Hydrogels with Appropriate/Tunable Properties for Biomedical Applications

Hydrogels with Appropriate/Tunable Properties for Biomedical Applications

Guest Editors

Yazhong Bu
Yanyu Yang
Feifei Sun

MDPI

Basel • Beijing • Wuhan • Barcelona • Belgrade • Novi Sad • Cluj • Manchester

Guest Editors

Yazhong Bu
School of Basic Medical Sciences
Xi'an Jiaotong University
Xi'an
China

Yanyu Yang
College of Materials Science and Engineering
Zhengzhou University
Zhengzhou
China

Feifei Sun
Institute of Process Engineering
Chinese Academy of Sciences
Beijing
China

Editorial Office
MDPI AG
Grosspeteranlage 5
4052 Basel, Switzerland

This is a reprint of the Special Issue, published open access by the journal *Gels* (ISSN 2310-2861), freely accessible at: https://www.mdpi.com/journal/gels/special_issues/Hydrogel_biomedical.

For citation purposes, cite each article independently as indicated on the article page online and as indicated below:

Lastname, A.A.; Lastname, B.B. Article Title. *Journal Name* **Year**, *Volume Number*, Page Range.

ISBN 978-3-7258-6388-4 (Hbk)
ISBN 978-3-7258-6389-1 (PDF)
https://doi.org/10.3390/books978-3-7258-6389-1

Contents

About the Editors vii

Yazhong Bu, Yanyu Yang and Feifei Sun
Editorial for the Special Issue "Hydrogels with Appropriate/Tunable Properties for Biomedical Applications"
Reprinted from: *Gels* **2025**, *11*, 277, https://doi.org/10.3390/gels11040277 1

Seung Hyeon Weon, Jiwoo Han, Yong-Keun Choi, Saerom Park and Sang Hyun Lee
Development of Blended Biopolymer-Based Photocatalytic Hydrogel Beads for Adsorption and Photodegradation of Dyes
Reprinted from: *Gels* **2023**, *9*, 630, https://doi.org/10.3390/gels9080630 6

Richard Heger, Natalia Zinkovska, Monika Trudicova, Martin Kadlec, Miloslav Pekar and Jiri Smilek
Lecithin as an Effective Modifier of the Transport Properties of Variously Crosslinked Hydrogels
Reprinted from: *Gels* **2023**, *9*, 367, https://doi.org/10.3390/gels9050367 24

Zewu Jiang, Qingqing Sun, Qian Li and Xiaomeng Li
Fabrication of Silk Hydrogel Scaffolds with Aligned Porous Structures and Tunable Mechanical Properties
Reprinted from: *Gels* **2023**, *9*, 181, https://doi.org/10.3390/gels9030181 42

Seiichiro Ishihara, Haruna Kurosawa and Hisashi Haga
Stiffness-Modulation of Collagen Gels by Genipin-Crosslinking for Cell Culture
Reprinted from: *Gels* **2023**, *9*, 148, https://doi.org/10.3390/gels9020148 57

Hongwen Yu, Jiaqi Song, Xianpeng Zhang, Kuo Jiang, Hong Fan, Yibing Li, et al.
Hydroxyapatite-Tethered Peptide Hydrogel Promotes Osteogenesis
Reprinted from: *Gels* **2022**, *8*, 804, https://doi.org/10.3390/gels8120804 68

Shuai Chang, Jiedong Wang, Nanfang Xu, Shaobo Wang, Hong Cai, Zhongjun Liu and Xing Wang
Facile Construction of Hybrid Hydrogels with High Strength and Biocompatibility for Cranial Bone Regeneration
Reprinted from: *Gels* **2022**, *8*, 745, https://doi.org/10.3390/gels8110745 83

Gang Zhang, Hai Qiu, Khalil I. Elkhodary, Shan Tang and Dan Peng
Modeling Tunable Fracture in Hydrogel Shell Structures for Biomedical Applications
Reprinted from: *Gels* **2022**, *8*, 515, https://doi.org/10.3390/gels8080515 96

Iulia Matei, Ana-Maria Ariciu, Elena Irina Popescu, Sorin Mocanu, Alexandru Vincentiu Florian Neculae, Florenta Savonea and Gabriela Ionita
Evaluation of the Accessibility of Molecules in Hydrogels Using a Scale of Spin Probes
Reprinted from: *Gels* **2022**, *8*, 428, https://doi.org/10.3390/gels8070428 117

Quan Diao, Hongyan Liu and Yanyu Yang
A Highly Mechanical, Conductive, and Cryophylactic Double Network Hydrogel for Flexible and Low-Temperature Tolerant Strain Sensors
Reprinted from: *Gels* **2022**, *8*, 424, https://doi.org/10.3390/gels8070424 127

Thi Ha My Phan, Yu-Hsun Yang, Yi-Jen Tsai, Fang-Yu Chung, Tooru Ooya, Shiho Kawasaki and Jeng-Shiung Jan
Synthesis and Hydrogelation of Star-Shaped Graft Copolypetides with Asymmetric Topology
Reprinted from: *Gels* **2022**, *8*, 366, https://doi.org/10.3390/gels8060366 139

Muhammad Farhan Ali Khan, Asim Ur.Rehman, Haidar Howari, Aiyeshah Alhodaib, Faiz Ullah, Zia ul Mustafa, et al.
Hydrogel Containing Solid Lipid Nanoparticles Loaded with Argan Oil and Simvastatin: Preparation, In Vitro and Ex Vivo Assessment
Reprinted from: *Gels* **2022**, *8*, 277, https://doi.org/10.3390/gels8050277 . . . . **153**

Haonan Wang, Qian Li, Yongchao Jiang and Xiaofeng Wang
Functional Hydrogels with Chondroitin Sulfate Release Properties Regulate the Angiogenesis Behaviors of Endothelial Cells
Reprinted from: *Gels* **2022**, *8*, 261, https://doi.org/10.3390/gels8050261 . . . . **169**

Theerasarn Pianpanit and Kiadtisak Saenboonruang
High-Energy Photon Attenuation Properties of Lead-Free and Self-Healing Poly (Vinyl Alcohol) (PVA) Hydrogels: Numerical Determination and Simulation
Reprinted from: *Gels* **2022**, *8*, 197, https://doi.org/10.3390/gels8040197 . . . . **182**

Fuminori Okazaki, Yusuke Hattori, Tetsuo Sasaki and Makoto Otsuka
Mechanochemical Effect on Controlled Drug Release of Konjac Glucomannan Matrix Tablets during Dry Grinding
Reprinted from: *Gels* **2022**, *8*, 181, https://doi.org/10.3390/gels8030181 . . . . **201**

Xingfan Li, Linyan Duan, Mingyue Kong, Xuejun Wen, Fangxia Guan and Shanshan Ma
Applications and Mechanisms of Stimuli-Responsive Hydrogels in Traumatic Brain Injury
Reprinted from: *Gels* **2022**, *8*, 482, https://doi.org/10.3390/gels8080482 . . . . **213**

Shuai Chang, Shaobo Wang, Zhongjun Liu and Xing Wang
Advances of Stimulus-Responsive Hydrogels for Bone Defects Repair in Tissue Engineering
Reprinted from: *Gels* **2022**, *8*, 389, https://doi.org/10.3390/gels8060389 . . . . **229**

Cosimo Ligorio, Judith A. Hoyland and Alberto Saiani
Self-Assembling Peptide Hydrogels as Functional Tools to Tackle Intervertebral Disc Degeneration
Reprinted from: *Gels* **2022**, *8*, 211, https://doi.org/10.3390/gels8040211 . . . . **244**

About the Editors

Yazhong Bu

Yazhong Bu received his bachelor's degree from Renmin University of China in 2013 and obtained his Ph.D. from the Institute of Chemistry, Chinese Academy of Sciences in 2018. From 2018 to 2020, he conducted postdoctoral research at the University of Galway, Ireland, as a Marie Skłodowska-Curie Fellow. He is currently a research fellow at the School of Basic Medical Sciences, Xi'an Jiaotong University, and is a distinguished professor in the Department of Burns, Plastic and Wound Repair Surgery at the Second Affiliated Hospital of Xi'an Jiaotong University. His research focuses on the design and application of biomaterials, including tissue engineering scaffolds, hydrogels, bioadhesives, and drug delivery systems.

Yanyu Yang

Yanyu Yang received her PhD degree in polymer chemistry and physics from the Institute of Chemistry of the Chinese Academy of Sciences in 2017. Currently, she works as a professor at the College of Materials Science and Engineering at Zhengzhou University. Her research is devoted to constructing smart hydrogel materials for flexible electronics and biomaterials.

Feifei Sun

Feifei Sun is an Associate Researcher specializing in chemical and biomolecular materials. She completed her postdoctoral research at the Institute of Chemistry, Chinese Academy of Sciences, and holds a PhD from China University of Petroleum (Beijing) and Georgia Institute of Technology. Her research focuses on advanced materials for wound closure, tissue adhesion, bone repair, and tumor inhibition.

gels

Editorial

Editorial for the Special Issue "Hydrogels with Appropriate/Tunable Properties for Biomedical Applications"

Yazhong Bu [1,*], Yanyu Yang [2,*] and Feifei Sun [3,*]

1 Institute of Medical Engineering, Department of Biophysics, School of Basic Medical Sciences, Health Science Center, Xi'an Jiaotong University, Xi'an 710061, China
2 College of Materials Science and Engineering, Zhengzhou University, Zhengzhou 450001, China
3 CAS Key Laboratory for Biomedical Effects of Nanomaterials & Nanosafety, National Center for Nanoscience and Technology, Beijing 100190, China
* Correspondence: yazhongbu@xjtu.edu.cn (Y.B.); yyyang@zzu.edu.cn (Y.Y.); feisun@iccas.ac.cn (F.S.)

Hydrogels are widely explored in biomedical fields, due to their porosity, high water content, and soft consistency, closely mimicking natural living tissue conditions. Efforts have been made to enhance their properties to increase their potential in biomedical applications. These enhancements encompass high mechanical strength, controllable degradation, bioadhesion capability, and responsiveness to stimulation, among other attributes.

Suitable properties for applications are very important in hydrogel preparation. While extensive research has aimed to imbue hydrogels with diverse properties, there appears to be a disproportionate emphasis on intensifying these properties. As a result, the fabrication of hydrogels with tailored properties is crucial for broadening their utility in biomedical applications. The focus of this Special Issue is to present the latest advancements in developing hydrogels with varied properties for biomedical applications. Both hydrogels with remarkably robust properties and those with adaptable characteristics are included in this Special Issue. It is believed that creating hydrogels with properties finely suited to specific applications will expedite the clinical implementation of these materials.

The brief introduction of the related articles are as follows.

(1) The article "Development of Blended Biopolymer-Based Photocatalytic Hydrogel Beads for Adsorption and Photodegradation of Dyes" by Weon et al. [1] develops blended biopolymer-based photocatalytic hydrogel beads to enhance adsorption and photodegradation capacities. Modified cellulose/TiO_2 beads significantly improved dye removal efficiency, with microbeads outperforming millimeter-sized beads. Incorporating Fe_2O_3 produced magnetic microbeads with tunable properties and rapid dye removal, highlighting their potential for environmental and biomedical applications.
(2) The article "Lecithin as an Effective Modifier of the Transport Properties of Variously Crosslinked Hydrogels" by Heger et al. [2] explores how lecithin, an amphiphile, modifies hydrogel transport properties crucial for drug delivery applications. By self-assembling within the hydrogel, lecithin alters its structure and reduces diffusion coefficients, as shown through dye-diffusion experiments and microscopy. Results highlight lecithin's potential to control transport properties, particularly in xerogels, enhancing hydrogel utility.
(3) The article "Fabrication of Silk Hydrogel Scaffolds with Aligned Porous Structures and Tunable Mechanical Properties" by Jing et al. [3] describes the development of silk fibroin glycidyl methacrylate (SF-GMA) hydrogels with aligned porous structures and enhanced toughness using directional freezing and in situ photo-crosslinking. These hydrogels exhibit improved mechanical properties, variable viscoelasticity, and

 https://doi.org/10.3390/gels11040277

excellent biocompatibility, making them suitable for applications in cell culture and tissue engineering.

(4) The article "Stiffness Modulation of Collagen Gels by Genipin-Crosslinking for Cell Culture" by Ishihara et al. [4] presents a method to modulate collagen gel stiffness (0.0292–12.5 kPa) using low-cytotoxic genipin as a crosslinker. This approach maintains cell viability while facilitating studies on stiffness-dependent phenomena, such as cancer cell-spreading and lineage-specific stromal cell differentiation. It provides an effective tool for investigating stiffness-controlled ECM behaviors in vitro.

(5) The article "Hydroxyapatite-Tethered Peptide Hydrogel Promotes Osteogenesis" by Yu et al. [5] introduces a peptide-based hydrogel (FmocFFRR/HAp) designed to stabilize hydroxyapatite (HAp) for bone regenerative engineering. By leveraging self-assembly, the hydrogel achieves stable HAp dispersion, improved mechanical properties, and strong biocompatibility. It supports 3D cell culture and effectively enhances osteoblast differentiation, showcasing its suitability for osteoinductive applications in bone tissue engineering.

(6) The article entitled "Facile Construction of Hybrid Hydrogels with High Strength and Biocompatibility for Cranial Bone Regeneration" by Chang et al. [6] introduces an osteogenic hybrid hydrogel between the amine-functionalized bioactive glass (ABG) and 4-armed poly(ethylene glycol) succinimidyl glutarate–gelatin network (SGgel) for cranial bone regeneration. Relying on the rapid ammonolysis reaction, the integration between the SGgel network and ABG moieties within a nano-scale level enabled the hybrid hydrogel to form adhesion to tissue, maintain the durable osteogenesis and accelerate bone regeneration, which may represent a promising strategy to design therapeutic scaffolds for tissue engineering in clinical applications.

(7) The article entitled "Modeling Tunable Fracture in Hydrogel Shell Structures for Biomedical Applications" by Zhang et al. [7] proposes a mixed graph-finite element method (FEM) phase-field approach to model the fracture of curved shells composed of hydrogels, for biomedical applications. Used in combination with experimental material testing, the method opens a new pathway to the efficient modeling of fracture in biomedical devices with surfaces of arbitrary curvature, helping in the design of devices with tunable fracture properties.

(8) The article entitled "Evaluation of the Accessibility of Molecules in Hydrogels Using a Scale of Spin Probes" by Matei et al. [8] explores, by means of electron paramagnetic resonance (EPR) spectroscopy, the accessibility of a series of spin probes, covering a scale of molecular weights in the range of 200–60,000 Da, in a variety of hydrogels: covalent network, ionotropic, interpenetrating polymer network (IPN), and semi-IPN. EPR spectroscopy can be an alternate method to evaluate the mesh size of gel systems and to provide information on local interactions inside gels.

(9) The article entitled "A Highly Mechanical, Conductive, and Cryophylactic Double-Network Hydrogel for Flexible and Low-Temperature Tolerant Strain Sensors" by Diao et al. [9] prepares a chitosan-poly (acrylic acid-co-acrylamide) double network (DN) hydrogel by immersing the chitosan-poly (acrylic acid-co-acrylamide) composite hydrogel into $Fe_2(SO_4)_3$ solution. Due to the formation of an energy dissipative chitosan physical network and the excellent conductivity and freeze-resistance, the DN hydrogel possesses excellent tensile, compression properties, remarkable sensitivity, and reliability in detecting stretching and bending deformations even at a lower temperature (−20 °C).

(10) The article entitled "Synthesis and Hydrogelation of Star-Shaped Graft Copolypetides with Asymmetric Topology" by Phan et al. [10] synthesizes the star-shaped poly(L-lysine) with various arm numbers by using asymmetric polyglycerol dendrimers

(PGDs) as the initiators and 1,1,3,3-tetramethylguanidine (TMG) as an activator for OH groups, followed by deprotection and grafting with indole or phenyl group on the side chain. By means of non-covalent interactions, the packing of the grafting moiety facilitates the polypeptide segments to adopt more ordered conformations and triggers the spontaneous hydrogelation. It is reported that the grafted moiety and polypeptide topology possessed the potential ability to modulate the polypeptide hydrogelation and hydrogel characteristics.

(11) The article entitled "Hydrogel-Containing Solid Lipid Nanoparticles Loaded with Argan Oil and Simvastatin: Preparation, In Vitro, and Ex Vivo Assessment" by Khan et al. [11] formulates and optimizes argan oil-loaded transdermal hydrogel containing lipid nanoparticles. They demonstrate that the hydrogel plays a crucial role in controlling the burst release and imparting the effect of argan oil as a hypolipidemic agent and permeation enhancer.

(12) The article entitled "Functional Hydrogels with Chondroitin Sulfate Release Properties Regulating the Angiogenesis Behaviors of Endothelial Cells" by Wang et al. [12] develops functional hydrogels with properties that mimicked the ECM structure using gelatin and chondroitin sulfate. It is proved that their functional hydrogels can regulate the angiogenesis behaviors of endothelial cells, thus, possessing potential in accelerating tissue regeneration.

(13) The article entitled "High-Energy Photon Attenuation Properties of Lead-Free and Self-Healing Poly (Vinyl Alcohol) (PVA) Hydrogels: Numerical Determination and Simulation" by Pianpanit et al. [13] numerically determines high-energy photon shielding properties of self-healing poly(vinyl alcohol) (PVA) hydrogels containing lead-free, heavy-metal compounds, namely, bismuth oxide (Bi_2O_3), tungsten oxide (WO_3), and barium sulfate ($BaSO_4$), through XCOM software packages (version 12). This work promisingly implies the potential of PVA hydrogels to be used as novel and potent X-ray and gamma-shielding materials with additional self-healing and nonlead properties.

(14) The article entitled "Mechanochemical Effect on Controlled Drug Release of Konjac Glucomannan Matrix Tablets during Dry Grinding" by Okazaki et al. [14] designs a controlled drug-release preparation based on a safe natural material, a Konjac glucomannan (KGM) mixture containing 16.0 w/w% calcium hydroxide ($Ca(OH)_2$). They explore how the degree of mechanochemical treatment would influence the drug release, which suggests that drug release from the KGM matrix tablet can be freely controlled by the degree of mechanochemical treatment.

(15) The Review entitled "Applications and Mechanisms of Stimuli-Responsive Hydrogels in Traumatic Brain Injury" by Li et al. [15] reviews the common types of stimuli-responsive hydrogels and their applications in traumatic brain injury (TBI) and further analyzes the therapeutic effects of hydrogels in TBI, such as pro-neurogenesis, anti-inflammatory, anti-apoptosis, anti-oxidation, and pro-angiogenesis. This study may provide strategies for the treatment of TBI by using stimuli-responsive hydrogels.

(16) The review entitled "Advances of Stimulus-Responsive Hydrogels for Bone Defects Repair in Tissue Engineering" by Chang et al. [16] focuses on the classification, design concepts, and research progress of stimulus-responsive hydrogels based on different types of external environmental stimuli, aiming at introducing new ideas and methods for repairing complex bone defects.

(17) The review entitled "Self-Assembling Peptide Hydrogels as Functional Tools to Tackle Intervertebral Disk Degeneration" by Ligorio et al. [17] describes the pathogenesis of intervertebral disk (IVD) degeneration, lists biomaterials requirements to attempt IVD repair, and focuses on current peptide hydrogel materials exploited for this purpose.

The properties of hydrogels need to be tuned to enable their application in biomedical fields. This is a necessary research focus in the continuous pursuit of hydrogel's performance breakthroughs. It implies that hydrogels must be tailored with appropriate functions based on their actual application scenarios. We appreciate the valuable contributions of the authors to this topic and hope that their work will continue to receive increased attention in the future.

Author Contributions: Y.B., conceptualization; writing—original draft preparation; writing—review and editing. Y.Y., conceptualization; writing—original draft preparation; writing—review and editing. F.S., conceptualization; writing—original draft preparation; writing—review and editing. All authors have read and agreed to the published version of the manuscript.

Funding: We would like to thank the National Natural Science Foundation of China (52103184, 52173144, and 51803188), Funding for Basic Scientific Research and Young Talent Support Plan of Xi'an Jiaotong University (xzy012022038).

Acknowledgments: We sincerely thank the editorial office, especially the assistant editor, for their kind assistance in the preparation and editing of this Special Issue.

Conflicts of Interest: The authors declare no conflicts of interest.

References

1. Weon, S.H.; Han, J.; Choi, Y.-K.; Park, S.; Lee, S.H. Development of Blended Biopolymer-Based Photocatalytic Hydrogel Beads for Adsorption and Photodegradation of Dyes. *Gels* **2023**, *9*, 630. [CrossRef] [PubMed]
2. Heger, R.; Zinkovska, N.; Trudicova, M.; Kadlec, M.; Pekar, M.; Smilek, J. Lecithin as an Effective Modifier of the Transport Properties of Variously Crosslinked Hydrogels. *Gels* **2023**, *9*, 367. [CrossRef] [PubMed]
3. Jiang, Z.; Sun, Q.; Li, Q.; Li, X. Fabrication of Silk Hydrogel Scaffolds with Aligned Porous Structures and Tunable Mechanical Properties. *Gels* **2023**, *9*, 181. [CrossRef] [PubMed]
4. Ishihara, S.; Kurosawa, H.; Haga, H. Stiffness-Modulation of Collagen Gels by Genipin-Crosslinking for Cell Culture. *Gels* **2023**, *9*, 148. [CrossRef] [PubMed]
5. Yu, H.; Song, J.; Zhang, X.; Jiang, K.; Fan, H.; Li, Y.; Zhao, Y.; Liu, S.; Hao, D.; Li, G. Hydroxyapatite-Tethered Peptide Hydrogel Promotes Osteogenesis. *Gels* **2022**, *8*, 804. [CrossRef] [PubMed]
6. Chang, S.; Wang, J.; Xu, N.; Wang, S.; Cai, H.; Liu, Z.; Wang, X. Facile Construction of Hybrid Hydrogels with High Strength and Biocompatibility for Cranial Bone Regeneration. *Gels* **2022**, *8*, 745. [CrossRef] [PubMed]
7. Zhang, G.; Qiu, H.; Elkhodary, K.I.; Tang, S.; Peng, D. Modeling Tunable Fracture in Hydrogel Shell Structures for Biomedical Applications. *Gels* **2022**, *8*, 515. [CrossRef] [PubMed]
8. Matei, I.; Ariciu, A.-M.; Popescu, E.I.; Mocanu, S.; Neculae, A.V.F.; Savonea, F.; Ionita, G. Evaluation of the Accessibility of Molecules in Hydrogels Using a Scale of Spin Probes. *Gels* **2022**, *8*, 428. [CrossRef] [PubMed]
9. Diao, Q.; Liu, H.; Yang, Y. A Highly Mechanical, Conductive, and Cryophylactic Double Network Hydrogel for Flexible and Low-Temperature Tolerant Strain Sensors. *Gels* **2022**, *8*, 424. [CrossRef] [PubMed]
10. Phan, T.H.M.; Yang, Y.-H.; Tsai, Y.-J.; Chung, F.-Y.; Ooya, T.; Kawasaki, S.; Jan, J.-S. Synthesis and Hydrogelation of Star-Shaped Graft Copolypetides with Asymmetric Topology. *Gels* **2022**, *8*, 366. [CrossRef] [PubMed]
11. Khan, M.F.A.; Ur. Rehman, A.; Howari, H.; Alhodaib, A.; Ullah, F.; Mustafa, Z.U.; Elaissari, A.; Ahmed, N. Hydrogel Containing Solid Lipid Nanoparticles Loaded with Argan Oil and Simvastatin: Preparation, In Vitro and Ex Vivo Assessment. *Gels* **2022**, *8*, 277. [CrossRef] [PubMed]
12. Wang, H.; Li, Q.; Jiang, Y.; Wang, X. Functional Hydrogels with Chondroitin Sulfate Release Properties Regulate the Angiogenesis Behaviors of Endothelial Cells. *Gels* **2022**, *8*, 261. [CrossRef] [PubMed]
13. Pianpanit, T.; Saenboonruang, K. High-Energy Photon Attenuation Properties of Lead-Free and Self-Healing Poly (Vinyl Alcohol) (PVA) Hydrogels: Numerical Determination and Simulation. *Gels* **2022**, *8*, 197. [CrossRef] [PubMed]
14. Okazaki, F.; Hattori, Y.; Sasaki, T.; Otsuka, M. Mechanochemical Effect on Controlled Drug Release of Konjac Glucomannan Matrix Tablets during Dry Grinding. *Gels* **2022**, *8*, 181. [CrossRef] [PubMed]
15. Li, X.; Duan, L.; Kong, M.; Wen, X.; Guan, F.; Ma, S. Applications and Mechanisms of Stimuli-Responsive Hydrogels in Traumatic Brain Injury. *Gels* **2022**, *8*, 482. [CrossRef]

16. Chang, S.; Wang, S.; Liu, Z.; Wang, X. Advances of Stimulus-Responsive Hydrogels for Bone Defects Repair in Tissue Engineering. *Gels* **2022**, *8*, 389. [CrossRef]
17. Ligorio, C.; Hoyland, J.A.; Saiani, A. Self-Assembling Peptide Hydrogels as Functional Tools to Tackle Intervertebral Disc Degeneration. *Gels* **2022**, *8*, 211. [CrossRef]

gels

MDPI

Article

Development of Blended Biopolymer-Based Photocatalytic Hydrogel Beads for Adsorption and Photodegradation of Dyes

Seung Hyeon Weon [1,†], Jiwoo Han [1,†], Yong-Keun Choi [1,2], Saerom Park [1,2,*] and Sang Hyun Lee [1,*]

[1] Department of Biological Engineering, Konkuk University, Seoul 05029, Republic of Korea; shweon99@gmail.com (S.H.W.); jw991021@naver.com (J.H.); dragonrt@konkuk.ac.kr (Y.-K.C.)
[2] R&D Center, ChoiLab Inc., Seoul 01811, Republic of Korea
* Correspondence: angel4y@naver.com (S.P.); sanghlee@konkuk.ac.kr (S.H.L.); Tel.: +82-10-4038-4656 (S.P.); +82-2-2049-6269 (S.H.L.)
† These authors contributed equally to this work.

Abstract: Blended biopolymer-based photocatalytic hydrogel beads were synthesized by dissolving the biopolymers in 1-ethyl-3-methylimidazolium acetate ([Emim][Ac]), adding TiO_2, and reconstituting the beads with ethanol. The incorporation of modifying biopolymer significantly enhanced the adsorption capacity of the cellulose/TiO_2 beads. Cellulose/carrageenan/TiO_2 beads exhibited a 7.0-fold increase in adsorption capacity for methylene blue (MB). In contrast, cellulose/chitosan/TiO_2 beads showed a 4.8-fold increase in adsorption capacity for methyl orange (MO) compared with cellulose/TiO_2 beads. In addition, cellulose/TiO_2 microbeads were prepared through the sol–gel transition of the [Emim][Ac]-in-oil emulsion to enhance photodegradation activity. These microbeads displayed a 4.6-fold higher adsorption capacity and 2.8-fold higher photodegradation activity for MB than the millimeter-sized beads. Furthermore, they exhibited superior dye removal efficiencies for various dyes such as Congo red, MO, MB, crystal violet, and rhodamine B, surpassing the performance of larger beads. To expand the industrial applicability of the microbeads, biopolymer/TiO_2 magnetic microbeads were developed by incorporating Fe_2O_3. These magnetic microbeads outperformed millimeter-sized beads regarding the efficiency and time required for MB removal from aqueous solutions. Furthermore, the physicochemical properties of magnetic microbeads can be easily controlled by adjusting the type of biopolymer modifier, the TiO_2 and magnetic particle content, and the ratio of each component based on the target molecule. Therefore, biopolymer-based photocatalytic magnetic microbeads have great potential not only in environmental fields but also in biomedical fields.

Keywords: biopolymer; magnetic microbead; adsorption; photodegradation; photocatalyst

1. Introduction

Photocatalytic oxidation systems employing photocatalysts such as TiO_2, ZnO, SnO_2, and WO_3 [1] have emerged as effective methods for degrading and mineralizing organic contaminants in wastewater. Under ultraviolet (UV) irradiation, these systems oxidize a wide range of organic and inorganic pollutants into harmless byproducts such as CO_2 and H_2O [2]. Among these photocatalysts, titanium dioxide (TiO_2) is a semiconductor that occurs in three crystalline forms: anatase, rutile, and brookite [3]. When TiO_2 is illuminated, it generates hydroxyl radicals on its surface, which possess strong oxidizing abilities and can degrade contaminants [3]. Due to its exceptional properties, including photoactivity, biological and chemical stability, non-toxicity, water insolubility, wide pH tolerance, and low cost, TiO_2 has become one of the most extensively utilized and researched photocatalysts for the degradation of organic pollutants (e.g., phenols, dyes, herbicides, pesticides, and antibiotics) in aqueous environments [3–6]. However, heterogeneous photocatalytic processes involving TiO_2 in sewage have the drawback of requiring a post-treatment step to recover the suspended photocatalysts, which can be laborious because of the nanoscale size of TiO_2 particles. Therefore, it should be incorporated into support materials to achieve efficient photodegradation of pollutants using TiO_2.

Biopolymer-based hydrogels, which are three-dimensional structures composed of hydrophilic biopolymer networks that retain large amounts of water, have attracted considerable attention as effective adsorbents. They are known for their high water retention capacity and low cost [7]. Recently, cellulose-based hydrogels have drawn particular interest in biomedical fields, such as drug delivery, tissue engineering scaffolds, biosensors, and adsorbents [8], owing to their excellent mechanical resistance and inherent eco-friendly properties, such as biodegradability and biocompatibility [9]. Furthermore, the cellulose hydrogel network containing −OH and −COOH groups can serve as an excellent support material, offering a robust platform for achieving highly dispersed TiO_2 particles. However, the low solubility of cellulose in conventional organic solvents hinders the preparation of unmodified cellulose-based hydrogels by dissolving cellulose and regenerating it using antisolvents. In recent years, the development of new cellulose-dissolving solvents, such as ionic liquids (ILs) and deep eutectic solvents, such as alkylammonium hydroxide and alkylphosphonium hydroxide, has enabled the creation of various cellulose-based composite hydrogels [10].

Blending two or more biopolymers facilitates the facile preparation of biopolymer materials with novel physicochemical properties [11]. Furthermore, the properties of biopolymer materials can be controlled [12–14]. However, blending biopolymers through co-dissolution and regeneration poses challenges because of their low solubility in conventional solvents and different dissolution conditions. Recent advances in the ability of ILs to dissolve biopolymers have facilitated the preparation of blended biopolymer-based materials [15,16]. ILs exhibit high dissolving power for a wide range of biopolymers, including polysaccharides (e.g., cellulose, starch, dextran, agarose, and gums), proteins (e.g., gelatin, collagen, and silk), polynucleotides, and phenolic biopolymers (e.g., lignin), with or without co-solvents [15–21]. Wendler et al. and Liu et al. [12,13] reported the preparation of polysaccharide-blended materials using ILs that exhibit various characteristics depending on the biopolymer composition. Our group [14] obtained different biopolymer-blended hydrogel films by mixing cellulose as a support compound with polysaccharides, proteins, or phenolic biopolymers as a functional modifier. The resulting films exhibited distinct adsorption properties based on the type of biopolymer used. For example, the adsorption capacity of a cellulose/chitosan film for methyl orange was 9.1-fold higher, that of a cellulose/silk film for crystal violet was 7.9-fold higher, and that of a cellulose/chitosan film for lysozyme was 7.9-fold higher than that of cellulose film.

Recent advances in biomedicine have led to notable developments in the use of cellulose-based composite hydrogel photocatalysts. These photocatalysts have proven to be mild, efficient, and environmentally friendly, making them suitable for antibiotic degradation and sterilizing microorganisms [22]. Amaly et al. and Yue et al. [23,24] developed a cellulose composite aerogel doped with montmorillonite and a nanocellulose-based CdS/MoS_2/montmorillonite composite for tetracycline degradation. Zhang et al. [25] reported on a cellulose/TiO_2/β-CD composite hydrogel, demonstrating its effectiveness as a photo-activated antibacterial agent. These findings collectively highlight the promising potential of cellulose-based hydrogel photocatalysts for various biomedical applications. Cellulose-based TiO_2 composite hydrogels in various forms have been prepared. Film forms of cellulose/carrageenan/TiO_2 [26], cellulose/N-doped TiO_2 [27], and cellulose/graphene oxide (GO)/TiO_2 [28] composites were prepared to degrade methylene blue (MB). Wittmar et al. [29] fabricated cellulose/TiO_2/Fe_3O_4 macrospheres using a mixture of [Emim][Ac] and dimethylsulfoxide (DMSO) to degrade rhodamine B. Monolithic forms of cellulose/TiO_2 [30], cellulose/carboxymethyl cellulose/TiO_2/Fe_3O_4 [31], and cellulose/Cu_2O/TiO_2/reduced graphene oxide (rGO) [32] composites were prepared and used for the degradation of MB and methyl orange (MO).

Micrometer-sized cellulose hydrogel beads have good potential for developing adsorbents, drug delivery carriers, chromatographic resins, and enzyme supports because of their large surface areas and hydrophilic properties [33]. Conventionally, cellulose hydrogel microspheres are prepared using a solvent-in-oil emulsion by the sol–gel transition method. Recently, we prepared cellulose hydrogel microspheres using [Emim][Ac] as a solvent for enzyme support [9]. The lipase immobilized on the cellulose hydrogel microspheres exhibited a significantly higher loading efficiency, immobilization yield, and specificity constant than that immobilized on microcrystalline cellulose or millimeter-sized hydrogel beads because of the larger surface area and favorable interactions of the cellulose microspheres. Additionally, we developed cellulose/biopolymer/Fe_3O_4 hydrogel microspheres to enhance enzyme adsorption capacity [34]. Liu et al. prepared cellulose/chitosan/Fe_3O_4 hydrogel microspheres for the immobilization of glucose oxidase [35], and Xue et al. investigated the immobilization of lysozyme on cellulose/Fe_3O_4 microspheres [36]. Thus, due to their favorable interactions with biomolecules, cellulose-based hydrogel microspheres exhibit significant potential as biomedical materials. However, incorporating TiO_2 as a photocatalyst into cellulose-based microbeads remains unexplored.

In this study, we developed blended biopolymer-based photocatalytic hydrogel beads through the sol–gel transition of biopolymer solutions co-dissolved in [Emim][Ac]. We evaluated the adsorption capacity and photodegradation activity of the blended biopolymer/TiO_2 hydrogel beads using various model dyes with different charges and hydrophobicities. Carrageenan, chitosan, and carbon nanotubes (CNT) were incorporated to modify the surface characteristics of the cellulose/TiO_2 hydrogel beads. Furthermore, we prepared biopolymer/TiO_2 magnetic hydrogel microbeads using an [Emim][Ac]-in-oil emulsion to enhance the efficiency of dye removal from aqueous solutions (Figure 1). The effects of the biopolymer charge, TiO_2 content, magnetic particle content, and ratio of each component on the adsorption capacity and photodegradation activity of the composite hydrogel microbeads were investigated. We developed efficient and versatile photocatalytic hydrogel microbeads for environmental and biomedical applications by exploring the potential of blended biopolymers and incorporating TiO_2 into hydrogel beads.

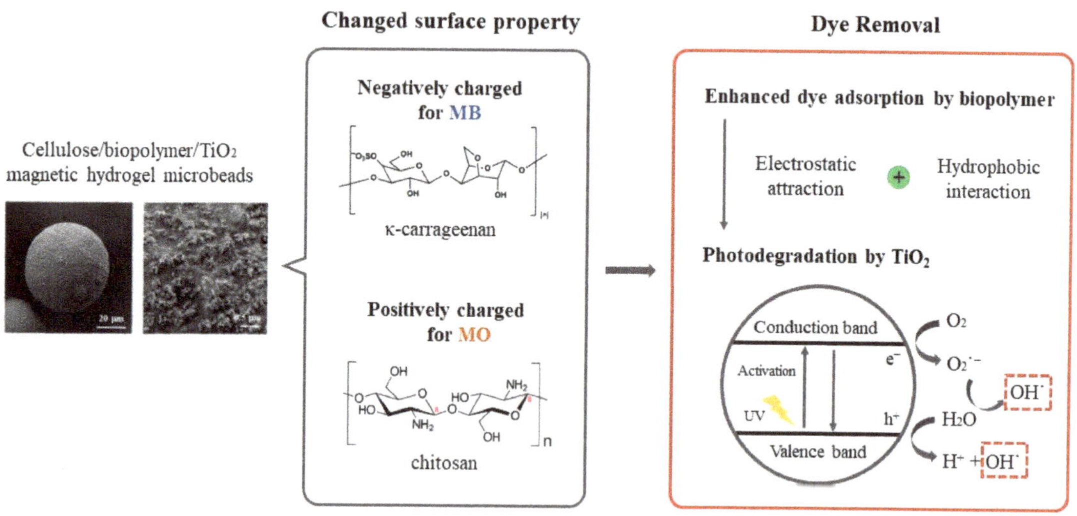

Figure 1. Schematic representation of the preparation of cellulose/biopolymer/TiO_2 composite hydrogel beads and their dye removal mechanism.

2. Results and Discussion

2.1. Preparation of Biopolymer/TiO_2 Hydrogel Beads

In this study, biopolymer-based photocatalytic hydrogel beads were prepared via the co-dissolution of cellulose and a biopolymer (chitosan or carrageenan) using [Emim][Ac] as the biopolymer solvent and the dispersion of TiO_2, followed by reconstitution with ethanol. Cellulose was used as the main supporting material for composite hydrogel beads because of its excellent mechanical strength and chemical stability. Chitosan and carrageenan were selected as modifiers to increase the positive and negative surface charges of the composite hydrogel beads, respectively. TiO_2 nanoparticles were used as photocatalysts. The capacities of chitosan and carrageenan as charge modifiers were compared with those of CNT, which were used to increase the surface area of composite materials and enhance the photocatalytic activity of TiO_2.

Millimeter-sized (ranging from 1.8 to 2.2 mm) composite hydrogel beads with the regular spherical shape of biopolymer/TiO_2 were prepared using syringe extrusion and sol–gel transition by ethanol as an anti-solvent (Figure 2). Depending on the additives used, the surfaces of the prepared biopolymer/TiO_2 composite beads exhibited no significant differences. All composite beads had nanosized bumpy surfaces due to the presence of TiO_2. In contrast, cellulose, cellulose/chitosan and cellulose/carrageenan beads without TiO_2 had smooth surfaces (Figure S1). The finely protruding surfaces of all the composite beads indicated that TiO_2 was fully incorporated into the biopolymers in the composite beads during the sol–gel transition process. The X-ray diffraction (XRD) patterns of the cellulose beads, TiO_2, and cellulose/TiO_2 composite beads are shown in Figure S2. The XRD patterns revealed the crystal structure of TiO_2 and confirmed its presence in the cellulose/TiO_2 composite beads. The TiO_2 nanoparticle can exist in three crystalline forms: anatase, rutile, and brookite. The main peak at 25.3° corresponded to the anatase form, which exhibited photocatalytic activity. These peaks were observed in the XRD patterns of cellulose/TiO_2 composite beads. The XRD profiles of the cellulose and cellulose/TiO_2 beads exhibited broad amorphous peaks, indicating a disruption in the crystalline nature of cellulose within the reconstituted beads. This disruption can be attributed to the breakage of internal hydrogen bonds in cellulose by [Emim][Ac].

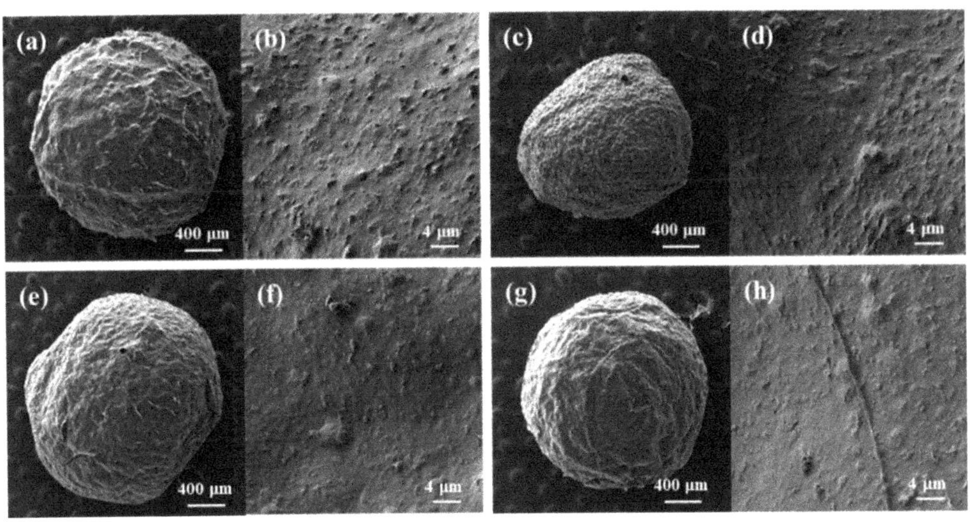

Figure 2. SEM images of freeze-dried biopolymer/TiO_2 hydrogel beads. (**a**,**b**): cellulose/TiO_2; (**c**,**d**): cellulose/CNT/TiO_2; (**e**,**f**): cellulose/chitosan/TiO_2; (**g**,**h**): cellulose/carrageenan/TiO_2.

2.2. Dye Removal of Biopolymer/TiO_2 Hydrogel Beads through Adsorption and Photodegradation

The biopolymer/TiO_2 hydrogel beads can remove various compounds from aqueous solutions via adsorption and photodegradation. To investigate the removal capacity of the target molecules of the biopolymer/TiO_2, MB and MO were chosen as model dyes for positively and negatively charged molecules, respectively. The adsorption capacity (q_e, mg/g) of the composite hydrogel beads was measured at 16 h, when the adsorption process reached equilibrium under dark conditions. The photocatalytic activity (k, min^{-1}) of composite hydrogel beads was measured at an initial rate under ultraviolet (UV) irradiation after the adsorption process reached equilibrium under dark conditions. The dye removal efficiency (%) of the composite hydrogel beads was measured after 16 h of adsorption in the dark and after 6 h of UV irradiation.

Table 1 lists the MB adsorption capacity and photodegradation activity of the composite hydrogel beads. The adsorption capacity of the cellulose/carrageenan/TiO_2 beads was 9.74 mg/g, which was 7.0 times higher than that of the cellulose/TiO_2 beads. Carrageenan, with negatively charged sulfate groups ($-OSO_3^-$), induced electrostatic attraction with positively charged MB. In contrast, the adsorption capacity of the cellulose/chitosan/TiO_2 beads was 8.2 times lower than that of the cellulose/TiO_2 beads. Chitosan, with positively charged amine groups ($-NH_3^+$), exhibited electrostatic repulsion with MB [10,34]. These results indicated that blending biopolymers with cellulose using [Emim][Ac] efficiently changed the surface charge of the cellulose/TiO_2 hydrogel beads without chemical modification. Traditional chemical modification methods can alter the biocompatibility and biodegradability of cellulose. The adsorption capacity of the cellulose/CNT/TiO_2 beads was 4.9-fold higher than that of the cellulose/TiO_2 beads. The enhanced adsorption capacity of the composite beads containing CNT can be explained by their large surface area and the hydrophobicity of CNT [37]. The highest adsorption capacity was obtained with cellulose/carrageenan/TiO_2.

Table 1. Adsorption and photodegradation capacity of biopolymer/TiO_2 hydrogel beads in aqueous methylene blue (MB) solution.

Biopolymers	Adsorption	Photodegradation	Dye Removal (%) [a]
	q_e (mg/g) [a]	k ($\times10^{-3}$/min) [a]	
Cellulose	1.40 ± 0.06	1.17 ± 0.06	40.1
Cellulose/CNT	6.79 ± 0.04	1.22 ± 0.13	41.6
Cellulose/Chitosan	0.17 ± 0.07	1.05 ± 0.04	19.6
Cellulose/Carrageenan	9.74 ± 0.06	0.98 ± 0.05	67.4

[a] Reaction conditions: 10 mg beads, 50 mg/L MB solution, and 16 h of adsorption in the dark, followed by 6 h of ultraviolet (UV) irradiation.

The photodegradation activity of the cellulose/TiO_2 beads was similar to that of cellulose/CNT/TiO_2. In contrast, cellulose/carrageenan/TiO_2 beads showed the lowest photocatalytic activity. The lowest k value of the cellulose/carrageenan/TiO_2 beads may be attributed to the inhibition of UV light transmission into the TiO_2 particles on the internal side of the beads due to the large number of MB molecules adsorbed on the surface of the composite beads. The inhibition of UV light transmission can also occur in cellulose/CNT/TiO_2 beads, which exhibit a high MB adsorption capacity. However, the enhanced photocatalytic activity of TiO_2 in the presence of CNT has been reported [38], which may compensate for the decreased photocatalytic activity caused by the inhibition of UV transmission. The photocatalytic activity of the cellulose/chitosan/TiO_2 beads was similar to that of the cellulose/TiO_2 beads. However, their MB adsorption capacities were deficient. These results indicate that the blended modifier did not significantly influence the photocatalytic activity of the composite beads for MB; however, a large amount of MB adsorbed on the composite surface could decrease the k value. The highest dye removal efficiency of 67.4% was obtained with the cellulose/carrageenan/TiO_2 beads, which showed a significantly enhanced MB adsorption capacity. In contrast, the dye removal efficiency of cellulose/chitosan/TiO_2, which showed the lowest MB adsorption capacity, was only 19.6%. The dye removal efficiency of the cellulose/CNT/TiO_2 beads was similar to that of the cellulose/TiO_2 beads.

Table 2 lists the MO adsorption capacity and photodegradation activity of the composite hydrogel beads. The adsorption capacity of cellulose/chitosan/TiO_2 beads was 0.96 mg/g, which was 4.8 times higher than that of cellulose/TiO_2 beads, whereas that of cellulose/carrageenan/TiO_2 beads was 2.0 times lower than that of cellulose/TiO_2 beads. These results can be explained by the attraction between positively charged chitosan and negatively charged MO and the repulsion between negatively charged carrageenan and MO. Therefore, chitosan and carrageenan can be successfully used as positive and negative surface modifiers, respectively, for the cellulose/TiO_2 composite beads. The cellulose/CNT/TiO_2 beads showed the highest adsorption capacity for MO, and their q_e value was 8.2-fold higher than that of the cellulose/TiO_2 beads. The enhanced adsorption capacity of the composite beads containing CNT may be explained by the large surface area of CNT and the hydrophobic interactions of CNT with MO, which was more hydrophobic than MB.

Table 2. Adsorption and photodegradation capacities of various biopolymer/TiO_2 hydrogel beads in aqueous methyl orange (MO) solution.

Biopolymers	Adsorption	Photodegradation	Dye Removal (%) [a]
	q_e (mg/g) [a]	k ($\times 10^{-3}$/min) [a]	
Cellulose	0.20 ± 0.05	1.85 ± 0.09	54.8
Cellulose/CNT	1.63 ± 0.03	0.36 ± 0.11	38.6
Cellulose/Chitosan	0.96 ± 0.15	1.40 ± 0.10	58.0
Cellulose/Carrageenan	0.10 ± 0.05	1.70 ± 0.12	45.4

[a] Reaction conditions: 5 mg beads, 10 mg/L MO solution, and 16 h of adsorption in the dark, followed by 6 h of UV irradiation.

The MO photodegradation activity of the composite hydrogel beads decreased as their adsorption capacity increased. The cellulose/CNT/TiO_2 beads, which exhibited the highest MO adsorption capacity, exhibited the lowest photodegradation activity. The cellulose/carrageenan/TiO_2 beads showed photocatalytic activity similar to that of cellulose/TiO_2. However, they showed the lowest adsorption capacity for MO. These results can be explained by the limitations of UV light transmission due to the adsorption of MO molecules on the surface of the composite beads. The cellulose/chitosan/TiO_2 beads exhibited the highest dye removal efficiency (58%). Although the cellulose/chitosan/TiO_2 beads demonstrated lower photodegradation activity than the cellulose/TiO_2 beads, their higher adsorption capacity could explain this discrepancy. In contrast, the cellulose/CNT/TiO_2 beads, despite having the highest adsorption capacity, displayed the lowest dye removal efficiency (39%). This can be attributed to their low photodegradation activity.

Cellulose/TiO_2 composite hydrogel beads exhibited enhanced MB and MO dye removal efficiencies when blended with carrageenan and chitosan, respectively. However, their enhanced dye removal efficiency could be mostly caused by the increased adsorption capacity, because their photodegradation activity was lower than that of the cellulose/TiO_2 beads. When dye molecules are adsorbed in large quantities on the surfaces of the composite beads, they can impede the transmission of UV light into the TiO_2 particles present within the beads. This problem can be addressed by using smaller-sized biopolymer/TiO_2 beads, which can significantly enhance the efficiency of dye removal.

2.3. *Preparation and Dye Removal of Cellulose/TiO_2 Hydrogel Microbeads*

To overcome the inhibition of light transmission caused by high dye concentrations in millimeter-sized cellulose/TiO_2 beads, we prepared micrometer-sized cellulose/TiO_2 beads through a sol–gel transition, using an [Emim][Ac]-in-oil emulsion. The adsorption capacity of the cellulose/TiO_2 microbeads was 6.41 mg/g (Table 3), whereas the q_e

value of the millimeter-sized cellulose/TiO_2 beads was 1.40 mg/g under the same adsorption conditions (Table 1). Microbeads showed 4.6-fold higher adsorption capacity than millimeter-sized beads. In addition, the adsorption of the microbeads reached equilibrium in 1 h, whereas that of the millimeter beads required 16 h to reach equilibrium. The photodegradation activity of the cellulose/TiO_2 microbeads was 2.8-fold higher than that of the millimeter-sized cellulose/TiO_2 beads (Tables 1 and 3). Microbeads showed much higher photodegradation activity than millimeter-sized beads, although the amount of MB adsorbed on the microbeads was much higher than that adsorbed on the millimeter-sized beads. These results clearly indicated that the size reduction in the cellulose/TiO_2 hydrogel beads could overcome the inhibition of light transmission caused by MB adsorbed in large quantities on the surfaces of the composite beads. The dye removal efficiency of the cellulose/TiO_2 microbeads was 95%, which was much higher (40%) than that of millimeter-sized cellulose/TiO_2 beads (Tables 1 and 3).

Table 3. Effect of magnetic particle content in cellulose/TiO_2 magnetic hydrogel microbeads on the adsorption and photodegradation capacity in aqueous methylene blue (MB) solution.

TiO_2 (%):Fe_2O_3 (%)	Adsorption	Photodegradation	Dye Removal (%) [a]
	q_e (mg/g) [a]	k ($\times 10^{-3}$/min) [a]	
0.5:0	6.41 ± 0.11	3.23 ± 0.31	95.0
0.5:0.1	3.85 ± 0.56	1.30 ± 0.22	56.0
0.5:0.5	1.62 ± 0.15	1.36 ± 0.07	47.7
0.5:0.75	2.90 ± 0.18	0.67 ± 0.10	34.8
0.5:1.0	7.84 ± 0.14	0.33 ± 0.03	59.3

[a] Reaction conditions: 10 mg beads, 50 mg/L MB solution, and 1 h of adsorption in the dark, followed by 6 h of UV irradiation.

The dye removal capacity of the cellulose/TiO_2 microbeads was measured using various dyes and compared with that of millimeter-scale cellulose/TiO_2 beads. The adsorption capacities (q_e) of the millimeter-sized cellulose/TiO_2 beads for Congo red (CR), MO, MB, crystal violet (CV), and rhodamine B (RB) were 3.5, 0.2, 0.6, 2.1, and 0.3 mg/g, respectively. Conversely, cellulose/TiO_2 microbeads exhibited q_e values of 6.8, 0.3, 3.0, 2.8, and 1.2 mg/g for CR, MO, MB, CV, and RB, respectively. The adsorption capacity of the microbeads was significantly higher, ranging from 1.3 to 5.0 times greater than that of the millimeter-sized beads. The dye removal efficiency of the cellulose/TiO_2 microbeads was much higher than that of the millimeter-sized cellulose/TiO_2 beads for all tested dyes (Figure 3). For example, microbeads showed 95% dye removal for MB, whereas millimeter-sized beads showed only 40% dye removal. These results indicated that cellulose/TiO_2 microbeads with a larger specific surface area than millimeter-sized beads could efficiently remove some dyes from aqueous solutions through higher adsorption capacity and photodegradation activity. This phenomenon applies to other chemicals and biomolecules. Kim et al. [39] and Jo et al. [9] reported that small-sized cellulose-based materials can enhance protein adsorption compared with larger-sized materials because of their improved specific surface area. Although cellulose/TiO_2 microbeads exhibit superior adsorption capacity and photodegradation activity compared with millimeter-sized cellulose/TiO_2 beads, because of their small size, industrial applications of microbeads may be constrained by the challenges associated with their recovery and reusability. The development of magnetic microbeads that can be easily separated using a magnet has the potential to overcome the limitations of industrial applications of microbeads.

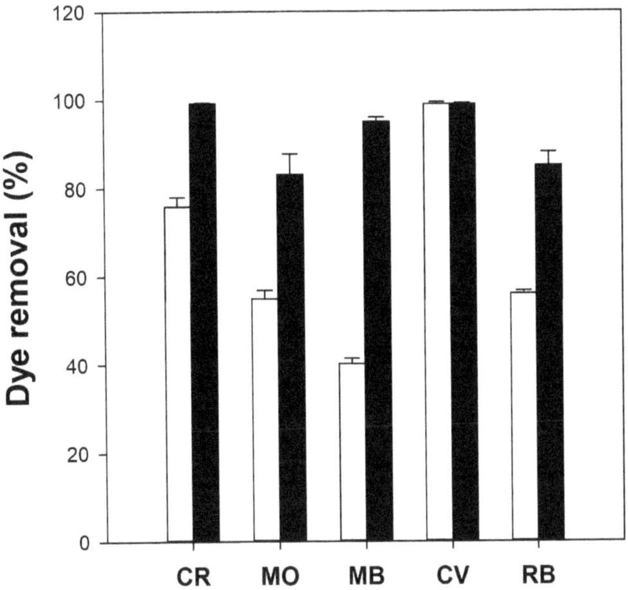

Figure 3. Dye removal efficiency (%) of millimeter-size (white bars) and micrometer-size (black bars) cellulose/TiO_2 hydrogel beads. Reaction conditions: 5 mg beads, 10 mg/L dye (Congo red [CR], methyl orange [MO], methylene blue [MB], crystal violet [CV] and rhodamine B [RB]) solution, and 16 h of adsorption in the dark, followed by 6 h of UV irradiation.

2.4. Preparation and Dye Removal of Cellulose/TiO_2 Magnetic Hydrogel Microbeads

To improve the recovery and reusability of the microbeads, we prepared cellulose/TiO_2/Fe_2O_3 microbeads by dispersing both TiO_2 and iron (III) oxide in an [Emim][Ac]-in-oil emulsion. The cellulose/TiO_2 magnetic hydrogel microbeads, containing more than 0.5% Fe_2O_3 in the bead-forming solution, were easily collected within 10 min using a magnet (Figure S3).

Figure 4 shows the surface morphology of cellulose/TiO_2 magnetic microbeads with the increasing content of Fe_2O_3. The surfaces of the cellulose/TiO_2 magnetic microbeads appeared rough, with clustered and aggregated formations scattered across various locations. This tendency became more pronounced as the amount of Fe_2O_3 increased at a fixed TiO_2 content (0.5%) (Figure 4a–c). The increase in the amount of used Fe_2O_3 and the consequent increase in the amount of exposed Fe_2O_3 on the surface led to a decrease in the photodegradation activity of TiO_2 by inhibiting light transmission. However, Fe_2O_3 exposed on the surface of cellulose/TiO_2 magnetic microbeads can act as both an adsorbent and a photocatalyst [40].

Table 3 shows the effect of the magnetic particle content of the cellulose/TiO_2 magnetic microbeads on the adsorption capacity and photodegradation activity of MB. Both adsorption capacity and photocatalytic activity were significantly influenced by the magnetic particle content. Adding Fe_2O_3 up to 0.5% decreased the adsorption capacity. In contrast, microbeads prepared with more than 0.5% showed higher q_e values than those prepared with 0.5% Fe_2O_3. The q_e value of the microbeads prepared with 1% Fe_2O_3 was 1.2 times higher than that of cellulose/TiO_2 microbeads without Fe_2O_3. The surfaces of the cellulose/TiO_2 magnetic microbeads prepared using 1% Fe_2O_3 were covered with aggregated Fe_2O_3 (Figure 4). The Fe_2O_3 aggregates exposed on the surface of the magnetic microbeads may have increased the adsorption capacity for MB because Fe_2O_3 has been reported as a good adsorbent for MB [40]. The photodegradation activity of the cellulose/TiO_2 magnetic microbeads decreased with increasing the content of Fe_2O_3. The k value of the cellulose/TiO_2 microbeads prepared with only 0.1% Fe_2O_3 was 2.5-fold lower than that of the cellulose/TiO_2 microbeads prepared without Fe_2O_3. Although the photocatalytic activity of F_2O_3 for MB has been reported [40], adding Fe_2O_3 decreased the photodegradation activity of cellulose/TiO_2 magnetic microbeads. Including Fe_2O_3 in the cellulose/TiO_2 magnetic microbeads did not significantly improve their adsorption capacity or photodegradation activity. Consequently, the subsequent experiments used the minimum amount of Fe_2O_3 required to serve as magnetic particles. Adding a minimum of 0.5% Fe_2O_3 to the gel-forming solution enabled rapid separation using a magnet.

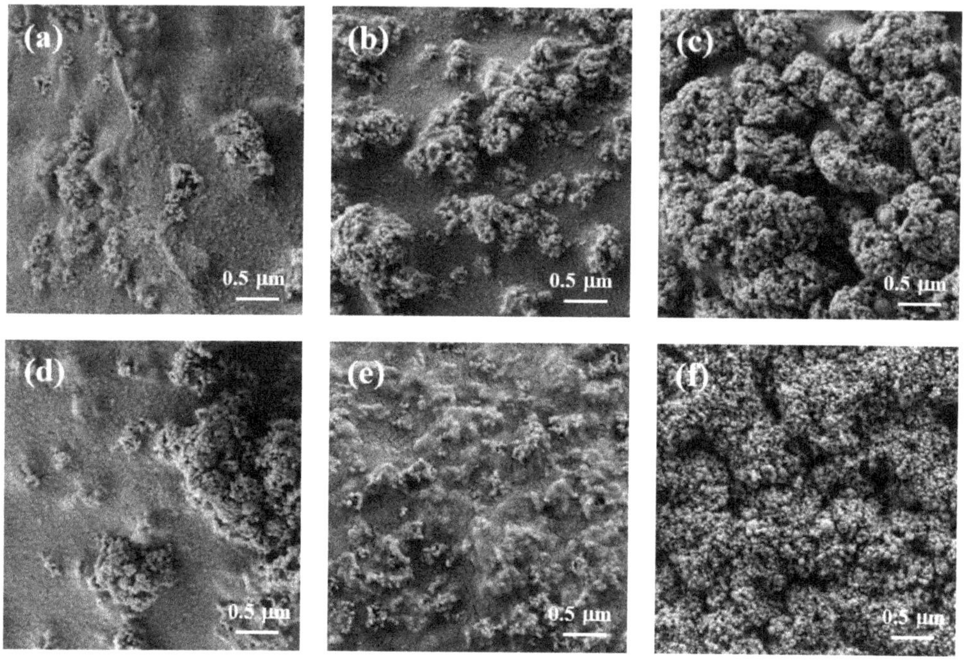

Figure 4. SEM images of freeze-dried cellulose/TiO_2 magnetic microbeads. (**a**) 0.5% TiO_2; (**b**) 0.5% TiO_2 and 0.1% Fe_2O_3; (**c**) 0.5% TiO_2 and 1% Fe_2O_3; (**d**) 0.5% TiO_2 and 0.5% Fe_2O_3; (**e**) 3.5% TiO_2 and 0.5% Fe_2O_3; (**f**) 7% TiO_2 and 0.5% Fe_2O_3.

To investigate the effect of the TiO_2 content of the cellulose/TiO_2 magnetic microbeads on MB adsorption capacity and photodegradation activity, the TiO_2 content was varied from 0.5% to 7% at a fixed content of Fe_2O_3 (0.5%). The surface morphology and mean diameter of the prepared cellulose/TiO_2 magnetic microbeads changed significantly depending on the content of TiO_2. The surfaces of the cellulose/TiO_2 magnetic microbeads had a rough texture with scattered aggregated particles. This tendency became more pronounced as the amount of TiO_2 increased at a fixed Fe_2O_3 content (0.5%) (Figure 4d–f). The surfaces of cellulose/TiO_2 magnetic microbeads, prepared with 7% TiO_2, were covered entirely with aggregated TiO_2 nanoparticles. As the amount of TiO_2 increased, the amount of exposed TiO_2 on the surface increased, which could have led to an increase in the photodegradation activity of TiO_2. However, the adsorption capacity of the cellulose/TiO_2 magnetic microbeads for MB could be decreased by decreasing the interactions between the cellulose and dye molecules because of the decreased cellulose content on the surface. The mean diameter of the cellulose/TiO_2 magnetic microbeads increased proportionally with TiO_2 content. The microbead diameters ranged from 22.3 mm to 52.6 mm, depending on the TiO_2 content, which varied from 0.5% to 7.0%. A strong correlation exists between the mean diameter of the microbeads and the TiO_2 content (r^2 = 0.960, n = 5).

Table 4 shows the effect of TiO_2 content in the cellulose/TiO_2 magnetic microbeads on the adsorption capacity and photodegradation activity of MB. When the TiO_2 content was 3.5%, the adsorption capacity and photodegradation activity were the highest. The q_e and k values increased as the TiO_2 content increased from 0.5% to 3.5%. However, beyond 3.5% TiO_2 content, both values decreased with a further increase in TiO_2 content up to 7%. As TiO_2 content increased, the mean diameter of the microbeads increased, whereas their specific surface area decreased. In addition, a noticeable transformation of the surface morphology was observed when the TiO_2 content exceeded 3.5%. Larger bead size and altered surface morphology may impede the interactions between cellulose and MB and hinder light transmission. To assess the efficiency of TiO_2 as a photocatalyst in the cellulose/TiO_2 magnetic microbeads, the specific rate constant was calculated based on the same TiO_2 content (k/mg TiO_2). Microbeads prepared with 3.5% TiO_2 exhibited the highest photodegradation activity, whereas those prepared with 0.5% TiO_2 showed the highest specific rate constant. The specific rate constant of the microbeads prepared with 3.5% TiO_2 was 3.2 times lower than that of those prepared with 0.5% TiO_2. These results could be attributed to several factors, including increased bead size, alterations in surface morphology, and TiO_2 aggregation. The increased bead size resulting from the higher TiO_2 content led to a decrease in the specific surface area of the microbeads. Furthermore, as the TiO_2 content increased, a significant amount of TiO_2 aggregated on the microbead surfaces. These aggregated forms of TiO_2 may have diminished the photocatalytic activity due to overlapping, thereby reducing overall efficiency. Wittmar and Ulbricht [29] reported the limitation of increasing the TiO_2 content in cellulose films due to the agglomeration of TiO_2.

Table 4. Effect of TiO_2 content in cellulose/TiO_2 magnetic hydrogel microbeads on the adsorption and photodegradation capacity in an aqueous methylene blue (MB) solution.

TiO_2 (%):Fe_2O_3 (%)	Mean Diameter (μm)	Adsorption	Photodegradation		Dye Removal (%) [a]
		q_e (mg/g) [a]	k ($\times10^{-3}$/min) [a]	k/mg TiO_2 ($\times10^{-3}$/min/mg) [a]	
0.5:0.5	22.3 ± 0.2	1.62 ± 0.15	1.36 ± 0.07	2.18 ± 0.11	47.7
2.0:0.5	25.9 ± 0.1	4.93 ± 0.21	1.49 ± 0.15	0.71 ± 0.07	66.8
3.5:0.5	34.6 ± 0.2	4.97 ± 0.42	2.18 ± 0.01	0.69 ± 0.00	77.4
5.0:0.5	37.7 ± 0.2	2.75 ± 0.29	1.88 ± 0.03	0.47 ± 0.01	59.5
7.0:0.5	52.6 ± 0.0	2.32 ± 0.20	1.23 ± 0.15	0.25 ± 0.03	48.1

[a] Reaction conditions: 10 mg beads, 50 mg/L MB solution, and 1 h of adsorption in the dark, followed by 6 h of UV irradiation.

2.5. Preparation and Dye Removal of Biopolymer/TiO_2 Magnetic Hydrogel Microbeads

The preparation conditions for the cellulose/TiO_2 magnetic microbeads were optimized for MB removal. However, the adsorption capacity and photodegradation activity of cellulose/TiO_2 magnetic microbeads for other molecules could be improved by blending specific biopolymers as modifiers for the target molecules. Chitosan, CNT, and carrageenan were added to cellulose/TiO_2 magnetic microbeads to improve the dye removal efficiency of MB and MO. Figure 5 shows the surfaces of freeze-dried biopolymer/TiO_2 magnetic microbeads. The surfaces of all the microbeads exhibited roughness, with scattered aggregated formations occurring in various areas due to the presence of exposed TiO_2 and Fe_2O_3. However, no significant differences were observed among the biopolymer/TiO_2 magnetic microbeads. The surface of the magnetic microbeads was rougher than that of the millimeter-sized beads (Figure 2), which could be attributed to the presence of exposed TiO_2 and Fe_2O_3 on the surface of the magnetic microbeads.

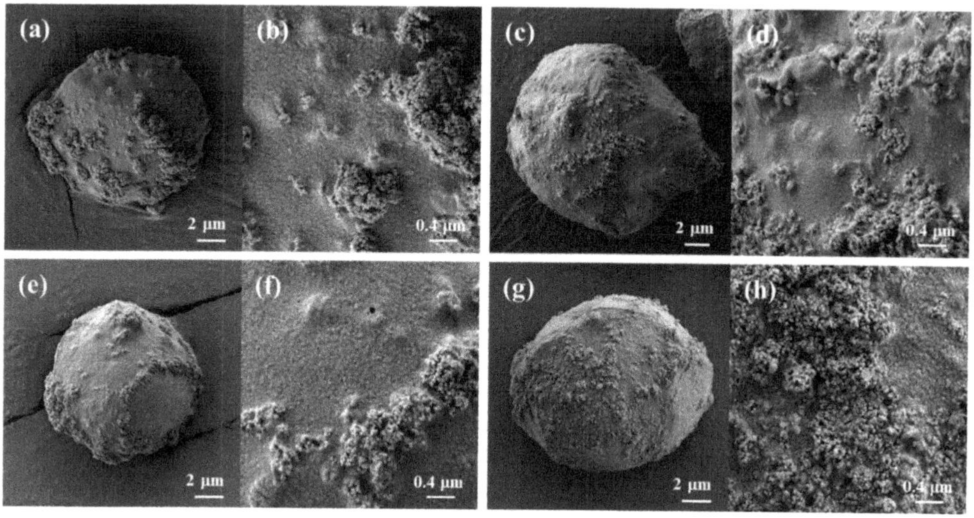

Figure 5. SEM images of freeze-dried biopolymer/TiO_2 magnetic microbeads. (**a**,**b**): cellulose/TiO_2/Fe_2O_3; (**c**,**d**): cellulose/CNT/TiO_2/Fe_2O_3; (**e**,**f**): cellulose/chitosan/TiO_2/Fe_2O_3; (**g**,**h**): cellulose/carrageenan/TiO_2/Fe_2O_3.

Table 5 shows the MB adsorption and photodegradation capacities of the biopolymer/TiO_2 magnetic microbeads for MB. The adsorption capacity of the negatively charged cellulose/carrageenan microbeads was 5.8 times higher than that of cellulose microbeads. In contrast, the adsorption capacity of the cellulose-chitosan microbeads was 1.1 times lower than that of the cellulose microbeads. Although the repulsion between the positively charged chitosan and MB significantly decreased the adsorption capacity for MB in millimeter-sized beads (Table 1), the TiO_2 and Fe_2O_3 exposed on the surface of the cellulose/chitosan magnetic microbeads may have increased the adsorption capacity for MB. The cellulose/CNT/TiO_2 magnetic microbeads exhibited the highest adsorption capacity, due to the large surface area and hydrophobicity of the CNT. Adding CNT, chitosan, and carrageenan to the cellulose/TiO_2 magnetic microbeads decreased the photodegradation activity. The lower photodegradation activities of the cellulose/CNT and cellulose/carrageenan magnetic microbeads may have been caused by the inhibition of UV light transmission by the MB molecules adsorbed in large quantities on the surface of the microbeads. Although microbeads partly overcome the inhibition of UV transmission, the much higher adsorption rate of microbeads compared with millimeter-sized beads may limit the photodegradation activity of magnetic microbeads. The cellulose/carrageenan magnetic microbeads showed the highest dye removal efficiency of 74.8%. In contrast, cellulose/chitosan showed the lowest dye removal efficiency (32.0%). The dye removal efficiency of the cellulose/CNT magnetic microbeads was 65.4%, which was 1.4 times higher than that of the cellulose magnetic microbeads. This may have been caused by the 5.8-fold enhanced adsorption capacity of the cellulose/CNT microbeads compared with that of the cellulose microbeads.

Table 5. Adsorption and photodegradation capacities of various biopolymer/TiO_2 magnetic microbeads in aqueous methylene blue (MB) solution.

Biopolymers	Adsorption	Photodegradation	Dye Removal (%) [a]
	q_e (mg/g) [a]	k (10^{-3}/min) [a]	
Cellulose	1.62 ± 0.15	1.36 ± 0.07	47.7
Cellulose/CNT	10.34 ± 0.09	0.68 ± 0.04	65.4
Cellulose/Chitosan	1.51 ± 0.13	0.26 ± 0.17	32.0
Cellulose/Carrageenan	9.46 ± 0.06	0.53 ± 0.09	74.8

[a] Reaction conditions: 10 mg beads, 50 mg/L MB solution, and 1 h of adsorption in the dark, followed by 6 h of UV irradiation.

The MB removal capacities of the biopolymer/TiO_2 magnetic microbeads (Table 5) were compared with those of millimeter-sized biopolymer/TiO_2 beads (Table 1) under the same reaction conditions, except for the adsorption time. The magnetic microbeads and millimeter-sized beads required 1 h and 16 h to reach equilibrium, respectively. Except for cellulose/carrageenan, the adsorption capacities of the magnetic microbeads were higher than those of the millimeter-sized beads. The adsorption capacity of the cellulose/carrageenan microbeads was not significantly different from that of the millimeter-sized beads. The presence of TiO_2 and Fe_2O_3 on the surface of the cellulose/carrageenan microbeads could have impeded the interactions between carrageenan and MB. These results indicated that magnetic microbeads could adsorb MB more efficiently and in a shorter time than millimeter-sized beads. The photodegradation activities of the microbeads, except for the cellulose beads, were lower than those of the millimeter-sized beads. The cellulose microbeads showed a 1.2-fold higher k value than that of the millimeter-sized cellulose beads. The lower photodegradation activity of the microbeads, except for the cellulose microbeads, may have been caused by the presence of Fe_2O_3. Although it has been reported that Fe_2O_3 can act as a photocatalyst [40], the addition of Fe_2O_3 to cellulose/TiO_2 microbeads decreased the photodegradation activity (Table 3). The dye removal efficiencies of all microbeads were higher than those of the millimeter-sized beads. These results indicated that the magnetic microbeads removed MB from the aqueous solution more efficiently and in a shorter time than the millimeter-sized beads.

Table 6 shows the adsorption capacity and photodegradation activity of biopolymer/TiO_2 magnetic microbeads for MO. The adsorption capacity of the positively charged cellulose/chitosan microbeads was 4.5 times higher than that of the cellulose microbeads. In contrast, the adsorption capacity of the cellulose/carrageenan microbeads was 5.7 times lower than that of the cellulose microbeads. The cellulose/CNT/TiO_2 magnetic microbeads exhibited the highest adsorption capacity, owing to the large surface area and hydrophobicity of the CNT. The addition of chitosan and carrageenan to the cellulose/TiO_2 magnetic microbeads increased their photodegradation activity. The decreased photodegradation activity of the cellulose/CNT magnetic microbeads may have been caused by the inhibition of UV light transmission by MO molecules adsorbed in large quantities on the surface of the microbeads. The cellulose/chitosan magnetic microbeads had the highest dye removal efficiency (57.3%). In contrast, the cellulose/CNT magnetic microbeads showed the lowest dye removal (30.2%).

Table 6. Adsorption and photodegradation capacities of various biopolymer/TiO_2 magnetic microbeads in aqueous methyl orange (MO) solution.

Biopolymers	Adsorption	Photodegradation	Dye Removal (%) [a]
	q_e (mg/g) [a]	k (10^{-3}/min) [a]	
Cellulose	0.57 ± 0.15	1.28 ± 0.07	42.6
Cellulose/CNT	3.87 ± 0.12	0.62 ± 0.10	30.2
Cellulose/Chitosan	2.55 ± 0.35	1.58 ± 0.05	57.3
Cellulose/Carrageenan	0.10 ± 0.05	1.52 ± 0.10	48.3

[a] Reaction conditions: 5 mg beads, 10 mg/L MO solution, and 1 h of adsorption in the dark, followed by 6 h of UV irradiation.

The MO removal capacities of the biopolymer/TiO_2 magnetic microbeads (Table 6) were compared with those of the millimeter-sized biopolymer/TiO_2 beads (Table 2) under the same reaction conditions, except for the adsorption time. The magnetic microbeads and millimeter-sized beads required 1 h and 16 h to reach equilibrium, respectively. The adsorption capacities of the magnetic microbeads, except for the cellulose/carrageenan beads, were higher than those of the millimeter-sized beads. The cellulose/carrageenan magnetic microbeads exhibited an adsorption capacity similar to that of millimeter-sized beads. The photodegradation activity of the cellulose/chitosan microbeads was 1.1-fold higher than that of the millimeter-sized beads. The k values of the cellulose and cellulose/carrageenan microbeads were lower than those of the millimeter-sized beads. The dye removal efficiencies of the magnetic microbeads, except for the cellulose microbeads, were similar to those of the millimeter-sized beads. Although the biopolymer/TiO_2 magnetic microbeads showed a much higher dye removal efficiency for MB than the millimeter-sized beads, their dye removal efficiency for MO was similar to or lower than that of the millimeter-sized beads. This could have been due to the presence of Fe_2O_3 because cellulose/TiO_2 microbeads without Fe_2O_3 showed a much higher dye removal efficiency for MO than millimeter-sized beads (Figure 3). The morphological differences between the microbeads and millimeter-sized beads changed their adsorption capacity and photodegradation activity. In addition, the surfaces of the microbeads were covered with more TiO_2 and Fe_2O_3 than those of the millimeter-sized beads (Figures 2 and 5). The surface of the microbeads, prepared through the sol–gel transition of the [Emim][Ac]-in-oil emulsion, can be changed using biopolymers as additives because the charge of the added biopolymers can influence the formation of the emulsion.

Table 7 compares the recently reported dye removal efficiencies achieved by cellulose/TiO_2-based photocatalytic hydrogels. Most studies have been conducted using hydrogels in the form of films or monoliths; studies on photocatalytic hydrogels using cellulose microbeads or blended biopolymers are scarce. A direct comparison is not feasible because of variations in the dye concentrations and hydrogel compositions used in each study. However, the blended biopolymer-based photocatalytic microbeads developed in this study exhibited comparable dye removal efficiencies to other forms of hydrogels. Microbead-shaped photocatalytic hydrogels have potential advantages for applications in the biomedical field, such as protein support and photo-antibacterial agents, thus expanding the scope of photocatalytic hydrogel applications.

Table 7. Dye removal by various forms of cellulose/TiO_2-based photocatalytic hydrogels.

Catalysts	Form	Dye (Concentration)	Time (h)	Dye Removal (%)	Ref.
Cellulose/TiO_2	Microbeads	MB (10 mg/L)	6	95	this study
Cellulose/TiO_2	Microbeads	MO (10 mg/L)	6	83	this study
Cellulose/TiO_2	Microbeads	RB (10 mg/L)	6	85	this study
Cellulose/carrageenan/TiO_2	Hydrogel film	MB (60 mg/L)	5	83	[26]
Cellulose/N-doped TiO_2	Hydrogel film	MB (40 mg/L)	6	96	[27]
Cellulose/GO/TiO_2	Hydrogel film	MB (10 mg/L)	2	93	[28]
Cellulose/TiO_2/Fe_3O_4	Macrospheres	RB (12 mg/L)	1	24	[29]
Cellulose/TiO_2	Monolith	MB (12 mg/L)	0.7	99	[30]
Cellulose/CMC/TiO_2/Fe_3O_4	Monolith	MB (20 mg/L)	1	98	[31]
Cellulose/Cu_2O/TiO_2/rGO	Monolith	MO (20 mg/L)	2	85	[32]

3. Conclusions

In this study, we successfully synthesized blended biopolymer-based photocatalytic hydrogel beads by co-dissolving biopolymers in [Emim][Ac], followed by the dispersion of TiO_2 and reconstitution with ethanol. Incorporating carrageenan and chitosan into the cellulose/TiO_2 hydrogel beads significantly enhanced their MB and MO adsorption capacities. However, many dye molecules adsorbed on the surface of the millimeter-sized composite beads hindered UV light transmission, resulting in reduced photodegradation activity.

To overcome this limitation, we developed cellulose/TiO_2 hydrogel microbeads through the sol–gel transition of an [Emim][Ac]-in-oil emulsion. These microbeads exhibited 4.6- and 2.8-fold higher adsorption capacities and photodegradation activities, respectively, than millimeter-sized beads. Furthermore, the cellulose/TiO_2 microbeads demonstrated superior dye removal efficiency for a wide range of dyes, including CR, MO, MB, CV, and RB, surpassing the performance of millimeter-sized beads. This demonstrated that the reduction in the size of the cellulose/TiO_2 hydrogel beads successfully overcame the limitation of light transmission, leading to improved photocatalytic performance.

Moreover, we expanded the industrial applicability of the cellulose/TiO_2 microbeads by incorporating their magnetic properties. By optimizing the content and ratio of TiO_2 and Fe_2O_3, we could fine-tune the adsorption capacity and photodegradation activity of the cellulose/TiO_2 magnetic microbeads. In addition, we explored the potential of blended, biopolymer-based, photocatalytic magnetic microbeads to further enhance the dye removal efficiency of cellulose/TiO_2 magnetic microbeads. The cellulose/carrageenan/TiO_2 magnetic microbeads exhibited significantly higher dye removal efficiencies than the cellulose/TiO_2 magnetic microbeads for MB.

The microbeads outperformed the millimeter-sized beads regarding the efficiency and time required for dye removal from aqueous solutions. Furthermore, the physicochemical properties of the biopolymer-based magnetic microbeads can be easily controlled by altering the type of modifier, adjusting the TiO_2 and magnetic particle contents, and varying the ratio of each component based on the target molecule. These findings highlight the potential of biopolymer-based photocatalytic magnetic microbeads for diverse applications in environmental fields and biomedical areas, such as tissue engineering, drug delivery systems, and biosensors.

4. Materials and Methods

4.1. Materials

Microcrystalline cellulose, titanium oxide (TiO_2), iron (III) oxide (Fe_2O_3), chitosan (low molecular weight), κ-carrageenan, methyl orange (MO), and 1-ethyl-3-methylimidazolium acetate ([Emim][Ac]) were purchased from Sigma-Aldrich (St. Louis, MO, USA). Span 80, Congo red (CR), methylene blue (MB), crystal violet (CV), rhodamine B (RB), ethanol, isopropanol, hexane, and high-performance liquid chromatography (HPLC) grade water were purchased from Samchun Pure Chemical Co., Ltd. (Gyeonggi-do, Republic of Korea).

Multi-walled carbon nanotubes (CMP-310F) were obtained from Hanwha Nanotech Co., Ltd. Incheon, Republic of Korea). All other chemicals used in this study were of analytical grade and used without further purification.

4.2. Preparation of Biopolymer/TiO_2 Hydrogel Beads

Millimeter-sized biopolymer/TiO_2 hydrogel beads were prepared using the sol–gel transition method with [Emim][Ac] as the solvent (Figure 6). Cellulose (7% w/v), TiO_2 (0.5% w/v), and 0.5% (w/v)-modifying additives (CNT, chitosan, or κ-carrageenan) were added to [Emim][Ac] and mixed in a mortar. The blended solutions were then incubated at 100 °C, with stirring for 3 h. For the fabrication of millimeter-sized hydrogel beads, 1 mL of the prepared mixture solution was added dropwise into ethanol using a 1 mL syringe with a 26-gauge needle and syringe pump (LSP01-2A; Longer Pump, Baoding city, China) at a rate of 50 μL/min, with stirring. The hydrogel beads were cured in ethanol for 1 h, washed with deionized water (DW), and then washed again with DW until the removal of [Emim][Ac] was confirmed by measuring the optical density of the washing solution at 211 nm [15].

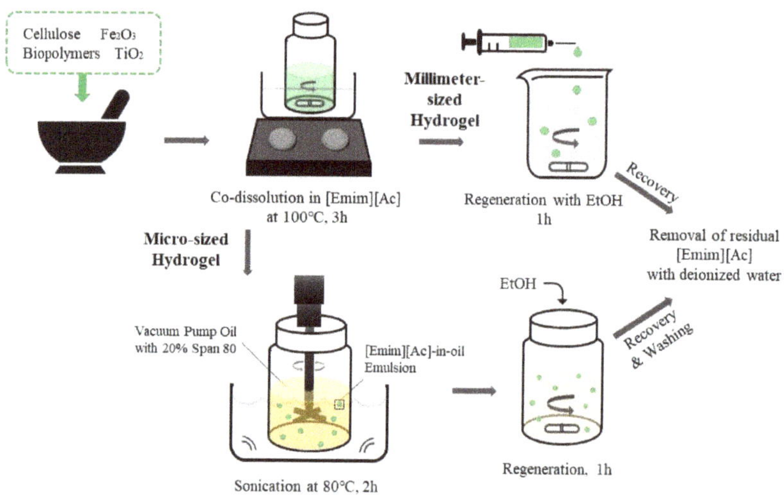

Figure 6. Schematic representation of the preparation process of biopolymer/TiO_2 composite hydrogel beads.

4.3. Preparation of Biopolymer/TiO_2 Hydrogel Microbeads

Biopolymer/TiO_2 hydrogel microbeads were prepared via the sol–gel transition method using an [Emim][Ac]-in-oil emulsion (Figure 6) [34]. Cellulose (7% w/v), TiO_2 (0.5%, 2%, 3.5%, 5%, or 7% w/v), Fe_2O_3 magnetic particles (0%, 0.1%, 0.5%, 0.75%, or 1% w/v), and 0.5% (w/v) modifying additives (CNT, chitosan, or carrageenan) were added to [Emim][Ac] and mixed in a mortar. The blended solutions were then incubated at 100 °C, with stirring for 3 h. To fabricate the microbeads, 3 mL of the prepared solution was added to 30 mL of vacuum pump oil containing 20% (v/v) Span 80, with mechanical stirring at 80 °C, followed by sonication in a sonic bath for 2 h. The resulting emulsion was slowly cooled to room temperature. Ethanol was added to the [Emim][Ac]-in-oil emulsion under stirring, and the mixture was stirred for an additional 1 h to regenerate cellulose and biopolymers. The resulting hydrogel microbeads were harvested by centrifugation, and the liquid phase was removed. The collected hydrogel microbeads were washed in the following order: n-hexane, isopropanol, aqueous ethanol, and water to remove vacuum oil, surfactant, and [Emim][Ac]. The prepared hydrogel microbeads were stored in HPLC-grade water until further use.

4.4. Characterization of Biopolymer/TiO_2 Hydrogel Beads

The dry weight of the biopolymer/TiO_2 hydrogel beads was measured with 30 beads of millimeter size or 2 mL of dispersed microbead solution after drying at 60 °C. The mean diameter of the hydrogel microbeads was measured using a particle size analyzer (Mastersizer 2000; Malvern, UK). To investigate the surface of the biopolymer/TiO_2 composite hydrogel beads, they were frozen overnight at −70 °C and then dried at −80 °C under vacuum for 24 h. All freeze-dried samples were sputter-coated with gold before surface observation under a scanning electron microscope (Carl Zeiss, Oberkochen, Germany). The XRD patterns of the hydrogel beads were analyzed after grinding the freeze-dried samples using a homogenizer. The samples were scanned using a D8 Advance Diffractometer (Bruker, Madison, WI, USA) at 40 kV and 40 mA. The 2θ angle was scanned from 10° to 40° at a width of 0.1°.

4.5. Adsorption and Photodegradation of Dyes

To determine the adsorption capacity and photodegradation activity of the biopolymer/TiO_2 composite hydrogel beads, the hydrogel beads (5 mg or 10 mg as dry weight) were immersed in 4 mL of dye solutions with shaking at 80 rpm under dark conditions at 25 °C until equilibrium was reached. After the adsorption process reached equilibrium, the hydrogel beads were exposed to UV light (20 W BL lamp; ALIM, Republic of Korea) for 6 h. Aliquots were harvested periodically, diluted with water, and centrifuged to obtain the supernatants. The dye concentrations in the sample solutions were calculated by measuring absorbance using a spectrophotometer. The extinction coefficients of MB, MO, CR, CV, and RB were 74,000 (664 nm), 21,600 (464 nm), 270,000 (499 nm), 87,000 (590 nm), and 106,000 (555 nm) cm^{-1} M^{-1}, respectively. All experiments were conducted three times, and the results were averaged.

The adsorption capacity of the biopolymer/TiO_2 hydrogel beads (q_e, mg/g) was calculated using the following equation:

$$q_e = \frac{(C_0 - C_e)}{m} \times V, \tag{1}$$

where C_0 is the initial dye concentration (mg/L), C_e is the equilibrium concentration of the dye solution (mg/L), m is the weight of the beads (g), and V is the volume of the adsorption solution (L).

The photocatalytic activities of the biopolymer/TiO_2 hydrogel beads (k, min^{-1}) were determined using the following equation, fitted with the change in dye concentration (Figure S4):

$$\ln\left(\frac{C_0}{C_t}\right) = kt, \tag{2}$$

where C_0 is the initial concentration of the dye (mg/L), C_t is the remaining concentration of the dye solution (mg/L), k is the reaction rate constant (min^{-1}), and t is the time (min).

Supplementary Materials: The following supporting information can be downloaded from https://www.mdpi.com/article/10.3390/gels9080630/s1, Figure S1: SEM images of freeze-dried biopolymers/TiO_2 beads. (a,b): cellulose; (c,d): cellulose/TiO_2; (e,f): cellulose/chitosan; (g,h): cellulose/chitosan/TiO_2; (i,j): cellulose/carrageenan; (k,l): cellulose/carrageenan/TiO_2. Figure S2: X-ray diffraction patterns of (a) cellulose beads, (b) TiO_2, and (c) cellulose/TiO_2 beads. The contents of cellulose and TiO_2 in the hydrogel bead-forming solutions were 7% and 0.5%, respectively. Figure S3: Recovery of cellulose/TiO_2/Fe_2O_3 hydrogel microbeads using a neodymium magnet. Figure S4: Plotting of "ln (C_0/C_t) vs. t" to determine the photodegradation activity of cellulose/TiO_2 millimeter-sized hydrogel beads (•) and cellulose/TiO_2 hydrogel microbeads (○).

Author Contributions: Conceptualization, S.P. and S.H.L.; Methodology, S.H.W. and J.H.; Validation, Y.-K.C.; Formal analysis, S.P. and S.H.W.; Investigation, S.P. and J.H.; Resources, S.H.L.; Writing—original draft preparation, S.H.W. and J.H.; Writing—review and editing, S.P. and S.H.L. All authors have read and agreed to the published version of the manuscript.

Funding: This study was supported by a National Research Foundation of Korea (NRF) grant funded by the Korean government (MSIT) [2021R1F1A1063364], a grant from the Development of Technology for Biomaterialization of Marine Fishery Byproducts of the Korea Institute of Marine Science & Technology Promotion (KIMST) funded by the Ministry of Oceans and Fisheries [KIMST-20220128], the Korea Water Cluster as a 2023 Demand-based Carbon-neutrality Water-tech Demonstration Supporting Project [B0080612001292], and a grant [21163MFDS501] from the Ministry of Food and Drug Safety in 2021. This study was partially supported by the LCA Expert Training Graduate Program for Knowledge-Based Environmental Services of the Ministry of Environment, Korea.

Institutional Review Board Statement: Not applicable.

Informed Consent Statement: Not applicable.

Data Availability Statement: Not applicable.

Conflicts of Interest: The authors declare no conflict of interest.

References

1. Zhang, H.; Chen, G.; Bahnemann, D.W. Photoelectrocatalytic materials for environmental applications. *J. Mater. Chem.* **2009**, *19*, 5089–5121. [CrossRef]
2. Amir, M.N.I.; Julkapli, N.M.; Hamid, S.B.A. Effective adsorption and photodegradation of methyl orange by TiO_2-chitosan supported glass plate photocatalysis. *Mater. Technol.* **2017**, *32*, 256–264. [CrossRef]
3. Khataee, A.R.; Kasiri, M.B. Photocatalytic degradation of organic dyes in the presence of nanostructured titanium dioxide: Influence of the chemical structure of dyes. *J. Mol. Catal. A-Chem.* **2010**, *328*, 8–26. [CrossRef]
4. Nawi, M.A.; Jawad, A.H.; Sabar, S.; Ngah, W.S.W. Immobilized bilayer TiO_2/chitosan system for the removal of phenol under irradiation by a 45 watt compact fluorescent lamp. *Desalination* **2011**, *280*, 288–296. [CrossRef]
5. Hosseini, N.; Toosi, M.R. Combined adsorption process and photocatalytic degradation of some commercial herbicides over N-doped TiO_2 particles supported on recyclable magnetic hexagonal mesoporous silica. *Sep. Sci. Technol.* **2019**, *54*, 1697–1709. [CrossRef]
6. Fiorenza, R.; Mauro, A.D.; Cantarella, M.; Iaria, C.; Scalisi, E.M.; Brundo, M.V.; Gulino, A.; Spitaleri, L.; Nicotra, G.; Dattilo, S.; et al. Preferential removal of pesticides from water by molecular imprinting on TiO_2 photocatalysts. *Chem. Eng. J.* **2020**, *379*, 122309. [CrossRef]
7. Tran, V.V.; Park, D.; Lee, Y.C. Hydrogel applications for adsorption of contaminants in water and wastewater treatment. *Environ. Sci. Pollut. Res.* **2018**, *25*, 24569–24599. [CrossRef]
8. Deng, J.; Liang, W. Fang J Liquid crystal droplet-embedded biopolymer hydrogel sheets for biosensor applications. *ACS Appl. Mater. Interfaces* **2016**, *8*, 3928–3932. [CrossRef]
9. Jo, S.; Park, S.; Oh, Y.; Hong, J.; Kim, H.J.; Kim, K.J.; Oh, K.K.; Lee, S.H. Development of cellulose hydrogel microspheres for lipase immobilization. *Biotechnol. Bioprocess Eng.* **2019**, *24*, 145–154. [CrossRef]
10. Park, S.; Oh, Y.; Jung, D.; Lee, S.H. Effect of cellulose solvents on the characteristics of cellulose/Fe_2O_3 hydrogel microspheres as enzyme supports. *Polymers* **2020**, *12*, 1869. [CrossRef]
11. Chang, C.; Zhang, L. Cellulose-based hydrogels: Present status and application prospects. *Carbohydr. Polym.* **2011**, *84*, 40–53. [CrossRef]
12. Wendler, F.; Persin, Z.; Stana-Kleinschek, K.; Reischl, M.; Ribitsch, V.; Bohn, A.; Fink, H.-P.; Meister, F. Morphology of polysaccharide blend fibers shaped from NaOH, N-methylmorpholine-N-oxide and 1-ethyl-3-methylimidazolium acetate. *Cellulose* **2011**, *18*, 1165–1178. [CrossRef]
13. Liu, Z.; Huang, H. Preparation and characterization of cellulose composite hydrogels from tea residue and carbohydrate additives. *Carbohydr. Polym.* **2016**, *147*, 226–233. [CrossRef]
14. Park, S.; Oh, Y.; Yun, J.; Yoo, E.; Jung, D.; Park, K.S.; Oh, K.K.; Lee, S.H. Characterization of blended cellulose/biopolymer films prepared using ionic liquid. *Cellulose* **2020**, *27*, 5101–5119. [CrossRef]
15. Kim, M.H.; An, S.; Won, K.; Kim, H.J.; Lee, S.H. Entrapment of enzymes into cellulose–biopolymer composite hydrogel beads using biocompatible ionic liquid. *J. Mol. Catal. B-Enzym.* **2012**, *75*, 68–72. [CrossRef]
16. Park, S.; Kim, S.H.; Won, K.; Choi, J.W.; Kim, Y.H.; Kim, H.J.; Yang, Y.; Lee, S.H. Wood mimetic hydrogel beads for enzyme immobilization. *Carbohydr. Polym.* **2015**, *115*, 223–229. [CrossRef] [PubMed]
17. Prasad, K.; Kaneko, Y.; Kadokawa, J. Novel gelling systems of κ-, ι- and λ-carrageenans and their composite gels with cellulose using ionic liquid. *Macromol. Biosci.* **2009**, *9*, 376–382. [CrossRef]
18. Idris, A.; Vijayaraghavan, R.; Rana, U.A.; Patti, A.F.; MacFarlane, D.R. Dissolution and regeneration of wool keratin in ionic liquids. *Green Chem.* **2014**, *16*, 2857–2864. [CrossRef]
19. Simmons, T.J.; Lee, S.H.; Miao, J.; Miyauchi, M.; Park, T.; Bale, S.S.; Pangule, R.; Bult, J.; Martin, J.G.; Dordick, J.S.; et al. Preparation of synthetic wood composites using ionic liquids. *Wood Sci. Technol.* **2011**, *45*, 719–733. [CrossRef]
20. Wang, J.; Wei, L.; Ma, Y.; Li, K.; Li, M.; Yu, Y.; Wang, L.; Qiu, H. Collagen cellulose hydrogel beads reconstituted from ionic liquid solution for Cu(II) adsorption. *Carbohydr. Polym.* **2013**, *98*, 736–743. [CrossRef]

21. Silva, S.S.; Santos, T.C.; Cerqueira, M.T.; Marques, A.P.; Reys, L.L.; Silva, T.H.; Caridade, S.G.; Mano, J.F.; Reis, R.L. The use of ionic liquids in the processing of chitosan/silk hydrogels for biomedical applications. *Green Chem.* **2012**, *14*, 1463–1470. [CrossRef]
22. Yang, J.; Liu, D.; Song, X.; Zhao, Y.; Wang, Y.; Rao, L.; Fu, L.; Wang, Z.; Yang, X.; Li, Y.; et al. Recent progress of cellulose-based hydrogel photocatalysts and their applications. *Gels* **2022**, *8*, 270. [CrossRef] [PubMed]
23. Amaly, N.; El-Moghazy, A.Y.; Nitin, N.; Sun, G.; Pandey, P.K. Synergistic adsorption-photocatalytic degradation of tetracycline by microcrystalline cellulose composite aerogel dopped with montmorillonite hosted methylene blue. *Chem. Eng. J.* **2022**, *430*, 133077. [CrossRef]
24. Yue, Y.; Shen, S.; Cheng, W.; Han, G.; Wu, Q.; Jiang, J. Construction of mechanically robust and recyclable photocatalytic hydrogel based on nanocellulose-supported CdS/MoS_2/Montmorillonite hybrid for antibiotic degradation. *Colloids Surf. A Physicochem. Eng. Asp.* **2022**, *636*, 128035. [CrossRef]
25. Zhang, H.; Zhu, J.; Hu, Y.; Chen, A.; Zhou, L.; Gao, H.; Liu, Y.; Liu, S. Study on photocatalytic antibacterial and sustained-release properties of cellulose/TiO_2/b-CD composite hydrogel. *J. Nanomater.* **2019**, *2019*, 2326042. [CrossRef]
26. Jo, S.; Oh, Y.; Park, S.; Kan, E.; Lee, S.H. Cellulose/carrageenan/TiO_2 nanocomposite for adsorption and photodegradation of cationic dye. *Biotechnol. Bioprocess Eng.* **2017**, *22*, 734–738. [CrossRef]
27. Mohamed, M.A.; Salleh, W.N.W.; Jaafar, J.; Ismail, A.F.; Mutalib, M.A.; Jamil, S.M. Incorporation of N-doped TiO_2 nanorods in regenerated cellulose thin films fabricated from recycled newspaper as a green portable photocatalyst. *Carbohydr. Polym.* **2015**, *133*, 429–437. [CrossRef]
28. Chen, Y.; Xiang, Z.; Wang, D.; Kang, J.; Qi, H. Effective photocatalytic degradation and physical adsorption of methylene blue using cellulose/GO/TiO_2 hydrogels. *RSC Adv.* **2020**, *10*, 23936. [CrossRef]
29. Wittmar, A.S.M.; Fu, Q.; Ulbricht, M. Photocatalytic and magnetic porous cellulose macrospheres for water purification. *Cellulose* **2019**, *26*, 4563–4578. [CrossRef]
30. Sun, X.; Wang, K.; Shu, Y.; Zou, F.; Zhang, B.; Sun, G.; Uyama, H.; Wang, X. One-pot route towards active TiO_2 doped hierarchically porous cellulose: Highly efficient photocatalysts for methylene blue degradation. *Materials* **2017**, *10*, 373. [CrossRef]
31. Zhu, J.-L.; Wang, M.-L.; Shi, S.-C.; Ren, J.-X.; Huang, H.-D.; Lin, W.; Li, Z.-M. In-situ constructing robust cellulose nanocomposite hydrogel network with well-dispersed dual catalysts for the efficient, stable and recyclable photo-Fenton degradation. *Cellulose* **2022**, *29*, 1929–1942. [CrossRef]
32. Zheng, A.L.T.; Sabidi, S.; Ohno, T.; Maeda, T.; Andou, Y. Cu_2O/TiO_2 decorated on cellulose nanofiber/reduced graphene hydrogel for enhanced photocatalytic activity and its antibacterial applications. *Chemosphere* **2022**, *286*, 131731. [CrossRef] [PubMed]
33. Gennari, A.; Führ, A.J.; Volpato, G.; de Souza, C.F.V. Magnetic cellulose: Versatile support for enzyme immobilization-A Review. *Carbohydr. Polym.* **2020**, *246*, 116646. [CrossRef] [PubMed]
34. Park, S.; Oh, Y.; Yun, J.; Yoo, E.; Jung, D.; Oh, K.K.; Lee, S.H. Cellulose/biopolymer/Fe_3O_4 hydrogel microbeads for dye and protein adsorption. *Cellulose* **2020**, *27*, 2757–2773. [CrossRef]
35. Liu, Z.; Wang, H.; Li, B.; Jiang, Y.; Yu, G.; Mu, X. Biocompatible magnetic cellulose-chitosan hybrid gel microspheres reconstituted from ionic liquids for enzyme immobilization. *J. Mater. Chem.* **2012**, *22*, 15085–15091. [CrossRef]
36. Xue, F.; Chen, Q.; Li, Y.; Liu, E.; Li, D. Immobilized lysozyme onto 1,2,3,4-butanetetracarboxylic (BTCA)-modified magnetic cellulose microsphere for improving bio-catalytic stability and activities. *Enzyme Microb. Technol.* **2019**, *131*, 109425. [CrossRef]
37. Ren, X.; Chen, C.; Nagatsu, M.; Wang, W. Carbon nanotubes as adsorbents in environmental pollution management: A review. *Chem. Eng. J.* **2011**, *170*, 395–410. [CrossRef]
38. Lam, S.M.; Sin, J.C.; Abdullah, A.Z.; Mohamed, A.R. Photocatalytic TiO_2/carbon nanotube nanocomposites for environmental applications: An overview and recent developments. *Fuller. Nanotub.* **2014**, *22*, 471–509. [CrossRef]
39. Kim, H.J.; Park, S.; Kim, S.H.; Kim, J.H.; Yu, H.; Kim, H.J.; Yang, Y.; Kan, E.; Kim, Y.H.; Lee, S.H. Biocompatible cellulose nanocrystals as supports to immobilize lipase. *J. Mol. Catal. B-Enzym.* **2015**, *122*, 170–178. [CrossRef]
40. Fatimah, I.; Purwiandono, G.; Hidayat, A.; Sagadevan, S.; Kamari, A. Mechanistic insight into the adsorption and photocatalytic activity of a magnetically separable γ-Fe_2O_3/Montmorillonite nanocomposite for rhodamine B removal. *Chem. Phys. Lett.* **2022**, *792*, 139410. [CrossRef]

gels

MDPI

Article

Lecithin as an Effective Modifier of the Transport Properties of Variously Crosslinked Hydrogels

Richard Heger *, Natalia Zinkovska, Monika Trudicova, Martin Kadlec, Miloslav Pekar * and Jiri Smilek *

Institute of Physical and Applied Chemistry, Faculty of Chemistry, Brno University of Technology, 61200 Brno, Czech Republic; xczinkovska@fch.vut.cz (N.Z.); xctrudicova@fch.vut.cz (M.T.); martin.kadlec@vut.cz (M.K.)

* Correspondence: xchegerr@fch.vut.cz (R.H.); pekar@fch.vut.cz (M.P.); smilek@fch.vut.cz (J.S.)

Abstract: Transport properties are one of the most crucial assets of hydrogel samples, influencing their main application potential, i.e., as drug carriers. Depending on the type of drug or the application itself, it is very important to be able to control these transport properties in an appropriate manner. This study seeks to modify these properties by adding amphiphiles, specifically lecithin. Through its self-assembly, lecithin modifies the inner structure of the hydrogel, which affects its properties, especially the transport ones. In the proposed paper, these properties are studied mainly using various probes (organic dyes) to effectively simulate drugs in simple release diffusion experiments controlled by UV-Vis spectrophotometry. Scanning electron microscopy was used to help characterize the diffusion systems. The effects of lecithin and its concentrations, as well as the effects of variously charged model drugs, were discussed. Lecithin decreases the values of the diffusion coefficient independently of the dye used and the type of crosslinking. The ability to influence transport properties is better observed in xerogel samples. The results, complementing previously published conclusions, showed that lecithin can alter a hydrogel's structure and therefore its transport properties.

Keywords: lecithin; hydrogel; diffusion; scanning electron microscopy; extracellular matrix; transport properties; model drugs

1. Introduction

Hydrogels are hydrophilic polymeric materials with a three-dimensional structure (or 3D networks of hydrophilic polymers) capable of swelling in aqueous solutions or biological fluids and also possessing unique properties, such as biocompatibility, biodegradability, superabsorbency, viscoelasticity, softness, and responsiveness to external stimuli (pH and temperature) [1–3]. The presence of these properties makes hydrogels a promising material in biomedicine, tissue engineering, and especially the preparation of drug-delivery systems with controlled release [4,5].

In the case of designing systems for the controlled release and delivery of drugs, special attention should be given to the mechanism of transportation and its correlation with the microstructure of hydrogels. The physical and mechanical properties of such systems can easily be improved (1) by adding a structure modifier; (2) by means of binary systems composed of two or more mixed polymers, such as interpenetrating polymer networks; (3) by using self-reinforced composite materials composed of fibers embedded in a matrix of the same polymer; and (4) by means of the sol-gel reaction to produce silica reinforcement [6]. Another alternative way to enhance the mechanical properties of hydrogel materials is to incorporate nanomaterials, such as silica, graphene, graphene nanotubes, carbon nanotubes, and plant fiber-based nanofibers, etc. For example, Mallikarjun B. Jalageri and G. C. Mohan Kumar, in their article, described a new material for designing cartilage scaffolds based on polyvinyl alcohol/polyvinyl pyrrolidone hydrogel reinforced by hydroxyapatite [7]. The addition of hydroxyapatite nanorods (HNr) improves the mechanical properties and morphological characteristics of PVA/PVP hydrogels. Soher N.

Jayash et al. prepared novel hybrid hydrogels based on chitosan and silica using the sol-gel technique [8]. Such hydrogel systems exhibited improved degradation and mechanical properties, no significant cytotoxic effects, and high loading efficiencies. Eric A. Appel and coauthors developed self-assembled hydrogels prepared by means of non-covalent interactions between hydroxypropylmethylcellulose derivatives (HPMC-x) and core-shell nanoparticles (NPs) [9]. This approach enabled the creation of materials with a dual loading of hydrophobic and hydrophilic molecules and high biocompatibility, which are appropriate for controlled drug delivery applications.

The characterization of hydrogels with a modified internal structure is often based on the analysis of their mechanical properties as well as their biocompatibility and biodegradability. In our previous paper [10], we provided a detailed and comprehensive analysis of the impact of lecithin (a naturally occurring mixture of phosphatides) on the morphological and rheological properties of hydrogels with different crosslinking mechanisms. The kinetics of drying and further swelling of these systems were studied as well. In addition to this valuable information, it is also essential to study the transport behavior of these materials in light of their potential application as controlled release systems.

Transport properties generally include viscosity, thermal conductivity, and the diffusion coefficient (diffusivity), which indicates the rate at which specific (per unit volume) momentum, heat, or mass are transferred [11]. Studying diffusion in hydrogels is extremely important in order to predict diffusant transport with respect to their potential application as drug delivery systems and also in order to extract information on polymer network structures [12]. Therefore, it should be kept in mind that the choice of an appropriate diffusion measurement method is essential since there are a huge number of different methods and instrumental techniques, as well as their combinations, and none of them is universal. Furthermore, experimental methods can be divided into a few main categories. The first (traditional) comprises concentration gradient methods (steady-state), in which a diffusional steady-state flux is set up through a gel diaphragm separating two liquid-filled, stirred compartments, and the effective diffusion coefficient is calculated. The second comprises concentration gradient methods (non-steady-state). This category includes, for example, uptake/release from a hydrogel immersed in a stirred solution [13]. In this case, the decrease or increase in the amount of diffusant in the outer solution is measured, and subsequently, the diffusion coefficient is calculated. The next category covers advanced instrumental techniques, such as diffusion-sensitive NMR, which enables the evaluation of solute self-diffusion [14–16]; fluorescence recovery after photobleaching (FRAP), which involves tracking the intensity distribution around the photobleached spot over time, as well as the use of powerful mathematical tools to characterize the diffusion coefficients of diffusants [17]; and fluorescence correlation spectroscopy (FCS), which allows the direct measurement of the translational diffusion of solutes [18,19]. The last category includes computer simulation studies on solute diffusion in gels, such as coarse-grained (CG) simulations, which are widely used to simulate polymer networks and also to predict the impact of formulation and processing parameters on the kinetics of drug release [20,21].

In this article, we present a study of the transport properties of hydrogel and xerogel systems determined on the basis of the given release experiments. In general, we investigated the diffusion of organic dyes (rhodamine 6G as a model of positively charged molecules; eosin B and amido black 10 B as models of negatively charged molecules) and calculated the diffusion coefficients for hydrogel systems with different weight concentrations of lecithin as a structure modifier.

2. Results and Discussion

According to previously published data and to extend the results obtained in the previous article [10], the same types of crosslinking and materials were chosen. For the formation of hydrogel matrices, the following materials were used: the linear polysaccharide agarose, as an example of physical crosslinking; sodium alginate crosslinked by calcium chloride, as an example of an ionic type of crosslinking; and poly(vinyl alcohol)-chitosan crosslinked by epichlorohydrin, as an example of a hydrogel with chemical crosslinking. Crosslinking is one of the main factors that influences the transport properties of the hydrogel (Figure S1), and thus it is important to take this factor into account when trying to characterize the influence of lecithin on the different hydrogel and xerogel samples.

For each type of crosslinking, samples differing in lecithin concentration were investigated. Lecithin concentrations were selected according to a previous study, and the samples were labeled according to these concentrations ("0.5", "1", and "2"); the reference sample without lecithin addition was marked as "R".

The probes that were chosen to study the transport properties of the hydrogel samples differed in some of their physicochemical properties (electric charge and molecular weight) as well as in some of their visual properties, such as their absorption maximum. Specifically, the probes chosen were the positively charged rhodamine 6G (479.02 $g{\cdot}mol^{-1}$) and the negatively-charged dyes eosin B (580.09 $g{\cdot}mol^{-1}$) and amido black 10B (616.50 $g{\cdot}mol^{-1}$).

Crucial transport parameters were obtained for hydrogel and xerogel samples placed in the container filled with distilled water, and the amount of released dye was controlled and measured by a UV-Vis spectrophotometer set up for hydrogels. In the case of xerogels, the same optical probes were used, but due to the very fast release of dye from the xerogel samples, more frequent measurements were recorded (also using UV-Vis). The supporting method was scanning electron microscopy (SEM), from which images of the studied hydrogels were obtained and further compared with previously published xerogel images.

2.1. *Physical Crosslinking*

To characterize the transport properties of the physically crosslinked hydrogel matrix, agarose hydrogels modified by different additions of lecithin were studied. Agarose was chosen because, in the context of amphiphilic modifications, we try to approach the functionality and properties of extracellular matrix (ECM), and agarose hydrogels are frequently used in vitro models across numerous disciplines [22]. Hydrogel samples, as well as dried xerogel samples, were prepared with a chosen dye already incorporated into the structure (the original hydrogel samples were prepared using a solution of the dye). A schematic figure of the preparation procedure can be seen in the Supplementary materials, Figure S2.

2.1.1. Dye-Release Experiments

The effective diffusion coefficient (D_{eff}) was the crucial transport parameter that provided us with information about the transport properties of the studied hydrogels. Specifically, it is a value that describes how each individual dye is released from the hydrogel. The procedure for obtaining D_{eff} and the method to calculate it are described in Section 4.2.

Rhodamine 6G was chosen as the only positively charged probe due to its suitable molecular weight (comparable with that of cationic drugs, e.g., loperamide or ebastine) and previous experience with this organic dye [23]. The results obtained for agarose hydrogel samples modified by lecithin suggest that lecithin addition influences the transport properties of agarose hydrogels by decreasing the value of the diffusion coefficient (as seen in Table 1 and Figure 1a), which means that, in our study, the release of the model organic probe was significantly slowed down. The large deviations between individual measurements could have been caused by the use of different probes or their calibration. However, in each set of measurements, a trend of decreasing diffusion coefficient with increasing lecithin concentration was evident. The reference sample immediately released more dye than all the other samples, while the samples with lecithin released the positively charged dye significantly more slowly. This could be the result of the modified internal structure of these hydrogels due to the self-assembly of lecithin inside the hydrogel. The presence of lecithin, as well as interactions between the positively charged rhodamine 6G and the negatively charged functional groups of the lecithin molecule, enable the dye inside the hydrogel structure to become partially blocked and thus slow down or delay its release.

This would also be confirmed by a number of publications dealing with obstructions inside agarose hydrogels [24,25].

Table 1. Effective diffusion coefficients calculated from dye-release experiments for physically crosslinked agarose hydrogels and xerogels with different concentrations of lecithin and the positively charged dye rhodamine 6G.

Rhodamine 6G	Hydrogel $D_{\text{eff}} \times 10^{-10}$ ($m^2 \cdot s^{-1}$)	Xerogel $D_{\text{eff}} \times 10^{-10}$ ($m^2 \cdot s^{-1}$)
R (0 wt.%)	2.7 ± 0.2	6.4 ± 0.9
0.5 wt.%	1.2 ± 0.1	—
1 wt.%	1.0 ± 0.1	—
2 wt.%	1.2 ± 0.1	—

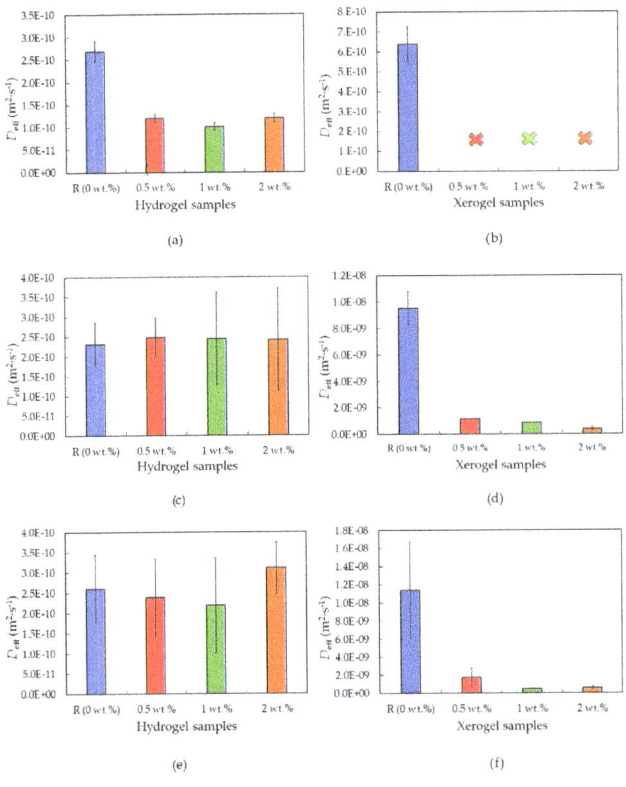

Figure 1. Effective diffusion coefficients for physically crosslinked agarose hydrogel and xerogel samples with different lecithin concentrations (0, 0.5, 1, and 2 wt.%); rhodamine 6G hydrogels (**a**), rhodamine 6G xerogels (**b**); eosin B hydrogels (**c**); eosin B xerogels (**d**); amido black 10B hydrogels (**e**); and amido black 10B xerogels (**f**).

What happens to the hydrogel after evaporation of the dispersion medium, and how does this affect the other substances in the gel? To answer this key question, the same hydrogel samples as used in the previous experiment were dried prior to characterization. In this case, however, the dye release from agarose was very quick, and it was not possible to measure the release over longer intervals as was performed for the dye release from non-dried hydrogel samples. For this reason, a fiber UV-Vis spectrometer was used, allowing the measurement of UV-Vis spectra every twenty seconds. In this setup, the absorbance was measured for approximately the first hour, which, for agarose xerogel samples, was enough time for around twenty percent of the dye to be released from the sample (for the reference sample, see Figure 1b). From the measured absorbance, the concentration was calculated.

The results from the dye-release experiments obtained for agarose xerogels loaded with rhodamine 6G confirmed the previously mentioned conclusions in Section 2.1.1. Therefore, the rhodamine 6G interacted with the lecithin modifier. This was particularly evident from the fact that it was not possible to obtain adequate values for the effective diffusion coefficient after the addition of lecithin. This hypothesis is also supported by the graphical representation of the release of the dye from the xerogel system (Figure 2a). These findings from Figure 1b show that the diffusion coefficient for the reference sample without lecithin was the only value obtainable for xerogels from the diffusion coefficient calculations (Table 1). For the lecithin additions, the amount of released dye, and thus the diffusion coefficient, was not measurable, and with increasing time, deformation of the samples occurred, which would again support the assumption of interactions between lecithin, rhodamine 6G, and agarose.

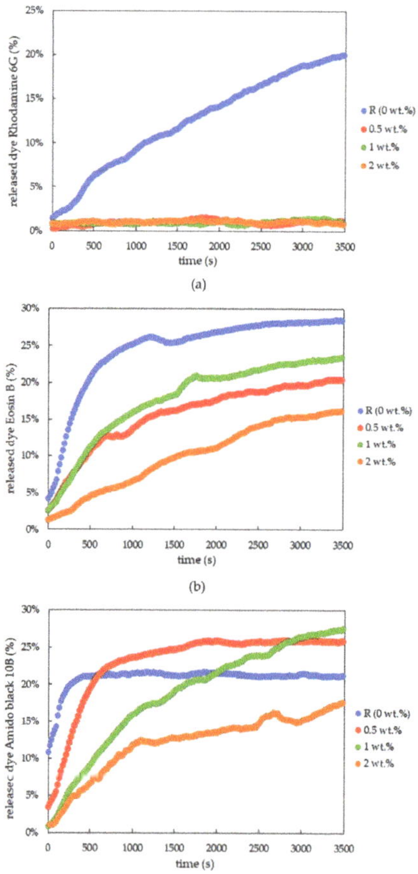

Figure 2. Time development of dye-release experiments for rhodamine 6G (**a**), eosin B (**b**), and amido black 10B (**c**), controlled by the optical probes for physically crosslinked agarose xerogels with different concentrations of lecithin.

Other drug models used were negatively charged eosin B and amido black 10B. These were chosen to study oppositely charged models and to avoid interactions with the modifier. The second studied model dye was the negatively charged eosin B. The characterization of the dye-release process in physically crosslinked hydrogels and xerogels tuned by the addition of lecithin was analogous to rhodamine 6G measurements and has already been described in Section 2.1.1. At the start of the experiment, the dye-release process was similar for all hydrogel samples. With the passage of time, the release of dye accelerated (leading to a higher concentration of dye outside of the sample) in the reference sample and in samples with smaller lecithin concentrations (Table 2). The associated diffusion coefficients were ultimately almost identical for all samples, which is explained by the fact that eosin B and hydrogel matrix interactions with each other should not occur (on the basis that eosin B can react with positively charged components [26]), and thus the diffusion of the negatively charged eosin B is not slowed down inside the hydrogel. The results of FTIR analysis using PCA showed that the main differences between the agarose gels without lecithin and gels with lecithin spectrally correspond to the characteristic bands of lecithin (Figures S3 and S4).

Table 2. Effective diffusion coefficients calculated from dye-release experiments for physically crosslinked agarose hydrogels and xerogels with different concentrations of lecithin and the negatively charged dye eosin B.

Eosin B	Hydrogel $D_{eff} \times 10^{-10}$ ($m^2 \cdot s^{-1}$)	Xerogel $D_{eff} \times 10^{-10}$ ($m^2 \cdot s^{-1}$)
R (0 wt.%)	2.3 ± 0.6	95 ± 13
0.5 wt.%	2.5 ± 0.5	$12 \pm 2 \cdot 10^{-2}$
1 wt.%	2.5 ± 1.2	$9 \pm 3 \cdot 10^{-7}$
2 wt.%	2.4 ± 1.3	4 ± 2

The results for physically crosslinked agarose xerogels and eosin B are presented in Table 2 and Figure 1d. The diffusion coefficient is greater for the reference sample than for samples with lecithin additions (0.5 and 1 wt.%), with the coefficient being the lowest for the sample with the highest lecithin concentration (2 wt.%). This suggests that lecithin heavily influences the coefficient and that even small additions decrease the coefficient significantly. The greater the concentration of lecithin, the lower the D_{eff} value, while the differences between the D_{eff} values themselves were not so substantial. However, if we ignore the reference sample, the magnitude of the diffusion coefficient decreases linearly with increasing lecithin concentration. It can be predicted that the value would decrease even lower to the point where the addition of the modifier was too great, and the dye release was almost negligible or even ceased altogether. This hypothesis is also supported by the time development of the release of eosin B (Figure 2b), which shows the fastest initial release in the reference sample. For samples with different concentrations of lecithin, it is clear that the lecithin inside the system caused a gradual slowing of the release. Samples with smaller lecithin additions (0.5 and 1 wt.%) were almost indistinguishable from each other, and the time of release was very similar. The highest lecithin addition (2 wt.%) led to the biggest retardation.

The third and last studied model dye was negatively charged amido black 10B, and the corresponding results for hydrogels are summarized in Table 3 and Figure 1e. These exhibit the same trend as the results obtained for the other negatively charged dye, eosin B (Table 2 and Figure 1c); thus, the kinetics of the released dye are very similar to each other, which is also confirmed by the calculation of very similar diffusion coefficients. Thus, it can be said that for a negatively charged model drug, the lecithin within the hydrogel does not heavily influence its release time, and the drug is easily released.

Table 3. Effective diffusion coefficients calculated from release dye experiments for physically crosslinked agarose hydrogels and xerogels with different concentrations of lecithin and the negatively charged amido black 10B.

Amido Black 10B	Hydrogel $D_{eff} \times 10^{-10}$ ($m^2 \cdot s^{-1}$)	Xerogel $D_{eff} \times 10^{-10}$ ($m^2 \cdot s^{-1}$)
R (0 wt.%)	2.6 ± 0.8	110 ± 53
0.5 wt.%	2.4 ± 1.0	18 ± 11
1 wt.%	2.2 ± 1.2	5.0 ± 0.4
2 wt.%	3.1 ± 0.1	6 ± 1

The conclusions for amido black 10B within physically-crosslinked agarose xerogels are similar to those for eosin B (Table 3 and Figure 1f). The diffusion coefficient decreases with increasing lecithin concentration in hydrogel samples in comparison with the reference sample, in which no additional factors influence the transport of amido black 10B dye. As seen from Table 3, the higher the concentration of lecithin, the lower the value of the diffusion coefficient obtained. In the case of the highest lecithin concentration (2 wt.%), dye release was substantially slowed down, probably due to the specific lecithin structures formed inside the hydrogel structure. This is also supported by the timing of the release (Figure 2c). These results are very similar to those obtained for eosin B, with the only difference being the smallest addition of lecithin (0.5 wt.%), which is more similar to the reference sample.

2.1.2. Morphological Characterization

The morphological characterizations of physically crosslinked agarose hydrogels complement the already-published SEM images of xerogel samples used in this study (SEM images already published in [10]). As stated in Section 4, hydrogel samples were shock-frozen in liquid nitrogen and immediately underwent lyophilization. Subsequently, the hydrogel structures were analyzed by SEM. The shock-freezing of samples was used to minimize the alteration of hydrogel structure caused by ice crystal growth, which can create some undesirable artifacts. Thus, the structures of freeze-dried hydrogels were similar to their structures in their native state [27].

The images obtained from SEM show differences in hydrogel structures between samples with different lecithin concentrations (Figure 3) and how these structures are potentially affected by the lecithin addition. With increasing concentrations of lecithin, the structures of hydrogels seem more altered and collapsed. Lecithin, either lecithin alone or some structure formed from it, was probably trapped in the pores of the hydrogel network, causing them to enlarge or making them more sensitive to changes caused by the growth of ice crystals. This corresponds with results from the dye-release experiments, where lecithin slowed down diffusion in hydrogels due to specific interactions between them or simply by its presence in a particular place (i.e., "obstruction effect"). At a magnification of 5000× (Figure 4), it is possible to observe pores in the structure of hydrogels (except in the sample with the highest concentration of lecithin, where the structure collapsed). These pores correspond in size to the pores in the structure of agarose hydrogels found in previous work [27].

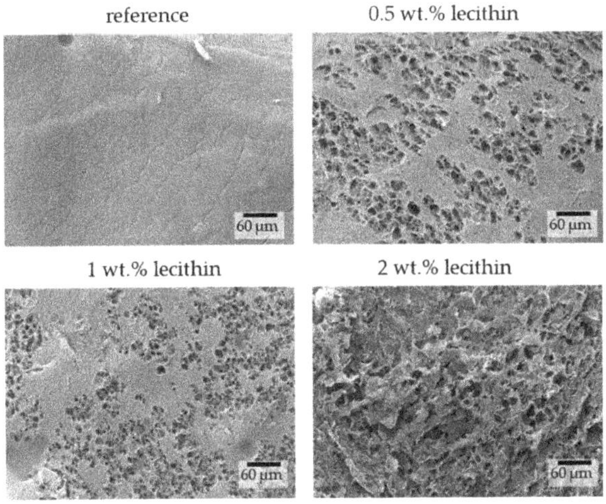

Figure 3. Surface morphologies of physically crosslinked agarose hydrogels with the addition of lecithin revealed by SEM. Original magnification: 500×.

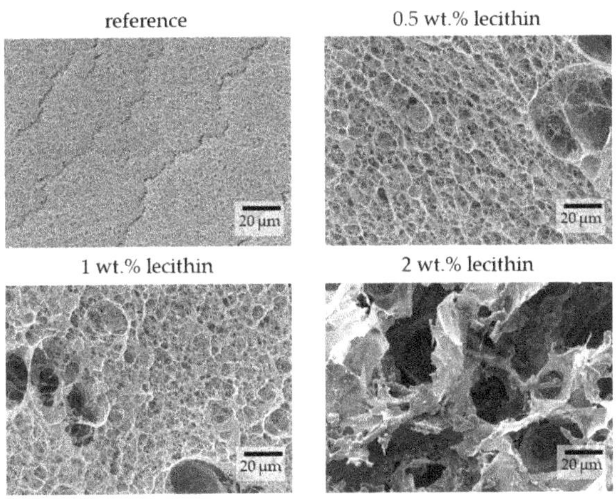

Figure 4. Surface morphologies of physically crosslinked agarose hydrogels with the addition of lecithin revealed by SEM. Original magnification: 5000×.

2.2. *Ionic Crosslinking*

As an example of an ionically crosslinked hydrogel matrix, sodium alginate crosslinked by calcium ions (2:1 wt. ratio) was studied. Bond formation resulted from the interaction of negatively charged poly(guluronic) acid units of alginate with divalent Ca^{2+} ions. Alginate has been extensively used as a hydrogel-synthetic ECM and thus is an ideal representative of an ionically crosslinked hydrogel for this study [28]. Structural modifications were evoked by lecithin additions, and the resulting transport properties were observed and compared by means of diffusion dye probes. A schematic figure of the preparation procedure can be seen in Supplementary Materials, Figure S5.

2.2.1. Dye-Release Experiments

The first studied model dye was positively charged rhodamine 6G, which was released from an ionically crosslinked alginate hydrogel. Lecithin, which acts as an internal structure modifier, was added to the hydrogel systems in different concentrations. The amount was the key factor that influenced both the hydrogel and xerogel structures and their related transport properties, which can be seen in Table 4 and Figure 5a.

Table 4. Effective diffusion coefficients calculated from dye-release experiments for ionically crosslinked alginate hydrogels and xerogels with different concentrations of lecithin and the positively charged dye rhodamine 6G.

Rhodamine 6G	Hydrogel $D_{eff} \times 10^{-10}$ ($m^2 \cdot s^{-1}$)	Xerogel $D_{eff} \times 10^{-10}$ ($m^2 \cdot s^{-1}$)
R (0 wt.%)	2.9 ± 0.2	22 ± 2
0.5 wt.%	1.5 ± 0.1	—
1 wt.%	1.2 ± 0.1	—
2 wt.%	1.2 ± 0.1	—

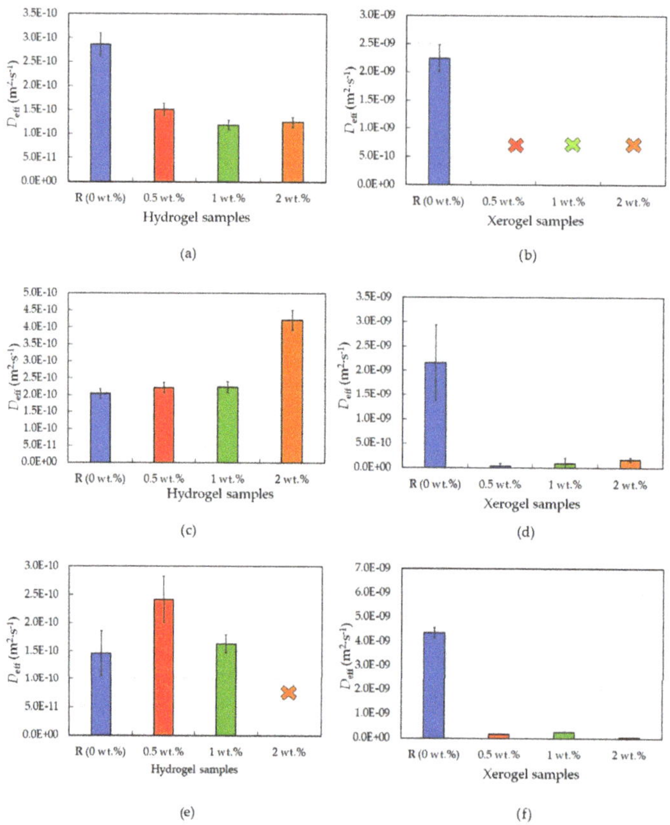

Figure 5. Effective diffusion coefficients for ionically-crosslinked alginate hydrogel and xerogel samples with different lecithin concentrations (0, 0.5, 1, and 2 wt.%); rhodamine 6G hydrogels (**a**), rhodamine 6G xerogels (**b**); eosin B hydrogels (**c**); eosin B xerogels (**d**); amido black 10B hydrogels (**e**); and amido black 10B xerogels (**f**).

The amount of released dye was the highest for the reference sample, as this hydrogel did not contain any structures or artifacts that would serve as obstacles or as storage moieties for the dye, and since the alginate hydrogel complex should act as a neutral system [29], the dye could be released out of the hydrogel without any problems. The kinetics of dye release slowed down with increasing lecithin concentration, i.e., with an increasing number of lecithin structures acting as obstacles. The associated effective diffusion coefficient (D_{eff}) was the highest for the reference sample and decreased with increasing lecithin concentration. The differences in D_{eff} values for the samples with distinct additions of lecithin were negligible and thus confirmed the insignificant influence of lecithin concentration on the final value of the diffusion coefficient (in comparison with the reference sample).

Xerogel samples for ionically crosslinked alginate yielded very similar results to those for physically crosslinked agarose samples. In the case of lecithin additions, the concentration of dye was undetectable, probably because of the blockage of positively charged dye release caused by lecithin throughout the whole duration of the experiment. The time of development of the dye release (Figure 6a) was almost identical to that for physically crosslinked hydrogels. The positively charged dye was almost immeasurable outside of the hydrogel, thus confirming interactions with the dye as well as the blockage caused by the lecithin additions [30].

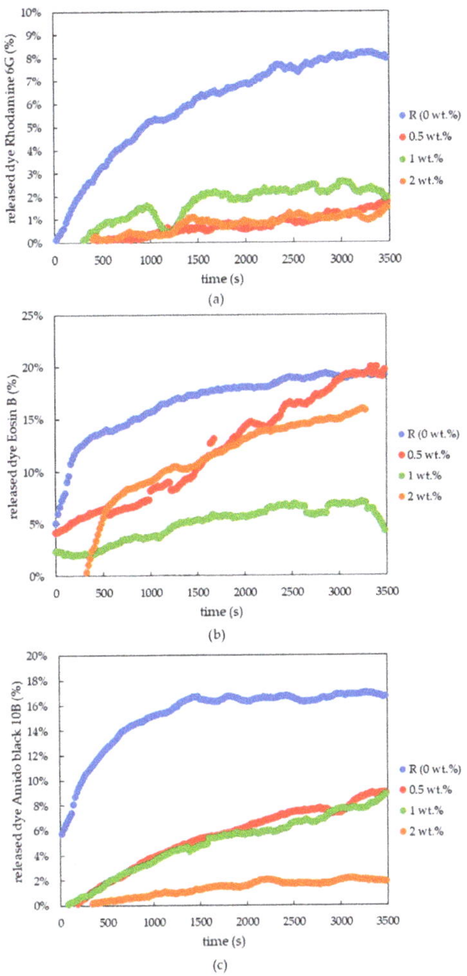

Figure 6. Time development of dye release for rhodamine 6G (**a**), eosin B (**b**), and amido black 10B (**c**), determined by optical probes for ionically crosslinked alginate xerogels with different concentrations of lecithin.

Negatively charged dyes were also studied. As is obvious from Table 5, the increasing lecithin concentration slowed down the release of the negatively charged dye eosin B from ionically crosslinked alginate hydrogels. The diffusion coefficient was then very similar for each sample, regardless of the amount of lecithin within each hydrogel. The sample with the highest lecithin concentration (2 wt.%) presented high deviation values, which were caused by the high content of lecithin released from the sample with the dye.

Table 5. Effective diffusion coefficients calculated from dye-release experiments for ionically crosslinked alginate hydrogels and xerogels with different concentrations of lecithin and the negatively charged eosin B.

Eosin B	Hydrogel $D_{eff} \times 10^{-10}$ ($m^2 \cdot s^{-1}$)	Xerogel $D_{eff} \times 10^{-10}$ ($m^2 \cdot s^{-1}$)
R (0 wt.%)	2.0 ± 0.1	22 ± 8
0.5 wt.%	2.2 ± 0.2	0.4 ± 0.6
1 wt.%	2.2 ± 0.2	0.9 ± 1.2
2 wt.%	4.2 ± 0.3	1.7 ± 0.4

The results for ionically crosslinked xerogels with negatively charged eosin B followed those for physically crosslinked xerogels with the same dye, showing that the reference sample had a much bigger diffusion coefficient than the other samples, their lower coefficients resulting from lecithin addition. According to this, it could be concluded that crosslinking does not have a considerable influence on the diffusion coefficients of xerogels for this type of model drug (dye). The timing of the release of this dye indicates that lecithin slowed down the release since alginate does not interact with eosin B [31]. As mentioned in the previous paragraph, the sample with the highest lecithin concentration (2 wt.%) initially had the slowest release, but, with time, it was possible to observe (Figure 6b) its acceleration, which was caused by the release of lecithin.

The release of the second negatively charged dye, amido black 10B, from ionically crosslinked alginate was significantly slowed down by lecithin additions. The associated diffusion coefficient increased with increasing concentration up to the 1 wt.% sample. During the experiment, the dye release from the samples with the two highest lecithin concentrations was either immeasurable or exhibited big deviations. This was caused by disintegration, which led to the release of both the dye and the lecithin. A similar order of magnitude for the diffusion coefficients (Figure 5c and Table 5) can be observed for the other negatively charged dye, eosin B; however, the differences between each sample for amido black 10b are obvious, which cannot be said for the eosin B samples. In comparison with physically crosslinked agarose hydrogels, again for amido black 10B (Figure 5e and Table 6), the coefficient has values in the same order of magnitude.

Table 6. Effective diffusion coefficients calculated from dye-release experiments for ionically crosslinked alginate hydrogels and xerogels with different concentrations of lecithin and the negatively charged amido black 10B.

Amido Black 10B	Hydrogel $D_{eff} \times 10^{-10}$ ($m^2 \cdot s^{-1}$)	Xerogel $D_{eff} \times 10^{-10}$ ($m^2 \cdot s^{-1}$)
R (0 wt.%)	1.5 ± 0.4	44 ± 2
0.5 wt.%	2.4 ± 0.4	1.8 ± 0.1
1 wt.%	1.7 ± 0.2	2.6 ± 0.1
2 wt.%	—	0.45 ± 0.02

On the other hand, ionically crosslinked xerogels with negatively charged amido black 10B show the exact same trend as physically crosslinked agarose hydrogels. The value of the diffusion coefficient decreases with increasing lecithin concentration. The results of FTIR analysis using PCA showed that the main differences between the alginate gels without and with lecithin spectrally correspond to the characteristic bands of lecithin (Figures S3 and S6).

2.2.2. Morphological Characterization

Prior to SEM observation, ionically crosslinked alginate hydrogels were shock-frozen in liquid nitrogen and subsequently lyophilized, as were the agarose hydrogels. As stated before, this procedure was used to minimize the alteration of hydrogel structure; however, for these types of hydrogels, it was not effective. As can be seen from Figure 7, the structure of all prepared hydrogels appears similar, leading to the conclusion that they are altered. This was most likely a result of the preparation of the samples. Since these hydrogels were prepared in a smaller volume than the agarose hydrogels, they may have been partially unfrozen prior to lyophilization, which would have caused ice crystal growth and thus the presence of structural artifacts [32]. These results therefore represent artifact-biased images of hydrogels in their native form.

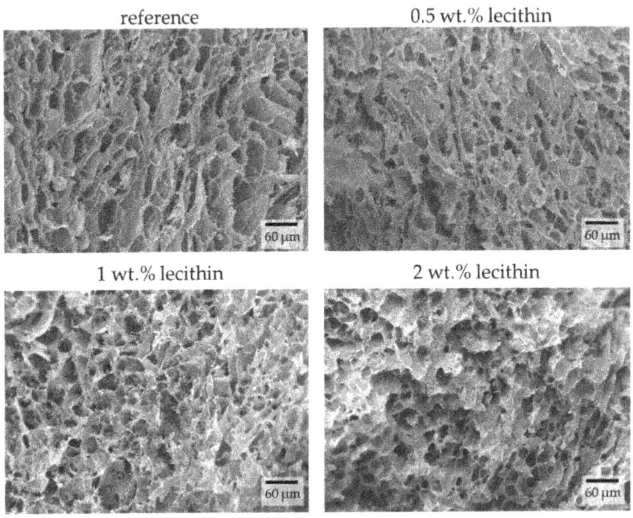

Figure 7. Surface morphologies of ionically crosslinked alginate hydrogels with the addition of lecithin revealed by SEM. Original magnification: 500×.

2.3. Chemical Crosslinking

A mixture of PVA (poly(vinyl alcohol)) with chitosan crosslinked by epichlorohydrin was studied as an example of a chemically crosslinked hydrogel. As well as other types of crosslinking, the internal structure and subsequent transport properties of samples were modified by the addition of different amounts of amphiphilic lecithin. Chemically crosslinked samples were studied only in their xerogel form since the preparation method finishes with mixtures in their dried state; thus, rehydrated samples would yield the same results (Section 4.1). The only difference was in the morphological characterization, in which samples were rehydrated to observe their structures in hydrogel form. The samples were chosen for their close similarity to ECM and their ideal biomedical properties [33]. A schematic figure of the preparation procedure can be seen in the Supplementary Materials, Figure S7.

2.3.1. Dye-Release Experiments

The main evaluated parameters were the diffusion coefficients and the amount of released dye. Three different dyes were used to study the transport properties of chemically crosslinked PVA-chitosan samples: positively charged rhodamine 6G, negatively charged eosin B, and amido black 10B. The negatively charged dyes were held inside the gel very strongly, which was because of the small pores in the highly organized structure of the chemically crosslinked gel as well as the interaction of the positively charged chitosan with the negatively charged dye. The only dye that was measurable outside the gel was released immediately after the experiment started; neither long nor short experiments, the latter focused on the first hour of the experiment, were able to provide results that could be analyzed. Only a residual amount of both of these dyes was released from the xerogels; therefore, no diffusion coefficient values could be calculated.

Only experiments with the positively charged dye, rhodamine 6G, yielded results sufficient for analyses and for obtaining information about diffusion coefficients. As can be seen in Figure 8a, the release of the dye was very fast, and the longer experiment setting was not appropriate for this type of gel. This burst release was studied on similar materials with the purpose of slowing the release [34]. Only experiments focused on the first hour of diffusion were able to yield results. Lecithin influenced the amount of dye released (Figure 8b), with the amount of released dye decreasing with increasing lecithin concentration. The same can be said about the diffusion coefficient (Figure 9 and Table 7), where the results obtained for the highest lecithin concentration (2 wt.%) are unequivocal since this lecithin concentration led to damage to the PVA-chitosan xerogel. The results of FTIR analysis using PCA showed that the main differences between the PVA-chitosan gels with and without spectroscopy correspond to the characteristic bands of lecithin (Figures S3 and S8).

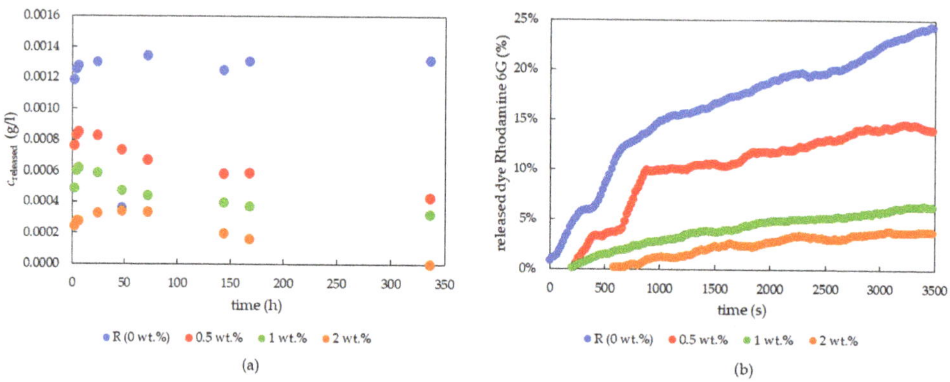

Figure 8. The concentration of rhodamine 6G released from xerogel samples over a longer period of time (**a**) and the time progress of dye-release experiments using rhodamine 6G for the first 60 min, determined by the optical probe for chemically cross-linked PVA-chitosan (**b**).

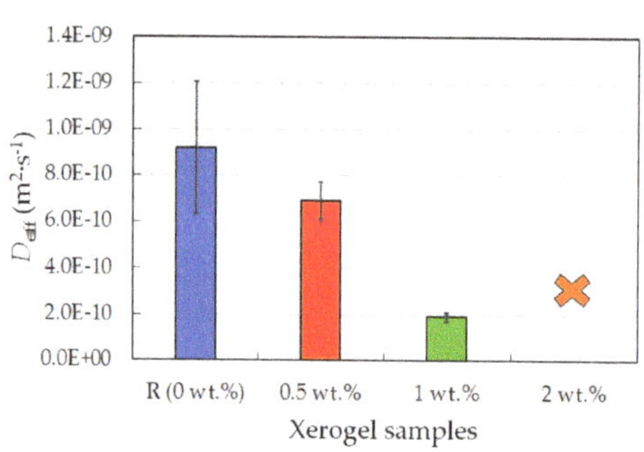

Figure 9. Effective diffusion coefficients for rhodamine 6G for chemically crosslinked PVA-chitosan xerogel samples with different lecithin concentrations (0, 0.5, 1, and 2 wt.%).

Table 7. Effective diffusion coefficient (D_{eff}) calculated from dye-release experiments for chemically crosslinked PVA-chitosan xerogels with different concentrations of lecithin and the positively charged dye rhodamine 6G.

Xerogel Sample (c of Lecithin)	R (0 wt.%)	0.5 wt.%	1 wt.%	2 wt.%
$D_{eff} \times 10^{-10}$ ($m^2 \cdot s^{-1}$)	9 ± 3	7 ± 1	1.9 ± 0.2	—

2.3.2. Morphological Characterization

The chemically crosslinked hydrogels were used in their native (hydrated) form for the preparation of samples for morphological study. The prepared xerogels were left to swell in water and were subsequently shock frozen in liquid nitrogen and freeze-dried. The images obtained using SEM show the layered structure of the hydrogels. The only observable difference between each sample is the smoothness of the cross-sectional view. With increasing lecithin content, the interlayer roughness increases (Figure 10).

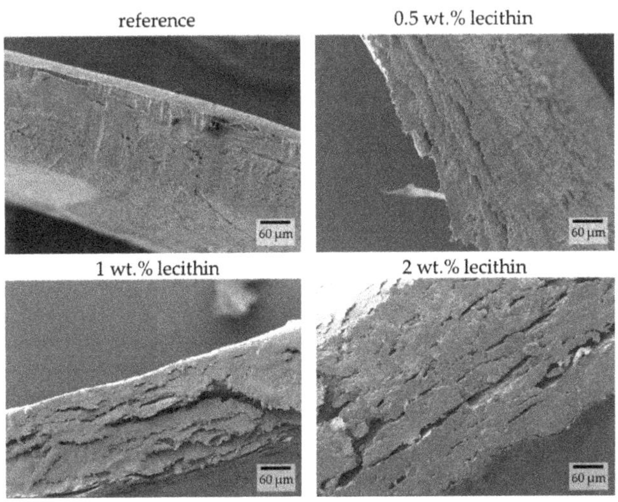

Figure 10. Cross-sectional view of chemically crosslinked PVA-chitosan hydrogels with the addition of lecithin revealed by SEM. Original magnification: 500×.

These results correspond with already-published data for xerogels of these chemically crosslinked hydrogels [10]. Since the structures of xerogels and hydrogels for this type of material are very similar, it does not matter which preparation procedure is used to observe the structure using SEM. Furthermore, since the hydrogel structure is smooth and no pores are visible, these observations correspond with results for dye release, in which only a small amount of dye is released.

3. Conclusions

This work focused on the description of the transport properties of differently crosslinked hydrogels and xerogels (physically crosslinked agarose, ionically crosslinked alginate, and chemically crosslinked PVA-chitosan), which were altered with different concentrations of amphiphilic lecithin as a structure modifier to influence the transport properties of the gels. For characterization of the transport properties of the above-mentioned hydrogel systems, various model probes with different charges were used (positively charged rhodamine 6G, negatively charged eosin B, and amido black 10B).

The positively charged dye interacted with the lecithin modifier, which meant that for the physically and ionically crosslinked hydrogels, the dye was held within the hydrogel and released very little dye or none at all, regardless of the quantity of the modifier. When the charge of the dyes did not interact with the modifier (in the case of negatively charged dyes), the quantity of lecithin had a significant impact on the speed and amount of dye release. Which meant that the diffusion coefficient was lower for the xerogel samples with lecithin compared to the reference samples. In summary, therefore, we can state that the addition of lecithin to physically or ionically crosslinked hydrogels slows down the release of the model drug and affects the amount of drug released. Chemically crosslinked hydrogels appear to be a system that is also affected by lecithin, but not as significantly as in other types of crosslinking.

This work closely follows already-published results dealing mainly with the mechanical properties of the same types of hydrogels and modifications used in this work. Together with our former work, it is possible to compare different morphological characterizations of gels in their hydrogel and xerogel forms.

4. Materials and Methods

Differently crosslinked hydrogel matrices and the effect of lecithin on their properties were studied. Three different crosslinking mechanisms were studied: physical, ionic, and chemical. Agarose E was purchased from Condalab (Madrid, Spain), calcium chloride from Lach-Ner (Neratovice, Czech Republic), and sodium alginate, poly(vinyl alcohol), chitosan, epichlorohydrin, and L-α-Phosphatidylcholine (lecithin) were all purchased from Sigma-Aldrich (Prague, Czech Republic). Dyes used for diffusion experiments differed in charge: positively charged rhodamine 6G (Sigma-Aldrich, Prague, Czech Republic), negatively charged amido black 10B (Merck, Darmstadt, Germany), and eosin B (Sigma-Aldrich, Prague, Czech Republic) were used.

4.1. Sample Preparations

Physically crosslinked hydrogels were prepared using agarose. For every sample, a mixture of agarose (1 wt.%) and lecithin (0, 0.5, 1, or 2 wt.%) was weighed in a container. A liquid dispersion medium was then added to the mixture. The dispersion medium was either pure deionized water (ELGA, PURELAB Classic) or, if needed, a solution of appropriate dye (rhodamine 6G—0.01 $g \cdot l^{-1}$, eosin B—0.015 $g \cdot l^{-1}$, or amido black 10B—0.01 $g \cdot l^{-1}$). The sample thus prepared was mixed and dissolved at an elevated temperature. It was then left to stand for 24 h to gel prior to further study [10,23].

Ionically-crosslinked hydrogels were prepared by sodium alginate (2 wt.%) crosslinked by calcium ions of calcium chloride (0.1 $mol \cdot dm^3$) at a two-to-one weight ratio. Lecithin was added to the alginate solution and stirred for 24 h. A solution of calcium chloride (if needed, the calcium chloride was dissolved in a solution of an appropriate dye) was sprayed on the layer of sodium alginate with lecithin in order to create a homogeneous hydrogel [10,28]. The sample was then left to stand for 24 h in order for bond creation to finish and for the sample to relax. The hydrogels were then ready for further investigation.

For the preparation of chemically crosslinked hydrogels, lecithin was added to the PVA solution and stirred for 24 h [10], after which the procedure closely followed that described by Garnica-Palafox et al. [33].

4.2. Diffusion Release Experiments

The most important information concerning prospective drug carriers, which were mimicked by hydrogel samples in this study, is their ability to release the drug at a optimal speed. The diffusion/dye-release experiments were the core technique for this study. Samples with different crosslinking mechanisms were studied, as well as lecithin's ability to influence this characteristic. Release characteristics were obtained for samples in their hydrogel form as well as for samples that were dried into their xerogel form. As model drugs released from the matrices, positively-charged rhodamine 6G, negatively charged eosin B, and amido black 10B were used.

The dye was added at the beginning of sample preparation. Thus, samples were prepared with the dye already in them. If the sample was supposed to be in xerogel form, hydrogels were dried to a constant weight in the laboratory oven at 45 °C. Hydrogel and xerogel samples were cut into the same-sized pieces and put into flasks with a defined volume of deionized water (hydrogels with 20 mL and xerogels with 40 mL) which was constantly stirred. The water was then routinely measured in a UV/Vis spectrophotometer every ten minutes to obtain the absorbance value for the released dye. The experiment ended when all the dye was released, the amount led to a constant weight, or the sample was damaged or destroyed.

The most fundamental parameter that can express (and consequently be used to compare) the transport properties of hydrogels with inner structural changes resulting from the addition of a modifier, such as lecithin, is the effective diffusion coefficient, expressed as the amount of dye released from the gel per $m^2 \cdot s^{-1}$.

It must be noted that calculated diffusion coefficients are not absolute values but only relative (or effective) ones. This means that porosity and tortuosity are hidden in the values of diffusion coefficients. The determination of absolute values of diffusion coefficients requires deeper research focused specifically on the diffusion phenomena of each specific hydrogel. This was not the principal aim of this publication. Therefore, we used effective values for the diffusion coefficients as proxies for comparing the transport properties of the studied hydrogels as a function of the additive concentration and/or type of crosslinking.

The time of development of the mass of released dye can be expressed as follows:

$$\frac{n_t}{n_{rov}} = \frac{4}{\delta} \cdot \sqrt{\left(\frac{D_{gel}}{\pi}\right)} \cdot \sqrt{t} \tag{1}$$

$$k = \frac{4}{\delta} \cdot \sqrt{\left(\frac{D_{gel}}{\pi}\right)} \tag{2}$$

$$D_{gel} = \left(k \cdot \frac{4}{\delta}\right)^2 \cdot \pi \tag{3}$$

The total amount of released dye at a specific time is expressed as the ratio of the amount of substance at a given time (n_t) divided by the amount of substance at equilibrium (n_{rov}). The effective diffusion coefficients were calculated from the slope of the linear regression of this ratio of the amount of substance as a function of the square root of time. The dimensions of the releasing medium (i.e., hydrogels) were considered as circles of diameter (δ). From this simple calculation, it is obvious that the release experiments must be performed until the equilibrium state is achieved (when the concentration of released dye does not change over time) [35].

During the measurements as well as after evaluation of the results from UV-Vis spectrophotometry, it was found that the release of the dyes from the xerogel samples was too fast for the experimental setup. For this reason, the xerogels were measured again using optical probes and OceanView 2.0 software (OceanInsight, Orlando, FL, USA), which allowed us to obtain UV-Vis absorbance every 20 s for the first hour of the experiment.

4.3. Scanning Electron Microscopy

Morphological characterization of the structures of all prepared hydrogel samples was performed using a ZEISS EVO LS 10 (Carl Zeiss AG, Germany) scanning electron microscope. The samples in their native (i.e., water-filled) form were shock-frozen in liquid nitrogen and then immediately freeze-dried. A few small specimens were taken from each studied hydrogel to maintain objective observation. Before observation, the specimens were gold-coated in a sputtering device (POLARON, Quorum Technologies, Lewes, UK).

The surface morphologies of every sample were recorded. Observations were realized in secondary electron mode, and the accelerating voltage was set to 5 kV to avoid charging the sample.

Supplementary Materials: The following supporting information can be downloaded at: https://www.mdpi.com/article/10.3390/gels9050367/s1, Reference [36] is cited in the Supplementary Material. Figure S1: Influence of crosslinking on the release of rhodamine 6G; Figure S2: Preparation procedure of physically crosslinked agarose hydrogels; Figure S3: Average ATR-FTIR spectra of: (a) dried gel samples and (b) powder lecithin; Figure S4: Results of principal component analysis of the ATR-FTIR spectra of dried reference agarose hydrogel and agarose hydrogel with lecithin; Figure S5: Preparation procedure of ionically crosslinked alginate hydrogels; Figure S6: Results of principal component analysis of the ATR-FTIR spectra of dried reference alginate hydrogel and alginate hydrogel with lecithin, respectively; S7: Preparation procedure of chemically crosslinked PVA-chitosan hydrogels; Figure S8: Results of principal component analysis of the ATR-FTIR spectra of dried reference poly(vinyl alcohol)-chitosan hydrogel and poly(vinyl alcohol)-chitosan hydrogel with lecithin, respectively.

Author Contributions: Conceptualization, R.H. and J.S.; methodology, R.H., N.Z. and M.T.; validation, R.H., N.Z. and M.T.; formal analysis, R.H. and J.S.; investigation, R.H., N.Z., M.K. and M.T.; data curation, R.H., N.Z. and M.T.; writing—original draft preparation, R.H., N.Z., M.T. and M.K.; writing—review and editing, J.S. and M.P.; visualization, R.H., M.T. and N.Z.; supervision, J.S.; project administration, R.H.; funding acquisition, R.H. All authors have read and agreed to the published version of the manuscript.

Funding: This research received no external funding.

Institutional Review Board Statement: Not applicable.

Informed Consent Statement: Not applicable.

Data Availability Statement: The data used in this study are available on request from the corresponding author.

Acknowledgments: The authors wish to warmly thank Petr Sedlacek (BUT CZ) for the FTIR measurements and analyses.

Conflicts of Interest: The authors declare no conflict of interest.

References

1. Bashir, S.; Hina, M.; Iqbal, J.; Rajpar, A.H.; Mujtaba, M.A.; Alghamdi, N.A.; Wageh, S.; Ramesh, K.; Ramesh, S. Fundamental Concepts of Hydrogels: Synthesis, Properties, and Their Applications. *Polymers* **2020**, *12*, 2702. [CrossRef] [PubMed]
2. Miyata, T.; Uragami, T.; Nakamae, K. Biomolecule-sensitive hydrogels. *Adv. Drug. Deliv. Rev.* **2002**, *54*, 79–98. [CrossRef] [PubMed]
3. Ahmed, E.M. Hydrogel: Preparation, characterization, and applications: A review. *J. Adv. Res.* **2015**, *6*, 105–121. [CrossRef]
4. Li, J.; Mooney, D.J. Designing hydrogels for controlled drug delivery. *Nat. Rev. Mater.* **2016**, *1*, 16071. [CrossRef]
5. Nezhad-Mokhtari, P.; Ghorbani, M.; Roshangar, L.; Soleimani Rad, J. A review on the construction of hydrogel scaffolds by various chemically techniques for tissue engineering. *Eur. Polym. J.* **2019**, *117*, 64–76. [CrossRef]
6. Serrano-Aroca, Á. Enhancement of Hydrogels' Properties for Biomedical Applications: Latest Achievements. In *Hydrogels*; IntechOpen: London, UK, 2018; pp. 91–120.
7. Jalageri, M.B.; Mohan Kumar, G.C. Hydroxyapatite Reinforced Polyvinyl Alcohol/Polyvinyl Pyrrolidone Based Hydrogel for Cartilage Replacement. *Gels* **2022**, *8*, 555. [CrossRef]
8. Jayash, S.N.; Cooper, P.R.; Shelton, R.M.; Kuehne, S.A.; Poologasundarampillai, G. Novel Chitosan-Silica Hybrid Hydrogels for Cell Encapsulation and Drug Delivery. *Int. J. Mol. Sci.* **2021**, *22*, 12267. [CrossRef]
9. Appel, E.A.; Tibbitt, M.W.; Webber, M.J.; Mattix, B.A.; Veiseh, O.; Langer, R. Self-assembled hydrogels utilizing polymer–nanoparticle interactions. *Nat. Commun.* **2015**, *6*, 6295. [CrossRef]
10. Heger, R.; Kadlec, M.; Trudicova, M.; Zinkovska, N.; Hajzler, J.; Pekar, M.; Smilek, J. Novel Hydrogel Material with Tailored Internal Architecture Modified by "Bio" Amphiphilic Components—Design and Analysis by a Physico-Chemical Approach. *Gels* **2022**, *8*, 115. [CrossRef]
11. Riazi, M.R. Chapter 8—Applications: Estimation of Transport Properties. In *Characterization and Properties of Petroleum Fractions*; ASTM International: West Conshohocken, PA, USA, 2007; pp. 329–336.
12. Kuo, C.K.; Ma, P.X. Ionically crosslinked alginate hydrogels as scaffolds for tissue engineering: Part 1. Structure, gelation rate and mechanical properties. *Biomaterials* **2001**, *22*, 511–521. [CrossRef]
13. Rahman, F.; Rafiquee, M.Z.A.; Aazam, E.S.; Alshabi, A.M.; Iqubal, S.M.S.; Khan, A.A.; Mohammed, T.; Dawoud, A.; Shaikh, I.A.; Maqbul, M.S.; et al. Determination of Rate of Release of Dye from the Hydrogels using Spectrophotometer Studies. *Asian. J. Pharm.* **2020**, *14*, 507–512. [CrossRef]

14. Feng, Y.; Taraban, M.; Yu, Y.B. The effect of ionic strength on the mechanical, structural and transport properties of peptide hydrogels. *Soft Matter* **2012**, *8*, 11723. [CrossRef] [PubMed]
15. Price, W.S. Pulsed-field gradient nuclear magnetic resonance as a tool for studying translational diffusion: Part 1. Basic theory. *Concepts Magn. Reson.* **1997**, *9*, 299–336. [CrossRef]
16. Evans, S.M.; Litzenberger, A.L.; Ellenberger, A.E.; Maneval, J.E.; Jablonski, E.L.; Vogel, B.M. A microfluidic method to measure small molecule diffusion in hydrogels. *Mater. Sci. Eng. C Mater. Biol. Appl.* **2014**, *35*, 322–334. [CrossRef]
17. Richbourg, N.R.; Peppas, N.A. High-Throughput FRAP Analysis of Solute Diffusion in Hydrogels. *Macromolecules* **2021**, *54*, 10477–10486. [CrossRef]
18. Piechocki, K.; Koynov, K.; Piechocka, J.; Chamerski, K.; Filipecki, J.; Maczugowska, P.; Kozanecki, M. Small molecule diffusion in poly-(olygo ethylene glycol methacrylate) based hydrogels studied by fluorescence correlation spectroscopy. *Polymer* **2022**, *244*, 124628. [CrossRef]
19. Jain, E.; Flanagan, M.; Sheth, S.; Patel, S.; Gan, Q.; Patel, B.; Montaño, A.M.; Zustiak, S.P. Biodegradable polyethylene glycol hydrogels for sustained release and enhanced stability of rhGALNS enzyme. *Drug Deliv. Transl. Res.* **2020**, *10*, 1341–1352. [CrossRef]
20. Quesada-Pérez, M.; Martín-Molina, A. Solute diffusion in gels: Thirty years of simulations. *Adv. Colloid Interface Sci.* **2021**, *287*, 102320. [CrossRef]
21. Jang, S.S.; Goddard, W.A.; Kalani, M.Y.S. Mechanical and Transport Properties of the Poly(ethylene oxide)−Poly(acrylic acid) Double Network Hydrogel from Molecular Dynamic Simulations. *J. Phys. Chem. B* **2007**, *111*, 1729–1737. [CrossRef]
22. Pomfret, R.; Sillay, K.; Miranpuri, G. An Exploration of the Electrical Properties of Agarose Gel: Characterization of Concentration Using Nyquist Plot Phase Angle and the Implications of a More Comprehensive In Vitro Model of the Brain. *Ann. Neurosci.* **2013**, *20*, 99–107. [CrossRef]
23. Smilek, J.; Sedláček, P.; Kalina, M.; Klučáková, M. On the role of humic acids' carboxyl groups in the binding of charged organic compounds. *Chemosphere* **2015**, *138*, 503–510. [CrossRef] [PubMed]
24. Klučáková, M. Agarose Hydrogels Enriched by Humic Acids as the Complexation Agent. *Polymers* **2020**, *12*, 687. [CrossRef] [PubMed]
25. Fatin-Rouge, N.; Starchev, K.; Buffle, J. Size Effects on Diffusion Processes within Agarose Gels. *Biophys. J.* **2004**, *86*, 2710–2719. [CrossRef] [PubMed]
26. Lai, M.; Lü, B. Tissue Preparation for Microscopy and Histology. *Compr. Sampl. Sample Prep.* **2012**, *3*, 53–93. [CrossRef]
27. Trudicova, M.; Smilek, J.; Kalina, M.; Smilkova, M.; Adamkova, K.; Hrubanova, K.; Krzyzanek, V.; Sedlacek, P. Multiscale Experimental Evaluation of Agarose-Based Semi-Interpenetrating Polymer Network Hydrogels as Materials with Tunable Rheological and Transport Performance. *Polymers* **2020**, *12*, 2561. [CrossRef]
28. Rowley, J.A.; Madlambayan, G.; Mooney, D.J. Alginate hydrogels as synthetic extracellular matrix materials. *Biomaterials* **1999**, *20*, 45–53. [CrossRef]
29. Golmohamadi, M.; Wilkinson, K.J. Diffusion of ions in a calcium alginate hydrogel-structure is the primary factor controlling diffusion. *Carbohydr. Polym.* **2013**, *94*, 82–87. [CrossRef]
30. Cohen, S.; Bañó, M.C.; Chow, M.; Langer, R. Lipid-alginate interactions render changes in phospholipid bilayer permeability. *Biochim. Et Biophys. Acta (BBA)—Biomembr.* **1991**, *1063*, 95–102. [CrossRef]
31. Klak, M.-C.; Lefebvre, E.; Rémy, L.; Agniel, R.; Picard, J.; Giraudier, S.; Larreta-Garde, V. Gelatin-Alginate Gels and Their Enzymatic Modifications: Controlling the Delivery of Small Molecules. *Macromol. Biosci.* **2013**, *13*, 687–695. [CrossRef]
32. Kaberova, Z.; Karpushkin, E.; Nevoralová, M.; Vetrík, M.; Šlouf, M.; Dušková-Smrčková, M. Microscopic Structure of Swollen Hydrogels by Scanning Electron and Light Microscopies: Artifacts and Reality. *Polymers* **2020**, *12*, 578. [CrossRef]
33. Garnica-Palafox, I.M.; Sánchez-Arévalo, F.M.; Velasquillo, C.; García-Carvajal, Z.Y.; García-López, J.; Ortega-Sánchez, C.; Ibarra, C.; Luna-Bárcenas, G.; Solís-Arrieta, L. Mechanical and structural response of a hybrid hydrogel based on chitosan and poly(vinyl alcohol) cross-linked with epichlorohydrin for potential use in tissue engineering. *J. Biomater. Sci. Polym. Ed.* **2014**, *25*, 32–50. [CrossRef] [PubMed]
34. Hezaveh, H.; Muhamad, I.I. Controlled drug release via minimization of burst release in pH-response kappa-carrageenan/polyvinyl alcohol hydrogels. *Chem. Eng. Res. Des.* **2013**, *91*, 508–519. [CrossRef]
35. Crank, J. *The Mathematics of Diffusion*, 2nd ed.; Oxford University Press USA: New York, NY, USA, 1979.
36. Mlynáriková, K.; Samek, O.; Bernatová, S.; Růžička, F.; Ježek, J.; Hároniková, A.; Šiler, M.; Zemánek, P.; Holá, V. Influence of Culture Media on Microbial Fingerprints Using Raman Spectroscopy. *Sensors* **2015**, *15*, 29635–29647. [CrossRef] [PubMed]

gels

MDPI

Article

Fabrication of Silk Hydrogel Scaffolds with Aligned Porous Structures and Tunable Mechanical Properties

Zewu Jiang [1,2], Qingqing Sun [3], Qian Li [1,2,*] and Xiaomeng Li [1,2,*]

1 School of Mechanics and Safety Engineering, Zhengzhou University, Zhengzhou 450001, China
2 National Center for International Joint Research of Micro-Nano Moulding Technology, Zhengzhou University, Zhengzhou 450001, China
3 School of Materials Science and Engineering, Zhengzhou University, Zhengzhou 450001, China
* Correspondence: qianli@zzu.edu.cn (Q.L.); xiaomeng.li@zzu.edu.cn (X.L.)

Abstract: The effectiveness of cell culture and tissue regeneration largely depends on the structural and physiochemical characteristics of tissue-engineering scaffolds. Hydrogels are frequently employed in tissue engineering because of their high-water content and strong biocompatibility, making them the ideal scaffold materials for simulating tissue structures and properties. However, hydrogels created using traditional methods have low mechanical strength and a non-porous structure, which severely restrict their application. Herein, we successfully developed silk fibroin glycidyl methacrylate (SF-GMA) hydrogels with oriented porous structures and substantial toughness through directional freezing (DF) and in situ photo-crosslinking (DF-SF-GMA). The oriented porous structures in the DF-SF-GMA hydrogels were induced by directional ice templates and maintained after photo-crosslinking. The mechanical properties, particularly the toughness, of these scaffolds were enhanced compared to the traditional bulk hydrogels. Interestingly, the DF-SF-GMA hydrogels exhibit fast stress relaxation and variable viscoelasticity. The remarkable biocompatibility of the DF-SF-GMA hydrogels was further demonstrated in cell culture. Accordingly, this work reports a method to prepare tough SF hydrogels with aligned porous structures, which can be extensively applied to cell culture and tissue engineering.

Keywords: silk hydrogel scaffold; mechanical property; aligned porous structure; freeze-casting; photo-crosslinking; tissue engineering

1. Introduction

Tissue-engineering (TE) scaffold primarily provides the seeding cells with specific microenvironments to guide cell proliferation and differentiation and, thereby, enhances tissue regeneration [1–3]. Hydrogels have gained increasing attention in the field of TE due to their resemblance to the cellular microenvironment in vivo and tunable physicochemical features. Hydrogels based on natural materials, such as alginate [4], hyaluronic acid [5], gelatin [6], and silk fibroin (SF) [7], can exhibit good biocompatibility and biodegradability [8,9]. Through mild crosslinking and processing techniques, hydrogels with diverse shapes, structures, and functions can be created for biomedical applications [10].

SF has been widely used in biomedical fields, such as cell culture [11], drug delivery [12,13], and tissue engineering [14], due to its high biocompatibility, degradability, and absorbability [15], as well as the fact that it has been approved by the US Food and Drug Administration (FDA) [16]. Physical crosslinking is typically utilized to transform the random coil into the β-sheet conformation while fabricating a SF hydrogel for TE [17]. However, this process is difficult to control, and the mechanical properties of the prepared SF scaffolds frequently fail to fulfill the criteria [18]. *N*-(3-Dimethylaminopropyl)-*N*′-ethyl carbodiimide hydrochloride (EDC), *N*-Hydroxysuccinimide (NHS), and other crosslinking agents are frequently used for chemical crosslinking to form amide bonds [19,20]. Nevertheless, the residual crosslinking agents may have an impact on cell viability. SF hydrogels with tunable mechanical properties have recently been enzymatically crosslinked using horseradish peroxidase for encapsulating human ovarian stromal cells [21]. Similarly, the cascade enzymatic crosslinking approach is employed to make functional SF hydrogels with a three-dimensional porous microenvironment for wound healing [22]. However, the displayed 3D porous structure is formed prior to observation by freeze drying. SF hydrogels with stable mechanical properties and low β-sheet content are crosslinked through di-tyrosine bonds by visible light, which can support the long-term culture of chondrocytes [23]. Meanwhile, photo-crosslinked silk fibroin glycidyl methacrylate (SF-GMA) hydrogels have been extensively applied in biomedical engineering due to their mild reaction conditions, few by-products, easy control of the reaction process, and high reaction efficiency [24,25]. However, the non-porous structure, brittle mechanical properties, and low toughness of these SF hydrogel scaffolds limit their further applications [26–28].

Porous hydrogel scaffolds have been shown to enhance nutrient diffusion and cell migration and proliferation [29,30]. Scaffolds with aligned porous structures can further guide cell morphology and promote tissue regeneration [31,32]. There are numerous strategies for controlling the porosity and microarchitecture of TE hydrogels [33]. In particular, freeze casting can be utilized to regulate the microscopic molecular arrangement of such gels and control pore morphology by controlling the directional ice formation in the solute solution [34]. Ice crystals form and propagate through predetermined directions during freezing, and melted ice crystals form interconnected microchannels within the scaffold [35]. Using ice templates to create programmable microporous structures eliminates the need for organic solvents [36,37]. For example, polyvinyl alcohol (PVA) hydrogels with high strength, toughness, and fatigue resistance were generated using directional freezing and salting-out treatment [38]. Bai et al., adjusted the PDMS wedges on the surface of a cold source to control the rate of ice crystal formation in the horizontal and vertical directions and produce large-scale aligned porous materials [39]. Additionally, using this technique, the degradable silk nacre and controllable performance were successfully processed [40]. However, the porous scaffolds prepared by current freeze-casting technology generally maintain a microporous structure after the freeze-drying and crosslinking treatment [41,42], and the obtained materials are unsuitable for 3D cell encapsulation and soft tissue engineering.

Here, we combine freeze-casting and photopolymerization processes to develop a silk hydrogel scaffold with aligned porous structures and robust mechanical properties. As illustrated schematically (Figure 1), a silk fibroin glycidyl methacrylate (SF-GMA) hydrogel with an aligned porous structure was prepared after directional freezing and UV crosslinking during the melting process. The oriented interconnected structures formed by the growth of ice crystals can be observed through an SEM. The mechanical properties of this directional-freezing and photo-crosslinked SF-GMA scaffold (DF-SF-GMA) were thoroughly investigated. The enhanced mechanical strength and tunable viscoelasticity are achieved due to the aligned porous structure. In addition, the secondary structure of the silk fibroin during this process was explored. The toughness of the DF-SF-GMA hydrogels is improved due to the porous structure and the alteration of the secondary structure. Then, the cytotoxicity of the DF-SF-GMA hydrogels was also examined to assess the potential for tissue regeneration applications.

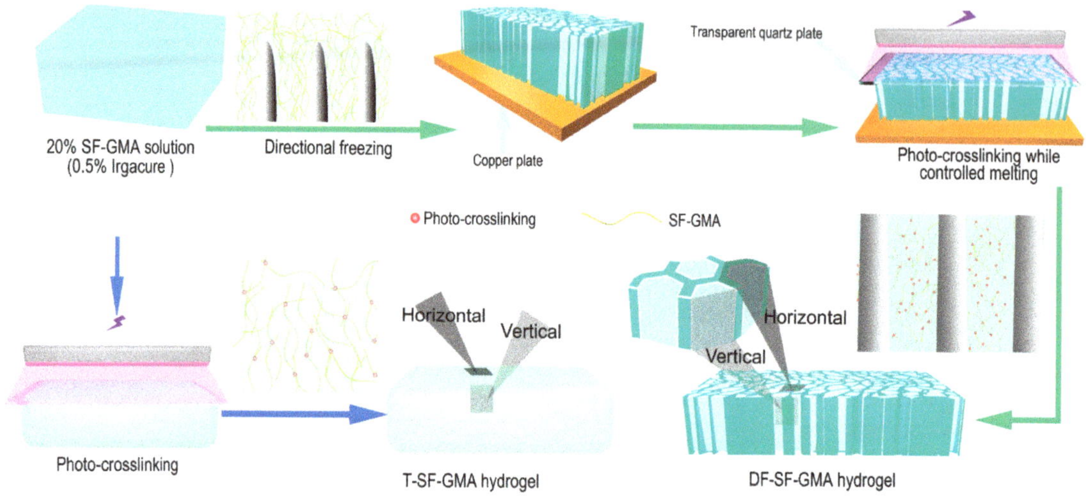

Figure 1. The schematic diagram for the fabrication of the glycidyl methacrylate (GMA) modified silk fibroin (SF) hydrogels with aligned porous structures via directional freezing (DF) and photo-crosslinking (DF-SF-GMA hydrogel) and the traditional SF hydrogel (T-SF-GMA hydrogel). The aligned porous structure of the DF-SF-GMA hydrogels is templated from ice crystal formation due to DF and is maintained after photo-crosslinking during controlled melting.

2. Results and Discussion

2.1. SF-GMA Characterization and Hydrogel Secondary Structure

The SF-GMA hydrogel was synthesized by methacrylate substitution of the SF primary amines (Figure 2A). To create a SF-GMA hydrogel with a high degree of functionalization, excess GMA was added to the regenerated SF solution. The success of grafting through the reaction of ring-opening of epoxy at the GMA and amino group on the molecular chain of SF was demonstrated using a proton nuclear magnetic resonance (^{1}H-NMR) spectroscopy (Figure 2B). It is evident that a characteristic resonance of the methacrylate vinyl group at δ = 5.8–6.2 ppm and a characteristic resonance of the methyl group of GMA at δ = 1.8 ppm appear in the SF-GMA macromer group. Moreover, there was a slight decrease in the lysine methylene signal at δ = 2.9 ppm after GMA modification, which all indicate the successful synthesis of the SF-GMA. The functionalization degree was calculated using the ^{1}H-NMR spectra, and the methacrylate degree of the SF-GMA is 36.4%.

The FTIR analysis was used to evaluate the secondary structures of the SF hydrogels (Figure 2C). The FTIR spectral regions between 1700–1600 cm^{-1} and 1600–1500 cm^{-1} represent the peptide backbone absorptions of amide I and amide II, respectively [43]. It indicates that all groups have a strong peak at 1654 cm^{-1}, which is assigned to the α-helix conformation. A prominent shoulder peak at 1627 cm^{-1} and a peak at 1515 cm^{-1} in amide II correspond to the β-sheet conformation that appears in the T-SF-GMA group, but it is not visible in the DF-SF-GMA group [44]. There is a strong peak at 1537 cm^{-1} in the directional-freezing groups, which is assigned to the random coil conformation [45].

To quantitatively analyze the content of the secondary structure of the SF-GMA hydrogels, the amide I region (1700–1600 cm^{-1}) in the FTIR spectra of the SF-GMA hydrogels was examined through Fourier self-deconvolution (FSD) (Figure S1A,B). The deconvolution of these spectra reveals 28.6% β-sheet, about 30.9% random coil, 25.9% α-helix, and 14.6% β-turn in the T-SF-GMA hydrogel (Figure 2D). Among the DF-SF-GMA groups, the content of their secondary structure is basically the same. Compared to the T group, the β-sheet content of the DF-SF-GMA groups decreases by about 7%, and the random coil content remains virtually unchanged, while the α-helix content and β-turn content increase by 3.3% and 3.6%, respectively. Importantly, these results show that the transformation from random coil to β-sheet structures is prevented during directional freezing and photo-crosslinking. It has been reported that a reduction in β-sheet normally improves the toughness of SF materials [46]. Similarly, the reduced β-sheet content caused by directional freezing enhances the toughness of the DF-SF-GMA hydrogels.

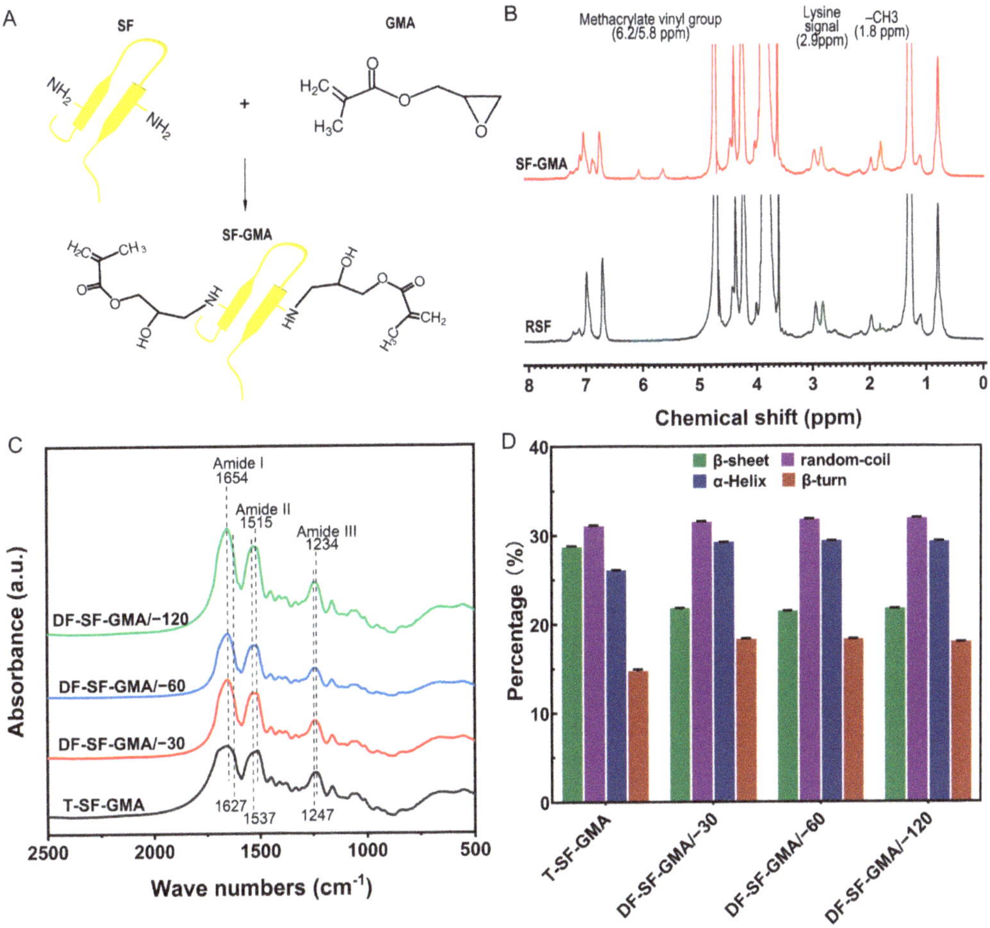

Figure 2. FTIR spectra and secondary structures of hydrogels. (**A**) Modification of SF molecule with GMA. SF was covalently grafted with GMA, which provides a vinyl double bond as a UV-crosslinking site. (**B**) ^{1}H-NMR spectra of the pristine RSF and the SF-GMA. The modification of lysine residues in SF with GMA was confirmed by the gradual decrease in the lysine signal and an increase in the methacrylate vinyl group and the methyl group signals. (**C**) FTIR spectra of the T-SF-GMA and DF-SF-GMA hydrogels. (**D**) Quantitative analysis of secondary structures in the T-SF-GMA and DF-SF-GMA hydrogels. Data are shown as mean ± S.D.; $n = 3$.

2.2. The Porous Architecture

The aligned porous lamellar structure in the SF hydrogels was constructed by the ice templates induced by directional freezing. As shown in Figure 3, the DF-SF-GMA hydrogels have a lamellar structure along the ice crystal growth direction. Moreover, the distance between the layers is larger as the freezing temperature increases. The hydrogel generated in the −30 °C plate exhibits the largest pore size, indicating less initial nucleation and a slower growth rate of ice crystals [47]. The morphology and structure of the control T-SF-GMA hydrogel are basically consistent with traditional bulk hydrogels [48]. The homogeneous pores in this hydrogel are formed by randomly growing ice crystals.

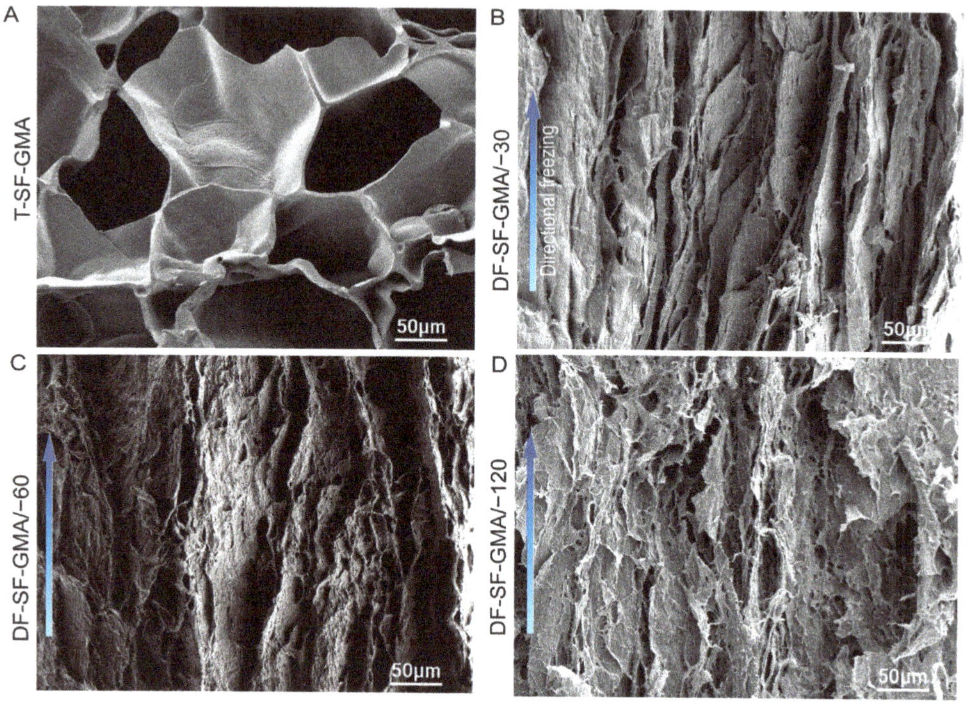

Figure 3. Characterization of the aligned porous structures of the DF-SF-GMA hydrogels. Scanning electron microscopy (SEM) images of the vertical section of the traditional hydrogel (T-SF-GMA) (**A**) and the hydrogels fabricated by directional freezing at −30 °C (DF-SF-GMA/−30) (**B**), −60 °C (DF-SF-GMA/−60) (**C**), and −120 °C (DF-SF-GMA/−120) (**D**). The porous structures in the T-SF-GMA hydrogel are induced by ice particles randomly formed during freezing. The SF molecules in the DF-SF-GMA hydrogels are excluded from the ice and aggregate between the aligned ice crystals and photo-crosslinked during melting. These aligned porous structures already exist prior to freeze drying.

2.3. Mechanical Properties of SF-GMA

The brittleness of traditional hydrogels is an inherent challenge in tissue engineering due to the architecture of hydrogels. DF-SF-GMA hydrogels with directional porous structures were fabricated in this study. The compression curves and the compressive modulus of these hydrogels were tested and determined to compare the different mechanical behaviors of the hydrogels with varied structures. The stress–strain curves show that the stress in the directional-freezing groups increases dramatically in the first 10% region (Figure 4A and Figure S2A), which may be attributed to the supporting of the directional lamellar structure formed during the ice crystal growth. Furthermore, the reinforcement in the directional-freezing hydrogel differs depending on the freezing temperatures, and the stress increases greater at a lower temperature. However, the stress of the control group increases gradually only from 0 kPa to around 1.7 kPa throughout the 10% strain.

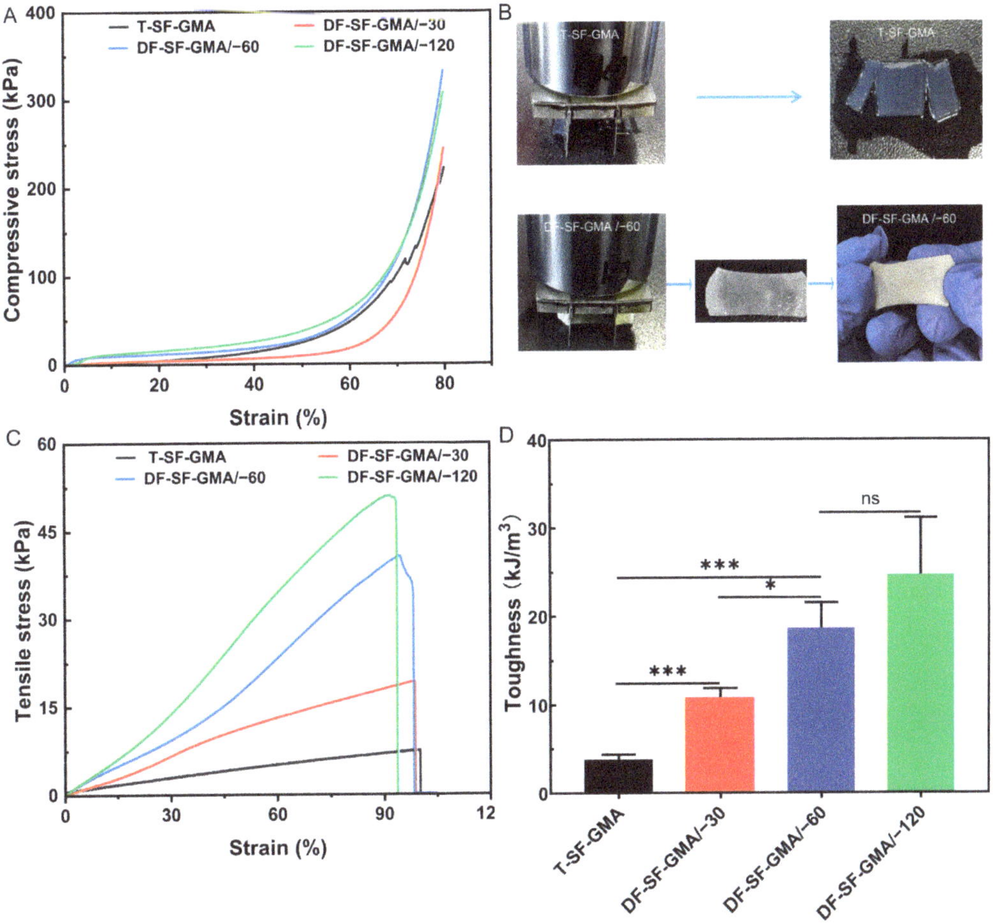

Figure 4. Physical properties of the DF-SF-GMA hydrogels. (**A**) Compression test for the T-SF-GMA and DF-SF-GMA hydrogels with varied freezing temperatures. (**B**) The T-SF-GMA and DF-SF-GMA/−60 were cut and compressed by a 1 kg weight for 1 min. The DF-SF-GMA/−60 hydrogel was still intact after removing the weight. (**C**) Tensile test for the SF-GMA hydrogels. (**D**) Toughness of the T-SF-GMA and DF-SF-GMA hydrogels (* $p < 0.05$; *** $p < 0.001$, $n = 3$).

In the control group, the stress drops vertically at a strain of 72% (Figures 4A and S2B), indicating a crack in the hydrogel. However, the DF-SF-GMA groups remain intact after compression. These results indicate that the toughness of SF hydrogels can be enhanced via directional freezing and photo-crosslinking, which may benefit from the oriented layered structure and the reduced content of β-sheet.

The compressive modulus of group T increases gradually at 3%, 25%, and 50% strain points (Figure S2C), which is consistent with the compressive properties of traditional photo-crosslinked hydrogels. Interestingly, the directional-freezing hydrogels show distinct characteristics. The compressive modulus of the DF-SF-GMA hydrogels produced at −60 °C and −120 °C shows a considerable enhancement at the strain of 3%, and then it falls at 25% and gradually increases again at 50%. These compressive behaviors are probably because of buckling. The lamellar materials formed by directional freezing play a supporting role in the early stage, buckling occurs as the strain increases, and then the modulus increases further with the densification of the materials.

Traditional hydrogels are not only easily fractured and brittle under pressure, but they are also incapable of stretching. The tensile mechanical test on the SF-GMA hydrogels was performed, and the tensile YounG′s modulus and toughness were calculated from the acquired tensile stress–strain curves (Figures 4C,D and S2D). The YounG′s modulus of the photo-crosslinked hydrogel after directional freezing is higher than that of the control group, which has YounG′s modulus of about 7.9 kPa. The tensile mechanical properties of the SF hydrogels increase steadily, reaching 39.5 kPa in the −120 °C group, which is about four times that of the control group. The tensile strength also exhibits a similar pattern, with 6.8 kPa and 53.1 kPa for the control group and the −120 °C group, respectively. There is no substantial difference in the elongation at the break between these groups, so the toughness also follows the same trend. The toughness of the directional-freezing groups is higher than that of the control group and the lower the freezing temperature, the greater the toughness. The space between the lamellar structures becomes tighter, and the pore size is smaller and more uniform as the temperature of the cold plate declines, further improving the mechanical properties of the SF-GMA hydrogel. These results indicate that the introduction of directional freezing has no effect on the elongation at the break of the SF-GMA hydrogel, and it can significantly boost the strength and toughness of SF hydrogels.

To visually investigate the toughness of the hydrogel, we placed a blade on the surface of the hydrogel and pressed a 1 kg weight on it for 1 min (Figure 4B). After cutting, the DF-SF-GMA hydrogel at −60 °C maintained good integrity and no evident incision was visible. The −30 °C and −120 °C groups also exhibited the same characteristics, but the control group was easily cut off. These results further indicate that the technique reported in this work can significantly enhance the toughness of the SF-GMA hydrogel.

2.4. Viscoelastic Properties of SF-GMA Hydrogels

Viscoelastic properties of hydrogel scaffolds have drawn much attention recently because many studies have proved that not only the stiffness of the surrounding matrix but also the viscoelastic mechanical properties, such as stress relaxation, can affect cell spreading, proliferation, and differentiation. Therefore, the viscoelastic properties of these novel scaffolds were also investigated. The stress relaxation tests were performed under 15%, 30%, and 60% strain (Figures 5A and S3A–C). The results of all the groups follow the same tendency. The T-SF-GMA hydrogel possesses the slowest stress relaxation speed, indicating high elasticity, which is consistent with previous studies. The hydrogels with aligned porous structures exhibit fast stress relaxation compared to the control group. Moreover, the degree of stress relaxation rises with decreasing freezing temperature under the same strain. The stress relaxation of the −60 °C and −120 °C groups could exceed 50% under the strain of 60%, and the time required for the −120 °C group to relax to 50% stress is around 5 s. These findings indicate that, with the introduction of directional freezing, the SF-GMA hydrogel has tunable viscoelasticity, which may be beneficial for cell culture and tissue regeneration.

Rheological analysis of the SF-GMA hydrogels was also performed to explore their viscoelasticity (Figure 5B). The loss modulus G'' and the storage modulus G' of the directional-freezing group are much larger than those of the control group, and the loss modulus G'' and the storage modulus G' rise as the temperature of the freezing plate decreases. These results further reveal that the introduction of directional freezing enhances the viscoelasticity of the SF-GMA hydrogels, with the augmentation being more pronounced at lower freezing temperatures.

To evaluate the effect of cyclic loading and the viscoelastic properties of the scaffolds, a cyclic compression test was also conducted (Figures 5C,D and S3D). The results show that the ratio of the hysteresis loop area of the traditional hydrogel is relatively small, and all the compression cycles essentially overlap, indicating the elasticity of the T-SF-GMA hydrogel. In the directional-freezing group, the proportion of the hysteretic area increases as the temperature of the cold plate decreases. The hysteretic area of the first compression cycle is larger than that of the last four compression cycles. This may be due to the fact that the oriented structure produced by directional freezing is crushed and water is squeezed out during the first compression cycle. Interestingly, there is a sudden change of stress in the DF-SF-GMA/−120 group, which could be attributed to the adhesion force of water and compression plates during the force unloading. These results illustrate that the DF-SF-GMA hydrogels display special architectures and tunable viscoelastic properties when compared to the traditional SF-GMA hydrogels.

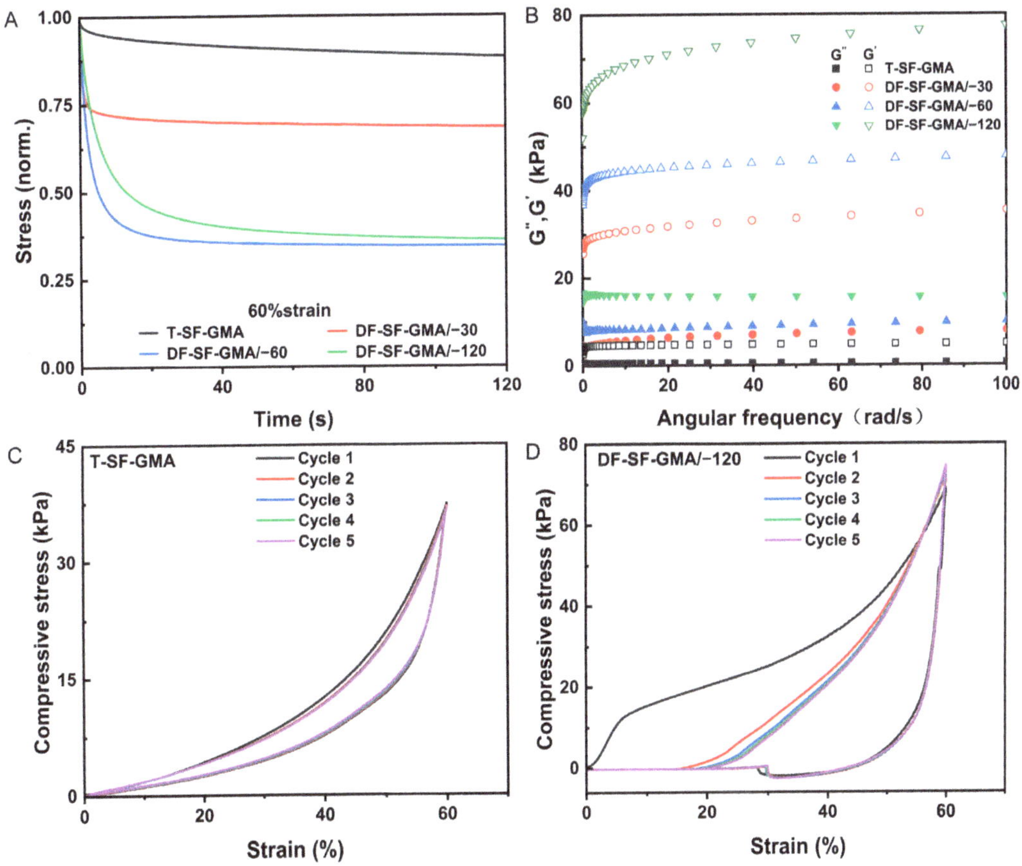

Figure 5. Viscoelastic properties of the T-SF-GMA and DF-SF-GMA hydrogels. (**A**) Stress relaxation test of the T-SF-GMA and DF-SF-GMA hydrogels at 60% strain. Stress is normalized by the initial stress. (**B**) Storage modulus and loss modulus of the T-SF-GMA and DF-SF-GMA hydrogels were tested using a rheometer at different frequencies. Cyclic compression test of the T-SF-GMA (**C**) and DF-SF-GMA/−120 (**D**) hydrogels under 60% strain.

The swelling of hydrogels is also an important concern for tissue regeneration applications, depending on the circumstances. The swelling and expansion ratios of the SF-GMA hydrogels were tested in water and PBS, respectively (Figure 6). The swelling ratio of the T-SF-GMA hydrogel is 240% in water and only about 6% in PBS. Interestingly, the −30 °C, −60 °C, and −120 °C groups show similar swelling and expansion ratios in water and PBS. The main reason for this behavior is that the SF-GMA solution concentration increased and resulted in a higher crosslinking density during the DF. Therefore, the swelling ratio and expansion degree of the DF-SF-GMA hydrogels are nearly the same whether in water or in the PBS solution. The swelling and expansion degree of the −30 °C group is slightly higher than that of the −60 °C and −120 °C groups, which may be affected by the crosslinking density and the large pore size. These findings suggest that the introduction of directional freezing could strengthen the swelling resistance of SF-GMA hydrogels.

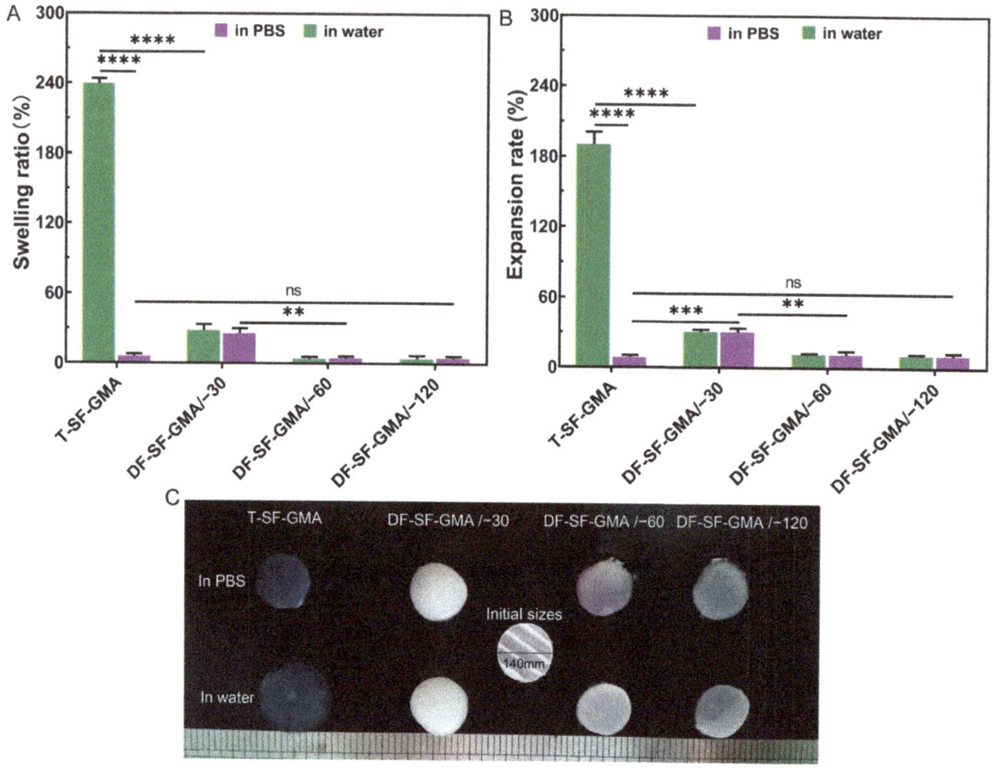

Figure 6. The (**A**) swelling ratio and (**B**) expansion rate of the T-SF-GMA and DF-SF-GMA hydrogels in PBS and milli-Q water. (**C**) Photomicrographs of all the hydrogels after swelling for 24 h (** $p < 0.01$; *** $p < 0.001$; and **** $p < 0.0001$, $n = 4$).

2.5. Cell Biocompatibility

To study the biocompatibility of the DF-SF-GMA hydrogel scaffolds and assess their potential for clinical usage in the future, we conducted in vitro cell culture experiments. Live/dead and cytoskeleton staining was performed after the cultivation of the cells in the scaffolds for three days. The cells in all groups maintain good viability, with nearly no dead cells seen (Figure 7A). It was found that the cell spreading in the −60 °C and −120 °C groups is better than that in the −30 °C and the control groups, which may be due to the various mechanical properties of the SF-GMA (Figure 7B). Meanwhile, a considerable degree of cell orientation is generated in the −60 °C and −120 °C groups, which is induced by the lamellar structure on the surface of the DF-SF-GMA hydrogels.

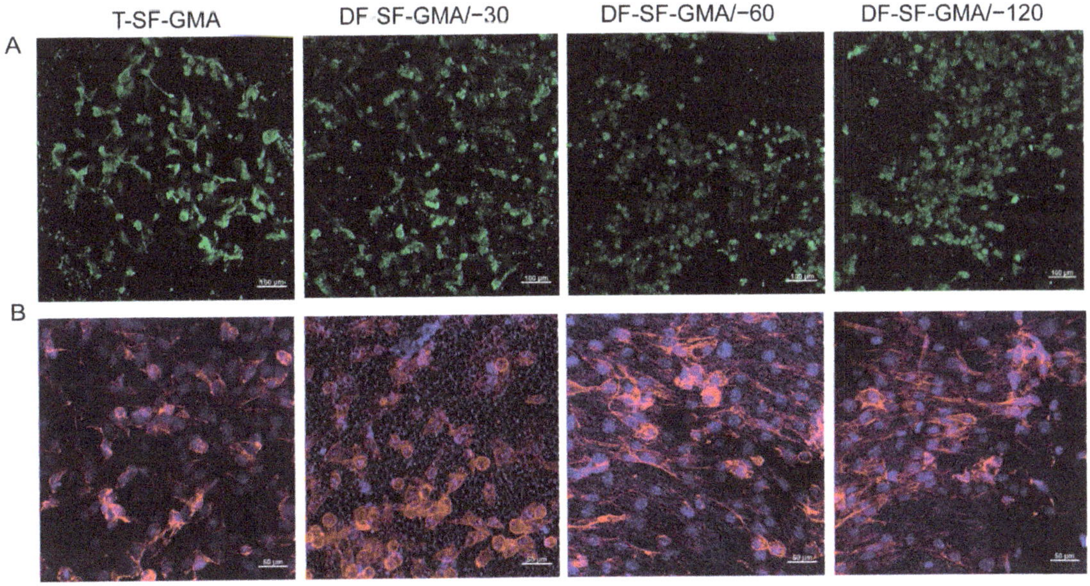

Figure 7. Cytocompatibility of the SF-GMA hydrogels. (**A**) Live/dead staining (Green: live cells; red: dead cells) and (**B**) F-actin staining (Blue: nucleus; red: F-actin) of cells cultured in the T-SF-GMA and DF-SF-GMA hydrogels for 3 days. All the cells exhibit excellent viabilities, and the cells cultured on the DF-SF-GMA/−60 and DF-SF-GMA/−120 hydrogels are aligned due to the surface morphology.

It has been proved that an orientated morphology can favor the growth and differentiation of some particular cells, such as endothelial cells [49], nerve [50], and muscle cells [51], and thus achieve a better outcome in tissue regeneration. Therefore, numerous studies have constructed scaffolds with oriented structures on the surface to stimulate cell alignment and tissue regeneration. For example, aligned electrospun fibers are employed to guide vascular endothelial cell alignment for vascular tissue regeneration [49]. Similarly, collagen scaffolds with a microgroove structure are created using liquid dispensing and freeze drying to mimic muscle basement membrane [32]. However, the majority of these research studies create oriented structures on the surface of scaffolds rather than inside three-dimensional scaffolds. Recently, an internal orientated and anisotropic structure inside a hydrogel is achieved using directional freezing technology, and the encapsulated cells show a high viability when combined with cryoprotective agents [52]. Although we have only so far cultivated cells on the surface of hydrogel scaffolds, cells could be encapsulated in the DF-SF-GMA hydrogels with cryoprotectants and employed for in vitro cell culture and in vivo tissue regeneration.

These results demonstrate the good biocompatibility of the DF-SF-GMA hydrogels produced via directional freezing and photo-crosslinking. Additionally, their tunable mechanical and structural properties can serve other functions of regulating cell behaviors and promoting tissue regeneration. Furthermore, this technology can be used not only for the construction of SF scaffolds but also for the manufacture of other material-based photo-crosslinking hydrogels.

3. Conclusions

In this study, DF-SF-GMA hydrogel for tissue-engineering scaffold is developed by directional freezing and in situ photopolymerization. The secondary structure, microstructure, mechanical properties, and biocompatibility of the DF-SF-GMA hydrogels are investigated. The results confirm that directional freezing at different temperatures can produce SF-GMA hydrogels with oriented lamellar structures. Directional freezing reduces the content of β-sheet and increases the content of α-helix and β-turn. The mechanical properties of the SF-GMA hydrogels can be tuned by adjusting the temperature of the cold plate for directional freezing. The prepared hydrogels outperform traditional hydrogels in mechanical strength and toughness. In addition, the DF-SF-GMA hydrogels have unique viscoelasticity and anti-swelling capability. Furthermore, these DF-SF-GMA hydrogel scaffolds also exhibit excellent biocompatibility. This work provides a new method for the preparation of hydrogels with aligned porous structures, which is expected to broaden the biomedical applications of hydrogels.

4. Materials and Methods

4.1. Reagents and Materials

Silkworm cocoons were obtained from the Northwest Silkworm Base (Shanxi, China). Lithium bromide (LiBr), sodium carbonate (Na_2CO_3), glycidyl methacrylate (GMA), and photoinitiator (Irgacure2959) were purchased from Aladdin (Aladdin, Shanghai, China). All reagents were used as received.

4.2. Synthesis of SF-GMA Macromer and ^{1}H NMR

The SF-GMA was synthesized by GMA modification according to a previous method. Briefly, the cocoons were boiled in a 0.05 M Na_2CO_3 solution for 30 min to remove sericin, and then they were washed several times with distilled water. Subsequently, 30 g of degummed and dried silk fibroin was dissolved in 150 mL of 9.3 M LiBr solution at 60 °C for 1 h, after which 9 mL of GMA solution was added to the mixture and stirred at 300 rpm for 3 h. The resulting solution was then dialyzed against milli-Q water for 4 days. Finally, the obtained methacrylate SF solution was concentrated to 20% using a 3500 Da dialysis bag against a PEG solution with a molecular weight of 8000.

The synthesis of the SF-GMA macromer was confirmed using ^{1}H NMR. After lyophilization, 5 mg of the regenerated silk fibroin (RSF) and SF-GMA was dissolved in 700 μL of deuterium oxide. The ^{1}H NMR spectra were collected using a Bruker DPX FT-NMR spectrometer with a frequency of 400 MHz. The degree of methacrylate substitution was determined using the formula of 1-(lysine integration signal of SF-GMA/lysine integration signal of unsubstituted RSF).

4.3. Preparation of SF-GMA Hydrogel

Irgacure2959 (0.5 w/v%) was added into the above-mentioned SF-GMA solution, and the solution was stirred slowly at 45 °C for one hour. As shown in Figure 1, the solution was directionally frozen on copper plates with different temperatures of −30 °C, −60 °C, and −120 °C. During directional freezing, an oriented lamellar porous structure was created [53,54]. After complete freezing, the quartz glass plate was covered on the scaffold, and they were subjected to ultraviolet light for 8 min. The solution was melted and crosslinked simultaneously during UV exposure, yielding directional-freezing, photo-crosslinked silk hydrogels (DF-SF-GMA). The bulk traditional silk hydrogel (T-SF-GMA), as the control group, was produced by direct photo-crosslinking for 8 min without directional freezing.

4.4. Structure Characterization

To analyze the internal microstructure of the silk fibroin hydrogels, the T-SF-GMA and the DF-SF-GMA hydrogels were observed using an SEM at a voltage of 15 kV. The prepared DF-SF-GMA and T-SF-GMA hydrogel samples were frozen at −80 °C and then lyophilized. After which, the samples were cut along the freezing direction, and the sections were coated in platinum using a sputtering system (SC7620, Quorum, San Diego, CA, USA).

4.5. FTIR Analysis

The hydrogels were ground with KBr after freeze drying to prepare the sample for Fourier-transform infrared spectrometer (FTIR, Thermo Scientific Nicolet iS20, Waltham, MA, USA). The spectra were recorded over the wavenumber range from 400 to 4000 cm^{-1} with a resolution of 4 cm^{-1}. Fourier self-deconvolution (FSD) of the infrared spectra covering the amide I at 1600–1700 cm^{-1} was performed using the PeakFit software. The FSD Spectra were curve-fitted to explore the relative areas of the amide I region components and the secondary structure of SF.

4.6. Mechanical Characteristics

To assess the mechanical properties of the hydrogels under physiological conditions, the DF-SF-GMA and T-SF-GMA hydrogels were punched into a disk with a diameter of 6 mm and a height of 2 mm and then swollen in a PBS solution at 37 °C for 24 h before the following mechanical testing (n = 3). For the compressive strain–stress test, the samples were subjected to a compression test at a rate of 1 mm/min. For the stress relaxation test, all samples were first compressed at a strain rate of 1 mm/min. The strain was held at 15%, 30%, and 60% strain, and the relaxation time was recorded to compare the stress relaxation of the hydrogels [55].

To measure the tensile properties, the hydrogel samples were cut into a rectangular shape and were tested at a strain rate of 3 mm/min. The toughness of the hydrogels was calculated according to the area of the tensile stress–strain curve. The rheological properties of the hydrogels were measured using a rheometer (TA Discovery DHR, Delaware, DE, USA) at 37°C, with an angular frequency sweep range of 0.1 to 100 rad/s and a shear strain of 1%.

4.7. Swelling Ratio Measurement

To test the swelling ratio, the hydrogels were swollen in PBS or water at 37 °C for 24 h. The weights m_1 and v_1 before swelling were recorded. After swelling, the samples were blotted with a filter paper to remove the residual water and weighted (m_2 and v_2). The swelling ratio Q and the swelling degree E were calculated according to the following formulas (n = 4):

$$Q = \frac{m_2 - m_1}{m_1} \times 100\%$$

$$E = \frac{v_2 - v_1}{v_1} \times 100\%$$

4.8. Biocompatibility

Human umbilical vein endothelial cells (HUVEC) were employed as the cell models to verify the biocompatibility of the DF-SF-GMA hydrogels. The HUVEC were cultured in 75 cm^2 tissue culture flasks in a culture medium (DMEM, Biological Industries Ltd., Shanghai, China.) supplemented with 10% fetal bovine serum (FBS, Biological Industries, Shanghai, China), 100 U mL^{-1} penicillin (Solarbio, Beijing, China), and 100 µg mL^{-1} streptomycin (Solarbio, Beijing, China) at 37 °C and 5% CO_2. The cell culture medium was refreshed every two days. After reaching 90% confluence, the cells were detached, suspended, and seeded into the hydrogels. Before cell seeding, the hydrogels were cut into small disks and sterilized in a 24-well plate. After incubation for 3 days, cell viability and cytoskeleton staining were performed using the live/dead kit and phalloidin staining. The fluorescence images were captured using a confocal microscope (LSM 880, Zeiss, Freiburg, Germany).

4.9. Statistical Analysis

All experiments were conducted independently three times with 3~6 samples to calculate the mean ± standard deviation (SD). The image data were processed using ImageJ. Statistical analysis was performed using Prism 6.0 (GraphPad Software, San Diego,

CA, USA) and Origin Pro to evaluate the significance of the experimental data, and *p*-values less than 0.05 were considered statistically significant.

Supplementary Materials: The following supporting information can be downloaded at: https://www.mdpi.com/article/10.3390/gels9030181/s1, Figure S1. The deconvolution of FTIR spectra for the amide I band for (A) T-SF-GMA and (B) DF-SF-GMA/−120 hydrogels; Figure S2. Physical properties of DF-SF-GMA hydrogels. The compressive stress-strain curves at early (A) and late stage (B). (C) The compressive modulus of all the hydrogels at different strains. The compressive modulus of DF-SF-GMA/−60 and DF-SF-GMA/−120 were much higher than other groups. (D) Tensile elastic modulus of T-SF-GMA and DF-SF-GMA hydrogels (* $p < 0.05$; ** $p < 0.01$; *** $p < 0.001$; and **** $p < 0.0001$, $n = 3$); Figure S3. Viscoelastic properties of T-SF-GMA and DF-SF-GMA hydrogels. Stress relaxation test of T-SF-GMA and DF-SF-GMA hydrogels at a strain of (A) 15% and (B) 30%. Stress is normalized by the initial stress. (C) The ratio of relaxed stress to initial stress at 60 s. (D) Cyclic compression test of DF-SF-GMA/−30 hydrogels under 60% strain.

Author Contributions: Conceptualization, X.L. and Q.L.; methodology, Z.J.; data curation, Z.J. and Q.S.; writing—original draft preparation, Z.J. and Q.S.; writing—review and editing, Q.S.; supervision, X.L. and Q.L.; project administration, X.L.; funding acquisition, X.L. All authors have read and agreed to the published version of the manuscript.

Funding: This research was funded by the National Natural Science Foundation of China, grant number 31900951, and the Joint Funds of the National Natural Science Foundation of China, grant number U1909219.

Institutional Review Board Statement: Not applicable.

Informed Consent Statement: Not applicable.

Data Availability Statement: Not applicable.

Conflicts of Interest: The authors declare no conflict of interest.

References

1. Griffith, L.G.; Naughton, G. Tissue engineering—Current challenges and expanding opportunities. *Science* **2002**, *295*, 1009–1014. [CrossRef]
2. Ikada, Y. Challenges in tissue engineering. *J. R. Soc. Interface* **2006**, *3*, 589–601. [CrossRef] [PubMed]
3. Zhu, J.; Marchant, R.E. Design properties of hydrogel tissue-engineering scaffolds. *Expert Rev. Med. Devices* **2011**, *8*, 607–626. [CrossRef]
4. Rastogi, P.; Kandasubramanian, B. Review of alginate-based hydrogel bioprinting for application in tissue engineering. *Biofabrication* **2019**, *11*, 042001. [CrossRef]
5. Burdick, J.A.; Prestwich, G.D. Hyaluronic Acid Hydrogels for Biomedical Applications. *Adv. Mater.* **2011**, *23*, H41–H56. [CrossRef]
6. Li, X.; Chen, S.; Li, J.; Wang, X.; Jing, Z.; Naoki, K.; Chen, G. 3D Culture of Chondrocytes in Gelatin Hydrogels with Different Stiffness. *Polymers* **2016**, *8*, 269. [CrossRef] [PubMed]
7. Chen, J.; Li, Y.; Liu, S.; Du, Y.; Zhang, S.; Wang, J. Freeze-casting osteochondral scaffolds: The presence of a nutrient-permeable film between the bone and cartilage defect reduces cartilage regeneration. *Acta Biomater.* **2022**, *154*, 168–179. [CrossRef] [PubMed]
8. Li, Y.; Yang, H.Y.; Lee, D.S. Advances in biodegradable and injectable hydrogels for biomedical applications. *J. Control. Release* **2021**, *330*, 151–160. [CrossRef]
9. Li, X.; Zhang, J.; Kawazoe, N.; Chen, G. Fabrication of Highly Crosslinked Gelatin Hydrogel and Its Influence on Chondrocyte Proliferation and Phenotype. *Polymers* **2017**, *9*, 309. [CrossRef]
10. Li, X.; Sun, Q.; Li, Q.; Kawazoe, N.; Chen, G. Functional Hydrogels with Tunable Structures and Properties for Tissue Engineering Applications. *Front. Chem.* **2018**, *6*, 499. [CrossRef]
11. Milazzo, L.; Vulcano, F.; Macioce, G.; Marziali, G.; Iosi, F.; Bertuccini, L.; Falchi, M.; Rech, F.; Giampaolo, A.; Pecci, R.; et al. Silk Fibroin Scaffolds as Biomaterials for 3D Mesenchymal Stromal Cells Cultures. *Appl. Sci.* **2021**, *11*, 11345. [CrossRef]
12. Ma, Y.; Duan, L.; Sun, J.; Gou, S.; Chen, F.; Liang, Y.; Dai, F.; Xiao, B. Oral nanotherapeutics based on *Antheraea pernyi* silk fibroin for synergistic treatment of ulcerative colitis. *Biomaterials* **2022**, *282*, 121410. [CrossRef]
13. Long, D.; Xiao, B.; Merlin, D. Genetically modified silk fibroin nanoparticles for drug delivery: Preparation strategies and application prospects. *Nanomedicine* **2020**, *15*, 1739–1742. [CrossRef]
14. Melke, J.; Midha, S.; Ghosh, S.; Ito, K.; Hofmann, S. Silk fibroin as biomaterial for bone tissue engineering. *Acta Biomater.* **2016**, *31*, 1–16. [CrossRef] [PubMed]

15. Suzuki, Y.; Aoki, A.; Nakazawa, Y.; Knight, D.P.; Asakura, T. Structural Analysis of the Synthetic Peptide (Ala-Gly-Ser-Gly-Ala-Gly)5, a Model for the Crystalline Domain of Bombyx mori Silk Fibroin, Studied with 13C CP/MAS NMR, REDOR, and Statistical Mechanical Calculations. *Macromolecules* **2010**, *43*, 9434–9440. [CrossRef]
16. Nguyen, T.P.; Nguyen, Q.V.; Nguyen, V.-H.; Le, T.-H.; Huynh, V.Q.N.; Vo, D.-V.N.; Trinh, Q.T.; Kim, S.Y.; Le, Q.V. Silk Fibroin-Based Biomaterials for Biomedical Applications: A Review. *Polymers* **2019**, *11*, 1933. [CrossRef]
17. Guan, Y.; You, H.; Cai, J.; Zhang, Q.; Yan, S.; You, R. Physically crosslinked silk fibroin/hyaluronic acid scaffolds. *Carbohydr. Polym.* **2020**, *239*, 116232. [CrossRef] [PubMed]
18. Han, H.; Ning, H.; Liu, S.; Lu, Q.; Fan, Z.; Lu, H.; Lu, G.; Kaplan, D.L. Silk Biomaterials with Vascularization Capacity. *Adv. Funct. Mater.* **2016**, *26*, 421–432. [CrossRef]
19. Yang, X.; Wang, X.; Yu, F.; Ma, L.; Pan, X.; Luo, G.; Lin, S.; Mo, X.; He, C.; Wang, H. Hyaluronic acid/EDC/NHS-crosslinked green electrospun silk fibroin nanofibrous scaffolds for tissue engineering. *RSC Adv.* **2016**, *6*, 99720–99728. [CrossRef]
20. Li, S.; Wu, P.; Ji, Z.; Zhang, Y.; Zhang, P.; He, Y.; Shen, Y. In vitro biocompatibility study of EDC/NHS cross-linked silk fibroin scaffold with olfactory ensheathing cells. *J. Biomater. Sci. Polym. Ed.* **2022**, *34*, 482–496. [CrossRef] [PubMed]
21. Jafari, H.; Dadashzadeh, A.; Moghassemi, S.; Zahedi, P.; Amorim, C.A.; Shavandi, A. Ovarian Cell Encapsulation in an Enzymatically Crosslinked Silk-Based Hydrogel with Tunable Mechanical Properties. *Gels* **2021**, *7*, 138. [CrossRef]
22. Shen, H.; Wang, P.; Han, X.; Ma, M.; Shang, Y.; Ju, Y.; Shen, S.; Yin, F.; Wang, Q. GOx/Hb Cascade Oxidized Crosslinking of Silk Fibroin for Tissue-Responsive Wound Repair. *Gels* **2022**, *8*, 56. [CrossRef]
23. Cui, X.; Soliman, B.G.; Alcala-Orozco, C.R.; Li, J.; Vis, M.A.M.; Santos, M.; Wise, S.G.; Levato, R.; Malda, J.; Woodfield, T.B.F.; et al. Rapid Photocrosslinking of Silk Hydrogels with High Cell Density and Enhanced Shape Fidelity. *Adv. Healthc. Mater.* **2020**, *9*, 1901667. [CrossRef] [PubMed]
24. Yue, K.; Trujillo-de Santiago, G.; Alvarez, M.M.; Tamayol, A.; Annabi, N.; Khademhosseini, A. Synthesis, properties, and biomedical applications of gelatin methacryloyl (GelMA) hydrogels. *Biomaterials* **2015**, *73*, 254–271. [CrossRef] [PubMed]
25. Liu, J.; Su, C.; Chen, Y.; Tian, S.; Lu, C.; Huang, W.; Lv, Q. Current Understanding of the Applications of Photocrosslinked Hydrogels in Biomedical Engineering. *Gels* **2022**, *8*, 216. [CrossRef]
26. Chen, G.; Ushida, T.; Tateishi, T. Scaffold Design for Tissue Engineering. *Macromol. Biosci.* **2002**, *2*, 67–77. [CrossRef]
27. Wang, Y.; Chen, Y.; Zheng, J.; Liu, L.; Zhang, Q. Three-Dimensional Printing Self-Healing Dynamic/Photocrosslinking Gelatin-Hyaluronic Acid Double-Network Hydrogel for Tissue Engineering. *ACS Omega* **2022**, *7*, 12076–12088. [CrossRef] [PubMed]
28. Sergeeva, A.; Vikulina, A.S.; Volodkin, D. Porous alginate scaffolds assembled using vaterite $CaCO_3$ crystals. *Micromachines* **2019**, *10*, 357. [CrossRef]
29. Li, X.; Chen, Y.; Kawazoe, N.; Chen, G. Influence of microporous gelatin hydrogels on chondrocyte functions. *J. Mater. Chem. B* **2017**, *5*, 5753–5762. [CrossRef] [PubMed]
30. Xie, Y.; Lee, K.; Wang, X.; Yoshitomi, T.; Kawazoe, N.; Yang, Y.; Chen, G. Interconnected collagen porous scaffolds prepared with sacrificial PLGA sponge templates for cartilage tissue engineering. *J. Mater. Chem. B* **2021**, *9*, 8491–8500. [CrossRef]
31. Francis, N.L.; Hunger, P.M.; Donius, A.E.; Riblett, B.W.; Zavaliangos, A.; Wegst, U.G.K.; Wheatley, M.A. An ice-templated, linearly aligned chitosan-alginate scaffold for neural tissue engineering. *J. Biomed. Mater. Res. Part A* **2013**, *101*, 3493–3503. [CrossRef]
32. Chen, S.; Nakamoto, T.; Kawazoe, N.; Chen, G. Engineering multi-layered skeletal muscle tissue by using 3D microgrooved collagen scaffolds. *Biomaterials* **2015**, *73*, 23–31. [CrossRef] [PubMed]
33. Annabi, N.; Nichol, J.W.; Zhong, X.; Ji, C.; Koshy, S.; Khademhosseini, A.; Dehghani, F. Controlling the porosity and microarchitecture of hydrogels for tissue engineering. *Tissue Eng. Part B Rev.* **2010**, *16*, 371–383. [CrossRef]
34. Zhang, H.; Cooper, A.I. Aligned Porous Structures by Directional Freezing. *Adv. Mater.* **2007**, *19*, 1529–1533. [CrossRef]
35. Yamamoto, T.; Randriantsilefisoa, R.; Sprecher, C.M.; D'Este, M. Fabrication of Collagen–Hyaluronic Acid Cryogels by Directional Freezing Mimicking Cartilage Arcade-like Structure. *Biomolecules* **2022**, *12*, 1809. [CrossRef]
36. Zhang, Q.; Lu, H.; Kawazoe, N.; Chen, G. Pore size effect of collagen scaffolds on cartilage regeneration. *Acta Biomater.* **2014**, *10*, 2005–2013. [CrossRef]
37. Chen, S.; Kawazoe, N.; Chen, G. Biomimetic Assembly of Vascular Endothelial Cells and Muscle Cells in Microgrooved Collagen Porous Scaffolds. *Tissue Eng. Part C Methods* **2017**, *23*, 367–376. [CrossRef] [PubMed]
38. Hua, M.; Wu, S.; Ma, Y.; Zhao, Y.; Chen, Z.; Frenkel, I.; Strzalka, J.; Zhou, H.; Zhu, X.; He, X. Strong tough hydrogels via the synergy of freeze-casting and salting out. *Nature* **2021**, *590*, 594–599. [CrossRef] [PubMed]
39. Bai, H.; Chen, Y.; Delattre, B.; Tomsia, A.P.; Ritchie, R.O. Bioinspired large-scale aligned porous materials assembled with dual temperature gradients. *Sci. Adv.* **2015**, *1*, e1500849. [CrossRef]
40. Xu, Z.; Wu, M.; Gao, W.; Bai, H. A sustainable single-component "Silk nacre". *Sci. Adv.* **2022**, *8*, eabo0946. [CrossRef]
41. Autissier, A.; Visage, C.L.; Pouzet, C.; Chaubet, F.; Letourneur, D. Fabrication of porous polysaccharide-based scaffolds using a combined freeze-drying/cross-linking process. *Acta Biomater.* **2010**, *6*, 3640–3648. [CrossRef]
42. Wu, X.; Liu, Y.; Li, X.; Wen, P.; Zhang, Y.; Long, Y.; Wang, X.; Guo, Y.; Xing, F.; Gao, J. Preparation of aligned porous gelatin scaffolds by unidirectional freeze-drying method. *Acta Biomater.* **2010**, *6*, 1167–1177. [CrossRef] [PubMed]
43. Lu, Q.; Hu, X.; Wang, X.; Kluge, J.A.; Lu, S.; Cebe, P.; Kaplan, D.L. Water-insoluble silk films with silk I structure. *Acta Biomater.* **2010**, *6*, 1380–1387. [CrossRef] [PubMed]
44. Zhang, X.; Xiao, L.; Ding, Z.; Lu, Q.; Kaplan, D.L. Engineered Tough Silk Hydrogels through Assembling β-Sheet Rich Nanofibers Based on a Solvent Replacement Strategy. *ACS Nano* **2022**, *16*, 10209–10218. [CrossRef] [PubMed]

45. Singh, S.; Cortes, G.; Kumar, U.; Sakthivel, T.S.; Niemiec, S.M.; Louiselle, A.E.; Azeltine-Bannerman, M.; Zgheib, C.; Liechty, K.W.; Seal, S. Silk fibroin nanofibrous mats for visible sensing of oxidative stress in cutaneous wounds. *Biomater. Sci.* **2020**, *8*, 5900–5910. [CrossRef]
46. Lyu, H.; Li, J.; Yuan, Z.; Liu, H.; Sun, Z.; Jiang, R.; Yu, X.; Hu, Y.; Pei, Y.; Ding, J.; et al. Supertough and highly stretchable silk protein-based films with controlled biodegradability. *Acta Biomater.* **2022**, *153*, 149–158. [CrossRef]
47. Pot, M.W.; Faraj, K.A.; Adawy, A.; van Enckevort, W.J.P.; van Moerkerk, H.T.B.; Vlieg, E.; Daamen, W.F.; van Kuppevelt, T.H. Versatile Wedge-Based System for the Construction of Unidirectional Collagen Scaffolds by Directional Freezing: Practical and Theoretical Considerations. *ACS Appl. Mater. Interfaces* **2015**, *7*, 8495–8505. [CrossRef] [PubMed]
48. Wang, Y.; Ma, M.; Wang, J.; Zhang, W.; Lu, W.; Gao, Y.; Zhang, B.; Guo, Y. Development of a Photo-Crosslinking, Biodegradable GelMA/PEGDA Hydrogel for Guided Bone Regeneration Materials. *Materials* **2018**, *11*, 1345. [CrossRef]
49. Yi, B.; Shen, Y.; Tang, H.; Wang, X.; Zhang, Y. Stiffness of the aligned fibers affects structural and functional integrity of the oriented endothelial cells. *Acta Biomater.* **2020**, *108*, 237–249. [CrossRef]
50. Ghaderinejad, P.; Najmoddin, N.; Bagher, Z.; Saeed, M.; Karimi, S.; Simorgh, S.; Pezeshki-Modaress, M. An injectable anisotropic alginate hydrogel containing oriented fibers for nerve tissue engineering. *Chem. Eng. J.* **2021**, *420*, 130465. [CrossRef]
51. Yi, B.; Shen, Y.; Tang, H.; Wang, X.; Li, B.; Zhang, Y. Stiffness of Aligned Fibers Regulates the Phenotypic Expression of Vascular Smooth Muscle Cells. *ACS Appl. Mater. Interfaces* **2019**, *11*, 6867–6880. [CrossRef]
52. Luo, Z.; Tang, G.; Ravanbakhsh, H.; Li, W.; Wang, M.; Kuang, X.; Garciamendez-Mijares, C.E.; Lian, L.; Yi, S.; Liao, J. Vertical extrusion cryo (bio) printing for anisotropic tissue manufacturing. *Adv. Mater.* **2022**, *34*, 2108931. [CrossRef] [PubMed]
53. Zhang, H.; Hussain, I.; Brust, M.; Butler, M.F.; Rannard, S.P.; Cooper, A.I. Aligned two- and three-dimensional structures by directional freezing of polymers and nanoparticles. *Nat. Mater.* **2005**, *4*, 787–793. [CrossRef] [PubMed]
54. Zhang, H. *Ice Templating and Freeze-Drying for Porous Materials and Their Applications*; John Wiley & Sons: New York, NY, USA, 2018.
55. Chaudhuri, O.; Gu, L.; Klumpers, D.; Darnell, M.; Bencherif, S.A.; Weaver, J.C.; Huebsch, N.; Lee, H.-P.; Lippens, E.; Duda, G.N.; et al. Hydrogels with tunable stress relaxation regulate stem cell fate and activity. *Nat. Mater.* **2016**, *15*, 326–334. [CrossRef] [PubMed]

gels

MDPI

Article

Stiffness-Modulation of Collagen Gels by Genipin-Crosslinking for Cell Culture

Seiichiro Ishihara [1,*], Haruna Kurosawa [2] and Hisashi Haga [1,*]

1 Department of Advanced Transdisciplinary Sciences, Faculty of Advanced Life Science, Hokkaido University, N10-W8, Kita-ku, Sapporo 060-0810, Japan
2 Division of Soft Matter, Graduate School of Life Science, Hokkaido University, N10-W8, Kita-ku, Sapporo 060-0810, Japan
* Correspondence: sishihara@sci.hokudai.ac.jp (S.I.); haga@sci.hokudai.ac.jp (H.H.)

Abstract: The stiffness of extracellular matrices (ECMs) is critical for cellular functions. Therefore, modulating the stiffness of ECMs in vitro is necessary to investigate the role of stiffness in cellular phenomena. Collagen gels are widely used for cell culture matrices in vitro. However, modulation of the stiffness in collagen gels for cell culture is challenging owing to the limited knowledge of the method to increase the stiffness while maintaining low cytotoxicity. Here, we established a novel method to modulate collagen gel stiffness from 0.0292 to 12.5 kPa with low cytotoxicity. We prepared collagens with genipin, a low-cytotoxic crosslinker of amines, at different concentrations and successfully modulated the stiffness of the gels. In addition, on 10 mM genipin-mixed collagen gels (approximately 12.5 kPa), H1299 human lung cancer cells showed spreading morphology and nuclear localization of yes-associated protein (YAP), typical phenomena of cells on stiff ECMs. Mouse mesenchymal stromal cells on 10 mM genipin-mixed collagen gels differentiated to vascular smooth muscle cells. On the other hand, the cells on 0 mM genipin-mixed collagen gels (approximately 0.0292 kPa) differentiated to visceral smooth muscle cells. Our new method provides a novel way to prepare stiffness-modulated collagen gels with low cytotoxicity in cell culture.

Keywords: collagen gel; extracellular matrix; stiffness; genipin; cell culture; cell morphology; YAP; mesenchymal stromal cells; cancer cells; differentiation

1. Introduction

Cells respond to mechanical stress and regulate their phenotype. In particular, once cells adhere to extracellular matrices (ECMs), they sense their stiffness and determine their fate [1]. For instance, fibroblasts easily migrate from soft ECMs to stiff ECMs, while it is difficult for cells to move from stiff ECMs to soft ECMs [2]. The morphology of fibroblasts is also regulated by ECM stiffness, as stiff ECMs induces spreading morphology [3]. In addition, epithelial cell clusters on stiff ECMs show random migration, whereas those on soft ECMs display collective migration [4]. Mesenchymal stromal/stem cells (MSCs) are multipotent mesenchymal cells that exist in many types of tissues [5] and regulate their differentiation potential by stiffness of the surrounding matrix. A previous study reported that soft ECMs, ECMs with moderate stiffness, and stiff ECMs trigger the neurogenic, myogenic, and osteogenic differentiation of MSCs, respectively [6]. Neural stem cells also respond to matrix stiffness and differentiate into neurons and astrocytes [7]. Furthermore, ECM stiffness is critical for cancer progression. Most solid cancers have stiff tissues, which affect the biochemical signaling and phenotypes of the cells in the tumors [8–10]. Stiff ECMs activate transcription factors, such as nuclear factor-κB (NF-κB), twist family bHLH transcription factor 1 (Twist1), and yes-associated protein (YAP) [11–13], thereby promoting malignant phenotypes in cancer cells. Stroma cells, such as cancer-associated fibroblasts (CAFs) in tumors, also respond to ECM stiffness and modulate cancer progression [14,15]. Therefore, the stiffness of the ECMs contributes to cellular functions and phenomena.

To evaluate the role of ECM stiffness in cell fate in vitro, we cultured the cells on ECMs with different stiffnesses. Collagens, one of the major components of ECMs [16], are widely used for cell culture. For example, the combination of a stiff substrate (collagen-coated glass or plastic dish) and a soft substrate (collagen gel) is one of the conventional methods used to determine the effect of ECM stiffness on cells [4,11,17]. In addition, by changing the collagen concentration, we can obtain gels with different stiffnesses, as higher collagen concentrations result in stiffer gel properties [8,18]. These methods are convenient; however, the exact concentration of collagen in these substrates cannot be controlled. Therefore, we cannot eliminate the biochemical effects of collagen molecules owing to differences in collagen concentration. Another way to prepare substrates with different stiffnesses is to use polyacrylamide gels with different crosslinker concentrations [19,20]. In this way, we can coat the gels with appropriate concentrations of ECMs; however, some cells can degrade the coated ECMs and, as a result, the artificial polyacrylamide gel surface is directly exposed to the cells. Another alternative method is crosslinking collagen gels with crosslinkers, such as N-(3-Dimethylaminopropyl)-N′-ethylcarbodiimide hydrochloride (EDC); however, using these crosslinkers poses a risk of cytotoxicity [21]. Although chitosan–hyaluronic acid dialdehyde hydrogels with different stiffnesses are used as substrates for cell culture [22], they are not derived from mammalian ECMs. Thus, a new way to control the stiffness of ECMs, such as collagen gels with the same concentration of ECMs and low cytotoxicity, is urgently needed.

Genipin is reported to be a crosslinker of amines in collagens [23,24] and a low cytotoxic reagent to cells [25]. Therefore, we used genipin to prepare collagen gels with different stiffness values for the cell culture. We successfully modulated the stiffness of collagen gels from 0.0292 to 12.5 kPa with low cytotoxicity. Moreover, we confirmed that the cells responded to the stiffness of these gels, as the H1299 lung cancer cells showed spreading morphology and nuclear localization of YAP, typical phenomena of cells on a stiff substrate. Furthermore, MSCs on stiff genipin-mixed collagen gels differentiated into vascular smooth muscle cells, whereas MSCs on soft collagen gels differentiated into visceral smooth muscle cells. These results suggest that our novel method for preparing stiffness-modulated collagen gels is a powerful tool for evaluating the effects of ECM stiffness on cultured cells in vitro.

2. Results and Discussion

2.1. Preparing Collagen Gels with Different Stiffness by Genipin-Crosslinking

First, we prepared collagen gels with genipin crosslinking to prepare gels with different stiffness values. We mixed a collagen solution and genipin-mixed HEPES (N-2-hydroxyethylpiperazine-N-2-ethane sulfonic acid) buffer and incubated the solution for gelation (Figure 1A). Collagen gels were mixed with 0, 0.01, 0.05, 0.1, 0.5, 1, or 10 mM genipin (final concentration). After washing the gels, we evaluated their stiffness (Young's modulus) using atomic force microscopy. The Young's moduli of the gels with 0, 0.01, 0.05, 0.1, 0.5, 1, and 10 mM genipin were 0.0292, 0.267, 0.678, 1.49, 3.36, 9.20, 12.5 kPa (mean value), respectively (Figure 1B). In addition, the genipin concentration in the media incubated for 24 h with 10 mM genipin-mixed collagen gel after washing was approximately 0.035 mM (determined by 595 nm absorbance). The cell viability of H1299 human lung cancer cell line and mouse mesenchymal stromal/stem cells (MSCs) was almost 100% with genipin concentrations less than 0.6 and 0.3 mM, respectively (Figure S1). Therefore, we confirmed that even at the highest genipin concentration in our experiments, the cytotoxicity of genipin derived from the genipin-mixed collagen gels was small enough to be ignored. As mentioned above, we successfully established a new method to prepare genipin-mixed collagen gels with different stiffnesses (0.0292–12.5 kPa) for cell culture.

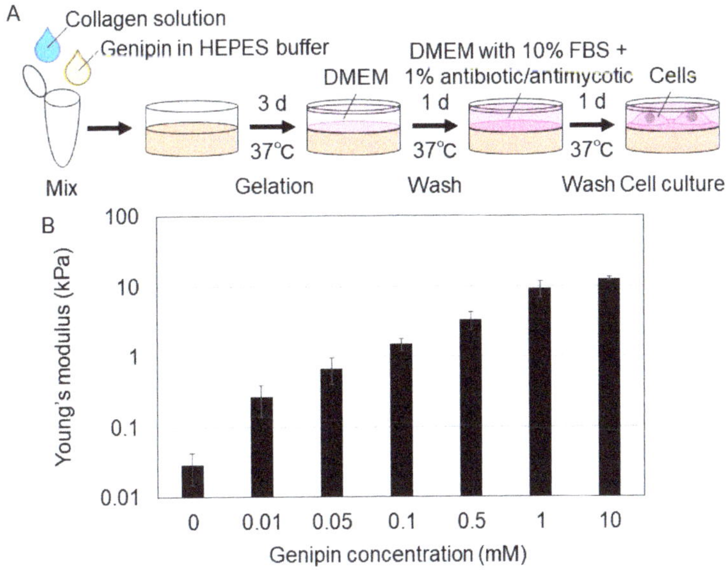

Figure 1. (**A**) The schematic procedure of preparing genipin-mixed collagen gels for cell culture. HEPES: N-2-hydroxyethylpiperazine-N-2-ethane sulfonic acid, DMEM: Dulbecco's modified Eagle's medium, FBS: fetal bovine serum. (**B**) Young's moduli of the 0, 0.01, 0.05, 0.1, 0.5, 1, or 10 mM genipin-mixed collagen gels. Mean ± S.D. N = 3 experiments.

2.2. Morphology of the Cells on Genipin-Mixed Collagen Gels

Next, we cultured H1299 cells on genipin-mixed collagen gels and investigated their phenotype. A previous study showed that cultured cells, including H1299 cells, display spreading morphology on a stiff substrate, whereas on a soft substrate, the cells maintained a round morphology [3,11]. Therefore, we cultured H1299 cells on genipin-mixed collagen gels with different stiffnesses and a collagen-coated glass dish (for the control of a stiff substrate). Our results confirmed that H1299 cells on 0 mM genipin-mixed collagen gels (0.0292 kPa gels) displayed a round shape; in contrast, the cells on 10 mM genipin-mixed collagen gels (12.5 kPa) and glass dishes showed spreading morphology (Figure 2A). The cells on 0.01 mM genipin-mixed collagen gels (0.267 kPa gels) induced moderate spreading (Figure 2A). Cell area analysis confirmed that cell spreading was significantly induced on stiff substrates (Figure 2B). These results showed that H1299 cells responded to the stiffness of genipin-mixed collagen gels and changed their morphology.

2.3. YAP Localization of the Cells on Genipin-Mixed Collagen Gels

YAP is a transcription factor that is localized to the nuclei of cells on a stiff substrate, whereas it diffuses or localizes to the cytoplasm of cells on a soft substrate [26]. Therefore, we cultured H1299 cells on genipin-mixed collagen gels and collagen-coated glass dishes and observed YAP localization using immunofluorescence staining. On 0 or 0.01 mM genipin-mixed collagen gels, YAP diffused in most cells (Figure 3A). In contrast, YAP localization to the nuclei was strongly induced in cells on 10 mM genipin-mixed collagen gels or glass substrates (Figure 3A). Statistical analysis of YAP nuclear localization showed that YAP was significantly localized to the nuclei of cells on 10 mM genipin-mixed collagen gels and glass dishes (Figure 3B). These results indicate that, similar to cell morphology, H1299 cells respond to the stiffness of genipin-mixed collagen gels and regulate YAP localization.

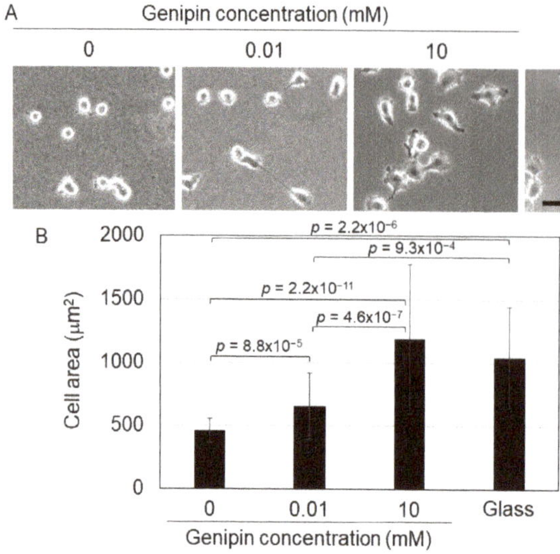

Figure 2. (**A**) Cell morphology of H1299 cells on the 0, 0.01, or 10 mM genipin-mixed collagen gels or collagen-coated glass substrates. Scale bar = 100 μm. (**B**) Cell area of H1299 cells shown in (**A**). Mean ± S.D. N = at least 23 cells in 3 independent experiments. p value was calculated using Welch's t-test with Bonferroni correction.

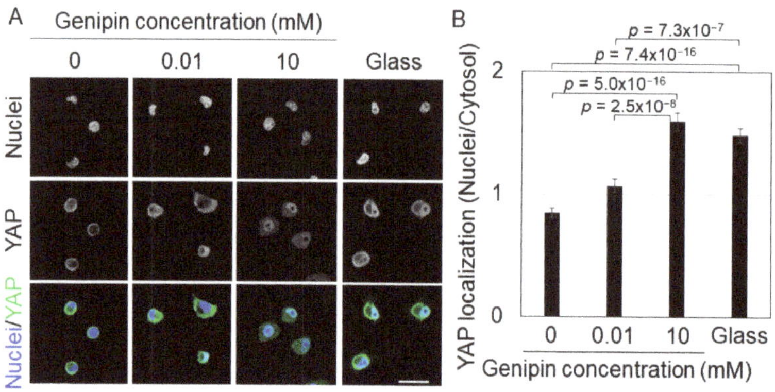

Figure 3. (**A**) Fluorescent staining of nuclei and YAP of H1299 cells on the 0, 0.01, or 10 mM genipin-mixed collagen gels or collagen-coated glass substrates. Scale bar = 50 μm. (**B**) YAP localization of H1299 cells shown in (**A**). Mean ± S.E. N = at least 84 cells in 4 independent experiments. p value was calculated using Welch's t-test with Bonferroni correction.

2.4. Differentiation of the MSCs on Genipin-Mixed Collagen Gels

Previous studies have shown that ECM stiffness is critical for the differentiation potential of MSCs [6,26]. Thus, we cultured MSCs on genipin-mixed collagen gels with different stiffnesses and evaluated their differentiation potential. First, we cultured MSCs on genipin-mixed collagen gels and collagen-coated plastic dishes (to control stiff substrates) under smooth muscle differentiation conditions. Our results showed that on 10 mM genipin-mixed collagen gels or plastic substrates, the cells showed spreading morphology, whereas on 0 mM genipin-mixed collagen gels, the cells did not (Figure 4A), suggesting that MSCs responded to the stiffness of each substrate. The qPCR results revealed that ACTA2 (αSMA, a marker of vascular smooth muscle [27]) expression was significantly higher in the cells on 10 mM genipin-mixed collagen gels or plastic dishes, on the other hand, ACTG2 (γSMA, a marker of visceral smooth muscle [27]) expression was significantly lower in the cells on 10 mM genipin-mixed collagen gels or plastic dishes (Figure 4B). We also confirmed that the smooth muscle marker smoothelin (SMTN) [27] was expressed in cells on all substrates, even though the expression level was significantly higher in cells on 0 mM genipin-mixed collagen gels (Figure S2).

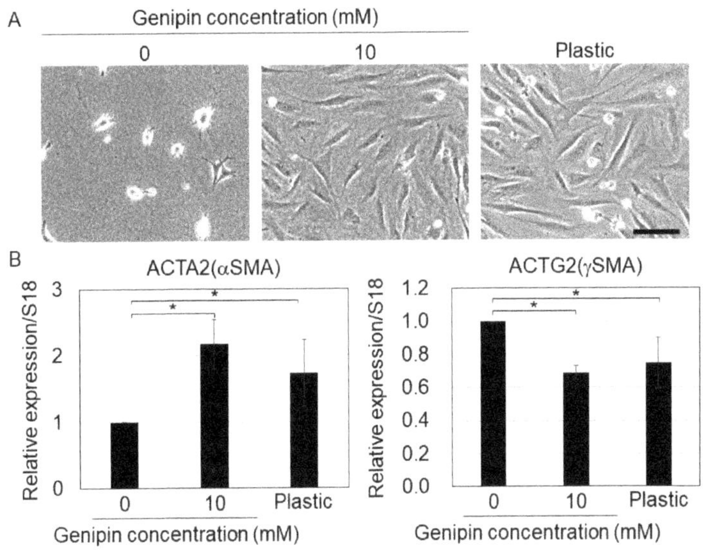

Figure 4. (**A**) Cell morphology of mouse mesenchymal stromal cells on the 0 or 10 mM genipin-mixed collagen gels or collagen-coated plastic substrates with smooth muscle-differentiation condition. Scale bar = 100 μm. (**B**) qPCR of ACTA2 and ACTG2 in (**A**). S18 was used as an internal control. Mean ± S.D. $N = 3$ independent experiments. * Statistical significance determined with 95% confidence interval with Bonferroni correction.

A previous study reported that in smooth muscle cells, high ACTA2 and low ACTG2 expressions are markers of vascular smooth muscle, whereas low ACTA2 and high ACTG2 expressions are markers of visceral smooth muscle [27]. Therefore, our results suggest that stiff substrates promote vascular smooth muscle differentiation and repress the visceral smooth muscle differentiation of MSCs. Although the mechanism that regulates the differentiation of MSCs into vascular or visceral smooth muscle is poorly understood, a previous study have shown that ACTA2 expression is upregulated by stiff substrates via YAP activation [15]. In addition, ACTA2-positive smooth muscle differentiation was reported to be promoted by the blockage of ERK signaling [28] and inhibited by miR-222-5p [29]. Therefore, stiff substrates may induce the differentiation of MSCs into vascular smooth muscles via the regulation of YAP, ERK, and miRNAs.

Next, we cultured MSCs on genipin-mixed collagen gels and collagen-coated plastic dishes under adipogenic conditions. On softer substrates, the cells stored larger lipid droplets (Figure 5A). A previous study found that lipid droplets grow large during the differentiation process [30], implying that soft substrates enhance the differentiation of MSCs into adipocytes. The qPCR results showed that the adipogenic markers (CEBPA and PPARG) [31] were downregulated in cells on stiff substrates (10 mM genipin-mixed collagen gels and plastic dishes). A previous study also showed that adipogenic differentiation is promoted in MSCs cultured on soft substrates [26]. Therefore, our results indicate that genipin-mixed collagen gels with different stiffnesses also regulate adipogenic differentiation of MSCs.

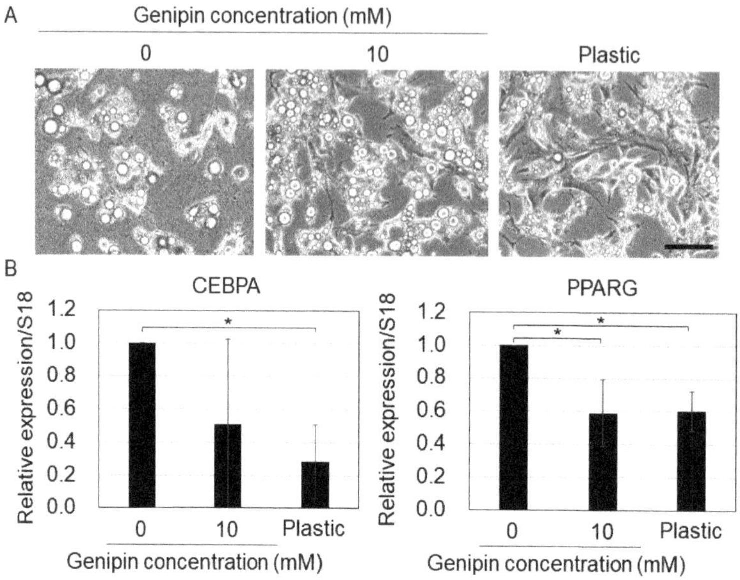

Figure 5. (**A**) Cell morphology of mouse mesenchymal stromal cells on the 0 or 10 mM genipin-mixed collagen gels or collagen-coated plastic substrates with adipogenic condition. Scale bar = 100 µm. (**B**) qPCR of CEBPA and PPARG in (**A**). S18 was used as an internal control. Mean ± S.D. $N = 3$ independent experiments. * Statistical significance determined with 95% confidence interval with Bonferroni correction.

2.5. Growth and Survival of the Cells on Genipin-Mixed Collagen Gels

We have previously shown that stiff ECMs promote growth of colon cancer cells [12] and MSCs in cancer-conditioned media [15]. In addition, hydrogel crosslinking affects gel porosity and cell proliferation [32]. Therefore, we examined the effect of hydrogel stiffness on the growth of cultured H1299 cells and MSCs by culturing them on genipin-mixed collagen gels of different stiffness. The results showed no significant differences in cell growth between the cells on 0 mM genipin-mixed collagen gels, 10 mM genipin-mixed collagen gels, and collagen-coated plastic substrates (Figure S3A,B). Cell survival was approximately 100% on all the substrates (Figure S3A,B), implying that cell growth reflects their proliferation. These results suggest that substrate stiffness and gel porosity are not critical for the proliferation of H1299 cells or MSCs.

These results demonstrate the successful preparation of collagen gels with different stiffness values, stable collagen concentrations, and low cytotoxicity. This method provides a useful system for culturing cells with different stiffness conditions and evaluating the effects of the mechanical environment on cell behavior.

3. Conclusions

In this study, we established a novel method for preparing cell culture substrates with different stiffness values. Our genipin-mixed collagen gels were substrates of modulated stiffness (0.0292–12.5 kPa) with the same concentration of collagen and low cytotoxicity. This is a good range of stiffness, as we can prepare substrates that mimic soft tissues such as brain tissue, moderate-stiff tissues such as lungs and livers, and abnormally stiff tissues such as tumors in vitro cell culture (Figure 6) [6,9]. Thus, we suggest that our method of preparing collagen gels with different stiffnesses is useful for evaluating the effects of in vivo tissue stiffness on development or cancer progression, contributing to the progress of regenerative medicine and the development of new cancer drugs.

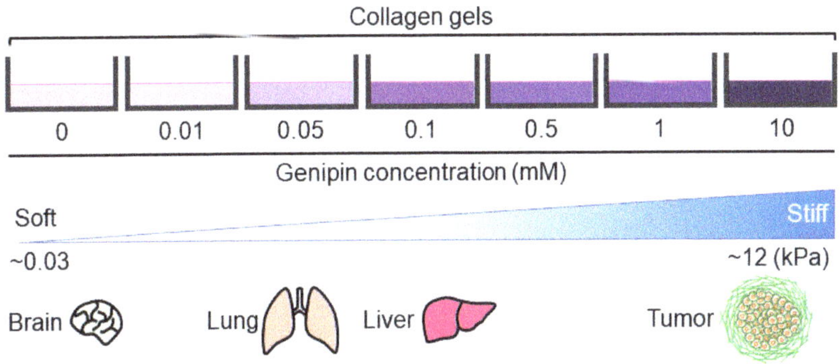

Figure 6. Stiffness-modulation of genipin-mixed collagen gels with the indicated genipin concentrations. The reported stiffness of brain, lung, liver, and tumor is also shown [6,9].

Previous studies have shown the potential of hydrogels as novel therapeutic agents. For instance, hydrogel microspheres are promising therapeutic material for cartilage repair [33], phytomedicine-loaded hydrogels are effective in killing gram-positive and gram-negative bacteria [34], and magnetic hydrogels may be useful for cancer therapy to induce local hyperthermia in tumors [35]. Genipin-mixed collagen gels may also be good therapeutic candidates for preparing biomimetic physical environments for specific tissues and transplanting gels with appropriate cells. Thus, we established a novel method to prepare stiffness-modulated collagen hydrogels for cell culture and potential medical applications.

4. Materials and Methods

4.1. Gel Preparation

Genipin-mixed collagen gels were prepared under sterile conditions (Figure 1A). First, HEPES buffer (100 mM HEPES, 16 g/L NaCl, 2.3 g/L Na_2HPO_4, 0.4 g/L KCl, 0.4 g/L KH_2PO_4, pH = 7.3–7.4) and 20 mM genipin (078-03021, Wako, Osaka, Japan) in HEPES buffer were mixed to prepare genipin premix solution (×2 genipin concentration of final collagen gel) on ice. Second, equal volumes of genipin premix solution and Atelocollagen Acidic Solution (5 mg/mL collagen concentration, IPC-50, KOKEN, Tokyo, Japan) were mixed well with gentle pipetting for 60 s on ice. The mixture was then quickly centrifuged to remove bubbles. The solution was poured on an iced dish and incubated at 37 °C in a humidified incubator with 5% CO_2 for 72 h for gelation. The gels were washed with Dulbecco's modified Eagle's medium (DMEM; D6046, Sigma-Aldrich, St. Louis, MO, USA or 08456-65, Nacalai Tesque Inc., Kyoto, japan) for 24 h at 37 °C in a humidified incubator with 5% CO_2. For MSC culture and growth and survival assay, the gels were washed with phosphate-buffered saline (PBS) for 1 h three times before washing with DMEM. Next, the gels were washed with DMEM supplemented with 10% fetal bovine serum (FBS; 172012, Sigma-Aldrich, St. Louis, MO, USA) and 1% antibiotic/antimycotic solution (A5955, Sigma-Aldrich, St. Louis, MO, USA) for 24 h at 37 °C in a humidified incubator with 5% CO_2. The gels were then used for cell culture or measurement of the Young's modulus using atomic force microscopy.

4.2. Atomic Force Microscopy (AFM)

Genipin-mixed collagen gels (500 μL) were prepared in a 35 mm plastic dish. AFM indentations for the measurement of surface stiffness were performed using Nanowizard 4 (JPK Instruments, Berlin, Germany) on a TE300 microscope (Nikon Instech, Tokyo, Japan). A 10 μm diameter bead (F8834, Life Technologies, Carlsbad, CA, USA) tip-bound silicon nitride cantilever (MLCT-E, Veeco, Woodbury, NY, USA) with a spring constant of ~0.1 N/m was used. The spring constants of the cantilevers were determined by thermal tuning using a simple harmonic oscillator model. The gels were indented with a calibrated forces of 0.2 nN (gels with 0 mM genipin), 0.5 nN (gels with 0.01, 0,05, or 0.1 mM genipin), 2 nN (gels with 0.5 or 1 mM genipin), or 5 nN (gels with 10 mM genipin) in a scan area of 1 μm^2 (4 pixels × 4 lines). The Hertz model of impact was used to determine stiffness (Young's modulus). Three points within the scanning area of each gel were selected, and their corresponding values were used to calculate the average Young's modulus of each gel. Only the force curves (indentation of at least 100 nm) that fit the Hertz model were used for the calculation. A Poisson's ratio of 0.5 was used to calculate the Young's modulus.

4.3. Cell Culture

The H1299 human lung cancer cell line (American Type Culture Collection) was cultured in H1299 culture media (DMEM supplemented with 10% FBS and 1% antibiotic/antimycotic solution). Mouse mesenchymal stromal/stem cells (MSCs) [15], kindly gifted by Dr. Suzanne Ponik (University of Wisconsin-Madison), were cultured in MSC culture media (DMEM supplemented with 20% FBS and 1% antibiotic/antimycotic solution). Cells were cultured at 37 °C in a humidified incubator with 5% CO_2. For collagen coating, plastic or glass dishes were treated with 0.5 mg/mL Atelocollagen Acidic Solution (IPC-50, KOKEN, Tokyo, Japan) or 0.3 mg/mL Cellmatrix Type I-C (Nitta Gelatin, Osaka, Japan; for growth and survival assay). Phase-contrast images were captured using a TE300 microscope (Nikon Instech, Tokyo, Japan) equipped with a 10× objective.

4.4. Cell Viability Assay

H1299 cells (1 × 10^5 cells) or MSCs (2 × 10^5 cells) were seeded on collagen-coated 35 mm plastic dishes and cultured for 24 h with genipin at the indicated concentrations. The cells were then treated with calcein AM (Dojindo, Mashiki, Japan) and incubated for 1 h. The bright field and fluorescent images of the cells were captured with a ZOE Fluorescent Cell Imager (Bio-Rad Laboratories Inc., Hercules, CA, USA) to count the number of live and total cells. Cell viability (%) was calculated as follows: (number of live cells)/(number of total cells) × 100.

4.5. Morphology Assay

H1299 cells (2 × 10^4 cells) were seeded on 200 µL of genipin-mixed collagen gels in 16 mm glass dishes or collagen-coated 16 mm glass dishes and cultured for 24 h. Then, phase-contrast images of the cells were captured with a TE300 microscope (Nikon Instech, Tokyo, Japan) equipped with a 10× objective. The cell area was calculated using ImageJ software.

4.6. Fluorescent Staining

H1299 cells (2 × 10^4 cells) were seeded on 200 µL of genipin-mixed collagen gels in 16 mm glass dishes or collagen-coated 16 mm glass dishes and cultured for 24 h. Then, the cells were fixed with 4% paraformaldehyde in PBS for 10 min at room temperature and washed three times with PBS. Next, the cells were permeabilized with 0.1% Triton X-100 in PBS for 15 min at room temperature and washed three times with PBS. The cells were blocked with 0.1 or 0.5% bovine serum albumin in PBS for 30–60 min at room temperature. Primary antibody solution (1:250 anti-YAP antibody; 14074S, Cell Signaling Technology Inc., Danvers, MA, USA) in PBS was added and incubated at 4 °C overnight. After three washes with PBS, the secondary antibody solution (1:250 Alexa Fluor 488 goat anti-rabbit; A27034, Invitrogen, Waltham, MA, USA and 1:10,000 Hoechst 33342; H1399, Invitrogen) in PBS was added and incubated at room temperature for 1 h. After three washes with PBS, fluorescent images were captured using an A1 confocal imaging system (Nikon Instech, Tokyo, Japan) with a 60× objective. To quantify the localization of YAP, the fluorescence intensity was calculated as the nuclear/cytosol ratio using ImageJ software.

4.7. Differentiation of Mesenchymal Stromal Cells

For smooth muscle differentiation, MSCs (5×10^4 cells) were seeded on 500 µL of genipin-mixed collagen gels in 35 mm plastic dishes or collagen-coated 35 mm plastic dishes and cultured for 24 h with MSC culture media. The medium was then changed to DMEM supplemented with 1% FBS and 1% antibiotic/antimycotic solution for 2 d of culture [36]. For adipogenic conditions, MSCs (8×10^4 cells) were seeded on 250 µL of genipin-mixed collagen gels in 12 well plates or collagen-coated 12 well plates and cultured for 24 h in MSC culture media. Adipogenic differentiation was then performed using the hMSC BulletKit (PT-3004, LONZA, Basel, Switzerland) for 12 d.

4.8. qPCR

RNA was extracted using TriPure isolation reagent (Sigma-Aldrich, St. Louis, MO, USA), followed by purification with the FastGene RNA Basic Kit (FG-80250, Nippon Genetics Co., Ltd., Tokyo, Japan). Reverse transcription was performed using the ReverTra Ace qPCR RT Kit (FSQ-201; TOYOBO, Osaka, Japan). qPCR with cDNA was performed using the KAPA SYBR Fast qPCR kit (Kapa Biosystems, Inc., Woburn, MA, USA) and the Applied Biosystems StepOnePlus Real-Time PCR System (Thermo Scientific, Waltham, MA, USA). The following primers (5′-3′) were used: S18 (forward: ACTTTTGGGGCCTTCGTGTC, reverse: GCAAAGGCCCAGAGACTCAT), ACTA2 (forward: CTATTCCTTCGTGACTACTGCCGAG; reverse: GTTATAGGTGGTTTCGTGGATGCCC), ACTG2 (forward: TCGAGTAGCACCAGAAGAGCAC, reverse: CGAATCCAGAACGATGCCTGTG), SMTN (forward: ATGGCAGACGAGGCTTTAGCTG, reverse: TCTCGCTGTTGAGAGTGTAGCC), CEBPA (forward: AACAGCTGAGCCGTGAACTG, reverse: TTTCAGGCCACACCGGAATC), and PPARG (forward: TGATCTTAACTGCCGGATCCAC, reverse: CCCAAACCTGATGGCATTGTG).

4.9. Growth and Survival Assay

H1299 cells and MSCs (2×10^4 cells) were seeded on 500 µL of genipin-mixed collagen gels in 35 mm plastic dishes or collagen-coated 35 mm plastic dishes and cultured for 48 h. Nuclei and live cells were stained with Hoechst 33342 (1:4000) and calcein AM (1:1000) in PBS for 30 min, followed by washing with PBS. The nuclei and live cells were imaged using a ZOE Fluorescent Cell Imager (Bio-Rad Laboratories Inc., Hercules, CA, USA). Relative cell growth was evaluated by counting the number of nuclei, with the mean in H1299 cells on 0 mM genipin-mixed collagen gels set as 1). The cell survival (%) was calculated as follows: (number of live cells)/(number of nuclei) $\times$ 100.

Supplementary Materials: The following supporting information can be downloaded from https://www.mdpi.com/article/10.3390/gels9020148/s1 Figure S1. Cell viability of H1299 and mouse mesenchymal stromal cells. Mean ± S.E. $N = 3$ independent experiments. Figure S2. qPCR of SMTN in mouse mesenchymal stromal cells on 0 or 10 mM genipin-mixed collagen gels or collagen-coated plastic substrates with smooth muscle-differentiation condition. S18 was used as an internal control. Mean ± S.D. $N = 3$ independent experiments. * Statistical significance was determined using a 95% confidence interval with Bonferroni correction. Figure S3. (A) Fluorescent staining of nuclei and live H1299 and mouse mesenchymal stromal cells on the 0 or 10 mM genipin-mixed collagen gels or collagen-coated plastic substrates. Scale bar = 100 µm. (B) Relative cell growth and survival of H1299 and mouse mesenchymal stromal cells in (A). Mean±S.D. $N = 3$ independent experiments. N.S.; no statistical significance with Welch's *t*-test (mesenchymal stromal cells, 10 mM vs plastic) or Student's *t*-test (the others) with Bonferroni correction.

Author Contributions: Conceptualization, S.I. and H.H.; methodology, S.I. and H.H.; investigation, S.I. and H.K.; formal analysis, S.I. and H.K.; writing—original draft preparation, S.I.; writing—review and editing, H.K. and H.H.; supervision, H.H.; project administration, H.H.; funding acquisition, S.I. and H.H. All authors have read and agreed to the published version of the manuscript.

Funding: This work was supported by Advanced Research and Development Programs for Medical Innovation by Japan Agency for Medical Research and Development (AMED, Grant Number

21gm0810007h0006 to H.H.); AMED under Grant Number JP22ym0126814 to H.H.; JSPS KAKENHI (Grant Numbers 21K07142 to H.H., 18K15232 to S.I, 21K07141 to S.I.); Grants-in-Aid for Regional R&D Proposal-Based Program from Northern Advancement Center for Science & Technology of Hokkaido Japan to S.I.; the research grant of Astellas Foundation for Research on Metabolic Disorders to S.I.; the research grant of The Uehara Memorial Foundation to S.I.; the research grant of The Yasuda Medical Foundation to S.I.; the research grant of The Akiyama Life Science Foundation to S.I.; Extramural Collaborative Research Grant of Cancer Research Institute, Kanazawa University to S.I.; periodic donations from M.D. Mariko Takamura to H.H.

Institutional Review Board Statement: Not applicable.

Informed Consent Statement: Not applicable.

Data Availability Statement: All data generated or analyzed during this study are included in this published article and its Supplementary Files.

Acknowledgments: The authors thank Hatsumi Sano for technical support.

Conflicts of Interest: The authors declare no conflict of interest.

References

1. Discher, D.E.; Janmey, P.; Wang, Y.L. Tissue cells feel and respond to the stiffness of their substrate. *Science* **2005**, *310*, 1139–1143. [CrossRef]
2. Lo, C.M.; Wang, H.B.; Dembo, M.; Wang, Y.L. Cell movement is guided by the rigidity of the substrate. *Biophys. J.* **2000**, *79*, 144–152. [CrossRef]
3. Yeung, T.; Georges, P.C.; Flanagan, L.A.; Marg, B.; Ortiz, M.; Funaki, M.; Zahir, N.; Ming, W.; Weaver, V.; Janmey, P.A. Effects of substrate stiffness on cell morphology, cytoskeletal structure, and adhesion. *Cell Motil. Cytoskelet.* **2005**, *60*, 24–34. [CrossRef] [PubMed]
4. Haga, H.; Irahara, C.; Kobayashi, R.; Nakagaki, T.; Kawabata, K. Collective movement of epithelial cells on a collagen gel substrate. *Biophys. J.* **2005**, *88*, 2250–2256. [CrossRef] [PubMed]
5. Meirelles, L.D.S.; Chagastelles, P.C.; Nardi, N.B. Mesehcnymal stem cells reside in virtually all post-natal organs and tissues. *J. Cell Sci.* **2006**, *119*, 2204–2213. [CrossRef] [PubMed]
6. Engler, A.J.; Sen, S.; Sweeney, H.L.; Discher, D.E. Matrix elasticity directs stem cell lineage specification. *Cell* **2006**, *126*, 677–689. [CrossRef]
7. Oyama, H.; Nukuda, A.; Ishihara, S.; Haga, H. Soft surfaces promote astrocytic differentiation of mouse embryonic neural stem cells via dephosphorylation of MRLC in the absence of serum. *Sci. Rep.* **2021**, *11*, 19574. [CrossRef]
8. Paszek, M.J.; Zahir, N.; Johnson, K.R.; Lakins, J.N.; Rozenberg, G.I.; Gefen, A.; Reinhart-King, C.A.; Margulies, S.S.; Dembo, M.; Boettiger, D.; et al. Tensional homeostasis and the malignant phenotype. *Cancer Cell* **2005**, *8*, 241–254. [CrossRef]
9. Butcher, D.T.; Alliston, T.; Weaver, V.M. A tense situation: Forcing tumour progression. *Nat. Rev. Cancer* **2009**, *9*, 108–122. [CrossRef]
10. Ishihara, S.; Haga, H. Matrix Stiffness Contributes to Cancer Progression by Regulating Transcription Factors. *Cancers* **2022**, *14*, 1049. [CrossRef]
11. Ishihara, S.; Yasuda, M.; Harada, I.; Mizutani, T.; Kawabata, K.; Haga, H. Substrate stiffness regulates temporary NF-κB activation via actomyosin contractions. *Exp. Cell Res.* **2013**, *319*, 2916–2927. [CrossRef]
12. Nukuda, A.; Sasaki, C.; Ishihara, S.; Mizutani, T.; Nakamura, K.; Ayabe, T.; Kawabata, K.; Haga, H. Stiff substrates increase YAP-signaling-mediated matrix metalloproteinase-7 expression. *Oncogenesis* **2015**, *4*, e165. [CrossRef]
13. Wei, S.C.; Fattet, L.; Tsai, J.H.; Guo, Y.; Pai, V.H.; Majeski, H.E.; Chen, A.C.; Sah, R.L.; Taylor, S.S.; Engler, A.J.; et al. Matrix stiffness drives epithelial-mesenchymal transition and tumour metastasis through a TWIST1-G3BP2 mechanotransduction pathway. *Nat. Cell Biol.* **2015**, *17*, 678–688. [CrossRef] [PubMed]
14. Calvo, F.; Ege, N.; Grande-Garcia, A.; Hooper, S.; Jenkins, R.P.; Chaudhry, S.I.; Harrington, K.; Williamson, P.; Moeendarbary, E.; Charras, G.; et al. Mechanotransduction and YAP-dependent matrix remodelling is required for the generation and maintenance of cancer-associated fibroblasts. *Nat. Cell Biol.* **2013**, *15*, 637–646. [CrossRef]
15. Ishihara, S.; Inman, D.R.; Li, W.J.; Ponik, S.M.; Keely, P.J. Mechano-Signal Transduction in Mesenchymal Stem Cells Induces Prosaposin Secretion to Drive the Proliferation of Breast Cancer Cells. *Cancer Res.* **2017**, *77*, 6179–6189. [CrossRef]
16. Shoulders, M.D.; Raines, R.T. Collagen structure and stability. *Annu. Rev. Biochem.* **2009**, *78*, 929–958. [CrossRef] [PubMed]
17. Ishihara, S.; Mizutani, T.; Kawabata, K.; Haga, H. An improved method for western blotting when extracting proteins from mammalian cells cultured on a collagen gel under serum-free conditions. *Cytotechnology* **2016**, *68*, 25–32. [CrossRef] [PubMed]
18. Provenzano, P.P.; Inman, D.R.; Eliceiri, K.W.; Keely, P.J. Matrix density-induced mechanoregulation of breast cell phenotype, signaling and gene expression through a FAK-ERK linkage. *Oncogene* **2009**, *28*, 4326–4343. [CrossRef]
19. Pelham, R.J.; Wang, Y. Cell locomotion and focal adhesions are regulated by substrate flexibility. *Proc. Natl. Acad. Sci. USA* **1997**, *94*, 13661–13665. [CrossRef]

20. Yip, A.K.; Iwasaki, K.; Ursekar, C.; Machiyama, H.; Saxena, M.; Chen, H.; Harada, I.; Chiam, K.H.; Sawada, Y. Cellular response to substrate rigidity is governed by either stress or strain. *Biophys. J.* **2013**, *104*, 19–29. [CrossRef]
21. Výborný, K.; Vallová, J.; Kočí, Z.; Kekulová, K.; Jiráková, K.; Jendelová, P.; Hodan, J.; Kubinová, Š. Genipin and EDC crosslinking of extracellular matrix hydrogel derived from human umbilical cord for neural tissue repair. *Sci. Rep.* **2019**, *9*, 10674. [CrossRef] [PubMed]
22. Thomas, L.V.; Vg, R.; Nair, P.D. Effect of stiffness of chitosan-hyaluronic acid dialdehyde hydrogels on the viability and growth of encapsulated chondrocytes. *Int. J. Biol. Macromol.* **2017**, *104*, 1925–1935. [CrossRef] [PubMed]
23. Sung, H.W.; Chang, W.H.; Ma, C.Y.; Lee, M.H. Crosslinking of biological tissues using genipin and/or carbodiimide. *J. Biomed. Mater. Res. Part A* **2003**, *64*, 427–438. [CrossRef]
24. Sundararaghavan, H.G.; Monteiro, G.A.; Lapin, N.A.; Chabal, Y.J.; Miksan, J.R.; Shreiber, D.I. Genipin-induced changes in collagen gels: Correlation of mechanical properties to fluorescence. *J. Biomed. Mater. Res. Part A* **2008**, *87*, 308–320. [CrossRef]
25. Sung, H.W.; Huang, R.N.; Huang, L.L.; Tsai, C.C.; Chiu, C.T. Feasibility study of a natural crosslinking reagent for biological tissue fixation. *J. Biomed. Mater. Res.* **1998**, *42*, 560–567. [CrossRef]
26. Dupont, S.; Morsut, L.; Aragona, M.; Enzo, E.; Giulitti, S.; Cordenonsi, M.; Zanconato, F.; Le Digabel, J.; Forcato, M.; Bicciato, S.; et al. Role of YAP/TAZ in mechanotransduction. *Nature* **2011**, *474*, 179–183. [CrossRef]
27. Jaslove, J.M.; Nelson, C.M. Smooth muscle: A stiff sculptor of epithelial shapes. *Philos. Trans. R. Soc. Lond. B Biol. Sci.* **2018**, *373*, 20170318. [CrossRef]
28. Tamama, K.; Sen, C.K.; Wells, A. Differentiation of bone marrow mesenchymal stem cells into the smooth muscle lineage by blocking ERK/MAPK signaling pathway. *Stem Cells Dev.* **2008**, *17*, 897–908. [CrossRef]
29. Gu, W.; Hong, X.; Le Bras, A.; Nowak, W.N.; Issa Bhaloo, S.; Deng, J.; Xie, Y.; Hu, Y.; Ruan, X.Z.; Xu, Q. Smooth muscle cells differentiated from mesenchymal stem cells are regulated by microRNAs and suitable for vascular tissue grafts. *J. Biol. Chem.* **2018**, *293*, 8089–8102. [CrossRef] [PubMed]
30. Nagayama, M.; Uchida, T.; Gohara, K. Temporal and spatial variations of lipid droplets during adipocyte division and differentiation. *J. Lipid Res.* **2007**, *48*, 9–18. [CrossRef]
31. Rosen, E.D.; Hsu, C.H.; Wang, X.; Sakai, S.; Freeman, M.W.; Gonzalez, F.J.; Spiegelman, B.M. C/EBPalpha induces adipogenesis through PPARgamma: A unified pathway. *Genes Dev.* **2002**, *16*, 22–26. [CrossRef] [PubMed]
32. Lai, J.Y.; Ma, D.H.; Lai, M.H.; Li, Y.T.; Chang, R.J.; Chen, L.M. Characterization of cross-linked porous gelatin carriers and their interaction with corneal endothelium: Biopolymer concentration effect. *PLoS ONE* **2013**, *8*, e54058. [CrossRef] [PubMed]
33. Lin, F.; Li, Y.; Cui, W. Injectable hydrogel microspheres in cartilage repair. *Biomed. Technol.* **2023**, *1*, 18–29. [CrossRef]
34. Das, P.; Ganguly, S.; Saravanan, A.; Margel, S.; Gedanken, A.; Srinivasan, S.; Rajabzadeh, A.R. Naturally Derived Carbon Dots In Situ Confined Self-Healing and Breathable Hydrogel Monolith for Anomalous Diffusion-Driven Phytomedicine Release. *ACS. Appl. Bio Mater.* **2022**, *5*, 5617–5633. [CrossRef]
35. Ganguly, S.; Margel, S. Design of Magnetic Hydrogels for Hyperthermia and Drug Delivery. *Polymers* **2021**, *13*, 4259. [CrossRef] [PubMed]
36. Park, J.S.; Chu, J.S.; Tsou, A.D.; Diop, R.; Tang, Z.; Wang, A.; Li, S. The effect of matrix stiffness on the differentiation of mesenchymal stem cells in response to TGF-β. *Biomaterials* **2011**, *32*, 3921–3930. [CrossRef]

gels

MDPI

Article

Hydroxyapatite-Tethered Peptide Hydrogel Promotes Osteogenesis

Hongwen Yu [1,2,3], Jiaqi Song [3], Xianpeng Zhang [3], Kuo Jiang [2], Hong Fan [2], Yibing Li [2], Yuanting Zhao [2], Shichang Liu [1,2,*], Dingjun Hao [1,2,*] and Guanying Li [3,*]

[1] The Second Clinical Medical School, Shaanxi University of Chinese Medicine, Xianyang 712046, China
[2] Honghui Hospital, Xi'an Jiaotong University, Xi'an 710054, China
[3] School of Basic Medical Sciences, Xi'an Jiaotong University, Xi'an 710061, China
* Correspondence: feitianyu1985@163.com (S.L.); hdj_hhyy@163.com (D.H.); guanyingli@xjtu.edu.cn (G.L.)

Abstract: Hydroxyapatite (HAp) as natural bone composition is highly osteoinductive. To harvest its osteoinductivity in bone regenerative engineering, the HAp-supporting hydrogel is urgently needed to minimize inhomogeneous aggregation of HAp. Here, we developed a HAp-stabilizing hydrogel based on peptide self-assembly. FmocFFRR was efficient for HAp-capping due to arginine-phosphate interaction. Tethering FmocFFRR on the HAp surface facilitated self-assembly to form FmocFFRR/HAp hybrid hydrogel, enabling stable dispersion of HAp in it. The molecular interactions between FmocFFRR and HAp particles were studied using microscopic and spectral characterizations. FmocFFRR/HAp hydrogel exhibited more enhanced mechanical properties than FmocFFRR. The biocompatibility of FmocFFRR/HAp hydrogel was verified using an ATP assay and live-dead staining assay. More importantly, FmocFFRR/HAp hydrogel not only enabled cell attachment on its surface, but also supported 3D cell culturing inside the hydrogel. Further, 3D culturing of MC3T3-E1 preosteoblasts inside FmocFFRR/HAp hydrogel significantly enhanced the expressions of osteogenesis markers, including alkaline phosphate (ALP), type-I collagen (COL1), and osteocalcin (OCN), demonstrating the promoting effect of osteoblast differentiation. These findings inspire its potential application in bone regenerative engineering.

Keywords: hydrogel; peptide assembly; hydroxyapatite; osteoinductive; mimetic bone matrix; bone regenerative engineering

1. Introduction

Despite the successful clinical usage of autografts and allografts for bone substitution, repair, and augmentation [1], they are both plagued with their high cost, limited availability, and secondary damage at the donor site. Bone regenerative engineering that uses a top-down approach to regenerate bone tissue has attracted increasing attention in repairing bone defects [2]. One of the key aspects of bone regenerative engineering is the osteoinductive properties of the regenerative scaffolds or incorporating materials [3]. Hydroxyapatite (HAp) constitutes the major inorganic component in natural bones. Nanostructured HAp particles possess high osteoinductivity [4–6] and have been widely used in bone regenerative engineering [7,8]. However, HAp is very stable with low solubility in physiological conditions. Nanosized HAp tends to aggregate due to its high surface energies [9], impeding its osteoinduction efficacy. What is worse, inhomogeneous aggregation of HAp is risky to bone malformation during bone regeneration [10].

Satisfying HAp stability in a three-dimensional (3D) scaffold greatly benefits from the advanced development of material design over the last decade [11,12]. Polymeric biomaterials including natural polymers [13–15] and synthetic polymers [10,16] have been extensively employed as HAp-supporting scaffolds. Compared with these polymeric scaffolds, supramolecular peptide hydrogels based on small molecular peptide assembly [17] show prominent biocompatibility and biodegradability without causing observable inflammation. Because of the noncovalent interactions that drive peptide assembly [18,19], peptide hydrogels are more adaptive and feasible as 3D scaffolds for tissue engineering than polymer-based scaffolds. More importantly, scaffolds embracing bioactive peptides as a mimetic extracellular matrix (ECM) [20] enable cell adhesion and regulate cell proliferation, cell motion, polarity, mechanotransduction, and ECM signaling cascades, which crucially affect cell differentiation during bone development. In this regard, self-assembling peptide hydrogels become promising candidates as osteoinductive scaffolds for bone regenerative engineering.

Fluorenyl-9-methoxycarbonyl diphenylalanine (FmocFF), as one of the smallest building blocks [21], could self-assemble into a self-supporting hydrogel. However, this basic self-assembled hydrogel lacks HAp-binding capacity and bioactivity allowing cell attachment. To address these issues, a co-assembly strategy using multicomponent was employed. Ghosh et al. [22] incorporated Fmoc-arginine (FmocR) with FmocFF to form HAp-decorated hydrogel. Wang et al. [23] introduced a cell adhesive ligand Fmoc-arginyl-glycyl-aspartate (FmocRGD) and developed FmocFF/FmocRGD hydrogel for three-dimensional (3D) culturing of mesenchymal stem cells. Recently, Vitale et al. [24] developed a three-component assembled hydrogel composed of FmocFF/FmocS/FmocRGD, where FmocS was Fmoc-serine. This hydrogel enabled HAp-binding and cell adhesion, and promoted osteoclast differentiation of RAW264.7 macrophages, showing potential in bone tissue engineering. One of the major drawbacks of multicomponent assembled hydrogel is that the HAp trapping manner, their mechanical properties, and their biological activities strongly depend on their formulation. Their osteoinductivity for osteoblast differentiation has not yet been explored.

To better harvest the osteoinductivity of HAp in bone regenerative engineering, in this work, we seek to construct Hap-stabilizing hydrogel composed of single peptide assembly for accurate guidance of Hap-induced osteogenesis. Recently, we reported a pentapeptide FmocFFRRR with high affinity to phosphate. The side chain of arginine residues has a pKa value of 12.48, indicating guanidine groups are protonated and possess positive charges under physiological condition. By adding phosphates, FmocFFRRR formed a self-assembling hydrogel, assisting the in situ mineralization of CaP [25]. When HAp that is rich in phosphate groups was added, however, FmocFFRRR was prone to form aggregation or microgels, probably due to the strong binding affinity. Herein, we compromised its HAp binding by reducing an arginine residue. The obtained tetrapeptide FmocFFRR possessed a balanced hydrophobic building block (FmocFF−) and hydrophilic residue (-RR), and self-assembled into nanofilaments in aqueous. We speculate that HAp is capable of tethering FmocFFRR via electrostatic phosphate–guanidine interaction [26]. Anchoring FmocFFRR on the HAp surface not only reduces the intermolecular distance of self-assembling building blocks that facilitates their self-assembly, but also promotes interfibrillar tethering, generating a HAp-trapping fibrous network for stable HAp dispersion (Figure 1). The mechanical properties, biocompatibility, and osteoinductive properties of HAp-tethered FmocFFRR assembling hydrogel were evaluated.

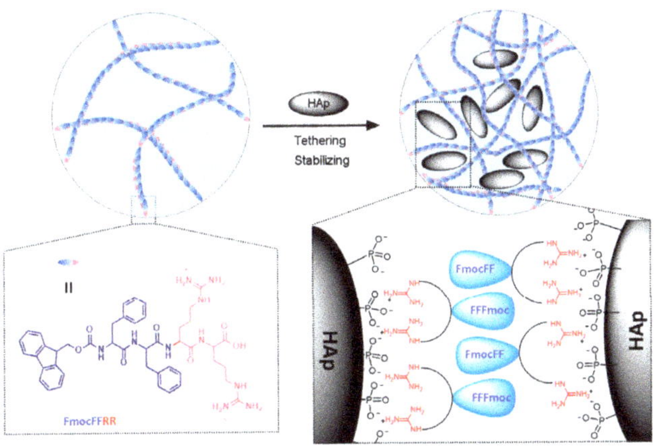

Figure 1. Schematic illustration of constructing FmocFFRR-tethering HAp hybrid hydrogel for HAp-stabilization.

2. Results and Discussion

2.1. Synthesis and Characterizations

FmocFFRR was synthesized by standard solid phase peptide synthesis (SPPS) using 2-chlorotritylchloride resin and Fmoc-protected amino acids (Scheme S1). After cleavage from resin using trifluoroacetic acid (TFA), the crude product was precipitated in ether and purified on a semi-prepared HPLC to obtain the target peptide with a 97.7% purity (Figure S1). Structural characterizations were carried out using a high-resolution mass spectrum (Figure S2) and ^{1}H NMR spectrum (Figure S3), confirming the production of target peptide FmocFFRR.

2.2. Formation of HAp-Anchoring Hydrogels

FmocFFRR dissolved well in water and formed viscous fluid at low concentrations (0.5, 1.0, and 2.0 mg/mL). When increasing its concentration (4.0 and 8.0 mg/mL), it formed a transparent, self-supporting hydrogel (Figure S4), suggesting the self-assembly of FmocFFRR. To investigate whether FmocFFRR self-assembly could immobilize HAp, we dispersed nanosized HAp particles into FmocFFRR fluids at different concentrations. As shown in Figure 2A, HAp rapidly precipitated within 30 min in the absence of FmocFFRR assembly. In very low concentrations of FmocFFRR (0.5 mg/mL), HAp particles could disperse well for hours, and partial precipitation was observed after 24 h. In FmocFFRR assembly higher than 0.5 mg/mL, HAp stood well dispersive and formed opaque mixtures that were stable even after 48 h (Figure S5), suggesting high stability of HAp on the FmocFFRR/HAp mixture. In addition, FmocFFRR at 2.0 mg/mL successfully gelated into self-supporting hydrogel after being tethered by HAp particles, compared to fluidic FmocFFRR in the absence of HAp (Figure 2B). It suggested that the addition of HAp particles significantly facilitated FmocFFRR assembly, because anchoring of FmocFFRR on the HAp surface could (1) neutralize its positive charges on the arginine residues overcoming charges repulsion, (2) reduce intermolecular distance of self-assembly building blocks, and (3) tether self-assembled filaments into the dense fibrous network. The stability of HAp dispersion was further quantified by measuring transmittance at 600 nm (Figure 2C). The results showed that the HAp/FmocFFRR mixture with a weight ratio higher than 2:0.75 kept opaque with low transmittance over 30 min period. Therefore, FmocFFRR assembly could efficiently stabilize HAp dispersion in the hydrogel.

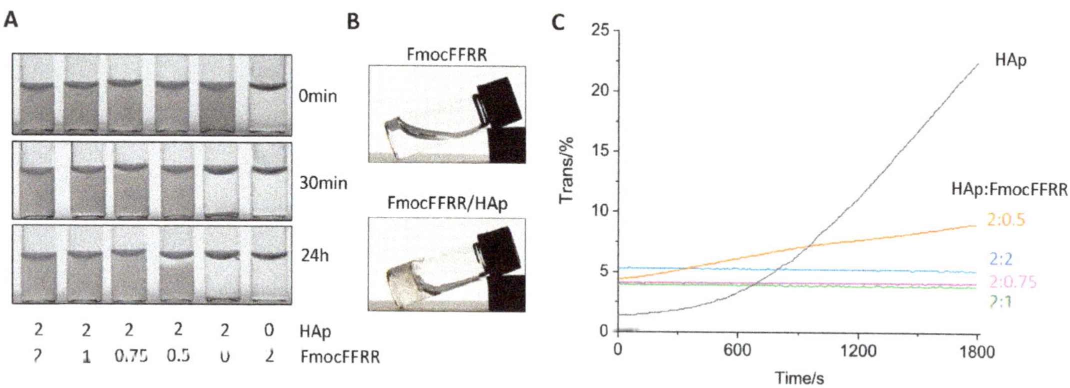

Figure 2. (**A**) Optical images of 2 mg/mL HAp particles dispersed in FmocFFRR solution in ddH_2O at different concentrations (0, 0.5, 0.75, 1.0, 2.0 mg/mL) for 0 min, 30 min, or 24 h. (**B**) Optical images of 2 mg/mL FmocFFRR in the absence or presence of HAp particles (4 mg/mL) in ddH_2O. (**C**) Transmittance curves of 2 mg/mL HAp dispersed in ddH_2O or in FmocFFRR assemblies at 0.5, 0.75, 1.0, or 2.0 mg/mL over a 30 min period. Transmittance was measured at 600 nm.

We analyzed the mechanical properties of HAp/FmocFFRR-tethering hydrogel using rheology. Oscillatory strain-sweep rheological analysis (Figure 3) showed that 4 mg/mL FmocFFRR assembly in the absence of HAp possessed low storage modulus (G′) and loss modulus (G″) values with a breaking strain at 47%, indicating the formation of a weak hydrogel. When mixing with 2 mg/mL HAp, FmocFFRR formed a strong gel that showed remarkably increased G′ and G″ values with a wide linear viscoelastic region (LVR). In a 4:4 (*w*/*w*) FmocFFRR/HAp hydrogel, both G′ (180 Pa) and G″ (closed to 10 Pa) values were similar, while in a 4:8 (*w*/*w*) FmocFFRR/HAp hydrogel, the G′ value was a two-fold increase and G″ value was a three-fold increase, compared to the 4:2 (*w*/*w*) FmocFFRR/HAp hydrogel. The incorporation of HAp into FmocFFRR assembly significantly enhanced its mechanical properties, in line with Ghosh's [22] and Vitale's findings [24]. It is reasonable to postulate that the interaction between FmocFFRR and HAp contributes to fibers crosslinking and strengthens the fibrous entanglements. When the composition of HAp increased to 16 mg/mL, however, the FmocFFRR/HAp hydrogel became weakened with reduced G′ and G″ values. It is not surprising, because oversaturation of HAp reinforces peptide binding to their surfaces, thus weakening fiber-fiber interaction. These results demonstrated a 1:2 (*w*/*w*) FmocFFRR/HAp formulate to fabricate a relatively strong hydrogel.

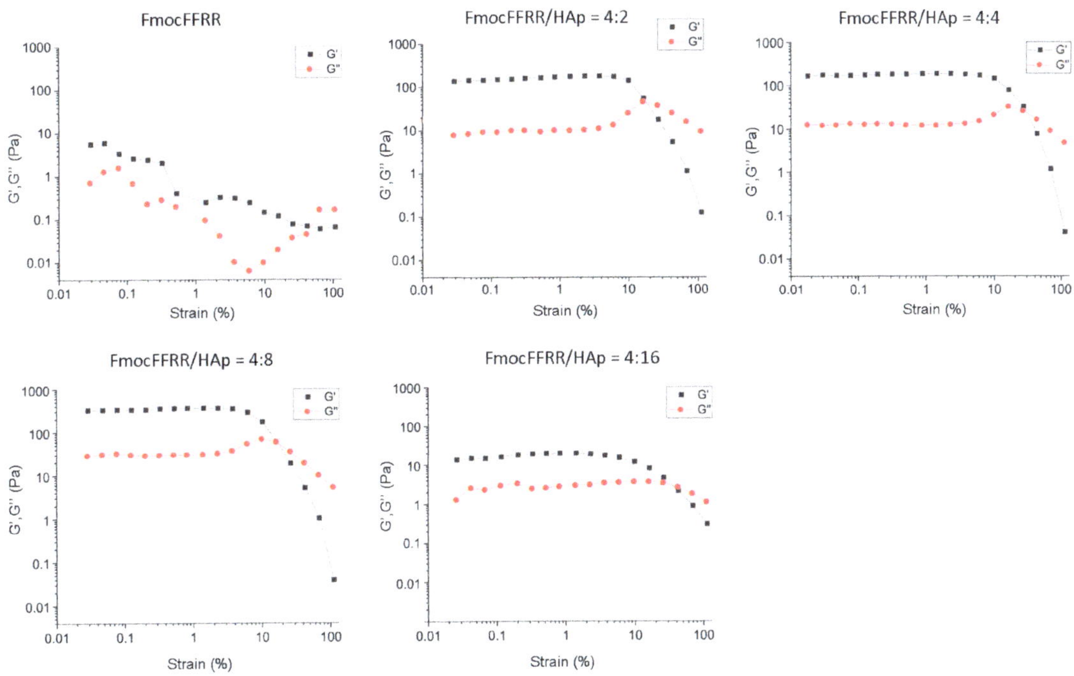

Figure 3. Strain amplitude sweep tests of 4 mg/mL FmocFFRR containing 0, 2, 4, 8, or 16 mg/mL HAp particles at a fixed angular frequency (1.0 rad/s) at room temperature.

2.3. Structural Characterizations of FmocFFRR/HAp Hydrogels

After formulating the compositions of the tough HAp/FmocFFRR hydrogel, we examined their microstructures using transmission electron microscopy (TEM). As shown in Figure 4A, HAp particles used in these tests showed nanorod-like structures. High-resolution TEM image revealed the crystalline structure of HAp nanoparticles with a lamellar distance of 0.69 nm, which was attributed to the crystal lattice *c* parameter [27,28]. FmocFFRR self-assembled into nanofibers with a diameter of 4 nm. When FmocFFRR was mixed with HAp, it formed a dense fibrous network. At a low concentration of mixed gel (HAp/FmocFFRR = 0.50/0.25, mg/mL), we could clearly observe individual HAp particles surrounded by the peptide assembling nanofibers (Figure 4B), indicating peptide bound to HAp. HAp clustering was observed at a high concentration of mixed gel (HAp/FmocFFRR = 5.0/2.5, mg/mL). However, the edge of HAp clusters was fused with fibrous bundles. These observations suggested HAp tethered FmocFFRR assembling nanofibers, which in turn stabilized HAp dispersion. X-ray diffraction (XRD) analysis (Figure 4C) demonstrated that the FmocFFRR/HAp mixture showed a similar XRD pattern compared with free HAp particles. This result suggested that the tether of FmocFFRR on the HAp surface did not affect its crystalline structure.

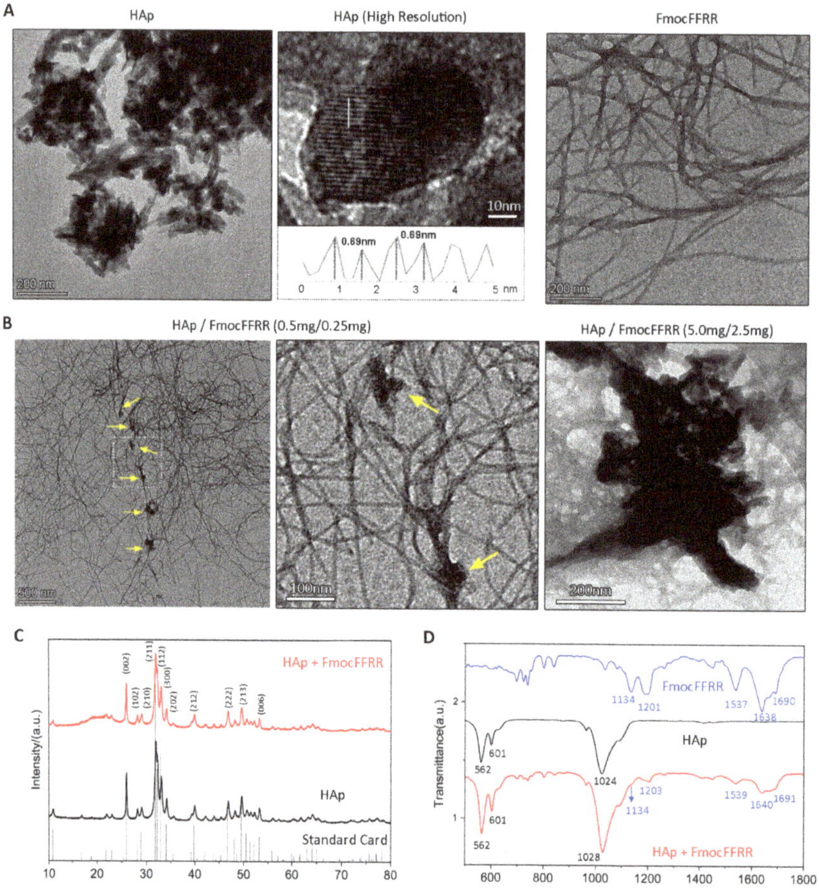

Figure 4. (**A**) TEM images of HAp particles (5.0 mg/mL) with high-resolution image (middle panel) showing lattice spaces. The gray value distribution curve of indicated white light was also showed to indicate the interlamellar distance. TEM image of FmocFFRR assembly (2.5 mg/mL) was shown at the right panel. (**B**) TEM images of HAp/FmocFFRR hydrogel at the concentration of 0.50/0.25 (mg/mL) or 5.0/2.5 (mg/mL) in Tris buffer. Yellow arrows indicated peptide-bound HAp particles. (**C**) XRD patterns of free HAp particles (black) and the lyophilized HAp/FmocFFRR sample (red). A standard card of the calculated XRD pattern of HAp was also shown for comparison. (**D**) FTIR spectra of FmocFFRR (blue), HAp (black), and HAp/FmocFFRR (red).

FT-IR spectra were further conducted to demonstrate the molecular interaction between HAp and FmocFFRR. The results (Figure 4D) showed that the phosphate stretching vibration at 1024 cm^{-1} of HAp slightly shifted to 1028 cm^{-1} after anchoring of FmocFFRR. Slight shifts of guanidine groups vibration (1537 cm^{-1}) and amide bands (1638 cm^{-1} and 1690 cm^{-1}) of FmocFFRR after HAp/FmocFFRR interaction were also observed. The intensity of the amide I band at 1690 cm^{-1} assigned as β-turn [29] of FmocFFRR assembly increased after HAp/FmocFFRR interaction compared to the amide II band internally at 1638 cm^{-1}, in accordant with our previous finding on FmocFFRRR/CaP minerals interaction [25]. These results suggested HAp/FmocFFRR interaction promoted the molecular assembly of FmocFFRR.

2.4. Biocompatibility of FmocFFRR/HAp Hydrogels

To evaluate the osteoinductive potency of FmocFFRR/HAp hydrogels, we first examined their biocompatibility. Preosteoblast MC3T3-E1 cells, a typical cell model for studying osteogenesis [30], were seeded on the culture plate that was previously coated with FmocFFRR/HAp hydrogel. The live-dead cell staining results (Figure S6) showed that a large number of MC3T3-E1 cells attached on the hydrogel surface without inducing much cell death, suggesting great cell adhesion of FmocFFRR/HAp hydrogel. MC3T3-E1 cells grew on a 96-well plate that was pretreated with free HAp particles, free FmocFFRR assemblies, and HAp/FmocFFRR hydrogels for 48 h, and the viability was further quantified using an ATP assay. As shown in Figure 5A, free HAp nanoparticles exhibited high cytotoxicity against MC3T3-E1 cells. To reason their cytotoxicity, we observed the microscopic images of HAp-treated MC3T3-E1 cells and found that HAp aggregates strongly attached to cell surfaces (Figure S7), which may affect cell functions and contribute to the cytotoxicity. FmocFFRR assemblies without tethering on HAp showed moderate cytotoxicity, with less than 50% cell viability at a concentration of 5mg/mL. Excitingly, when HAp and FmocFFRR mixed to form hydrogels, they showed significant biocompatibility with cell viability greater than 90%. The 1:2 (w/w) FmocFFRR/HAp formulating hydrogel at 5/10 mg/mL concentration exhibited higher cell viability than the non-treated cells, suggesting a promoting effect on cell proliferation. This was reasoned that the 1:2 (w/w) ratio of FmocFFRR to HAp formed the stable hydrogel (Figure 3) with a minimum amount of free HAp or free FmocFFRR compositions leaked into culture media. The cytotoxicity of 1:2 (w/w) FmocFFRR/HAp hydrogel at different amount concentrations was also conducted. The results (Figure 5B) demonstrated that at low content of FmocFFRR (0.1, 0.25, 0.5, and 1.0 mg/mL), the FmocFFRR/HAp mixture exhibited a cytotoxic effect with cell viability less than 40%. With a high content of FmocFFRR (2.5, 5.0 mg/mL) that enabled stable FmocFFRR/HAp hydrogel formation, the hydrogel was of great biocompatibility with cell viability greater than 80%. Non-covalent interactions, including π-π stacking, H-bonds, and electrostatic interactions, drove the formation of the FmocFFRR/HAp hybrid hydrogel. To evaluate whether FmocFFRR/HAp hydrogel would dissemble to leak FmocFFRR and HAp particles that show moderate cytotoxic effect in this study, we incubated the hydrogel with MC3T3-E1 cells which were attached to the culture plate. The results (Figure S8) showed that the leaching contents showed comparable vital cell numbers and a few numbers of dead cells compared to the controlled experiment.

To further investigate the capability of cell encapsulation, a mixture of FmocFFRR/HAp was in situ gelated with MC3T3-E1 cell suspension in a basal culture medium and kept in a cell-cultured condition overnight. Hoechst 33,258 staining indicated the existence of MC3T3-E1 cells inside the hydrogel (Figure S9). Live-dead staining of encapsulated cells was performed. The results (Figure 5C) suggested that most of the encapsulated cells inside the FmocFFRR/HAp hydrogel were alive (stained green) with a few cells stained red (indicating dead cells), comparable with controlled cells attached to the cell-cultured plate. Notably that portion of live cells encapsulated in the hydrogel remained round in shape, assigning to their 3D culturing environment. In contrast, cell suspension that was treated with free FmocFFRR was capable of attaching to the culture plate and showed little dead cells. However, the cell number was less than the non-treated group, suggesting FmocFFRR affected the proliferation of MC3T3-E1 cells. Cell suspension treated with free nanosized HAp particles showed large numbers of dead cells, suggesting the cytotoxic effect against the MC3T3-E1 cell. These data agreed with the results of the ATP assay (Figure 5A).

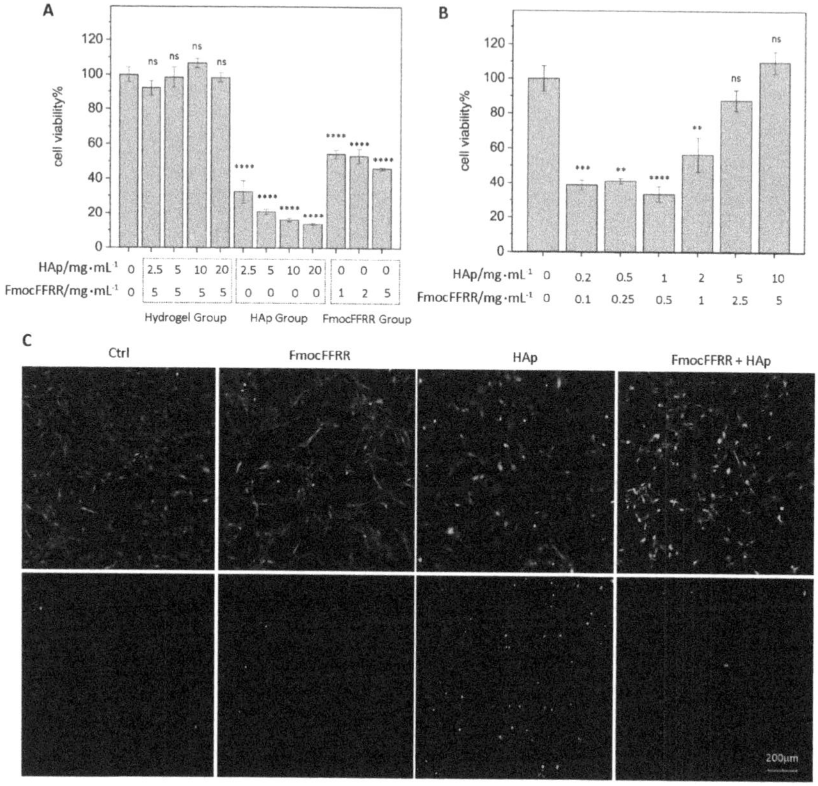

Figure 5. (**A**) MC3T3-E1 cell suspensions were mixed with freshly prepared FmocFFRR/HAp mixture for in situ gelations (hydrogel group), with free HAp particles (HAp group), or with FmocFFRR assemblies (FmocFFRR group). For the controlled groups, MC3T3-E1 cell suspensions were mixed with culture media. Cells were allowed for cultured for another 48 h and the cell viability was evaluated. (**B**) Cell viability of MC3T3-E1 cells after culturing in FmocFFRR/HAp hydrogels at different concentrations with a fixed 1:2 weight ratio. ns: non-significant; **: $p < 0.01$; ***: $p < 0.001$; ****: $p < 0.0001$. (**C**) Live-dead staining of MC3T3-E1 cells after culturing in FmocFFRR/HAp (5.0/10.0 mg/mL) hydrogel, or treated with free HAp particles (10.0 mg/mL), FmocFFRR assemblies (5.0 mg/mL) for 48 h. The controlled cells were seeded on a tissue-culture plate. Green channels indicated live cells; red channels indicated dead cells, respectively.

2.5. Osteoinductive Properties of FmocFFRR/HAp Hydrogels

MC3T3-E1 preosteoblasts exhibit a high level of osteoblast differentiation into mature osteocytes after growing in an osteoinduction medium that usually contains *L*-ascorbic acid and phosphate sources. When the MC3T3-E1 preosteoblasts were maintained in the FmocFFRR/HAp hydrogel for culturing in the basal DMEM media without supplementary *L*-ascorbic and phosphate, we observed that cell bodies elongated and started growing dendrites after 5 days, similar to those treated with osteoinduction medium (Figure 6). This finding implied that the FmocFFRR/HAp hydrogel potentially induced osteogenesis of preosteoblasts. MC3T3-E1 cells treated with free FmocFFRR exhibited irregular morphologies, meanwhile treatment of free HAp particles induced preosteoblasts body shrinkage and nuclear condensation, suggesting induced cytotoxicity accordantly with previous results. The controlled MC3T3-E1 cells remained in fibroblast shape without the treatment of osteoinduction medium.

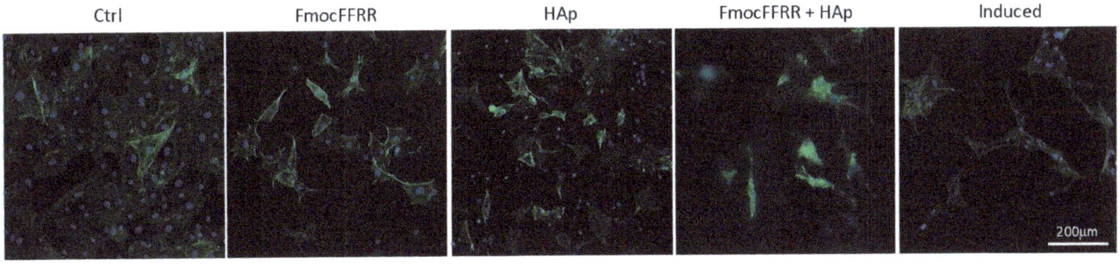

Figure 6. Morphological analysis of MC3T3-E1 cells after culturing in FmocFFRR/HAp (5.0/10.0 mg/mL) hydrogel, free HAp particles (10.0 mg/mL), FmocFFRR assemblies (5.0 mg/mL), or treated with osteoinduction medium (Induced) for 5days. F-actin filaments were stained with Alexa Fluor 488-conjugated phalloidins, and nuclear were stained with DAPI (blue).

Osteoblast differentiation of 3D culturing preosteoblasts in FmocFFRR/HAp hydrogel was further confirmed by assessing the expression of osteogenesis biomarkers. After 7 days culture of MC3T3-E1 cells in the FmocFFRR/HAp hydrogel without supplementary *L*-ascorbic and phosphate, the expression of alkaline phosphatase (ALP) and type-I collagen (COL1) as biomarkers for early-stage osteoblast differentiation [31] was evaluated using immunostaining fluorescent staining. Compared to control MC3T3-E1 cells that were cultured with basal culture medium, HAp-treated cells showed elevated expression of ALP (Figure S10), and FmocFFRR assembly-treated cells showed slightly increased expression of COL1 (Figure S11). However, HAp treatment or FmocFFRR assembly treatment of MC3T3-E1 cells induced a negligible change of osteocalcin (OCN) expression (Figure 7A), a commonly used biomarker for late-stage osteoblast differentiation [31]. These results suggested that sole treatment with FmocFFRR assemblies or HAp particles could induce the osteoblasts differentiation into the early stage for ECM secretion and maturation. In comparison, 3D culturing of MC3T3-E1 cells in FmocFFRR/HAp hydrogel significantly promoted the expression of osteogenesis markers including ALP, COL1, and OCN, suggesting a late stage of osteoblasts differentiation occurred for bone matrix mineralization. MC3T3-E1 cells that were induced with osteoinduction medium for 7 days exhibited increasing expressions of ALP, COL1, and OCN. When MC3T3-E1 cells were 3D cultured in FmocFFRR/HAp hydrogel, higher expression levels of COL-1 and OCN were observed compared to the induced cells by osteoinduction medium, revealing FmocFFRR/HAp hydrogel has a higher osteoinductivity than osteoinduction medium used in this study.

The RNA level of osteopontin (OPN) expression, another biomarker for late-stage osteoblast differentiation, was quantified by qRT-PCR. The results (Figure 7B) demonstrated that 3D culturing in hydrogel promoted the OPN level, while the treatment of HAp or FmocFFRR assembly reduced its expression. MC3T3-E1 cells cultured in hydrogel also exhibited increasing RNA levels of bone morphogenic protein 2 (BMP-2) (Figure 7C), which positively regulates osteogenesis processes [32]. Taking into consideration the morphological changes of preosteoblasts and the increasing level of both protein expressions and RNA expressions of osteoblast differentiation-related markers, we could conclude that FmocFFRR/HAp hydrogel capably supported 3D culturing of preosteoblasts, and promoted their osteoblast differentiation, exhibiting great potentials as a highly osteoinductive scaffold in bone regenerative engineering.

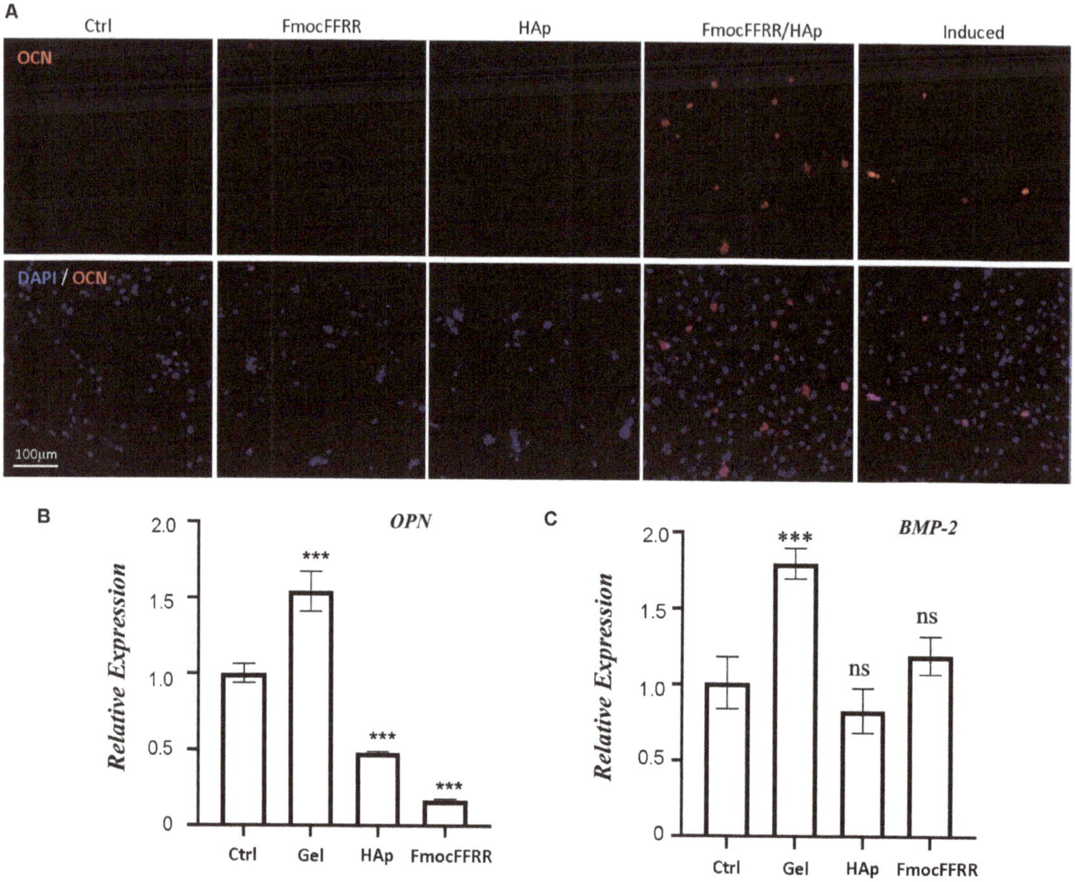

Figure 7. MC3T3-E1 cells were cultured with basal culture medium in FmocFFRR/HAp (5.0/10.0, mg/mL) hydrogel, free HAp particles (10.0 mg/mL), FmocFFRR assemblies (5.0 mg/mL), or treated with osteoinduction medium (Induced) for 7 days. (**A**) Protein expression of OCN was immunostained with an anti-OCN antibody. Cell nuclear was also stained with DAPI. Gene expression of osteogenesis markers OPN (**B**) and BMP-2 (**C**) was quantified using qRT-PCR. The expression of GAPDH was used as an internal standard. Data were shown as mean ± SD, $n = 6$; *** $p < 0.001$.

Despite the regulatory effects of both positively charged arginine and negatively charged aspartic acid or glutamic acid on HAp mineralization [33,34], recent material-design for HAp-supporting scaffolds focused on polycarboxylate-rich scaffold materials as HAp capping [10,35]. We demonstrated an arginine-containing peptide as an alternative to HAp-stabilizing materials. The HAp-anchorable peptide FmocFFRR was designed by tailoring an arginine residue to a self-assembling building block FmocFF. Anchoring FmocFFRR on the HAp surface through electrostatic guanidine–phosphate interaction facilitated the self-assembly of FmocFFRR, due to the neutralization of charge repulsion, the reduction of intermolecular distance between building blocks, and the tether of self-assembling filaments into the dense fibrous network. On the other hand, FmocFFRR assembling on the nanosized HAp surface reduced its surface energy and stabilized its dispersion in the scaffold.

Nanosized HAp particles are reported biocompatible [10,22,24]. In this study, however, we used commercially available HAp particles with an average size of 20 nm and found that they exhibited cytotoxic effect against MC3T3-E1 preosteoblasts. Strong attachment of them to the cell surfaces (Figure S7) is one of the factors contributing to their cytotoxicity. An assembling peptide FmocFFRR was not cytotoxic but showed moderate anti-proliferative effect (Figure 5C). Mixing FmocFFRR and HAp at a ratio of 1:2 (*w*/*w*) in a high concentration (FmocFFRR > 2.5 mg/mL) generated a stable hybrid hydrogel, which surprisingly turned over the cytotoxicity of single components and exhibited high biocompatibility. This action is similar in many cases of polymeric hydrogels. For example, methacryloyl monomer and initiators are known to be harmful. They polymerize in situ to form polymeric hydrogels with less cytotoxicity. Non-covalent interactions including π-π stacking, H-bonds, and electrostatic interactions are the major forces that drive the formation of FmocFFRR/HAp hydrogel. However, the electrostatic interaction between arginine and phosphate is very strong and considered "covalent-like" [26,36]. This feature endows FmocFFRR/HAp hydrogel with polymer-like properties showing high stability and biocompatibility.

Despite the cytotoxic effect of nanosized HAp particles we used in this study, they induced an increasing expression level of ALP (Figure S10). FmocFFRR itself promoted the secretion of COL1 (Figure S11), which participates in the bone ECM construction. MC3T3-E1 preosteoblasts cultured in the FmocFFRR/HAp hydrogel not only showed lifted levels of ALP and COL1, but also higher expression of OCN than that induced by osteoinduction medium, suggesting that FmocFFRR/HAp hydrogel is highly osteoinductive. The Uesugi group [37,38] reported positively charged self-assembling molecules could selectively bind to heparan sulfate and promote clustering of heparan sulfate proteoglycans (HSPGs) such as syndecan-4. It is reasonable to hypothesize that FmocFFRR binds to heparan sulfate. Therefore, FmocFFRR/HAp hydrogel may involve in heparan sulfate-growth factors related signaling for promoting osteoblast differentiation.

3. Conclusions

In summary, we demonstrated a HAp-supporting hydrogel based on peptide assembly. FmocFFRR bound to HAp through strong electrostatic interactions between arginine residues of FmocFFRR and phosphate groups on the surface of HAp. HAp-tethering FmocFFRR facilitated self-assembly and crosslinking to form FmocFFRR/HAp hybrid hydrogel, mimicking the natural bone ECM structure. The 1:2 (*w*/*w*) ratio of FmocFFRR/HAp mixture at a high concentration (FmocFFRR > 2.5 mg/mL) formed stable hydrogel with enhanced mechanical properties and high biocompatibility. Both the ATP assay and live-dead staining assay verified the formulating FmocFFRR/HAp hydrogel showed negligible cytotoxicity. More importantly, FmocFFRR/HAp hydrogel not only enabled cell attachment on its surface, but also supported 3D cell culturing inside the hydrogel to promote the osteoblast differentiation of MC3T3-E1 preosteoblasts, inspiring its potential application in bone regenerative engineering. These findings not only expand the diversity of osteoinductive hydrogels for osteogenesis, but also provides a bottom-up approach for fabricating ECM-mimicking hydrogel with high accuracy. We believe that this strategy can be programmed with multifunctionalities and broadens its appeal.

4. Materials and Methods

4.1. Materials and Instruments

All the solvents and chemicals were commercially available and used as received without further purification. HAp nanoparticles with an average size of 20 nm were purchased from Macklin Co., Ltd. (Shanghai, China). Mouse Embryonic Osteoblasts Cell (MC3T3-E1) was purchased from the American Type Culture Collection (ATCC). Dulbecco's modified Eagle medium (DMEM), fetal bovine serum (FBS), and penicillin-streptomycin liquid (100X) were purchased from Gibco (Waltham, MA, USA). The Calcein-AM/PI live and dead cell double stain kit was purchased from Solarbio Technology Co., Ltd. (Beijing, China). The primary antibody of OCN was purchased from Boster Biological Technology Co., Ltd. (Pleasanton, CA, USA); primary antibodies of ALP, and Col1 were purchased from Proteintech Group Inc, Ltd. (Chicago, IL, USA). Cy5-conjugated secondary antibody was purchased from Boster Biological Technology Co., Ltd. (Pleasanton, CA, USA). CellTiter-Glo® Luminescent Cell Viability Assay was purchased from Promega Co., Ltd. (Madison, WI, USA). PCR primers were purchased from Sangon Biotech Co, Ltd. (Shanghai, China). SPARK easy Cell RNA Kit, SPARK script || RT Plus kit, SYBR Green qPCR Master Mix were purchased from SparkJade Biological Technology Co., Ltd. (Shandong, China). NucBlue Live ReadyProbe, NucBlue Dead ReadyProbe, and ActinGreen 488 ReadyProbe reagents were purchased from Thermo, Ltd. (Waltham, MA, USA).

The mass spectrum of FmocFFRR was measured by Agilent-6125B ESI mass spectrometer (Agilent, Santa Clara, CA, USA), and the purity was checked on an EClassical® 3140 HPLC (Elite, China). ^{1}H spectrum was acquired on a 400 MHz JEOL spectrometer (JEOL, Tokyo, Japan). The ^{1}H NMR chemical shifts (δ) were given in ppm referring to internal standard tetramethylsilane (TMS). The transmittance of the FmocFFRR/HAp mixture was recorded on an Ultraviolet-visible Spectrophotometer (YOKE L6). The rheology data were collected using a HAAKE MARS rheometer (Thermo) with an environmental test chamber. TEM images were obtained using a Talos L120C transmission electron microscope. X-ray diffractometry utilized Bruker AXS D8 diffractometer (Bruker, Billerica, MA, USA) using a copper Kα (λ = 1.5406 Å) radiation source. FTIR spectra were recorded using a Bruker Vertex 70 FTIR spectrophotometer (Bruker, USA) in an attenuated total reflectance (ATR) mode and analyzed through OriginPro 2021 Software Student Version (Originlab, Northampton, MA, USA). Fluorescence images were captured on a DMi8 fluorescent inverted microscope (Leica, Wetzlar, Germany). qRT-PCR was performed on a QuantGene 9600 real-time PCR (Bioer, Hangzhou, China).

4.2. Synthesis

Peptide FmocFFRR was designed and synthesized by solid phase peptide synthesis (SPPS) using 2-chlorotritylchloride resin and Fmoc-protected amino acids (Scheme S1). The crude product was purified on an EClassical P3140 HPLC equipped with a SinoChrom C18 column (ODS-BP, 10 μm, 20.0 mm × 250 mm) to obtain white powders. Yield 52.1%. The purity was checked as 97.7%, using an EClassical P3140 HPLC equipped with a SinoChrom C18 column (ODS-BP, 5 μm, 4.6 mm × 250 mm). ESI-MS (m/z) calcd. for C45H54N10O7, 846.9900, found $[M + 2H]^{2+}$ 424.2444 and $[M + H]^+$ 847.4774. ^{1}H NMR (400 MHz, DMSO-d6) (ppm): δ 8.22 (dd, J = 14.8, 7.7 Hz, 2H), 8.09 (d, J = 8.0 Hz, 1H), 7.88 (d, J = 7.7 Hz, 2H), 7.65–7.51 (m, 4H), 7.41 (ddt, J = 7.4, 4.2, 2.2 Hz, 2H), 7.33–7.14 (m, 12H), 4.70–4.54 (m, 1H), 4.37–4.31 (m, 1H), 4.27–4.15 (m, 3H), 4.15–4.04 (m, 2H), 3.19–3.01 (m, 5H), 2.95–2.79 (m, 2H), 2.73–2.65 (m, 1H), 1.82–1.66 (m, 2H), 1.66–1.45 (m, 6H).

4.3. Preparation of FmocFFRR/HAp Hydrogels

To a freshly prepared FmocFFRR solution in ddH$_2$O was added HAp suspension in ddH$_2$O and mixed with a pipette. The mixture was allowed to stand still for aging at room temperature.

4.4. Transmittance Measurements

First, 1 mL of freshly prepared FmocFFRR solution in ddH$_2$O was placed in quartz, to which was added 1mL HAp suspension containing 4 mg HAp in ddH$_2$O and mixed with a pipette. The quartz was immediately placed in the sample chamber of the UV-vis absorbance spectrometer. The transmittance was recorded at 600 nm for a 30 min period. The final weight ratio of FmocFFRR/HAp was (mg/mg) 0/2, 0.5/2, 0.75/2, 1/2, and 2/2, respectively.

4.5. Rheological Analysis

The mixtures of FmocFFRR/HAp containing 4 mg/mL of FmocFFRR and different amounts of HAp (0, 2, 4, 8, 16 mg/mL) were prepared and aged for 1 h at room temperature. Samples were cast between a 25 mm diameter stainless steel upper plate and a lower Peltier plate with the gap between plates at 1.0 mm for all measurements. Oscillatory strain sweeps (0.01–100% strain) at a fixed frequency of 1 Hz were performed.

4.6. Transmission Electron Microscopy (TEM)

A drop of 5 μL sample was added to the discharged copper grids coated with thin carbon film for about 60 s. After removing the excess solution with filter paper, the grids were washed with pure water and stained with 1% uranyl acetate for 30 s. The grid was allowed to dry at room temperature and ready for TEM imaging.

4.7. XRD Analysis

XRD diffraction patterns of HAp particles and lyophilized FmocFFRR/HAp sample were measured on a BrukerAXS D8 diffractometer using a copper Kα (λ = 1.5406 Å) radiation source. Scans were performed over a 2θ range of 10–80°.

4.8. FTIR Analysis

FTIR spectra of HAp particles, lyophilized FmocFFRR, or FmocFFRR/HAp mixtures were recorded using a Bruker Vertex 70 FTIR spectrophotometer in an attenuated total reflectance (ATR) mode and analyzed through OriginPro 2021 Software Student Version (Originlab, Northampton, MA, USA). The wavenumber range is 4000–400 cm^{-1}.

4.9. Cell Culture

Pre-osteoblasts MC3T3-E1 cells were purchased from the American Type Culture Collection (ATCC, CRL-2593™) and cultured in DMEM medium supplemented with 10% (v/v) FBS and 1% (v/v) penicillin–streptomycin. Cells were maintained in a fully humidified incubator containing 5% CO_2 at 37 °C.

4.10. Cell Viability Analysis

DdH_2O used in cellular experiments was filtered through a 0.22 μm pore-size membrane. Then, 40 μL of freshly prepared HAp suspension (2.5, 5.0, 10, or 20 mg/mL), FmocFFRR assembly (1.0, 2.0, or 5.0 mg/mL), FmocFFRR/HAp mixture containing various HAp contents (5.0/2.5, 5.0/5.0, 5.0/10, or 5.0/20 mg/mL), or 1:2 (w/w) FmocFFRR/HAp mixture at different concentrations (0.1/0.2, 0.25/0.5, 0.5/1.0, 1.0/2.0, 2.5/5.0, or 5.0/10 mg/mL) in ddH_2O was added to a 96-well plate and was sterilized under UV light irradiation for 4 h at room temperature. Then, 100 μL of 1×10^5 cells/mL MC3T3-E1 cell suspension was seeded to each well and cultured for 48 h. The ddH_2O pre-coated wells were used as the control. The viability was assessed using a luminescence-based ATP assay following the manufacturer's instruction. Briefly, 100 μL assay reagent was added to each well and mix the contents vigorously for 5 min to induce cell lysis. After maintaining the plate in the incubator for 15 min, 100 μL mixture in each well was transferred into a black plate. The luminescence was recorded immediately using a SuPerMax 3100 plate reader (Flash, Shanghai, China).

4.11. Cell Staining

MC3T3-E1 cell suspension was mixed with HAp, FmocFFRR of FmocFFRR/HAp mixture in DMEM media and placed seeded on a 24-well plate. Cell viability was assessed by Calcein-AM/PI staining following the manufacturer's instruction. For staining the live cell nuclear of encapsulated MC3T3-E1 cells inside the FmocFFRR/HAp hydrogel, two drops of NucBlue Live ReadyProbe were added and stained at 37 °C for 30 min. After being washed with phosphate-buffered saline (PBS) three times, the samples were captured on a fluorescent inverted microscope. For staining F-actin of MC3T3-E1 cells, cells were fixed with 4% paraformaldehyde (PFA) in PBS for 15 min and permeabilized with 0.1% Triton X-100 in PBS for 5 min. Additionally, cells were treated with a drop of NucBlue Dead ReadyProbe and ActinGreen 488 ReadyProbe and stained at room temperature for 1 h. After being washed with phosphate-buffered saline (PBS) three times, the samples were captured on a fluorescent inverted microscope.

4.12. Immunofluorescent Staining

MC3T3-E1 cells were fixed with 4% PFA and permeabilized with 0.1% (*v*/*v*) Triton X-100 in PBS. After blocking with 5% (*w*/*v*) BSA for 1 h, the cells were incubated with an anti-osteocalcin antibody (1:100; PB1009), anti-collagen type-I antibody (1:100; 14695-1-AP), and anti-alkaline phosphatase antibody (1:100; ET1601-21) overnight at 4 °C. Subsequently, cells were washed three times with PBS and incubated with Cy3-coupled secondary antibody (1:1000; Abcam, UK) for 1 h, followed by incubation with DAPI for 15 min. Fluorescent signals were captured using a fluorescence microscope.

4.13. RNA Quantification

Total RNA was extracted using an RNA Purification Kit (SparkJade, China) and cDNA was obtained from 500 ng of total RNA using the Reverse Transcription Kit (SparkJade). qRT-PCR was performed using SYBR Green qPCR Master Mix (SparkJade). Relative gene expression was calculated using the $2^{-\Delta\Delta CT}$ method, and GAPDH was used as a reference for normalization. The primers were purchased from BioTNT, and primer sequences for BMP-2, OPN, and GADPH were used as follows:

hBMP-2: 5′-CAGAACACAAGTCAGTGGGAGAGC-3′ (F), 5′-GAGGTGCCACGATCCAGTCATTC-3′ (R);
hOPN: 5′-ATGGACGACGATGATGACGATGATG-3′ (F), 5′-ATGGCTGCCCTTTCCGTTGTTG-3′ (R);
hGADPH: 5′-TGAACGGGAAGCTCACTGG-3′ (F), 5′-TCCACCACCCTGTTGCTGTA-3′ (R).

4.14. Statistical Analysis

All results were presented as mean and standard deviation with 3–6 independent experiments. The statistics were analyzed using the SPSS software. The p value was calculated using Tukey's multiple comparison test. When $p > 0.05$, differences were considered non-significant (ns); when $p < 0.05$, differences were considered to be significant (* $p < 0.05$, ** $p < 0.01$, *** $p < 0.001$, **** $p < 0.0001$, respectively).

Supplementary Materials: The following supporting information can be downloaded at: https://www.mdpi.com/article/10.3390/gels8120804/s1, The synthesis, HPLC profile, mass spectrum and ^{1}H NMR spectrum of FmocFFRR, preparation of FmocFFRR hydrogel, cell attachment experiments, cell encapsulation experiments, and immunofluorescent staining of ALP and COL1 were listed in the supporting information.

Author Contributions: Conceptualization, S.L., D.H. and G.L.; Data curation, H.Y., J.S. and X.Z.; Formal analysis, H.Y.; Funding acquisition, S.L. and G.L.; Methodology, H.Y., J.S., X.Z., K.J., H.F., Y.L. and Y.Z.; Project administration, S.L., D.H. and G.L.; Supervision, S.L., D.H. and G.L.; Writing—original draft, H.Y. and G.L.; Writing—review and editing, H.Y., J.S., X.Z., K.J., H.F., Y.L., Y.Z., S.L., D.H. and G.L. All authors have read and agreed to the published version of the manuscript.

Funding: This research was funded by the National Natural Science Foundation of China (No. 81802690), the Fundamental Research Funds for the Central Universities (No. xzy012022127), and the "Yong Talent Support Plan" of Xi'an Jiaotong University.

Institutional Review Board Statement: Not applicable.

Informed Consent Statement: Not applicable.

Data Availability Statement: Not applicable.

Conflicts of Interest: The authors declare no conflict of interest.

References

1. Wang, W.; Yeung, K.W.K. Bone grafts and biomaterials substitutes for bone defect repair: A review. *Bioact. Mater.* **2017**, *2*, 224–247. [CrossRef] [PubMed]
2. Orciani, M.; Fini, M.; Di Primio, R.; Mattioli-Belmonte, M. Biofabrication and Bone Tissue Regeneration: Cell Source, Approaches, and Challenges. *Front. Bioeng. Biotechnol.* **2017**, *5*, 17. [CrossRef] [PubMed]

3. Henkel, J.; Woodruff, M.A.; Epari, D.R.; Steck, R.; Glatt, V.; Dickinson, I.C.; Choong PF, M.; Schuetz, M.A.; Hutmacher, D.W. Bone Regeneration Based on Tissue Engineering Conceptions—A 21st Century Perspective. *Bone Res.* **2013**, *1*, 216–248. [CrossRef] [PubMed]
4. Wang, Z.; Han, T.; Zhu, H.; Tang, J.; Guo, Y.; Jin, Y.; Wang, Y.; Chen, G.; Gu, N.; Wang, C. Potential Osteoinductive Effects of Hydroxyapatite Nanoparticles on Mesenchymal Stem Cells by Endothelial Cell Interaction. *Nanoscale Res. Lett.* **2021**, *16*, 67. [CrossRef]
5. Pupo, Y.M.; Leite, L.M.B.; Senegaglia, A.C.; Antunes, L.; Nadal, J.M.; de Lara, E.L.; Saito, R.E.; Antunes, S.R.M.; Lacerda, W.F.; Farago, P.V. Effect of Hydroxyapatite Microspheres, Amoxicillin–Hydroxyapatite and Collagen–Hydroxyapatite Composites on Human Dental Pulp-Derived Mesenchymal Stem Cells. *Materials* **2021**, *14*, 7515. [CrossRef]
6. Yu, F.; Lian, R.; Liu, L.; Liu, T.; Bi, C.; Hong, K.; Zhang, S.; Ren, J.; Wang, H.; Ouyang, N.; et al. Biomimetic Hydroxyapatite Nanorods Promote Bone Regeneration via Accelerating Osteogenesis of BMSCs through T Cell-Derived IL-22. *ACS Nano* **2022**, *16*, 755–770. [CrossRef]
7. Shi, H.; Zhou, Z.; Li, W.; Fan, Y.; Li, Z.; Wei, J. Hydroxyapatite Based Materials for Bone Tissue Engineering: A Brief and Comprehensive Introduction. *Crystals* **2021**, *11*, 149. [CrossRef]
8. Jeong, J.; Kim, J.H.; Shim, J.H.; Hwang, N.S.; Heo, C.Y. Bioactive calcium phosphate materials and applications in bone regeneration. *Biomater. Res.* **2019**, *23*, 4. [CrossRef]
9. Alrebaish, A.S.; Wilson, O.C., Jr. Mesocrystal aggregation of biological apatite nanocrystals. *Med. Devices Sens.* **2021**, *4*, e10155. [CrossRef]
10. Zhang, J.; Jia, J.; Kim, J.P.; Shen, H.; Yang, F.; Zhang, Q.; Xu, M.; Bi, W.; Wang, X.; Yang, J.; et al. Ionic Colloidal Molding as a Biomimetic Scaffolding Strategy for Uniform Bone Tissue Regeneration. *Adv. Mater.* **2017**, *29*, 1605546. [CrossRef]
11. Koons, G.L.; Diba, M.; Mikos, A.G. Materials design for bone-tissue engineering. *Nat. Rev. Mater.* **2020**, *5*, 584–603. [CrossRef]
12. Tang, G.; Liu, Z.; Liu, Y.; Yu, J.; Wang, X.; Tan, Z.; Ye, X. Recent Trends in the Development of Bone Regenerative Biomaterials. *Front. Cell Dev. Biol.* **2021**, *9*, 665813. [CrossRef] [PubMed]
13. Pu, X.; Tong, L.; Wang, X.; Liu, Q.; Chen, M.; Li, X.; Lu, G.; Lan, W.; Li, Q.; Liang, J.; et al. Bioinspired Hydrogel Anchoring 3DP GelMA/HAp Scaffolds Accelerates Bone Reconstruction. *ACS Appl. Mater. Interfaces* **2022**, *14*, 20591–20602. [CrossRef] [PubMed]
14. Choi, S.; Lee, J.S.; Shin, J.; Lee, M.S.; Kang, D.; Hwang, N.S.; Lee, H.; Yang, H.S.; Cho, S.-W. Osteoconductive hybrid hyaluronic acid hydrogel patch for effective bone formation. *J. Control. Release* **2020**, *327*, 571–583. [CrossRef] [PubMed]
15. Li, X.; Chen, M.; Wang, P.; Yao, Y.; Han, X.; Liang, J.; Jiang, Q.; Sun, Y.; Fan, Y.; Zhang, X. A highly interweaved HA-SS-nHAp/collagen hybrid fibering hydrogel enhances osteoinductivity and mineralization. *Nanoscale* **2020**, *12*, 12869–12882. [CrossRef] [PubMed]
16. El-Habashy, S.E.; Eltaher, H.M.; Gaballah, A.; Zaki, E.I.; Mehanna, R.A.; El-Kamel, A.H. Hybrid bioactive hydroxyapatite/polycaprolactone nanoparticles for enhanced osteogenesis. *Mater. Sci. Eng. C* **2021**, *119*, 111599. [CrossRef]
17. Hao, Z.; Li, H.; Wang, Y.; Hu, Y.; Chen, T.; Zhang, S.; Guo, X.; Cai, L.; Li, J. Supramolecular Peptide Nanofiber Hydrogels for Bone Tissue Engineering: From Multihierarchical Fabrications to Comprehensive Applications. *Adv. Sci.* **2022**, *9*, 2103820. [CrossRef]
18. Sheehan, F.; Sementa, D.; Jain, A.; Kumar, M.; Tayarani-Najjaran, M.; Kroiss, D.; Ulijn, R.V. Peptide-Based Supramolecular Systems Chemistry. *Chem. Rev.* **2021**, *121*, 13869–13914. [CrossRef]
19. Levin, A.; Hakala, T.A.; Schnaider, L.; Bernardes, G.J.L.; Gazit, E.; Knowles, T.P.J. Biomimetic peptide self-assembly for functional materials. *Nat. Rev. Chem.* **2020**, *4*, 615–634. [CrossRef]
20. Song, J.; Zhang, Q.; Li, G.; Zhang, Y. Constructing ECM-like Structure on the Plasma Membrane via Peptide Assembly to Regulate the Cellular Response. *Langmuir* **2022**, *38*, 8733–8747. [CrossRef]
21. Fleming, S.; Ulijn, R.V. Design of nanostructures based on aromatic peptide amphiphiles. *Chem. Soc. Rev.* **2014**, *43*, 8150–8177. [CrossRef] [PubMed]
22. Ghosh, M.; Halperin-Sternfeld, M.; Grigoriants, I.; Lee, J.; Nam, K.T.; Adler-Abramovich, L. Arginine-Presenting Peptide Hydrogels Decorated with Hydroxyapatite as Biomimetic Scaffolds for Bone Regeneration. *Biomacromolecules* **2017**, *18*, 3541–3550. [CrossRef] [PubMed]
23. Wang, Y.L.; Lin, S.P.; Nelli, S.R.; Zhan, F.K.; Cheng, H.; Lai, T.S.; Yeh, M.Y.; Lin, H.C.; Hung, S.C. Self-Assembled Peptide-Based Hydrogels as Scaffolds for Proliferation and Multi-Differentiation of Mesenchymal Stem Cells. *Macromol. Biosci.* **2017**, *17*, 1600192. [CrossRef] [PubMed]
24. Vitale, M.; Ligorio, C.; McAvan, B.; Hodson, N.W.; Allan, C.; Richardson, S.M.; Hoyland, J.A.; Bella, J. Hydroxyapatite-decorated Fmoc-hydrogel as a bone-mimicking substrate for osteoclast differentiation and culture. *Acta Biomater.* **2022**, *138*, 144–154. [CrossRef] [PubMed]
25. Shao, L.; Liu, S.; Li, G.; Zhang, Y. Capture Phosphates via Peptide Self-assembly to Construct Templates Assisting Mineralization. *ChemNanoMat* **2022**, *8*, e202100542. [CrossRef]
26. Woods, A.S.; Ferré, S. Amazing stability of the arginine-phosphate electrostatic interaction. *J. Proteome Res.* **2005**, *4*, 1397–1402. [CrossRef] [PubMed]
27. Lin, K.; Zhou, Y.; Zhou, Y.; Qu, H.; Chen, F.; Zhu, Y.; Chang, J. Biomimetic hydroxyapatite porous microspheres with co-substituted essential trace elements: Surfactant-free hydrothermal synthesis, enhanced degradation and drug release. *J. Mater. Chem.* **2011**, *21*, 16558–16565. [CrossRef]

28. Arkharova, N.A.; Suvorova, E.I.; Severin, A.V.; Khripunov, A.K.; Krasheninnikov, S.V.; Klechkovskaya, V.V. SEM and TEM for structure and properties characterization of bacterial cellulose/hydroxyapatite composites. *Scanning* **2016**, *38*, 757–765. [CrossRef]
29. Bagińska, K.; Makowska, J.; Wiczk, W.; Kasprzykowski, F.; ChmurzyńSKI, L. Conformational studies of alanine-rich peptide using CD and FTIR spectroscopy. *J. Pept. Sci.* **2008**, *14*, 283–289. [CrossRef]
30. Wang, Z.; Chen, L.; Wang, Y.; Chen, X.; Zhang, P. Improved Cell Adhesion and Osteogenesis of op-HA/PLGA Composite by Poly(dopamine)-Assisted Immobilization of Collagen Mimetic Peptide and Osteogenic Growth Peptide. *ACS Appl. Mater. Interfaces* **2016**, *8*, 26559–26569. [CrossRef]
31. Amarasekara, D.S.; Kim, S.; Rho, J. Regulation of Osteoblast Differentiation by Cytokine Networks. *Int. J. Mol. Sci.* **2021**, *22*, 2851. [CrossRef] [PubMed]
32. Salazar, V.S.; Gamer, L.W.; Rosen, V. BMP signalling in skeletal development, disease and repair. *Nat. Rev. Endocrinol.* **2016**, *12*, 203–221. [CrossRef] [PubMed]
33. Tavafoghi, M.; Cerruti, M. The role of amino acids in hydroxyapatite mineralization. *J. R. Soc. Interface* **2016**, *13*, 20160462. [CrossRef] [PubMed]
34. Yanyan, S.; Guangxin, W.; Guoqing, S.; Yaming, W.; Wuhui, L.; Osaka, A. Effects of amino acids on conversion of calcium carbonate to hydroxyapatite. *RSC Adv.* **2020**, *10*, 37005–37013. [CrossRef] [PubMed]
35. Xia, H.; Dong, L.; Hao, M.; Wei, Y.; Duan, J.; Chen, X.; Yu, L.; Li, H.; Sang, Y.; Liu, H. Osteogenic Property Regulation of Stem Cells by a Hydroxyapatite 3D-Hybrid Scaffold with Cancellous Bone Structure. *Front. Chem.* **2021**, *9*, 798299. [CrossRef]
36. Lenton, S.; Hervø-Hansen, S.; Popov, A.M.; Tully, M.D.; Lund, M.; Skepö, M. Impact of Arginine–Phosphate Interactions on the Reentrant Condensation of Disordered Proteins. *Biomacromolecules* **2021**, *22*, 1532–1544. [CrossRef]
37. Takashima, I.; Kusamori, K.; Hakariya, H.; Takashima, M.; Vu, T.H.; Mizukami, Y.; Noda, N.; Takayama, Y.; Katsuda, Y.; Sato, S.I.; et al. Multifunctionalization of Cells with a Self-Assembling Molecule to Enhance Cell Engraftment. *ACS Chem. Biol.* **2019**, *14*, 775–783. [CrossRef] [PubMed]
38. Takemoto, N.; Suehara, T.; Frisco, H.L.; Sato, S.; Sezaki, T.; Kusamori, K.; Kawazoe, Y.; Park, S.M.; Yamazoe, S.; Mizuhata, Y.; et al. Small-molecule-induced clustering of heparan sulfate promotes cell adhesion. *J. Am. Chem. Soc.* **2013**, *135*, 11032–11039. [CrossRef]

gels

MDPI

Article

Facile Construction of Hybrid Hydrogels with High Strength and Biocompatibility for Cranial Bone Regeneration

Shuai Chang [1,2,3,4,†], Jiedong Wang [1,2,3,4,†], Nanfang Xu [1,2,3], Shaobo Wang [1,2,3], Hong Cai [1,2,3,*], Zhongjun Liu [1,2,3,*] and Xing Wang [4,5,*]

1 Department of Orthopedics, Peking University Third Hospital, Beijing 100191, China
2 Beijing Key Laboratory of Spinal Disease Research, Peking University Third Hospital, Beijing 100191, China
3 Engineering Research Center of Bone and Joint Precision Medicine, Ministry of Education, Peking University Third Hospital, Beijing 100191, China
4 Beijing National Laboratory for Molecular Sciences, Institute of Chemistry, Chinese Academy of Sciences, Beijing 100190, China
5 University of Chinese Academy of Sciences, Beijing 100049, China
* Correspondence: hongcai@bjmu.edu.cn (H.C.); zjliu@bjmu.edu.cn (Z.L.); wangxing@iccas.ac.cn (X.W.)
† These authors contributed equally to this work.

Abstract: The significant efforts being made towards the utilization of artificial soft materials holds considerable promise for developing tissue engineering scaffolds for bone-related diseases in clinics. However, most of these biomaterials cannot simultaneously satisfy the multiple requirements of high mechanics, good compatibility, and biological osteogenesis. In this study, an osteogenic hybrid hydrogel between the amine-functionalized bioactive glass (ABG) and 4-armed poly(ethylene glycol) succinimidyl glutarate-gelatin network (SGgel) is introduced to flexibly adhere onto the defective tissue and to subsequently guide bone regeneration. Relying on the rapid ammonolysis reaction between amine groups (-NH_2) of gelatin and ABG components and N-hydroxysuccinimide (NHS)-ester of tetra-PEG-SG polymer, the hydrogel networks were formed within seconds, offering a multifunctional performance, including easy injection, favorable biocompatibility, biological and mechanical properties (compressive strength: 4.2 MPa; storage modulus: 10^4 kPa; adhesive strength: 56 kPa), which could facilitate the stem cell viability, proliferation, migration and differentiation into osteocytes. In addition, the integration between the SGgel network and ABG moieties within a nano-scale level enabled the hybrid hydrogel to form adhesion to tissue, maintain the durable osteogenesis and accelerate bone regeneration. Therefore, a robust approach to the simultaneously satisfying tough adhesion onto the tissue defects and high efficiency for bone regeneration on a mouse skull was achieved, which may represent a promising strategy to design therapeutic scaffolds for tissue engineering in clinical applications.

Keywords: bone tissue engineering; bioactive glass; hybrid hydrogel; high mechanics; injectability

1. Introduction

Bone tissue possesses unique self-repair and regeneration capacities for minor injuries, however, severe bone fractures and defects, as well as trauma and osteosarcoma, are still harmful diseases in orthopedics [1–3]. Although significant progress in surgical techniques has been made in clinic, delayed union or nonunion usually occurs in 10% of patients, resulting in disability and significant socioeconomic burdens [4–7]. Efforts are being made to improve the surgical treatment of autologous and allogeneic bone grafts for critical bone defects, however, these can cause clinical complications such as donor shortage, delayed operation time, limited donor volume, immune rejection of donor sites, and infection [8–11]. To tackle these shortcomings, the advent of the tissue engineering technique has provided creative ideas for promoting bone regeneration, on the basis of its unique advantages of exogenous progenitor cells and tailored release of biological agents, wherein the rapid development of biocompatible scaffolds with sufficient mechanical strength, matched degradation rate and promoted osteogenic activity has evolved as a primary topic in bone tissue engineering research [12–18].

Hydrogels comprising of 3D crosslinking polymer networks with multifariously physical structures and chemical properties have been widely explored as biological scaffolds to mimic the natural ECM environment when properly designed, thus facilitating the viability, growth, proliferation, and differentiation of the stem cells for tissue repair [19–26]. Nevertheless, most hydrogel scaffolds cannot simultaneously satisfy the multiple requirements of robust mechanics, cell compatibility and biological osteogenesis, thus hindering the clinical therapy of bone defects [27–30]. Moreover, an ideal scaffold should also possess a suitable porous architecture to offer sufficient biological immobilization space between the implanted scaffolds and regenerative tissues for adequate bone induction and angiogenesis.

Using high performance hydrogels as biological materials for bone tissue repair has led to great progress in enhancing the robustness of bone tissue in recent years. In general, the incorporation of bioactive components (e. g. POSS, BG, clay, HAp) is an effective method to improve the biological performance of bone constructs [31–38]. Wherein, bioactive glass (BG), mainly consisting of silicon, calcium, and phosphorus elements, has more capabilities on inducing the osteo-stimulation and bone mineralization for defect repair [39–41]. It was previously believed that, along with the implanted hydrogel scaffold degradation in vivo, the bioactive nanoparticles of hydroxyapatite or BG could be exposed-contacted into and subsequently bound onto the tissues during the degradation process in order to facilitate the angiogenesis, which was of vital importance to enhance bone regeneration [42,43]. Therefore, it is likely that the bioactive components may gradually degrade into the main biological components of the silicon, calcium and phosphorus elements, which could be absorbed in situ to promote bone mineralization. In particular, BG can be induced to produce hydroxyapatite layers by chemically binding to bone under physiological conditions, and analogously to the bone mineral stage, providing a favorable environment for the activity, growth, proliferation, and differentiation of osteoblast-related cells [44]. However, complex microenvironments in vivo can cause the explicit microphase separation between the inorganic phase and organic network, which endangers the long-term stability, durability, integrity, and osteogenic potentials of these nanocomposite materials, thus resulting in the disharmony between bone ingrowth and constructs degradation [45,46]. Therefore, the pursuit of advanced strategies for the achievement of long-term stability and maintenance of phase integration in vivo are urgently required in order to develop more intelligent scaffolds to guide bone regeneration.

In this work, in order to fabricate an advanced osteogenic scaffold that could simultaneously satisfy the multiple requirements of strong adhesion onto tissue defects, good compatibility, high mechanics and promote the sustainable osteogenesis effect, a class of highly integrated inorganic-inorganic hydrogel scaffold (SGgel@ABG) was developed via the rapid ammonolysis reaction between NHS-activated tetra-PEG-SG polymer and gelatin in the presence of an amine-terminated bioactive glass (ABG) particle in the one-pot strategy (Figure 1). On account of the powerful integration between the biological SGgel network and ABG moieties within a nano-scale level, the significant enhancement of the mechanics and osteogenic activity enabled SGgel@ABG composite hydrogel to form adhesion to the tissue and to maintain the durable osteogenesis and accelerate bone regeneration. This not only offered the surgeon the injectable and shapeless ability to fill in the gaps of different patterns, but also provided stable and sustained osteogenic activity that guided long-term bone growth in vivo, thereby indicating the potential therapeutic efficiency for the bone-related disease in clinic.

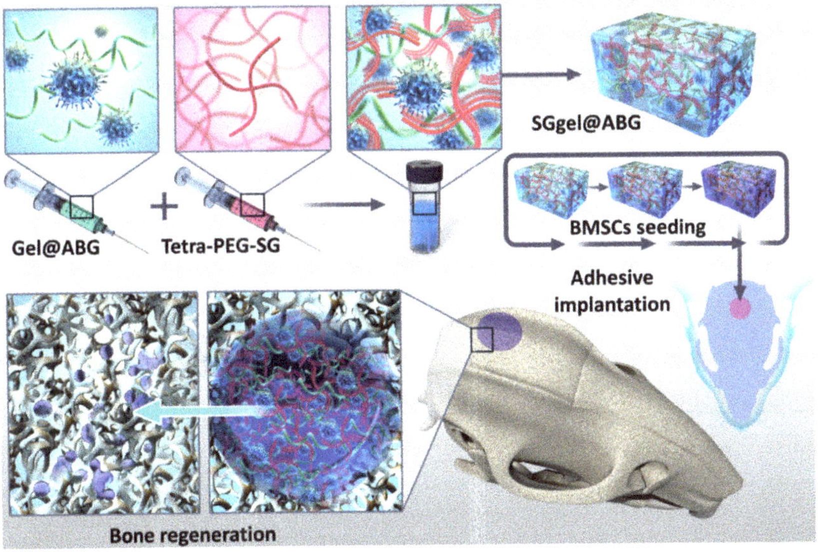

Figure 1. Schematic illustration of the fabrication procedures of SGgel@ABG composite hydrogel for calvaria bone defects repair.

2. Results and Discussion

2.1. Preparation and Characterization of SGgel and SGgel@ABG Hydrogels

The schematic illustration of the SGgel@ABG hybrid hydrogel was shown in Figure 2A. In order to avoid the misgivings of inhibition by anticoagulants and disease transfer, the tetra-PEG-SG was prepared using two steps, with high efficacy. All of the characteristic peaks in the ^{1}H NMR spectrum could be found and belonged to its explicit molecular structure and the clear integration peak ratio (a/b/c/d = 2:1:1:1), as shown in Figure 2B, revealing the precise preparation of tetra-PEG-SG. On the basis of efficient ammonolysis reactions between the active ester groups and the amine in mild conditions, the SGgel hydrogel could be quickly obtained via simultaneously mixing the polymeric gelatin and tetra-PEG-SG solutions using a dual syringe. It was mentioned that the high adhesive strength of SGgel hydrogel with the biological tissue was also achieved by forming the chemical linkage among the amine-terminated proteins tissue and the tetra-PEG-SG polymers, which demonstrated its facile and accessible usage for the bone diseases with unique advantages, such as accessible injection behavior, short gelation time and powerful tissue adhesion. Although the residual active ester group of the SGgel hydrogel can provide adhesive strength on the amine-abundant tissues or other adherends, its cohesion strength was too weak to maintain the tough and durable adhesion to bone tissue. Therefore, a conventional hybridization method with BG moieties could furnish the composite hydrogel with tough cohesion onto the tissue and a strong osteointegration effect. However, the weak interactions between the inorganic phases and polymeric materials within the hybrid hydrogels could impair the mechanical toughness under exposure onto the external force. Therefore, a nanomodification strategy for improving the interface bonding strength of the inorganic-organic phase was proposed to construct polymeric nanocomposites via the amine functionalization by the coupling agent of APTES. In regard to the functionalized BG with abundant amine groups, the obtained ABG was simultaneously integrated into an organic network via the physical aggregation effect and chemical ammonolysis reaction with the ester-active tetra-PEG polymer. This spontaneously strengthened the adhesion and cohesion interactions and formed a highly integrated inorganic-organic structure of the SGgel@ABG hydrogel. In addition, due to the introduction of the inorganic ABG component, the swelling rate of the SGgel@ABG hydrogel declined in comparison to the traditional PEG-based hydrogels. In addition, the less humid environment in the defected bone also hindered the swelling behavior of the hydrogels. Therefore, the involved ABG moieties in the hydrogels were not only beneficial for epithelialization and collagen synthesis in chronic healing, but also retained its bioactivity in stimulating osteogenesis and angiogenesis at the target site.

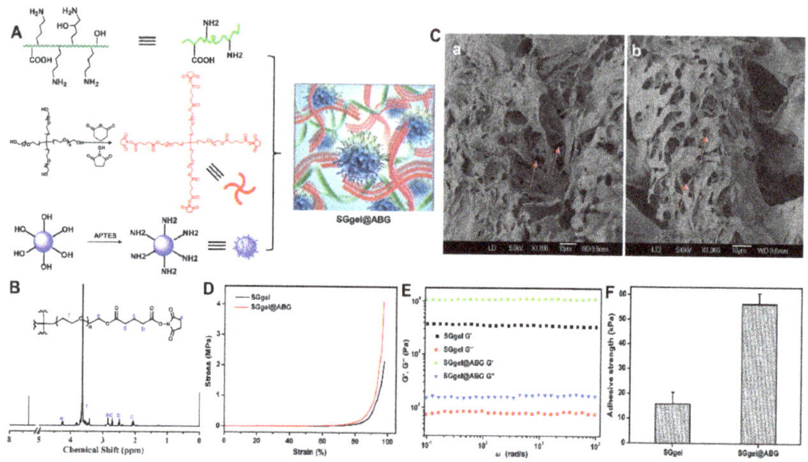

Figure 2. Structure and property characterizations. (**A**) Synthesis route of modified polymers and amine-functionalized ABG. (**B**) ^{1}H NMR spectrum of the tetra-PEG-SG polymer. (**C–F**) SEM images, compressive, rheology and adhesive profiles of (**a**) SGgel and (**b**) SGgel@ABG hydrogels. Red arrows represent the similar inner pores.

The SEM images showed the morphologies of the hydrogel scaffolds, as can be seen in Figure 2C, in which the SGgel and SGgel@ABG hydrogels displayed similar architectures and inner pores (red arrow). These composite hydrogels had large pore diameters and uniform porosity, enabling the entry of host cells and the exchange of nutrients and metabolic wastes. Under this circumstance, the prepared SGgel@ABG hybrid hydrogel may enable the sustained ABG release, cell infiltration and differentiation, as well as the internal and external substance exchange of hydrogels. The compressive behavior was a key parameter in estimating the stability, mechanical stability, and practicability of the composite scaffolds. As shown in Figure 2D, compared to the SGgel, the SGgel@ABG hydrogel exhibited higher mechanical performances as a result of the introduction of rigid ABG, uniform distribution and modulus enhancement. Moreover, rheology was also applied to verify the rationalized network and optimized incorporation of bioactive ABG moieties on improving the mechanical properties of SGgel@ABG hydrogel (Figure 2E). The storage modulus G′ (10^4 kPa), which is larger than G″, indicated an elastic state and rigid backbone, which is indicative of its potential values for the tissue regenerations.

Figure 2F showed that the SGgel@ABG hydrogel adhesive possessed a significant improvement in adhesive strength and tough cohesion compared to the SGgel hydrogel, which was attributed to the optimal involvement and uneven arrangement of ABG particles within the network, through both the physical absorption and chemical anchor. The rational design and ratio of the SGgel@ABG adhesive was relative to the cohesiveness and adhesiveness. The SGgel adhesives possessed weak mechanical strength due to its poor cohesiveness. It has been stated that, although the addition of ABG can significantly enhance the cohesiveness of the adhesive, it also depletes the NHS-ester groups for the adhesiveness in the SGgel. Therefore, over-dosed ABG could exhaust NHS-ester groups and decline the adhesiveness of the SGgel@ABG hydrogel adhesive. Taking both the cohesiveness and adhesiveness of hydrogel adhesive into consideration, it was concluded that the SGgel@ABG, containing 5% *w/v* of ABG, was an optimized formula with optimal structures and well-balanced properties. Therefore, with facile injection, tough adhesion and biological properties, this hydrogel adhesive may spark innovative clinical treatments to guide bone regeneration.

2.2. Cell Viability and Proliferation

Biomaterial compatibility is a prominent prerequisite for high-qualified cell carriers. The FDA-approved PEG and biocompatible gelatin are widely used in biomedical applications, such as in neural, articular cartilage tissue, bladder and bone repair. Therefore, cytotoxicity was first conducted to assess the safety of the SGgel and SGgel@ABG hydrogel scaffolds, by the BMSCs, using a live/dead assay. Figure 3A confirms the excellent cell biocompatibility with a clear observation of high BMSCs viability. Quantitatively, this ABG-conjoined SGgel could not only keep the favourable cell survival rate, but also significantly facilitated the cell growth after 5 days of in vitro culture, revealing its low toxicity on BMSCs. This is a preliminary indicator for the implantation of hydrogel scaffolds in the tissue engineering fields. The BMSCs proliferation experiment was further performed after the in vitro culture (1, 3 and 5 days) on each group. As expected, cell proliferation continued to increase in all of the groups, within the five days. As shown in Figure 3B, the density of BMSCs within the SGgel@ABG hydrogel scaffold was measured to be $(0.83 \pm 0.03) \times 10^4$ cells/mL on day 1, $(1.74 \pm 0.08) \times 10^4$ cells/mL on day 3 and $(3.51 \pm 0.1) \times 10^4$ cells/mL on day 5, higher than those of the SGgel hydrogel and control group, indicating the important roles of ABG substrate on improving cell proliferation after prolonged incubation.

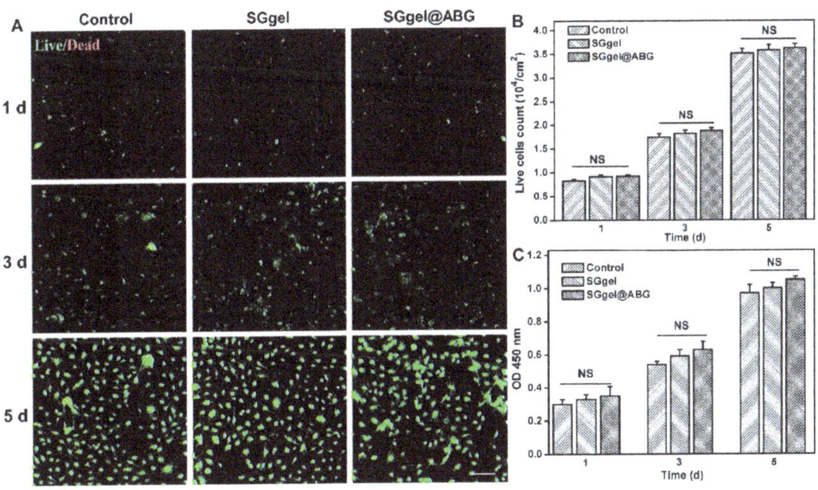

Figure 3. Cell cytotoxicity of SGgel and SGgel@ABG hybrid scaffolds in vitro. (**A**) Live/dead staining of BMSCs. Wherein, the green cells are the living BMSCs, and the red cells are the dead BMSCs. (**B**) Cell viability and (**C**) Cell proliferation of SGgel and SGgel@ABG hybrid scaffolds after the cultivation for the appointed time. NS, not significant.

To explore whether this hybrid hydrogel could support cell activity, growth, and proliferation, we performed the proliferation experiment and measured the OD values at 450 nm after the in vitro culture on the hydrogel scaffolds using a CCK8 assay. After the co-culture of the SGgel@ABG hydrogel, the cell proliferation rate was slightly improved initially, and significantly increased at days 3 and 5 of in vitro incubation (Figure 3C), revealing the satisfactory cell growth and proliferation behavior of the ABG-conjoined SGgel hybrid scaffold.

2.3. In Vitro Osteogenic Differentiation

In addition to the requirement good biocompatibility, ideal engineered scaffolds should also meet the needs of boosting progenitor cell's osteogenic differentiation. ALP staining, ARS staining and qPCR were applied to assess the osteoinductivity of the SGgel@ABG hydrogel scaffold. As shown in Figure 4A–C, ALP staining and ARS staining revealed that the SGgel@ABG group possessed the greatest staining intensity and ratios of staining positive areas, compared to the other groups. These results indicated that the ALP activity at day 14 and the mineralized nodules at day 21 were conspicuously enhanced, demonstrating that the BMSCs cultured on the SGgel@ABG hydrogel could be activated and could enhance osteogenic differentiation. As an early osteogenic marker, the increasing ALP expressions within the hydrogel scaffold had a certain guiding significance for the osteogenic differentiation (Figure 4B). OCN, Osterix and RUNX2 were another three crucial markers of the BMSCs that play an important role in regulating bone metabolism, and their upregulated levels illustrated high-grade mineralization. The relative mRNA expressions of osteogenic marker genes (OCN, ALP, Osterix and RUNX2) were analyzed at various stages of osteogenesis. RT-qPCR was used to examine their osteoinduction abilities. Compared to the control group, ALP expression was significantly upregulated at each time point for the SGgel and SGgel@ABG groups, indicating their favorable components, architectures and mechanics for ossification effects. Intriguingly, it was shown that the expression of the osteogenesis-related gene markers of the SGgel@ABG group were more conspicuously increased than those in the SGgel group (Figure 4D), indicating the promotion of cell differentiation by favorable ABG moieties, as well as its uniform distribution, controlled release and improved mechanical properties. The highest upregulated expression after 7 d further implied that the ABG-laden SGgel hydrogel scaffold was capable of impelling BMSCs osteogenic differentiation in vitro. Collectively, these findings reinforced the statement that the SGgel@ABG hydrogel scaffold could effectively promote the osteogenic differentiation that is important for therapeutic ossification and bone regeneration in vivo.

2.4. Hybrid Hydrogel Scaffolds Regenerate Bone Formation In Vivo

Inorganic BG is a series of bioactive, biodegradable and osteoconductive materials for bone tissue engineering, but its application in bone tissue engineering is limited due to its easy removal from the target site. In regard to the SGgel@ABG composite hydrogel with good mechanical, compatible and biological properties in vitro, we performed in vivo bone-defect experiments of this bioactive scaffold to demonstrate its strong bonding capacity, excellent cell ingrowth and regenerative bioactivity. A circular bone piece was fixed within freshly formed cranial defects and treated with the SGgel and SGgel@ABG hydrogels. The stability and bone regeneration of the circular bone plates were evaluated eight weeks after surgery. The micro-CT images exhibited the dislocation of the circular bone piece in the control group. This was the result of its lack of fracture fixation, sustainability and regenerative ability (Figure 5A). No similar phenotype was seen in the implanted hydrogel scaffolds, and the circular bone pieces were not fully healed in the SGgel group due to its insufficient nutrient and osteogenic supplies. By comparison, the SGgel@ABG group exhibited a good integration between the circular bone piece and its surrounding bone tissue. More importantly, its gap was conspicuously reduced at eight weeks, indicating the formation of vast new bones as a result of the bioactive ABG participant in the SGgel@ABG group. The microarchitectural parameters of the regenerated bone tissue within the cranial defect were confirmed by quantifying the bone volume within circular gap and the total volume within the circular gap (Figure 5B–D). The TV of the SGgel@ABG group was larger than that of the control and SGgel groups at eight weeks. The similar trend of BV/TV and BMD further indicated the achievement of the largest volume of new bone formation for the SGgel@ABG scaffold, which was consistent with the assumption of essential ABG introduction to satisfy the bioactive and mechanical support for new bone formation.

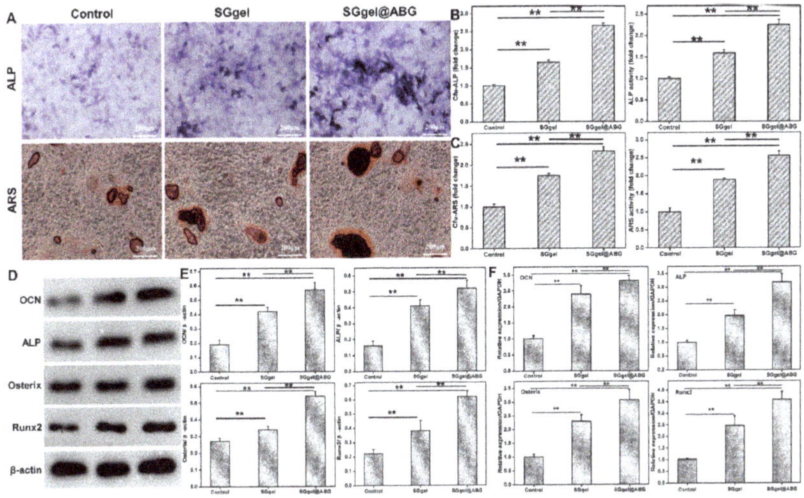

Figure 4. In vitro osteogenic differentiation of SGgel@ABG hydrogel. (**A–C**) ALP (14 d) and ARS staining (21 d) revealing the enhanced osteogenic differentiation of BMSCs. (**D**,**E**) Western blotting analysis and (**F**) qPCR quantification showing the highest osteogenic expression markers (OCN, ALP, Osterix and RUNX2) in the hydrogels. Statistically significant differences in comparison with untreated cells (control), SGgel hydrogel and SGgel@ABG hydrogel. ** $p < 0.01$.

A histologic analysis experiment was carried out to differentiate the bone and soft tissue within the defect after eight weeks (Figure 5E–G). This was consistent with the micro-CT as soft tissue formation was observed but no new bone formation was observed in the control group, which implied the uneliminated the defect. In contrast, the SGgel@ABG group showed many fibrous connective tissues that had filled most of the defected areas. H&E staining showed the maximum bone regeneration in the defect area in the SGgel@ABG group, while less marginal bone formation was achieved in the SGgel group. The microstructure of newly formed bone in SGgel@ABG was relatively orderly and the distribution of osteocytes was more uniform, revealing the greatest abundant bone regeneration in the SGgel@ABG group. In addition, mature bone tissue was also detected in the targeted site to reveal its bone regeneration ability. Masson trichrome staining showed that the SGgel@ABG group had significantly more mature neo-bone than the SGgel group. The latter had more immature collagen fibers and osteoid blue-stained areas, and the former revealed the best quality of newly formed bone, indicating that ABG moieties could promote bone tissue growth. Thereafter, the cranial repair results proved its potential application as a suitable tissue-engineered scaffold.

The repair mechanism of the SGgel@ABG composite hydrogel may be attributed to the following factors: (1) the component of osteoid and bone tissues. Biocompatible natural gelatin was favored for the chondrogenesis to improve proteoglycan/collagen secretion, and polymeric tetra-PEG-SG provided a dense crosslinking junction and sufficient adhesion to the tissue surface to maintain the durable mechanical support and long-lasting stability; (2) the SGgel@ABG composite hydrogel could offer a moisture environment to ensure the surrounding cells' survival, growth and proliferation, which played a role in the initial stage of bone regeneration; (3) in addition to the hydrogel degradation in vivo, ABG could be exposed-contacted into, and subsequently bound onto, the tissues to facilitate angiogenesis, which was of vital importance to enhance bone regeneration; (4) on account of the main biological components of silicon, calcium and phosphorus elements within the ABG, the gradual degradation and element release could be absorbed to promote bone mineralization. Consequently, in comparison to the previous attempts to fabricate tissue engineered hydrogel scaffolds for tissue regeneration, the current attempt presents a distinctive formula to offer a potential clinic alternative with good biocompatibility, facile injectability, usage flexibility and long-lasting osteogenic capacity.

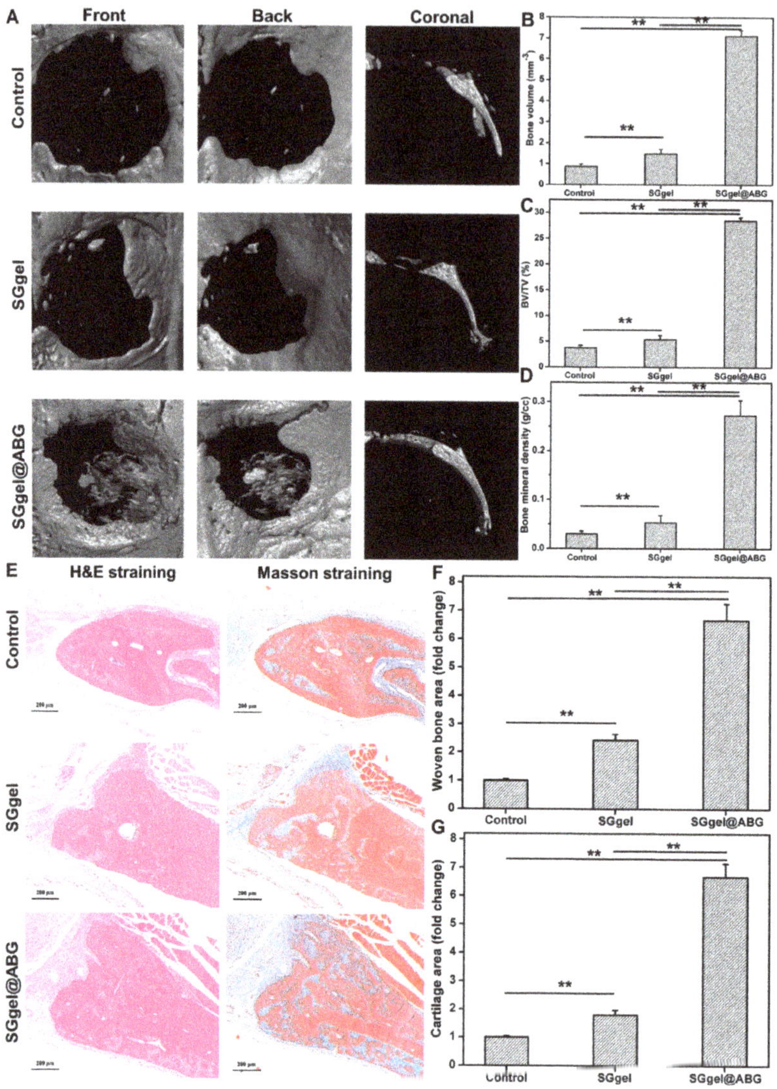

Figure 5. (**A**) 3D reconstruction of Micro-CT images of regenerated bone formation in rat cranium after the hydrogel implantation for 8 weeks with control group, SGgel group and SGgel@ABG group. (**B–D**) Quantitative analysis of BV, BV/TV and BMD of newly formed bone tissue. (**E**) H&E and Masson's trichrome staining. (**F**,**G**) Woven bone and cartilage areas were analyzed in defect bone region. ** $p < 0.01$.

3. Conclusions

In summary, we designed a novel SGgel@ABG hybrid hydrogel adhesive for local bone regeneration. The porous structure and favorable mechanics of the SGgel@ABG scaffold could enhance cell activity, growth, proliferation, and differentiation, while its strong adhesive strength and durable stability on bone tissue was essential for clinical bone repair. In addition, the amine-functionalized ABG moieties not only rationally provided the physical-chemical crosslinking junction on the improving the mechanical properties, but also significantly contributed to the bioactive osteogenic activity, which was found to effectively guide the osteogenesis, maintain the perdurability of bone formation, and accelerate the bone repair both in vitro and in vivo. This study demonstrated an adaptive strategy for producing bone scaffolds with a rational design and fabrication of advanced biomaterial hydrogels for clinical treatments.

4. Materials and Methods

4.1. Materials

Four-arm poly(ethylene glycol) (tetra-PEG-OH, M_w = 20 kDa, M_w/M_n = 1.03, SINOPEG), glutaric anhydride (98%, Energy Chemical), N-hydroxysuccinimide (NHS, 98%, Energy Chemical), 3-aminopropyltriethoxysilane (APTES, J and K, 98%), gelatin (80–100 kDa, J and K), and bioactive glass (BG, 10–30 μm, biomedical grade) was purchased from Shanghai Aladdin Bio-chem Technology Co., Ltd. (Shanghai, China). All other chemical and biological compounds were purchased from Beijing Chemical Works (Beijing, China) and used without any further treatments and purification steps.

4.2. Measurements

The nuclear magnetic resonance spectrum was obtained on a Bruker DRX-400 spectrometer. Scanning electron microscopy (SEM) images were measured at an acceleration voltage of 5 kV, on a JSM-6700F microscope. The sample was sputter-coated with a thin layer of Pt for 120 s before the measurement. The rheological measurements were carried out on a Thermo Haake Rheometer, equipped with cone-parallel plate geometry (diameter: 35 mm), at a gap of 0.5. The sample was spread on a parallel plate, followed by sealing with the oil to prevent water evaporation during the experiment, which was conducted at 25 °C with a constant strain of 0.05%, in the frequency range of 100–0.1 rad s^{-1}. The compressive testing was carried out by a universal Instron 3365 machine, at room temperature, and the hydrogels were cut into cylinders (diameter: 9 mm) with a velocity of 2 mm min^{-1}. The adhesion testing was also performed using Instron 3365 at room temperature. One side of fresh squared porcine skins (diameter: 16 mm × 16 mm) was adhered to a Perspex cylinder using an α-cyanoacrylate glue, while the other side was adhered using the hydrogel adhesive at room temperature. When the attaching porcine skins were torn apart under the driven force, the shear strength was measured with a force sensor of 100 N.

4.3. Synthesis of Tetra-PEG-SG Polymer

First, tetra-PEG-OH (4 g, 0.2 mmol), DMAP (244 mg, 2 mmol) and glutaric anhydride (228 mg, 2 mmol) were dissolved in 50 mL of anhydrous CH_2Cl_2. After stirring for 24 h, the solution was washed with a hydrochloric acid aqueous solution, saturated sodium chloride aqueous solution and DI-water, several times over, and dried over anhydrous $MgSO_4$. Next, the obtained product was further precipitated into diethyl ether for twice to obtain the tetra-arm PEG-glutaric acid polymer. Then, tetra-PEG-COOH (2 g, 0.1 mmol), NHS (460 mg, 4 mmol) and EDCI (768 mg, 4 mmol) were dissolved in 25 mL of anhydrous CH_2Cl_2. After stirring for 24 h, the solution was washed with a hydrochloric acid aqueous solution, saturated sodium chloride aqueous solution and DI-water, three times, and dried over anhydrous $MgSO_4$ to yield the white solid of tetra-PEG-SG.

4.4. Synthesis of ABG

The modification method used to prepare ABG particle was carried out by adding 5 mL APTES into 100 mL of hexane, containing 0.3 g of BG. After stirring for 24 h at 60 °C, the amine-terminated ABG was facilely yielded after being washed with the ethanol and DI water several times.

4.5. Preparation of SGgel and SGgel@ABG Hydrogels

The SGgel adhesive was prepared by mixing the same volume of solution A, containing 20 wt% of gelatin, and solution B, containing 15% of tetra-PEG-SG polymer. For the SGgel@ABG, 5 wt% of ABG was incorporated into solution A with stable sonication, 10 min before mixing solution A and B.

4.6. In Vitro Cytotoxicity Assay

A CCK-8 assay was used to evaluate the cytotoxicity of the hydrogel scaffold. After suspending it into a fresh cell culture medium and seeding it into a 48-well plate, the BMSCs with a density of 1×10^4 cells/100 µL were incubated for 24 h in a 5% CO_2 humidified incubator. The hydrogel was treated in fresh cell medium for another 24 h in order to obtain the extracts. Then, the treated cell medium was used to replace the fresh cell medium to obtain further cell cultures. The cells that were cultured in the fresh medium formed the control group. The cell viability was calculated via the equation:

$$\text{Cell viability (\%)} = [(A_{sample} - A_{blank})/(A_{control} - A_{blank})] \times 100\%$$

4.7. Live/Dead Staining Assay

The cell biocompatibilities were evaluated using a Live/Dead Viability Kit. The BMSCs on either a blank plate or hydrogels were stained with live/dead staining working solution (Calcein-AM/PI), and the stained cells were directly observed under the inverted optical microscope (Olympus, Japan). Calcein-AM generates a green fluorescence signal in living cells and PI only reaches the nuclei of dead cells to emit red fluorescence. The number of live cells was quantitatively analyzed using Image J software (Image J 1.52, Wayne Rasband, National Institute of Mental Health, Bethesda, Rockville, MD, USA).

4.8. Cell Proliferation Assay

Cell proliferation measurement was also carried out using a CCK-8 assay. Firstly, the BMSCs were incubated in a growth medium for 1 day. Next, the hydrogel samples were added into the medium and further incubated for another 1 day. After 5 days of cell incubation, 100 µL of fresh culture medium was used to replace the cell culture medium and added into 10 µL of CCK-8 for another 4 h. Then, the absorbance was measured at 450 nm and recorded, using a microplate reader to calculate the cell proliferation.

4.9. ALP Staining and ARS Staining

The BMSCs were seeded on the blank plate or the hydrogels pre-coated plate, with a density of 1×10^6 cells/well in osteogenic induction medium. On differentiation days 14 and 21, the cells were washed several times, fixed with 4% paraformaldehyde for 15 min, and stained for 30 min using the ALP staining kit (day 14) and ARS solution (day 21). An inverted optical microscope was used to capture the images. The areas of ALP positive colony forming unit (Cfu-ALP) and ARS positive colony forming unit (Cfu-ARS) were calculated using Image J software. The stained cells were lysed with 10% hexadecylpyridinium chloride monohydrate for 15 min and the absorbance of the collected supernatant at 540 nm was measured using a spectrophotometer. The relative ALP and ARS activities were normalized to total protein content, which was determined using a BCA kit.

4.10. Quantitative Real-Time Polymerase Chain Reaction (qRT-PCR) Analysis

The total mRNA extraction was performed using TRIzol and was then subjected to reverse transcription by PrimeScript RT Master Mix. The qRT-PCR was conducted using the ABI 7500 Real-time PCR System (Atlantic Avenue, New York, CA, USA). The expression of the target genes was measured using TB Green Premix Ex Taq and the expression of β-actin was used as an endogenous control.

4.11. In Vivo Experiments

In order to study the osteogenesis effects of hydrogel scaffolds in vivo, the animal experiments were carried out on 8-week-old male Sprague−Dawley (SD) rats using the calvarial defect model. These rats were anesthetized by an intraperitoneal injection of pentobarbital (Nembutal 3.5 mg/100 g), and the sagittal incisions were made on the skin and the muscle to the cranial roof. After elevating the skin-muscle-periosteum flap, a 5 mm thick circular defect was created in the center of each parietal bone with a ring drill of a corresponding external diameter. Next, the implanted hydrogel samples were injected into the rat defects and the incisions were closed with the resorbable sutures. The rats returned to normal function after surgery. A total of 30 rats were randomly divided into the graft study groups: (1) negative control group (defect only), (2) SGgel/BMSCs group, and (3) SGgel@ABG/BMSCs group. All surgical interventions and postoperative animal care procedures were approved by the Animal Research Committee of Peking University (LA2020465).

4.12. Microcomputed Tomography (Micro-CT) Analysis

To assess the quality of the new bone formation, skull bones possessing a defect region were extracted 8 weeks after the implantation. Micro-CT was performed to scan the skull defects using the SkyScan 1076 System (Bruker, Kontich, Belgium), under a standard condition (40 kV X-ray voltage; 250 μA electric current; 18 μm per pixel). DataViewer was used to reconstruct 3D images of micro-CT. A diameter of 5 mm, covering the bone defect region in the reconstructed 3D image, was defined as the region of interest (ROI), and the bone morphometric parameters, including bone volume (BV), bone volume/tissue volume (BV/TV) and bone mineral density (BMD), were analyzed in ROI.

4.13. Histological Observations

The skull samples were fixed in 4% paraformaldehyde for 2 d and subsequently decalcified with a daily change of 15% tetrasodium EDTA for 28 d. From the center of the defect region in the coronal plane, a decalcified skull sample was divided into 8 mm thick slices. Next, the samples were dehydrated in an ethanol series before being embedded in paraffin. From the center of the defect region in the coronal plane, decalcified skull samples were sliced into 5 μm thick sections for the histological examination and morphological analysis with Hematoxylin-Eosin (H&E) staining and Masson's trichrome staining.

4.14. Statistics Analysis

All the results were expressed in the form of mean and standard deviation and there were at least 3 independent experiments. For multi-group comparisons, the one-way ANOVA was assessed, followed by either the Student-Newman-Keuls or Dunnett's test, wherever appropriate; $p < 0.05$ was considered statistically significant.

Author Contributions: H.C., Z.L. and X.W. initiated the project. S.C., J.W., N.X., S.W., Z.L. and X.W. searched the database, wrote, and finalized the manuscript. H.C. and Z.L. gave important advice and helped revising the manuscript. All authors have read and agreed to the published version of the manuscript.

Funding: This work was supported by the Youth Innovation Promotion Association CAS (2019031), National Natural Science Foundation of China (51973226) and the Ministry of Science and Technology of the People's Republic of China (2016YFB1101501).

Institutional Review Board Statement: The animal study protocol was approved by the Animal Research Committee of Peking University (LA2020465).

Informed Consent Statement: Not applicable.

Data Availability Statement: Not applicable.

Conflicts of Interest: The authors declare that they have no conflict of interest.

References

1. Service, R.F. Tissue engineers build new bone. *Science* **2000**, *289*, 1498–1500. [CrossRef]
2. Senarath-Yapa, K.; Li, S.; Walmsley, G.G.; Zielins, E.; Paik, K.; Britto, J.A.; Grigoriadis, A.E.; Wan, D.C.; Liu, K.J.; Longaker, M.T.; et al. Small molecule inhibition of transforming growth factor beta signaling enables the endogenous regenerative potential of the mammalian calvarium. *Tissue Eng. Part A* **2016**, *22*, 707–720. [CrossRef]
3. Vi, L.; Baht, G.S.; Soderblom, E.J.; Whetstone, H.; Wei, Q.; Furman, B.; Puviindran, V.; Nadesan, P.; Foster, M.; Poon, R.; et al. Macrophage cells secrete factors including LRP1 that orchestrate the rejuvenation of bone repair in mice. *Nat. Commun.* **2018**, *9*, 5191. [CrossRef] [PubMed]
4. Fan, B.; Guo, Z.; Li, X.; Li, S.; Gao, P.; Xiao, X.; Wu, J.; Shen, C.; Jiao, Y.; Hou, W. Electroactive barium titanate coated titanium scaffold improves osteogenesis and osseointegration with low-intensity pulsed ultrasound for large segmental bone defects. *Bioact. Mater.* **2020**, *5*, 1087–1101. [CrossRef] [PubMed]
5. Zhang, Y.; Dou, X.; Zhang, L.; Wang, H.; Zhang, T.; Bai, R.; Sun, Q.; Wang, X.; Yu, T.; Wu, D.; et al. Facile fabrication of a biocompatible composite gel with sustained release of aspirin for bone regeneration. *Bioact. Mater.* **2022**, *11*, 130–139. [CrossRef]
6. Dimitriou, R.; Jones, E.; McGonagle, D.; Giannoudis, P.V. Bone regeneration: Current concepts and future directions. *BMC Med.* **2011**, *9*, 66. [CrossRef] [PubMed]
7. Gómez-Barrena, E.; Rosset, P.; Lozano, D.; Stanovici, J.; Ermthaller, C.; Gerbhard, F. Bone fracture healing: Cell therapy in delayed unions and nonunions. *Bone* **2018**, *70*, 93–101. [CrossRef]
8. Wang, W.; Yeung, K.W.K. Bone grafts and biomaterials substitutes for bone defect repair: A review. *Bioact. Mater.* **2017**, *2*, 224–247. [CrossRef]
9. Ho-Shui-Ling, A.; Bolander, J.; Rustom, L.E.; Johnson, A.W.; Luyten, F.P.; Picart, C. Bone regeneration strategies: Engineered scaffolds, bioactive molecules and stem cells current stage and future perspectives. *Biomaterials* **2018**, *180*, 143–162. [CrossRef]
10. Bose, S.; Sarkar, N. Natural medicinal compounds in bone tissue engineering. *Trends Biotechnol.* **2020**, *38*, 404–417. [CrossRef]
11. William, G., Jr.; Einhorn, T.A.; Koval, K.; McKee, M.; Smith, W.; Sanders, R.; Watson, T. Bone grafts and bone graft substitutes in orthopaedic trauma surgery. A critical analysis. *J. Bone Joint Surg. Am.* **2007**, *89*, 649–658. [CrossRef]
12. Senthil, R.; Kavukcu, S.B. Efficacy of glycoprotein-based nanocurcumin/silk fibroin electrospun scaffolds: Perspective for bone apatite formation. *Mater. Chem. Phys.* **2022**, *289*, 126444. [CrossRef]
13. Petite, H.; Viateau, V.; Bensaïd, W.; Meunier, A.; de Pollak, C.; Bourguignon, M.; Oudina, K.; Sedel, L.; Guillemin, G. Tissue engineered bone regeneration. *Nat. Biotechnol.* **2000**, *18*, 959. [CrossRef]
14. Li, R.; Sun, Y.; Cai, Z.; Li, Y.; Sun, J.; Bi, W.; Yang, F.; Zhou, Q.; Ye, T.; Yu, Y. Highly bioactive peptide-HA photo-crosslinking hydrogel for sustained promoting bone regeneration. *Chem. Eng. J.* **2021**, *415*, 129015. [CrossRef]
15. Armiento, A.R.; Hatt, L.P.; Rosenberg, G.S.; Thompson, K.; Stoddart, M.J. Functional biomaterials for bone regeneration: A lesson in complex biology. *Adv. Funct. Mater.* **2020**, *30*, 1909874. [CrossRef]
16. Xu, B.; Ye, J.; Fan, B.-S.; Wang, X.; Zhang, J.-Y.; Song, S.; Song, Y.; Jiang, W.-B.; Wang, X.; Yu, J.-K. Protein-spatiotemporal partition releasing gradient porous scaffolds and anti-inflammatory and antioxidant regulation remodel tissue engineered anisotropic meniscus. *Bioact. Mater.* **2023**, *20*, 194–207. [CrossRef] [PubMed]
17. Wang, S.J.; Jiang, D.; Zhang, Z.Z.; Chen, Y.R.; Yang, Z.D.; Zhang, J.Y.; Shi, J.; Wang, X.; Yu, J.K. Biomimetic nanosilica-collagen scaffolds for in situ bone regeneration: Toward a cell-free, one-step surgery. *Adv. Mater.* **2019**, *31*, 1904341. [CrossRef] [PubMed]
18. Ren, X.; Chen, X.; Geng, Z.; Su, J. (Bone-targeted biomaterials: Strategies and applications. *Chem. Eng. J.* **2022**, *446*, 137133. [CrossRef]
19. Yang, Y.; Zhou, M.; Peng, J.; Wang, X.; Liu, Y.; Wang, W.; Wu, D. Robust, anti-freezing and conductive bonding of chitosan-based double-network hydrogels for stable-performance flexible electronic. *Carbohyd. Polym.* **2022**, *276*, 118753. [CrossRef]
20. Lei, L.; Bai, Y.; Qin, X.; Liu, J.; Huang, W.; Lv, Q. Current understanding of hydrogel for drug release and tissue engineering. *Gels* **2022**, *8*, 301. [CrossRef] [PubMed]
21. Kim, H.D.; Lee, E.A.; An, Y.-H.; Kim, S.L.; Lee, S.S.; Yu, S.J.; Jang, H.L.; Nam, K.T.; Im, S.G.; Hwang, N.S. Chondroitin sulfate-based biomineralizing surface hydrogels for bone tissue engineering. *ACS Appl. Mater. Interfaces* **2017**, *9*, 21639–21650. [CrossRef]
22. Sun, J.-Y.; Zhao, X.; Illeperuma, W.R.K.; Chaudhuri, O.; Oh, K.H.; Mooney, D.J.; Vlassak, J.J.; Suo, Z. Highly stretchable and tough hydrogels. *Nature* **2012**, *489*, 133–136. [CrossRef] [PubMed]
23. Tsou, Y.-H.; Khoneisser, J.; Huang, P.-C.; Xu, X. Hydrogel as a bioactive material to regulate stem cell fate. *Bioact. Mater.* **2016**, *1*, 39–55. [CrossRef] [PubMed]
24. Yang, Y.; Wang, X.; Yang, F.; Wang, L.; Wu, D. Highly elastic and ultratough hybrid ionic-covalent hydrogels with tunable structures and mechanics. *Adv. Mater.* **2018**, *30*, 1707071. [CrossRef]

25. Oprita, E.I.; Iosageanu, A.; Craciunescu, O. Progress in composite hydrogels and scaffolds enriched with icariin for osteochondral defect healing. *Gels* **2022**, *8*, 648. [CrossRef] [PubMed]
26. Gull, N.; Khan, S.M.; Butt MT, Z., Zia, S.; Khalid, S.; Islam, A.; Sajid, I.; Khan, R.U.; King, M.W. Hybrid cross-linked hydrogels as a technology platform for in vitro release of cephradine. *Polym. Adv. Technol.* **2019**, *30*, 2414–2424. [CrossRef]
27. Zhu, T.; Wang, H.; Jing, Z.; Fan, D.; Liu, Z.; Wang, X.; Tian, Y. High efficacy of tetra-PEG hydrogel sealants for sutureless dural closure. *Bioact. Mater.* **2022**, *8*, 12–19. [CrossRef]
28. Shi, L.; Wang, F.; Zhu, W.; Xu, Z.; Fuchs, S.; Hilborn, J.; Zhu, L.; Ma, Q.; Wang, Y.; Weng, X.; et al. Self-healing silk fibroin-based hydrogel for bone regeneration: Dynamic metal-ligand self-assembly approach. *Adv. Funct. Mater.* **2017**, *27*, 1700591. [CrossRef]
29. Ge, J.; Li, Y.; Wang, M.; Gao, C.; Yang, S.; Lei, B. Engineering conductive antioxidative antibacterial nanocomposite hydrogel scaffolds with oriented channels promotes structure-functional skeletal muscle regeneration. *Chem. Eng. J.* **2021**, *425*, 130333. [CrossRef]
30. Dou, X.; Cao, Q.; Sun, F.; Wang, Y.; Wang, H.; Shen, H.; Yang, F.; Wang, X.; Wu, D. Synergistic control of dual cross-linking strategy toward tailor-made hydrogels. *Sci. China Chem.* **2020**, *63*, 1793–1798. [CrossRef]
31. Li, D.; Zhou, J.; Zhang, M.; Ma, Y.; Yang, Y.; Han, X.; Wang, X. Long-term delivery of alendronate through injectable tetra-peg hydrogel to promote osteoporosis therapy. *Biomater. Sci.* **2020**, *8*, 3138–3146. [CrossRef] [PubMed]
32. Wang, X.; Gao, P.; Yang, Y.; Guo, H.; Wu, D. Dynamic and programmable morphology and size evolution via a living hierarchical self-assembly strategy. *Nat. Commun.* **2018**, *9*, 2772. [CrossRef] [PubMed]
33. Che, L.; Wang, Y.; Sha, D.; Li, G.; Wei, Z.; Liu, C.; Yuan, Y.; Song, D. A biomimetic and bioactive scaffold with intelligently pulsatile teriparatide delivery for local and systemic osteoporosis regeneration. *Bioact. Mater.* **2023**, *19*, 75–87. [CrossRef]
34. Li, C.; Zhang, W.; Wang, R.; Du, X.-F.; Jiang, D.; Liu, B.; Nie, Y.; Liao, J.; Chen, Y.; Liang, X.; et al. Nanocomposite multifunctional hydrogel for suppressing osteosarcoma recurrence and enhancing bone regeneration. *Chem. Eng. J.* **2022**, *435*, 134896. [CrossRef]
35. Wang, X.; Gao, P.; Wang, J.; Yang, Y.; You, Y.; Wu, D. A robust strategy for precise fabrication of rigid-flexible coupling dendrimers toward self-coordinated hierarchical assembly. *CCS Chem.* **2020**, *2*, 1093–1104. [CrossRef]
36. Fan, L.F.; Wang, X.; Wu, D.C. Polyhedral oligomeric silsesquioxanes (POSS)-based hybrid materials: Molecular design, solution self-assembly and biomedical applications. *Chinese J. Chem.* **2021**, *39*, 757–774. [CrossRef]
37. Yu, T.; Zhang, L.; Dou, X.; Bai, R.; Wang, H.; Deng, J.; Zhang, Y.; Sun, Q.; Li, Q.; Wang, X.; et al. Mechanically robust hydrogels facilitating bone regeneration through epigenetic modulation. *Adv. Sci.* **2022**, *9*, e2203734. [CrossRef]
38. Fan, L.-F.; Hou, C.-L.; Wang, X.; Yan, L.-T.; Wu, D.-C. Tunable multiple morphological transformation of supramolecular hyperbranched polymers based on an A_2B_6-type POSS monomer. *Chinese J. Polym. Sci.* **2021**, *39*, 1562–1571. [CrossRef]
39. Granel, H.; Bossard, C.; Nucke, L.; Wauquier, F.; Rochefort, G.Y.; Guicheux, J.; Jallot, E.; Lao, J.; Wittrant, Y. Optimized bioactive glass: The quest for the bony graft. *Adv. Healthc. Mater.* **2019**, *8*, 1801542. [CrossRef]
40. Zhang, K.; Zhou, Y.; Xiao, C.; Zhao, W.; Wu, H.; Tang, J.; Li, Z.; Yu, S.; Li, X.; Min, L.; et al. Application of hydroxyapatite nanoparticles in tumor-associated bone segmental defect. *Sci. Adv.* **2019**, *5*, eaax6946. [CrossRef]
41. Bai, X.; Lü, S.; Liu, H.; Cao, Z.; Ning, P.; Wang, Z.; Gao, C.; Ni, B.; Ma, D.; Liu, M. Polysaccharides based injectable hydrogel compositing bio-glass for cranial bone repair. *Carbohyd. Polym.* **2017**, *175*, 557–564. [CrossRef] [PubMed]
42. Li, X.; Zou, Q.; Chen, H.; Li, W. In vivo changes of nanoapatite crystals during bone reconstruction and the differences with native bone apatite. *Sci. Adv.* **2019**, *5*, eaay6484. [CrossRef] [PubMed]
43. Xu, L.; Gao, S.; Zhou, R.; Zhou, F.; Qiao, Y.; Qiu, D. Bioactive pore-forming bone adhesives facilitating cell ingrowth for fracture healing. *Adv. Mater.* **2020**, *32*, 1907491. [CrossRef]
44. Tang, J.; Xi, K.; Chen, H.; Wang, L.; Li, D.; Xu, Y.; Xin, T.; Wu, L.; Zhou, Y.; Bian, J.; et al. Flexible osteogenic glue as an all-in-one solution to assist fracture fixation and healing. *Adv. Funct. Mater.* **2021**, *31*, 2102465. [CrossRef]
45. Zhang, K.; Lin, S.; Feng, Q.; Dong, C.; Yang, Y.; Li, G.; Bian, L. Nanocomposite hydrogels stabilized by self-assembled multivalent bisphosphonate-magnesium nanoparticles mediate sustained release of magnesium ion and promote in-situ bone regeneration. *Acta Biomater.* **2017**, *64*, 389. [CrossRef] [PubMed]
46. Tomlinson, S.K.; Ghita, O.R.; Hooper, R.M.; Evans, K.E. Monomer conversion and hardness of novel dental cements based on ethyl cyanoacrylate. *Dent. Mater.* **2007**, *23*, 799. [CrossRef] [PubMed]

gels

MDPI

Article

Modeling Tunable Fracture in Hydrogel Shell Structures for Biomedical Applications

Gang Zhang [1,2,†], Hai Qiu [3,†], Khalil I. Elkhodary [4], Shan Tang [5,6,*] and Dan Peng [7,*]

1 Hubei Provincial Key Laboratory of Chemical Equipment Intensification and Intrinsic Safety, Wuhan 430205, China
2 School of Mechanical and Electrical Engineering, Wuhan Institute of Technology, Wuhan 430200, China
3 School of Mechanical Engineering, Jiangsu University of Science and Technology, Zhenjiang 212003, China
4 The Department of Mechanical Engineering, The American University in Cairo, New Cairo 11835, Egypt
5 Department of Engineering Mechanics, Dalian University of Technology, Dalian 116024, China
6 State Key Laboratory of Structural Analysis for Industrial Equipment, International Research Center for Computational Mechanics, Department of Engineering Mechanics, Dalian University of Technology, Dalian 116024, China
7 Department of Neurology, The Second Hospital of Dalian Medical University, Dalian 116023, China
* Correspondence: shantang@dlut.edu.cn (S.T.); pengdanmed@163.com (D.P.)
† These authors contributed equally to this work.

Abstract: Hydrogels are nowadays widely used in various biomedical applications, and show great potential for the making of devices such as biosensors, drug- delivery vectors, carriers, or matrices for cell cultures in tissue engineering, etc. In these applications, due to the irregular complex surface of the human body or its organs/structures, the devices are often designed with a small thickness, and are required to be flexible when attached to biological surfaces. The devices will deform as driven by human motion and under external loading. In terms of mechanical modeling, most of these devices can be abstracted as shells. In this paper, we propose a mixed graph-finite element method (FEM) phase field approach to model the fracture of curved shells composed of hydrogels, for biomedical applications. We present herein examples for the fracture of a wearable biosensor, a membrane-coated drug, and a matrix for a cell culture, each made of a hydrogel. Used in combination with experimental material testing, our method opens a new pathway to the efficient modeling of fracture in biomedical devices with surfaces of arbitrary curvature, helping in the design of devices with tunable fracture properties.

Keywords: hydrogels; curved shell; biomedical devices; phase field

1. Introduction

Hydrogels are three-dimensional hydrophilic, cross-linked polymeric networks which can absorb water, or biological fluids, and swell with the water, without dissolving in it. Due to the advantages of high biocompatibility and tunable biodegradability, hydrogels are increasingly used for controllable devices in medical applications [1–7]. Various strategies have been developed to synthesize hydrogels. The polymeric network structure and its chemical composition can be adjusted to control the capability for water absorption, and to achieve desirable mechanical and functional properties, such as strength or conductivity. Filler materials, such as graphene, carbon nanotubes, carbon black, metallic nanowires, and nanoparticles can also be introduced to further tune their mechanical and conductive properties [8–11]. Introducing stimuli-responsive functional groups into the polymeric networks, hydrogel properties can also be controlled by environmental stimuli, including heat, light, magnetic fields, chemical agents, and acidity (pH) [12–16]. Due to the tunability of their material properties, hydrogels are becoming increasingly used to fabricate biomedical devices such as biosensors, capsules for drug delivery, and scaffolds for cell cultures, etc. As the human body and its organs/substructures often have irregular surfaces of complex shape, these devices are often designed with a small thickness, to better adapt to such biological surfaces. Drawing parallels to the more frequently encountered large-scale engineering structures, such as airplane wings and architectural structures, as shown in Figure 1, devices in biomedical engineering can also be considered as shell structures, in terms of their mechanical modeling, as shown in Figure 1b. This parallel is expedient as, in some biomedical applications, structural integrity of the devices should be maintained, and crack growth and propagation should be prevented, whereas in other applications cracks on a device can be harnessed to realize complex functions, such as camouflage [17]. Shell modeling of fracture may thus be called upon to cheaply and accurately explore such diverse applications for hydrogels, and to better tune their properties.

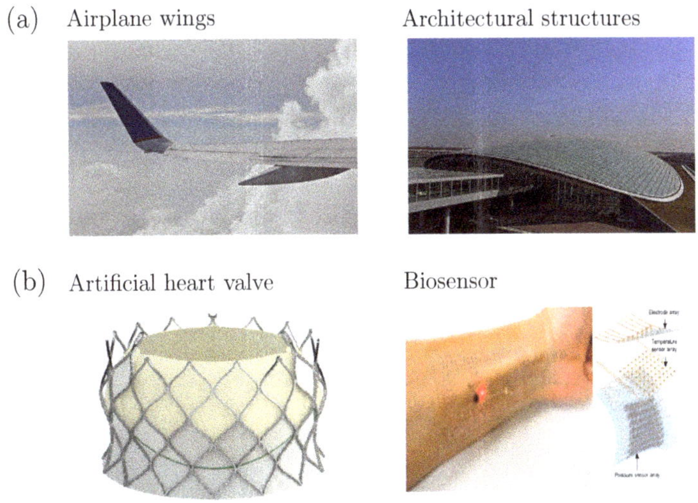

Figure 1. Shell structure. (**a**) Aircraft wing [18] and architectural structures [19]; (**b**) Artificial heart valve (reprinted with permission from Ref. [20]. Copyright 2017, Elsevier) and biosensor (reprinted with permission from Ref. [21]. Copyright 2020, Springer Nature).

To produce devices made from hydrogels with tunable fracture properties (fracture path, pattern, and toughness), trial-and-error via experimentation may be a possible route. However, the usual mechanical tests, such as puncturing, and uniaxial tension or compression, are typically used to measure the corresponding properties prior to crack growth, such as the Young's modulus and the fracture energy, whether directly or indirectly. Additional tests are frequently needed to further measure the fracture path, its evolving pattern, and fracture toughness. However, reproducing realistic loading and operating conditions for these devices that simulate the living body can be challenging. Furthermore, identifying the mechanisms that underly the evolving fracture may not be straightforward with such experiments. Thus, to reduce trial-and-error cost and complexity, to improve the efficiency of hydrogel development cycles, and to deepen our understanding, various computational simulations have been more recently performed. Fracture modeling of hydrogels, however, has proven challenging. Several examples are listed here for illustration of the state of the art. Mao et al. [22] developed a large deformation viscoelastic model for a double-network hydrogel, taking into account the debonding process for both polymeric networks and the swollen hydrogel. The final fracture of the double network hydrogel was not considered, however. Similarly, based on their previous work on the large deformation of nonlinear viscoelastic materials, which incorporates the kinetics of breaking physical crosslinks [23,24], Guo et al. [25] further proposed a novel time-integration scheme that investigated the stress and deformation fields near a crack tip for (single) edge-cracked specimens. Likewise, the rate-dependent fracture behavior of a polyvinyl alcohol (PVA) hydrogel was studied via experiments and numerical simulation. Liu et al. [26,27] developed a constitutive model that could accurately describe the mechanical behavior of double cross-linked hydrogels at different temperatures and loading rates. Mode-I crack tip stress fields, and the corresponding stress-intensity factors for the hydrogel were then analyzed through a finite element analysis. Shen and Vernerey [28] developed a framework to capture the interplay between energy dissipation and crack propagation in hydrogels constituted of transient networks. A cohesive zone model was then employed to relate the crack driving force to crack initiation (instability) and propagation velocity. Liu et al. [29,30] furthermore proposed a physically based poro-visco-hyperelastic model within the framework of the theory of porous media under finite strain to describe the solid-fluid-coupled material behavior of hydrogels. Their model was applied to study time-dependent fiber-reinforced hydrogel composites, and fatigue crack growth in them.

For these developed material models, it might seem straightforward to extend their modeling to shells of hydrogels. However, these models mostly focus on the underlying mechanisms of diffusion, bonding, and debonding in transient networks or double networks, and their coupling to applied deformation. They involve many material parameters, which may not be easily calibrated through experiments. Also, their programmatic implementation is relatively sophisticated. As such, they are not readily amenable for use with engineering design. Specifically for bio-device design, when its characteristic deformation time-scale is considerably smaller than the time-scale of water diffusion, the accounted-for diffusion mechanisms should be ignored [31], bringing the validity of some of these material models into question at small scales. We further note that a tensile stress state is considered primary in a typical hydrogel model. However, shell structures (curved bio-devices) may operate under severe loading conditions, where large deformation arises with shear or compression dominating the stress state. Typical hydrogel models might thus lead to unstable computations under such conditions.

It is also important to note the difficulty in modeling the entire fracture process computationally, while accounting for the path and pattern of fracture. Established methods for numerical fracture modeling in engineering design, such as the cohesive zone model and XFEM [32–35], as available today in commercial finite-element software, do not typically fair well when modeling complex fracture propagation modes, such as crack branching and merging. Moreover, elements of C^1 continuous basis functions are preferable when the modeling shells. However, most available elements in the commercial software are C^0.

It seems to us that in view of these many challenges, studies that model fracture in shells composed of hydrogels, especially for biomedical applications, are today rare. We thus aim to model the large deformation of hydrogel shells of arbitrary curvature using C^0 elements, as available in finite-element software, with application to bio-devices. In this paper, we adopt the phase-field model, which renders it possible to track crack initiation and propagation with greater ease. Phase field can be discretized via the graph method, which is widely used in the computational geometry, machine learning and computer science communities [36–38]. For hydrogels, we consider a nonlinear elastic model, in which the deformation energy is decomposed into tensile and compressive parts. This decomposition can enhance numerical stability when C^0 solid shell elements are used. Although the proposed model is phenomenological, it can relate to water content, the fillers, and the stimuli, by means of experimental data. This model aims to be is easier and more convenient for users who are not familiar with the relevant mathematics and theories in mechanics. It is consistent with the idea of data-driven computational mechanics, which has been developing very rapidly in recent years [39–47]. In Section 2, this material model with phase field degradation is presented. The phase field is discretized based on the graph method. In Section 3, a staggered solution method is proposed. We then model the fracture of a wearable biosensor, a membrane-coated drug, and a matrix for a cell culture, as illustrative examples that highlight the strength of our proposed method. Finally, we summarize our main results and conclude our work in Section 4.

2. Material Models for Fracture

To model the deformation and fracture of hydrogels, a phenomenological mechanical model with few parameters is proposed. To help preserve numerical stability, deformation is decomposed into tensile and compressive parts. We first define λ_i as the i-th principal stretch of the deformation gradient F_{iJ}. The first invariant of the right Cauchy–Green strain is $I_1 = \sum_{i=1}^{3} \lambda_i^2$, and the Jacobian of the deformation gradient is $J = \lambda_1\lambda_2\lambda_3$. The phase field p can be introduced to the free energy density W per the following,

$$W = \left[(1-\mathcal{K})p^2 + \mathcal{K}\right]W^+ + W^-, \tag{1}$$

to model the hyperelastic deformation of hydrogels with damage. Our model assumes that crack growth occurs faster than the rate of water diffusion in the hydrogels, so that their viscous behavior can be ignored [48,49]. For readers who are not familiar with phase field fracture modeling, the phase field p can be thought of as the damage variable. When it is set equal to 1, it signifies that the material has full integrity, and no damage. When it is set equal to zero, it signifies that the material is in a fully damaged state. $\mathcal{K}$ is a numerical constant that is correspondingly often set to a small value to enhance numerical stability, as was shown for the small deformation regime [50–53] and the large deformation regime [48,54]. W^+ and W^- are defined by

$$W^+ = G(\lambda_i^+, J^+) \quad \text{where } \lambda_i^+ = \begin{cases} \lambda_i & \text{for } \lambda_i > 1 \\ 1 & \text{for } \lambda_i \le 1 \end{cases} \qquad J^+ = \begin{cases} J & \text{If } J > 1 \\ 1 & \text{If } J \le 1 \end{cases}$$

and

$$W^- = G(\lambda_i^-, J^-) \quad \text{where } \lambda_i^- = \begin{cases} \lambda_i & \text{for } \lambda_i < 1 \\ 1 & \text{for } \lambda_i \ge 1 \end{cases} \qquad J^- = \begin{cases} J & \text{If } J < 1 \\ 1 & \text{If } J \ge 1 \end{cases} .$$

The free energy density function G takes the form

$$G(\lambda_i, J) = \frac{1}{2}\mu\sum_{i=1}^{3}\left(\lambda_i^2 - 1 - 2\ln\lambda_i\right) + \frac{1}{2}\chi(J-1)^2, \tag{2}$$

where μ and χ are the (initial) shear and Lamé constants, respectively. This is the neo-Hookean model, which is widely used to model soft solids, including hydrogels [55–57]. The first Piola–Kirchhoff (PK) stress can then be computed from

$$P_{iJ} = \frac{\partial W}{\partial F_{iJ}} = \left[(1-\mathcal{K})p^2 + \mathcal{K}\right]\frac{\partial W^+}{\partial F_{iJ}} + \frac{\partial W^-}{\partial F_{iJ}}. \tag{3}$$

2.1. Phase Field Discretization

The evolution of the phase field identifies the fracture process. A Lagrangian formulation is adopted to discretize the phase field equations. All integrations are performed on the undeformed configuration, and differentiation is computed with respect to its reference coordinates. We also assume, for simplicity of computational implementation, that the phase field variable at each Gaussian integration point will take the same value within an element (usually more than three Gaussian points are considered along the thickness direction of a shell element), which greatly reduces computational cost. Specifically, as the fracture process evolves, only a single phase field value for every element in the mesh is computed and stored, which is interpreted as an average phase field value for all the Gaussian points in that element. This assumption has been successfully employed by [58].

Consistent with this assumption, the energy density can be therefore averaged over the element's thickness,

$$\bar{W}^{+} = \frac{1}{N_G}\sum_{I=1}^{N_G} \bar{W}_I^{+},$$

where $\bar{W}_I^{+}$ is the energy density at the I-th Gaussian point, and N_G is the number of Gaussian points along the thickness direction of a shell element.

Now, a correspondence between a finite element mesh and a graph may be easily established. The finite element mesh for a curved surface can be considered as a weighted, undirected graph. A graph is here defined as the pair (V, E) where V denotes the set of n vertices (a vertex is equivalent to a node in finite-element terminology), and $E \subseteq V \times V$ the set of edges. Therefore, the cotangent formulation from discrete differential geometry can be leveraged to obtain the discrete phase field equations. At a vertex point i, this can be written as

$$\left[4\ell(1-\mathcal{K})\bar{W}^{+}/\mathcal{G}_c + 1\right]p_i + 4\ell^2\frac{1}{a_i}\sum_{j:(i,j)\in E} w_{ij}(p_i - p_j) = 1. \tag{4}$$

Note that i, j label the finite-element node numbers. $(i, j) \in E$ represents an edge in the graph. ℓ is the length scale parameter in the phase field model, which usually characterizes the width of a crack. a_i and w_{ij} are related to the area and the length of the element on the graph. $\mathcal{G}_c$ is correspondingly the fracture energy per unit area for an element. For the detailed derivation process, see Section 3 of Zhang et al. [54].

As we discussed in the introduction, many mathematical material models have been already proposed in the literature [22–30] to model the fracture of the gels. However, most models seem significantly more complex than what we propose herein. Our model for gels only requires three main parameters (Young's modulus, bulk modulus, and the fracture energy per unit area). The corresponding polymer structures that relate to the three parameters can be tuned to control the fracture pattern and the toughness of the bio-devices made by the gels discussed further in Section 2.3.

Finite-element modeling of the fracture of plates and shells has also attracted much research interest for a long time, with discrete and diffuse methods having been proposed. In discrete methods, cracks are taken as explicit discontinuities in the displacement field. Typical methods are the extended finite element method (XFEM) and the cohesive zone method, first proposed by [32,35,59]. It is difficult for these models to simulate complex fracture modes, such as crack branching and merging. In diffuse fracture methods, cracks are smeared out and represented directly by a damage variable in the constitutive formulation. A popular diffuse damage model is the phase-field approach [48,54,60,61]. However, discretizing a phase field in thin shells by using modern discrete differential geometry, as we pursued herein, is rare for fracture modeling. Specifically, our discretization process preserves the governing differential equations' intrinsic geometric structure when on non-flat manifolds (shells). Convergence is thus guaranteed even with C^0 continuity elements [62], which is a difficulty for traditional finite-element modeling. We thus showed how to employ our resulting modeling tool to accurately capture the fracture of bio-devices made of gels.

2.2. A Material Model for Heart Tissue

In addition to modeling the hydrogel, a bio-device may be for instance implanted on the heart's surface, as will be modeled in a later example. A model for cardiac tissue is thus also required. The Ogden model [63,64] is herein employed. The Ogden model shows great applicability to nonlinear elastic materials, including soft tissue, and retains a simple form. The form of the free energy density for the Odgen model is given by

$$W = \frac{2\mu_{Og}}{\alpha^2}(\bar{\lambda}_1^{\alpha} + \bar{\lambda}_2^{\alpha} + \bar{\lambda}_3^{\alpha} - 3) + \frac{1}{2}K_{Og}(J-1)^2, \tag{5}$$

where $\bar{\lambda}_i(i = 1 \cdots 3)$ is the deviatoric principal stretch, which is defined as $\bar{\lambda}_i = J^{-1/3}\lambda_i$. μ_{Og} and K_{Og} are the shear modulus and the bulk modulus respectively, and α is a material constant. The first PK stress for the Odgen model can be computed from

$$P_{iJ} = \frac{\partial W}{\partial F_{iJ}}, \tag{6}$$

similar to our model for the hydrogels.

Three material models (e.g., Mooney–Rivlin, Yeoh, Ogden model) are often used to describe the mechanical response of heart tissue. The Mooney–Rivlin model is only suitable for small deformation and cannot describe well the mechanical behavior of heart tissue under multiaxial loadings, ignoring the effect of fiber orientation. The Yeoh model is a special form of reduced-polynomial hyperelastic model, which can describe well the mechanical behavior of tissue during tensile and shear deformation. However, too many material parameters are required. Conversely, the Ogden model is expressed by elongation in three principal directions, using few parameters, and can be effectively and efficiently used to model the mechanical behavior of heart tissue [65,66].

2.3. Toward Tuned Fracture Properties of Devices

As discussed in our introduction, and summarized in Figure 2a, the water content, the content and distribution of fillers (e.g., graphene, carbon nanotubes, carbon black, metallic nanowires and nanoparticles), and the stimuli (heat, light, magnetic fields, chemical agents, and pH) tune the mechanical and fracture properties of hydrogels. In terms of mechanical modeling, careful constitutive models for each hydrogel in each situation should be built, if we wish to model the hydrogel accurately. However, these are not easy to build. Many such material models are complicated by their attempted coverage of the underlying mechanisms of deformation [22–28]. This complication leads to a plurality of material parameters that are often difficult to calibrate. Such in-depth modeling is not amenable for detailed fracture modeling (itself complex), limiting its scope of use in device design. Indeed, we developed in earlier work a material model for hydrogels that considers the contained water and its interaction with the polymeric network. We used that in-depth model to simulate the fracture of a hydrogel block under compressive loading, and found that its predicted fracture pattern is less accurate than that obtained by the phenomenological model that we propose in the present paper, when compared against experiments [48,54,67].

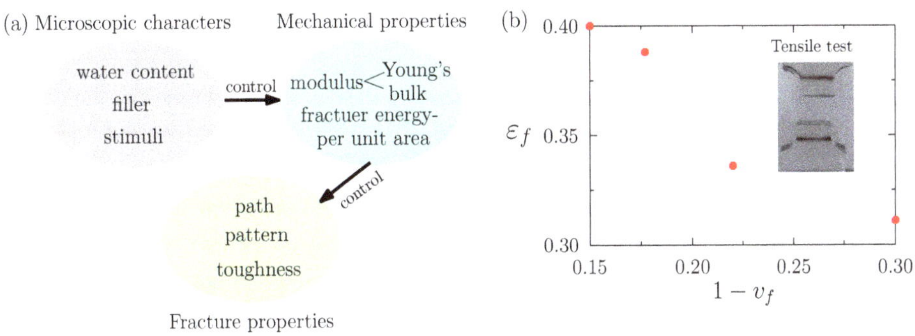

Figure 2. (**a**) Tuning the fracture property by microscopic characters such as water content, filler or stimuli for hydrogels. (**b**) Results of tensile fracture test for a brittle hydrogel (fracture strain vs. water volume fraction).

Our proposed material model herein is also easier for implementation and computation. In fact, water content, content and distribution of fillers, and environmental stimuli, can all be accounted for in the material's Young's modulus E, bulk modulus K, and fracture energy per unit area $\mathcal{G}_c$.

Here, we consider an example. We fabricated in our labs a poly-acrylamide hydrogel for tensile testing. The hydrogel, with different initial volume fractions of water (v_f), is prepared by first loading the glass container with an aqueous pre-gel solution and N,N′-methylene-bis-acrylamide. Gelation can be initiated by adding an ammonium persulfate aqueous solution as the initiator, and N,N,N′,N′-tetramethylenediamine as the catalyst, into the pre-gel solution. In our previous work, we have demonstrated the thermal stability and chemical stability of this fabricated gel, which we used to study experimentally the phenomena of surface morphology transition and fracture [48,68]. That is, we showed that gel structure and properties do not change significantly for a wide range of temperatures and under the action of its chemical environment. The fabrication method is reproducible, which are used in several our previous works [48,68]. In this paper, we do not try to show that our fabrication method is advantageous over other methods. We just take the fabricated gel as an example to demonstrate that the relevant mechanical parameters of the gel can be measured. With these measured parameters, it is possible for us to accurately simulate the fracture behavior of the hydrogel in biomedical applications and conveniently control the fracture, show in the numerical results section.

Via this method, three different hydrogels with different initial volume fractions of water are fabricated. Tensile testing until fracture is then carried out. The results are presented in Figure 2b. It can be seen that for an increasing initial volume fraction of water, the fracture strain is much larger. This finding implies that the fracture energy per unit area $\mathcal{G}_c$ increases with water content. This calibrated $\mathcal{G}_c$ can be therefore used to model the fracture process of a hydrogel device via our proposed computational framework. Finally, the water content can be altered when fabricating, to tune the fracture property of the device, based on the simulation results obtained. As such, if experiments are performed that vary the water content, the content and distribution of the filler, and the stimuli, one can with ease connect experimental data to the Young's modulus E, bulk modulus K, and/or the fracture energy per unit area $\mathcal{G}_c$ in our computational model for device design. There are also many ways to build this relationship through machine learning [69–71]. Therefore, our method can supplement and guide experiments to more easily tune the desired fracture properties of bio-devices.

The present work focused on the tunable fracture of bio-devices made of gels. Indeed, tunable hydrogels and their applications are a hot research area. A tunable hydrogel can be important in a variety of applications, including wound healing, hemostasis, drug delivery, cell encapsulation and release, 3D bioprinting, and tissue engineering. For example, Bratskaya et al. [72] discussed the applicability of a stimuli-responsive hydrogel of tunable dissolution rate under physiological conditions. Similarly, Danko et al. [73] prepared a polyelectrolyte hydrogel, the swelling and water state of which can be controlled by external stimulation. Their hydrogel shows a capability of free-floating on the water surface, indicating its potential use in floating pH detection devices. In the same vein, Jafari et al. [74] developed hydrogels of controllable gelation time and adjustable degradation rate. Their new hydrogel exhibited remarkable prospects for the reconstruction of artificial ovaries. However, most of these tunable gel properties are oriented toward the functionality of the hydrogels. At present, there are too few studies on tunable gel fracture in bio-devices, risking the reliability of these otherwise impressive advances.

3. Numerical Results

We apply a staggered scheme to solve the evolutionary equations of fracture, leveraging the finite-element method. In this scheme, at a given time step, the mechanical equilibrium equation is solved first to update the displacement. With the updated displacements, the phase field equation on all the nodes is then solved. Once both the displacement and the phase fields have converged, the computation will then proceed to the next time step. Thus, the original coupled problem is reduced to two smaller systems, which renders the algorithm more numerically stable, as per our numerical simulations.

To demonstrate the capabilities of our proposed approach for hydrogel device design in medical applications, the fracture modeling of a wearable biosensor, a membrane-coated drug, and a matrix for a cell culture, is carried out. We adopt nondimensional units in all our simulations.

3.1. A Wearable Biosensor

In vivo monitoring of heath is very important for many diseases. Biosensors with programmed diagnostic functions can measure, for instance, blood temperature, pressure, pH, glucose concentration, hormonal, and alcohol levels, and are becoming more and more popular. They can also can be used to detect human movement, or small toxic molecules [75–78]. Biosensors usually attach to specified positions in the human body, where important biomarkers (e.g., pressure, alcohol level) can be best collected. They can be attached externally (e.g., on the skin) or internally (e.g., on the surface of heart). Hydrogels have been taken as excellent candidates for such applications, specifically due to their biocompatibility, mechanical flexibility, and biodegradability. Such devices should be designed with good mechanical strength to resist fracture, and to maintain robust adhesion onto dynamic, curved surfaces (such as the skin, the heart, etc.).

Let us here consider the example of a biosensor for the heart, as shown in Figure 3 [79]. Note how the heart surface is curved and irregular (of varying curvature). The biosensor must be attached to this surface, which expands and contracts non-uniformly as the heart beats. How to ensure the integrity and the cohesive strength of the biosensor on this dynamic surface is a very important engineering design problem. An accurate evaluation of its integrity and bonding strength could help tune the functionality and longevity of the device, to minimize patient exposure to additional surgery.

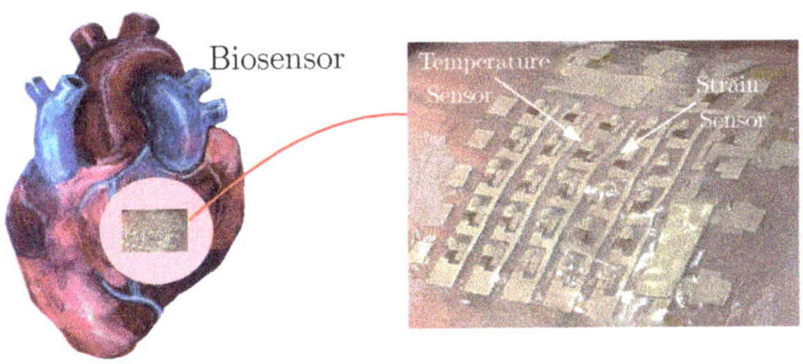

Figure 3. Schematic diagram of biosensor monitoring heart health [80].

To study fracture initiation and propagation in such biosensors, we establish two shell models, one for the heart, and another for the biosensor. Both the heart model and biosensor model are meshed, using a finite element mesher. The heart model contains 14,485 triangular shell elements, and 7594 nodes, with a finer mesh used in the area where it comes into contact with the biosensor. The mesh and coordinates are shown (see Figure 4a). An Odgen model is used to describe the mechanical response. Similarly, the biosensor contains 5242 triangular shell elements and 2732 nodes, and are attached to the heart model through a tie-binding contact technology [81]. Our phase field model is employed to describe the biosensor's mechanical behavior. The shear modulus and the Lamé constants of the biosensor model are taken as, $\mu = 1.0$ and $\chi/\mu = 2.17$, respectively. The diffuse length scale (for the crack width) in our phase field is set at $\ell = 1.0$, and the fracture energy density is varied between $\mathcal{G}_c/(\mu\ell) = 0.3 \times 10^{-4}$, 0.6×10^{-4} and 1.2×10^{-4}, respectively. The effective element size in the critical zone is $h_{ele}/\ell \approx 0.1$, which is less than ℓ. The elastic parameters for the Odgen material model are $\mu_{Og}/\mu = 0.1$, $K_{Og}/\mu = 1.0$, and $\alpha = -20.0$ [82]. The displacement along the radial direction is imposed at the wall of the large blood vessel, and the bottom part (apex to midline, as marked by black points) is fixed, as shown in Figure 4b. This boundary condition aims to only generate physiologically consistent maximum stretches in the atrium's tissue beneath the biosensor, in a given cardiac cycle.

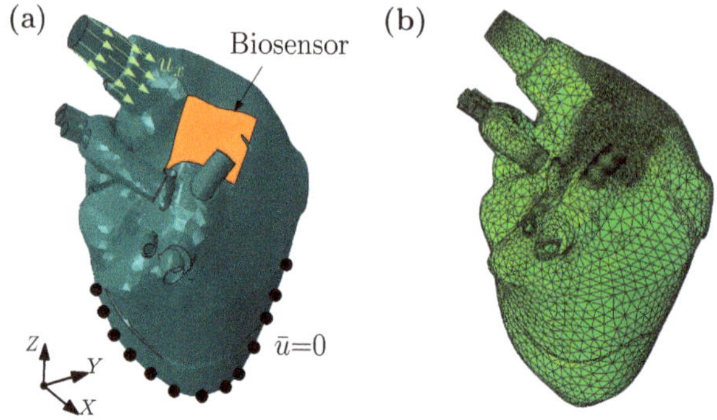

Figure 4. Fracture of a biosensor attached to the heart. (**a**) Geometric dimensions and boundary conditions of a heart model. (**b**) Mesh model of heart.

Figure 5 plots the contours of the effective Cauchy stress and the phase fields for $\mathcal{G}_c/(\mu\ell) = 0.3 \times 10^{-4}$, as generated under three levels of imposed displacement u_x/l. For an imposed displacement u_x/l = 0.728, the fracture phase field is essentially undeveloped (see Figure 5b). When u_x/l reaches 0.829, the fracture phase field begins to develop at the left corner of the biosensor. With increasing displacement, the crack propagates until full fracture of the biosensor. Figure 6 plots the contours of the phase field variable for $\mathcal{G}_c/(\mu\ell) = 0.3 \times 10^{-4}$, 0.6×10^{-4} and 1.2×10^{-4}, respectively. From the Figure 6, it can be seen that due to the different values of $\mathcal{G}_c/(\mu\ell)$, the damage modes in the biosensor's shell are different for the same displacement. With an increase to $\mathcal{G}_c$, the imposed displacement required for crack initiation, and for complete fracture, will also increase. When designing these sensors, an appropriate $\mathcal{G}_c$ may be obtained by tuning the water content or the volume fraction of the filler, which can prevent the biosensor from the premature damage during the operational life of the biosensor.

In summary, wearable flexible sensors [83–85] for human health monitoring [86–90] have shown new advances. These flexible sensors can better monitor subtle changes in the external environment, such as temperature [91–93], humidity [91,94], and deformation [95,96], and are evolving rather fast. For these kinds of sensors, various types of hydrogels are being actively developed, such as cellulose hydrogels [97], PNIPAM hydrogels [98], and double chain hydrogels [99]. These hydrogels are sensitive to mechanical signals and can be used as dynamic soft sensors that accurately detect small external forces and strains. However, it should be noted that there are too few computational studies to ensure the good mechanical strength of the resulting hydrogel devices, their ability to resist fracture, or their robust adhesion onto dynamic, deformable, curved surfaces (such as the skin, the heart, etc.). Our approach herein aims to serve as a first sure step in filling this knowledge gap.

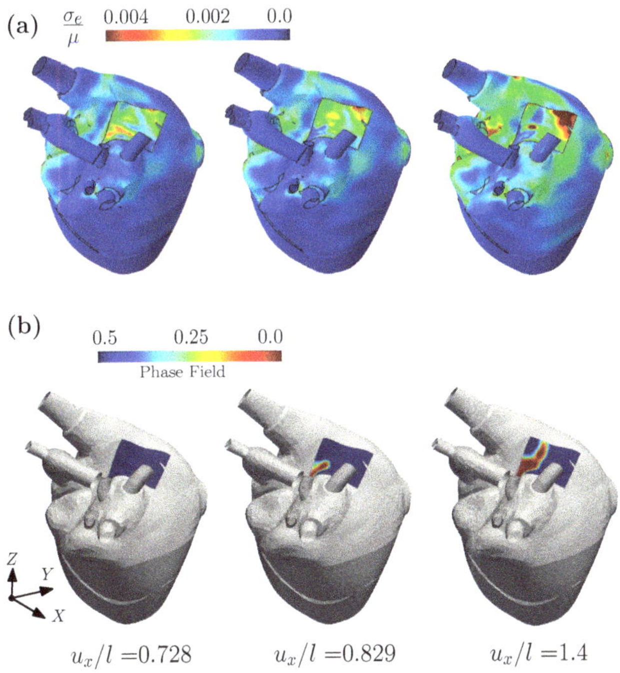

Figure 5. (**a**) The contour plots of normalized effective Cauchy stress and (**b**) the contour plots of phase field for a heart model and a biosensor shell, under different displacement.

3.2. Outer Membrane of Drug-Delivery Capsule

Hydrogels have received a great deal of research attention in the field of drug delivery. Indeed, there is much literature and many patents on the application of hydrogels to drug delivery [4,100,101]. Hydrogels usually have a high porosity, which can be controlled by the cross-linkers in the polymeric network, and its affinity to water. The porous structure allows the drug to be first loaded, then gradually released at a controlled rate, for sustained periods. The release of the drug can in fact be controlled rather flexibly by tuning the diffusion, swelling, chemical potential, or the environment (such as PH).

A diffusion-controlled drug release system can be represented by a reservoir delivery system, which is usually composed of a drug-containing core, coated with a hydrogel membrane, as with commonly available capsules (cylinders, slabs, or spheres). When this reservoir system is transported in the blood vessel, it is subjected to interaction with a dynamic fluid (blood), with compression and shear acting on the capsule's membrane, as shown in Figure 7. Although the porous hydrogel brings advantages to drug release, nevertheless, the same voids induce a lower mechanical strength, which may undermine the capsule's integrity and sustained functionality.

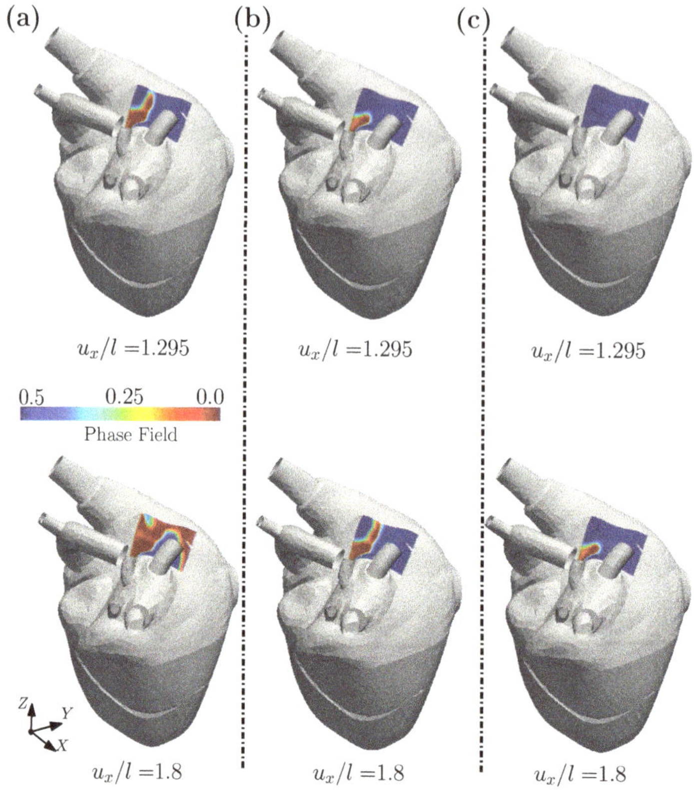

Figure 6. The contour plots of phase field for a biosensor shell with different $\mathcal{G}_c/(\mu\ell)$ = (**a**) 0.3×10^{-4}; (**b**) 0.6×10^{-4}; (**c**) 1.2×10^{-4}.

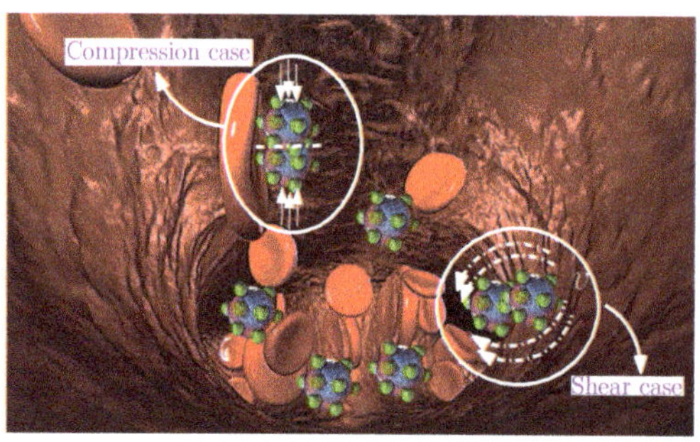

Figure 7. Schematic diagram of drug delivery in blood vessels [102].

We here examine the fracture of a spherical drug-delivery capsule. A spherical shell model is built with a radius $R = 1.0$, which contains multiple holes of radius $r/R = 0.05$ that facilitate the release of drugs. The holes are uniformly distributed on the surface of the sphere, as shown in Figure 8a. In this study, we consider two external loading conditions, as shown in Figure 8b,c respectively. The two loadings approximately consider the compression and shear forces acting on the capsule's membrane from the dynamic fluids in which the capsule is immersed. Considering the symmetry of the spherical shell and the imposed loading, only a quarter and an eighth of the full sphere are here analyzed, for the shear and compression-loading scenarios respectively, as noted in Figure 8b,c. The shear modulus and the Lamé constants of the membrane are $\mu = 1.0$ and $\chi/\mu = 2.17$, respectively. For the compression case, the model contains 11,108 triangular shell elements and 5700 nodes. For the shear case, the model contains 12,456 triangular shell elements and 6428 nodes. The two cases have the same shell radius, shell thickness, length scale ℓ, and fracture energy per unit area $\mathcal{G}_c$. The length scale for the phase field model is set at $\ell/R = 0.06$, and the fracture energy per unit area is taken as $\mathcal{G}_c/(\mu R) = 0.3 \times 10^{-3}$. The effective element size in the critical zone is less than ℓ. Figure 8d shows the normalized resultant reaction force vs. the imposed displacement for two cases, where brown and purple represent the compression case and the shear case respectively. It can be seen that the resultant force increases linearly at the beginning, followed by a softening phase, then suddenly dropping to a very low value with fracture.

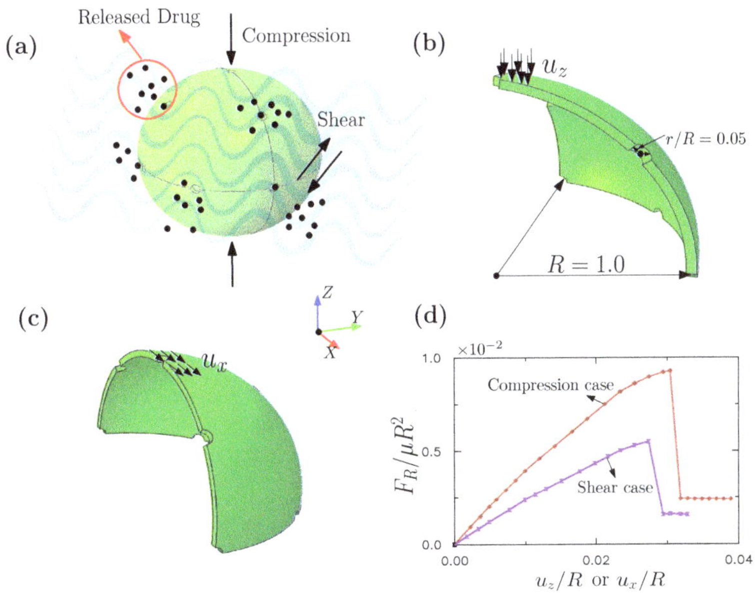

Figure 8. Fracture of a spherical shell subjected to compression and shear. (**a**) Spherical shell model for drug delivery. (**b**) One eighth spherical shell model. (**c**) A quarter spherical shell model. Geometric dimensions and boundary conditions of the model are given. (**d**) The reaction force vs. displacement curves.

Figure 9 shows the contours of effective stress and phase field, at different levels of imposed displacement for the compression case. The left column of the figure plots contours of the effective stress, while the right column of the figure shows the contours of the fracture phase field. At the imposed displacement of $u_z/R = 0.031$, the spherical shell is deformed, but the fracture phase field has not yet developed. At an imposed displacement $u_z/R = 0.032$, it can be seen that the fracture phase field begins to develop around hole A (marked in the figure) and rapidly expands along the Z direction (see Figure 9b). It should be noted that due to the symmetry of the shell, there is also a corresponding fracture phase field at a symmetrical position on the sphere (not shown in the figure). Similar to Figure 9, Figure 10 shows the contours of effective stress and fracture phase field for the case of shear. Different from the compression case, when the fracture phase field appears around point A it extends along a direction towards point B and point C (point A, B and C are marked in Figure 10b). Finally, noting that the velocity of blood varies with the different blood vessel diameters in the body, we can expect different loading conditions and force distributions on the capsule's surface (combinations between compression and shear). Our simulation thus shows that using hydrogels for the membrane of a capsules should be carefully designed to resist fracture under different external loading conditions, and may differ depending on the target organ and the hemodynamic pathway followed.

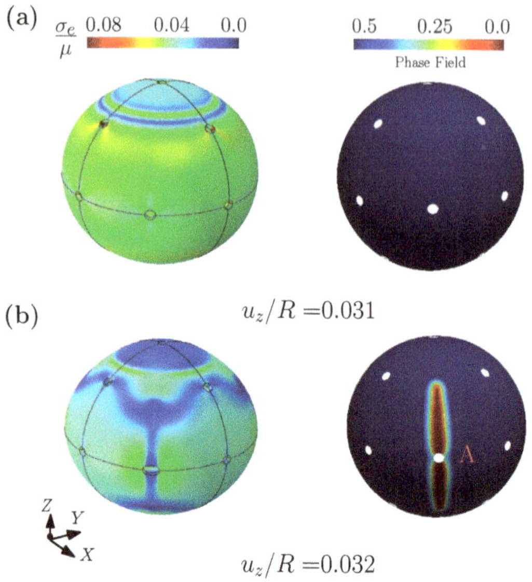

Figure 9. The contour plots of normalized effective Cauchy stress and phase field for spherical shell considering compression, under different displacement u_z/R =: (**a**) 0.031; (**b**) 0.032.

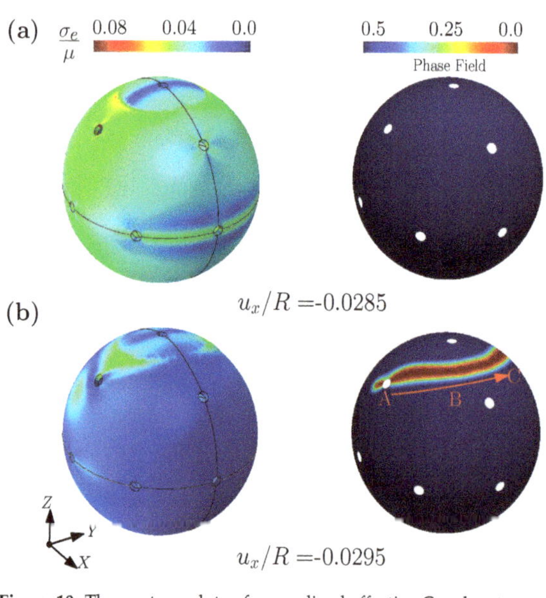

Figure 10. The contour plots of normalized effective Cauchy stress and phase field for spherical shell considering shear, under different displacement u_x/R =: (**a**) −0.0285; (**b**) −0.0295.

It can be summarized that hydrogels have broad application prospects for controllable drug delivery, by virtue of their unique programmability, designable responsiveness, biocompatibility, and biodegradability. These desirable properties can offset prevalent shortcomings in traditional drug-delivery materials, such as their high invasiveness, low efficiency, and intermittent local delivery [103,104]. To further promote the application of hydrogels in drug delivery, DNA hydrogels [105], and smart hydrogels [106] have also been recently developed. However, hydrogels are prone to fracture when interacting with the dynamic blood flow during the process of drug delivery [29,107,108], decreasing their success probability at the target. Our proposed method can capture this fracturing process during drug delivery so that the hydrogel device can be designed for improved success rates.

3.3. A Fibered Matrix for a Cell Culture

Hydrogel usage as scaffolds requires it to combine with biological cells to generate living tissue, both in vitro and in vivo. Such construction of three-dimensional regenerative tissue further requires that we create an environment that enables the biological cells to grow, and to interact with their surroundings. The porous microstructure of the hydrogels is critical for such cell growth. The matrix for a cell culture is often designed in the form of thin films composed of nanofibers or nanosheets. Again, the voids on these films often result in lowered mechanical strength, which can compromise scaffold functionality.

The fracture of porous sheets is here examined. A sheet can be modeled as a shell. A fibered topology is here taken as an example, as shown in Figure 11a. It can be fabricated by tetrazine-norbornene chemistry [109]. A finite-element model for this fibered shell structure has been here developed. The size and boundary conditions of the model are shown in Figure 11b. The length l, width h, and thickness t of the model are 4.0, 1.0 and 0.1, respectively. The mesh includes 10,330 triangular shell elements and 6032 nodes. The left and right edges of the model are fixed, and a displacement along the z direction is imposed at the middle of the model, which mimics the weight of a cell on the hydrogel. The shear modulus and Lamé constants of the fibered structure are taken as $\mu = 1.0$ and $\chi/\mu = 2.17$, respectively. The length scale is set at $\ell/h = 0.06$, and the fracture energy per unit area is varied between $\mathcal{G}_c/(\mu h) = 0.1 \times 10^{-3}$, 0.2×10^{-3} and 0.3×10^{-3}, respectively. The effective element size in the critical zone is less than ℓ.

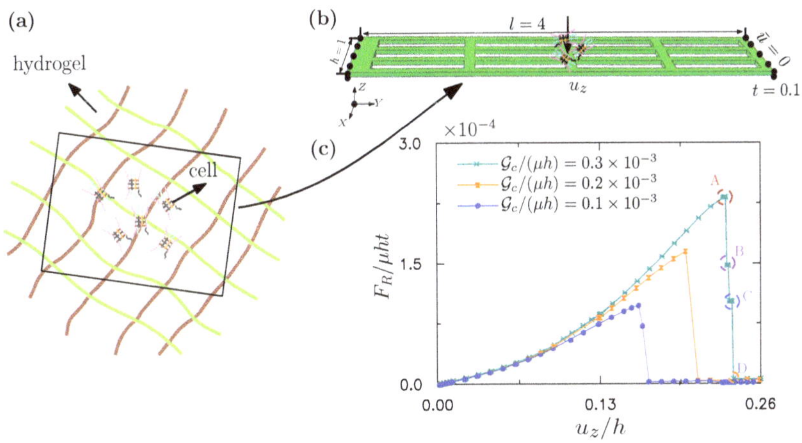

Figure 11. Fracture of hydrogel matrix for cell culture. (**a**) Schematics of cell culture on the hydrogel matrix. (**b**) Geometric dimensions and boundary conditions of the simplified model. (**c**) The normalized reaction force vs. the normalized displacement.

Figure 11c shows the reaction force F_R along the z direction as a function of displacement u_z. It can be seen from Figure 11c that the reaction force first increases, then suddenly drops. We here mark four points on the unloading curve of the fracture process. Point A indicates that the shell structure has just commenced to fracture, point B and C indicate increasing damage to the shell, prior to final fracture. The reaction force then drops to its lowest value (point D) when full damage to the shell structure has occurred.

Figure 12 plots the contours of effective stress and fracture phase fields for $\mathcal{G}_c/(\mu h) = 0.3 \times 10^{-3}$, under different levels of imposed displacement, viz. $u_z/h = 0.232$, 0.233, 0.235, and 0.238. For an imposed displacement of $u_z/h = 0.232$, the shell structure has undergone large deformation, but the fracture phase field is yet undeveloped (see Figure 12a). When the applied displacement reaches $u_z/h = 0.0233$, the fracture phase field begins to develop, which is consistent with the concentration of the effective stress, as shown in Figure 12b. With an increase to the imposed displacement, the number of the locations where the fracture initiates increases gradually, until final fracture has occurred. In addition, Figure 13 plots the contour of the fracture phase fields for three different levels of $\mathcal{G}_c/(\mu h) = 0.1 \times 10^{-3}$, 0.2×10^{-3} and 0.3×10^{-3}, under the same displacement. It can be seen from Figure 13 that due to the different values of $\mathcal{G}_c/(\mu h)$, the fracture modes of the fibered structure are different under the same level of displacement. With an increase to $\mathcal{G}_c$, fracture occurs only in the two fibers at the center. This example clearly shows that the fracture pattern can be controlled by $\mathcal{G}_c$, which can again be achieved by tuning the hydrogel's water content, etc.

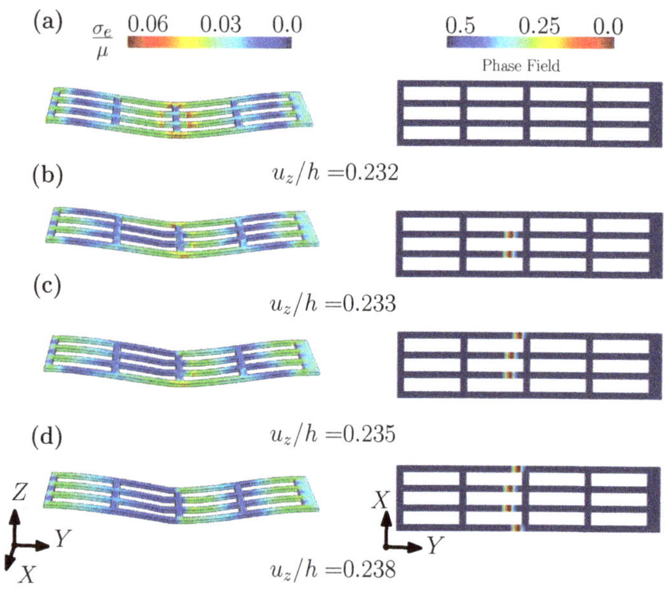

Figure 12. The contour plots of normalized effective Cauchy stress and phase field under different displacement u_z/h =: (**a**) 0.232; (**b**) 0.233; (**c**) 0.235; (**d**) 0.238.

The use of hydrogels as scaffolds has made further progress recently. For example, Liu et al. [110] prepared a thermosensitive hydrogel and used it for in situ embedding and three-dimensional culture of MCF-7 breast cancer cells. It was shown that this thermosensitive hydrogel can be used as a three-dimensional platform with high biocompatibility and easy functionalization, and has great potential in the biomedical field. Similarly, Jiao et al. [111] prepared sodium alginate-based composite hydrogels with good cell compatibility, which can be used for three-dimensional cell culture in vitro. Moreover, Nguyen et al. [112] developed a co-culture system of ECM degradation products and MSCs with macrophages. Bone marrow mesenchymal stem cells cultured on this hydrogel show immunomodulatory properties and can be used in regenerative medicine. It was shown that the hydrogel cell culture offers great prospects for the development of modern medicine. We deem that the numerical simulations of this section for hydrogel scaffolds indicate that our proposed method can be used to investigate the fracture (durability) of hydrogel scaffolds in new similar applications.

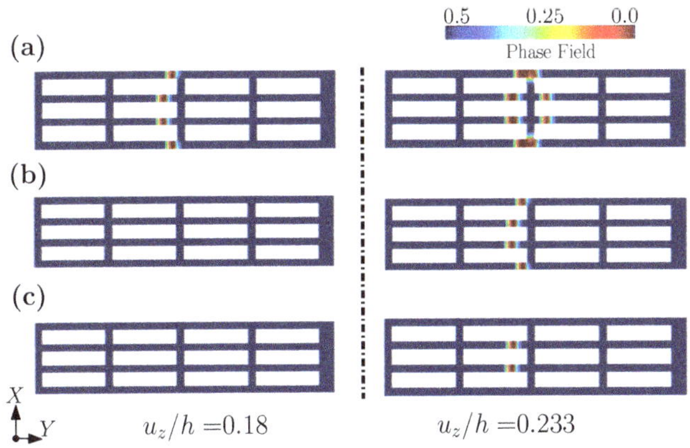

Figure 13. The contour plots of phase field with different $\mathcal{G}_c/(\mu h)$ =: (**a**) 0.1×10^{-3}; (**b**) 0.2×10^{-3}; (**c**) 0.3×10^{-3}.

4. Concluding Remarks

In recent years, research into hydrogels has made significant breakthroughs (e.g., self-healing hydrogels [88–90,113], smart hydrogels [98], cellulose hydrogels [97], double chain hydrogels [114], supramolecular hydrogels [115], etc.). However, most of the research focuses on hydrogel functionality, as for flexible wearable sensors, drug delivery, and medical engineering. The mechanical strength of most hydrogels today, however, is still far from optimal, being easy to damage (e.g., by accidental fracture, or premature degradation), as a result of undesirable or uncontrolled growth to their micro- or macrocracks [116]. Hydrogels have various other shortcomings to be overcome by researchers, such as their easy loss of sensor signal, poor antifreeze performance at low temperature, foreign body reaction, and undesired performance losses. At present, most functional hydrogels remain in the experimental stage and have not been put into use due to their immature production technologies. Much research work is thus still needed to promote functional hydrogels in our daily lives, realize their commercialization, and translate their fullest application potential. Mechanical modeling as we propose herein can serve an important role in this pathway of their development, in particular for bio-devices.

Hydrogels, with easily tuning properties, have been widely explored as prime candidates for modern medical devices, such as biosensors, drug delivery systems, and tissue engineering applications. Because it is today challenging to model the fracture of hydrogels, there are today few computational design tools that can investigate and control the fracture behavior of their medical devices, for instance fracture initiation, fracture path and propagation pattern, and resulting toughness. In this paper, a modeling framework is laid to examine the fracture of such devices across various biomedical applications, while leveraging simple phenomenological constitutive formulations. In our approach, the devices are treated as shell structures. Then, a phase field approach is adopted to model fracture initiation and propagation in those devices, while leveraging finite-element software. By harnessing discrete differential geometry, C^0 element in the finite-element simulations can be effectively used to obtain accurate results. Moreover, as the devices are treated as shells, computational cost is greatly reduced.

Three examples of fracture modeling have been here presented; a wearable biosensor, a membrane-coated drug, and a fibered matrix for a cell culture, all composed of hydrogels. In all these examples, by tuning the fracture energy per unit area $\mathcal{G}_c$, we can control fracture initiation location, and the pattern of cracking. As we showed with a simple experiment in our lab, $\mathcal{G}_c$ is strongly related to water content, as well as to fillers and environmental stimuli, offering a clear avenue to tune hydrogel devices computationally. This connection of medical device performance to water content, filler choice, or stimuli should thus be carried out in future research. We also plan to collaborate with more experts on experiments that can help us solve these challenging design issues.

Author Contributions: G.Z. and H.Q. envisaged, devised and implemented the numerical computations; K.I.E. provided some valuable advice and revised this paper; S.T. and D.P. guided the work and wrote the paper together. All authors have read and agreed to the published version of the manuscript.

Funding: This research was funded by the National Natural Science Foundation of China (grant number 12002246, 11872139) and 2020 annual Open Fund of Failure Mechanics & Engineering Disaster Prevention and Mitigation, Key Laboratory of Sichuan Province (Sichuan University) (2020JDS0022), funded By Open Research Fund Program of Hubei Provincial Key Laboratory of Chemical Equipment Intensification and Intrinsic Safety (2019KA03).

Institutional Review Board Statement: Not applicable.

Informed Consent Statement: Not applicable.

Data Availability Statement: Not applicable.

Acknowledgments: The financial support from the National Natural Science Foundation of China (grant number 12002246, 11872139). 2020 annual Open Fund of Failure Mechanics & Engineering Disaster Prevention and Mitigation, Key Laboratory of Sichuan Province (Sichuan University) (2020JDS0022), funded By Open Research Fund Program of Hubei Provincial Key Laboratory of Chemical Equipment Intensification and Intrinsic Safety (2019KA03).

Conflicts of Interest: The authors declare no conflict of interest.

References

1. Awasthi, S. A Review on Hydrogels and Ferrogels for Biomedical Applications. *JOM* **2021**, *73*, 2440–2451. [CrossRef]
2. Tavakoli, J.; Tang, Y. Hydrogel Based Sensors for Biomedical Applications: An Updated Review. *Polymers* **2017**, *9*, 364. [CrossRef] [PubMed]
3. Gibbs, D.M.R.; Black, C.R.M.; Dawson, J.I.; Oreffo, R.O.C. A review of hydrogel use in fracture healing and bone regeneration. *J. Tissue Eng. Regen. Med.* **2016**, *10*, 187–198. [CrossRef] [PubMed]
4. Calo, E.; Khutoryanskiy, V.V. Biomedical applications of hydrogels: A review of patents and commercial products. *Eur. Polym. J.* **2015**, *65*, 252–267. [CrossRef]
5. Chai, Q.; Jiao, Y.; Yu, X. Hydrogels for Biomedical Applications: Their Characteristics and the Mechanisms behind Them. *Gels* **2017**, *3*, 6. [CrossRef]
6. Ghorbanizamani, F.; Moulahoum, H.; Guler Celik, E.; Timur, S. Ionic liquids enhancement of hydrogels and impact on biosensing applications. *J. Mol. Liq.* **2022**, *357*, 119075. [CrossRef]
7. Tenje, M.; Cantoni, F.; Hernández, A.M.P.; Searle, S.S.; Johansson, S.; Barbe, L.; Antfolk, M.; Pohlit, H. A practical guide to microfabrication and patterning of hydrogels for biomimetic cell culture scaffolds. *Organs-on-a-Chip* **2020**, *2*, 100003. [CrossRef]
8. Zhao, J.; Wang, G.; Yang, R.; Lu, X.; Cheng, M.; He, C.; Xie, G.; Meng, J.; Shi, D.X.; Zhang, G. Tunable Piezoresistivity of Nanographene Films for Strain Sensing. *ACS Nano* **2015**, *9*, 1622–1629. [CrossRef]
9. Zhou, J.; Yu, H.; Xu, X.; Han, F.; Lubineau, G. Ultrasensitive, Stretchable Strain Sensors Based on Fragmented Carbon Nanotube Papers. *ACS Appl. Mater. Interfaces* **2017**, *9*, 4835–4842. [CrossRef]
10. Lim, G.H.; Lee, N.E.; Lim, B. Highly sensitive, tunable, and durable gold nanosheet strain sensors for human motion detection. *J. Mater. Chem. C* **2017**, *4*, 5642–5647. [CrossRef]
11. Yang, Z.; Wang, D.Y.; Pang, Y.; Li, Y.X.; Wang, Q.; Zhang, T.Y.; Wang, J.B.; Liu, X.; Yang, Y.Y.; Jian, J.M.; et al. Simultaneously Detecting Subtle and Intensive Human Motions Based on a Silver Nanoparticles Bridged Graphene Strain Sensor. *Acs Appl. Mater. Interfaces* **2018**, *10*, 3948–3954. [CrossRef] [PubMed]
12. Sorber, J.; Steiner, G.; Schulz, V.; Guenther, M.; Gerlach, G.; Salzer, R.; Arndt, K.F. Hydrogel-based piezoresistive pH sensors: Investigations using FT-IR attenuated total reflection spectroscopic imaging. *Anal. Chem.* **2008**, *80*, 2957–2962. [CrossRef] [PubMed]
13. Richter, A.; Bund, A.; Keller, M.; Arndt, K.F. Characterization of a microgravimetric sensor based on pH sensitive hydrogels. *Sens. Actuators B Chem.* **2004**, *99*, 579–585. [CrossRef]
14. Milo, S.; Thet, N.T.; Liu, D.; Nzakizwanayo, J.; Jones, B.V.; Jenkins, A.T.A. An in-situ infection detection sensor coating for urinary catheters. *Biosens. Bioelectron.* **2016**, *81*, 166–172. [CrossRef] [PubMed]

15. Wischerhoff, E.; Zacher, T.; Laschewsky, A.; Rekaï, E.D. Direct Observation of the Lower Critical Solution Temperature of Surface-Attached Thermo-Responsive Hydrogels by Surface Plasmon Resonance. *Angew. Chem. Int. Ed.* **2000**, *39*, 4602–4604. [CrossRef]
16. Lagus, T.P.; Edd, J.F. A review of the theory, methods and recent applications of high-throughput single-cell droplet microfluidics. *J. Phys. D Appl. Phys.* **2013**, *46*, 114005. [CrossRef]
17. Zeng, S.; Zhang, D.; Sun, L. Bio-inspired sensitive and reversible mechanochromisms via strain-dependent cracks and folds. *Nat. Commun.* **2016**, *7*, 11802. [CrossRef]
18. Johner. Wing. Available online: https://m.quanjing.com/imgbuy/qj9103024971.html (accessed on 8 June 2022).
19. Wu, X. T3 Terminal of Capital Airport. Available online: https://m.quanjing.com/imgbuy/QJ6325712014.html (accessed on 15 June 2022).
20. Reichenspurner, H.; Schaefer, D.; Linke, A.; Spence, M.; Sondergaard, L.; LeBreton, H.; Schymik, G.; Abdel-Wahab, M.; Leipsic, J.; Walters, D.; et al. Self-Expanding Transcatheter Aortic Valve System for Symptomatic High-Risk Patients With Severe Aortic Stenosis. *J. Am. Coll. Cardiol.* **2017**, *70*, 3127–3136. [CrossRef]
21. Han, M.; Chen, L.; Aras, K.; Liang, C.; Chen, X.; Zhao, H.; Li, K.; Ndeye Rokhaya, F.; Sun, B.; Kim, J.H.; et al. Catheter-integrated soft multilayer electronic arrays for multiplexed sensing and actuation during cardiac surgery. *Nat. Biomed. Eng.* **2020**, *4*, 1–13. [CrossRef]
22. Mao, Y.; Lin, S.; Zhao, X.; Anand, L. A large deformation viscoelastic model for double-network hydrogels. *J. Mech. Phys. Solids* **2017**, *100*, 103–130. [CrossRef]
23. Long, R.; Mayumi, K.; Creton, C.; Narita, T.; Hui, C.Y. Time Dependent Behavior of a Dual Cross-Link Self-Healing Gel: Theory and Experiments. *Macromolecules* **2014**, *47*, 7243–7250. [CrossRef]
24. Long, R.; Hui, C.Y. Fracture toughness of hydrogels: Measurement and interpretation. *Soft Matter* **2016**, *12*, 8069–8086. [CrossRef] [PubMed]
25. Guo, J.; Liu, M.; Zehnder, A.T.; Zhao, J.; Narita, T.; Creton, C.; Hui, C.Y. Fracture mechanics of a self-healing hydrogel with covalent and physical crosslinks: A numerical study. *J. Mech. Phys. Solids* **2018**, *120*, 79–95. [CrossRef]
26. Liu, M.; Guo, J.; Hui, C.Y.; Creton, C.; Narita, T.; Zehnder, A. Time-temperature equivalence in a PVA dual cross-link self-healing hydrogel. *J. Rheol.* **2018**, *62*, 991–1000. [CrossRef]
27. Liu, M.; Guo, J.; Hui, C.Y.; Zehnder, A. Crack tip stress based kinetic fracture model of a PVA dual-crosslink hydrogel. *Extrem. Mech. Lett.* **2019**, *29*, 100457. [CrossRef]
28. Shen, T.; Vernerey, F.J. Rate-dependent fracture of transient networks. *J. Mech. Phys. Solids* **2020**, *143*, 104028. [CrossRef]
29. Liu, D.; Ma, S.; Yuan, H.; Markert, B. Modelling and simulation of coupled fluid transport and time-dependent fracture in fibre-reinforced hydrogel composites. *Comput. Methods Appl. Mech. Eng.* **2022**, *390*, 114470. [CrossRef]
30. Liu, D.; Ma, S.; Yuan, H.; Markert, B. Computational modelling of poro-visco-hyperelastic effects on time-dependent fatigue crack growth of hydrogels. *Int. J. Plast.* **2022**, *155*, 103307. [CrossRef]
31. Wang, X.; Hong, W. Delayed fracture in gels. *Soft Matter* **2012**, *8*, 8171–8178. [CrossRef]
32. Belytschko, T.; Black, T. Elastic crack growth in finite elements with minimal remeshing. *Int. J. Numer. Methods Eng.* **1999**, *45*, 601–620. [CrossRef]
33. Dolbow, J.; Moës, N.; Belytschko, T. Modeling fracture in Mindlin-Reissner plates with the eXtended finite element method. *Int. J. Solids Struct.* **2000**, *37*, 7161–7183. [CrossRef]
34. Areias, P.; Belytschko, T. Non-linear analysis of shells with arbitrary evolving cracks using XFEM. *J. Mech. Phys. Solids* **2005**, *62*, 384–415. [CrossRef]
35. Moës, N.; Dolbow, J.; Belytschko, T. A finite element method for crack growth without remeshing. *Int. J. Numer. Methods Eng.* **1999**, *46*, 131–150. [CrossRef]
36. Grinspun, E. A discrete model of thin shells. In Proceedings of the ACM SIGGRAPH 2005 Courses, Los Angeles, CA, USA, 31 July–4 August 2005; pp. 14–19.
37. Desbrun, M.; Kanso, E.; Tong, Y. Discrete differential forms for computational modeling. In *Discrete Differential Geometry*; Springer: Berlin/Heidelberg, Germany, 2008; pp. 287–324.
38. Bronstein, M.M.; Bruna, J.; LeCun, Y.; Szlam, A.; Vandergheynst, P. Geometric deep learning: Going beyond euclidean data. *IEEE Signal Process. Mag.* **2017**, *34*, 18–42. [CrossRef]
39. Yang, H.; Guo, X.; Tang, S.; Liu, W. Derivation of heterogeneous material laws via data-driven principal component expansions. *Comput. Mech.* **2019**, *64*, 365–379. [CrossRef]
40. Tang, S.; Zhang, G.; Yang, H.; Li, Y.; Liu, W.K.; Guo, X. MAP123: A data-driven approach to use 1D data for 3D nonlinear elastic materials modeling. *Comput. Methods Appl. Mech. Eng.* **2019**, *357*, 112587. [CrossRef]
41. Kirchdoerfer, T.; Ortiz, M. Data-driven computational mechanics. *Comput. Methods Appl. Mech. Eng.* **2016**, *304*, 81–101. [CrossRef]
42. Kirchdoerfer, T.; Ortiz, M. Data Driven Computing with Noisy Material Data Sets. *Comput. Methods Appl. Mech. Eng.* **2017**, *326*, 622–641. [CrossRef]
43. Eggersmann, R.; Kirchdoerfer, T.; Reese, S.; Stainier, L.; Ortiz, M. Model-Free Data-Driven Inelasticity. *Comput. Methods Appl. Mech. Eng.* **2018**, *350*, 81–99. [CrossRef]
44. Abueidda, D.W.; Koric, S.; Sobh, N.A.; Sehitoglu, H. Deep learning for plasticity and thermo-viscoplasticity. *Int. J. Plast.* **2021**, *136*, 102852. [CrossRef]

45. Logarzo, H.J.; Capuano, G.; Rimoli, J.J. Smart constitutive laws: Inelastic homogenization through machine learning. *Comput. Methods Appl. Mech. Eng.* **2021**, *373*, 113482. [CrossRef]
46. Liu, Z.; Bessa, M.A.; Liu, W.K. Self-consistent clustering analysis: An efficient multi-scale scheme for inelastic heterogeneous materials. *Comput. Methods Appl. Mech. Eng.* **2016**, *306*, 319–341. [CrossRef]
47. Liu, Z.; Moore, J.A.; Liu, W.K. An extended micromechanics method for probing interphase properties in polymer nanocomposites. *J. Mech. Phys. Solids* **2016**, *95*, 663–680. [CrossRef]
48. Zhang, G.; Guo, T.F.; Zhou, Z.; Tang, S.; Guo, X. A phase-field model for fracture in water-containing soft solids. *Eng. Fract. Mech.* **2019**, *212*, 180–196. [CrossRef]
49. Tvergaard, V.; Hutchinson, J.W. The relation between crack growth resistance and fracture process parameters in elastic-plastic solids. *J. Mech. Phys. Solids* **1992**, *40*, 1377–1397. [CrossRef]
50. Bourdin, B.; Francfort, G.A.; Marigo, J.J. Numerical experiments in revisited brittle fracture. *J. Mech. Phys. Solids* **2000**, *48*, 797–826. [CrossRef]
51. Amor, H.; Marigo, J.J.; Maurini, C. Regularized formulation of the variational brittle fracture with unilateral contact: Numerical experiments. *J. Mech. Phys. Solids* **2009**, *57*, 1209–1229. [CrossRef]
52. Miehe, C.; Hofacker, M.; Welschinger, F. A phase field model for rate-independent crack propagation: Robust algorithmic implementation based on operator splits. *Comput. Methods Appl. Mech. Eng.* **2010**, *199*, 2765–2778. [CrossRef]
53. Borden, M.J.; Verhoosel, C.V.; Scott, M.A.; Hughes, T.J.; Landis, C.M. A phase-field description of dynamic brittle fracture. *Comput. Methods Appl. Mech. Eng.* **2012**, *217*, 77–95. [CrossRef]
54. Zhang, G.; Guo, T.F.; Elkhodary, K.I.; Tang, S.; Guo, X. Mixed Graph-FEM phase field modeling of fracture in plates and shells with nonlinearly elastic solids. *Comput. Methods Appl. Mech. Eng.* **2022**, *389*, 114282. [CrossRef]
55. Lin, J.; Zheng, S.Y.; Xiao, R.; Yin, J.; Wu, Z.L.; Zheng, Q.; Qian, J. Constitutive behaviors of tough physical hydrogels with dynamic metal-coordinated bonds. *J. Mech. Phys. Solids* **2020**, *139*, 103935. [CrossRef]
56. Krishnan, V.; Hui, C.Y.; Long, R. Finite Strain Crack Tip Fields in Soft Incompressible Elastic Solids. *Langmuir ACS J. Surfaces Colloids* **2008**, *24*, 14245–14253. [CrossRef] [PubMed]
57. Faghihi, S.; Karimi, A.; Jamadi, M.; Imani, R.; Salarian, R. Graphene oxide/poly(acrylic acid)/gelatin nanocomposite hydrogel: Experimental and numerical validation of hyperelastic model. *Mater. Sci. Eng. C* **2014**, *38*, 299–305. [CrossRef]
58. Pillai, U.; Triantafyllou, S.P.; Ashcroft, I.; Essa, Y.; de la Escalera, F.M. Phase-field modelling of brittle fracture in thin shell elements based on the MITC4+ approach. *Comput. Mech.* **2020**, *65*, 1413–1432. [CrossRef]
59. Needleman, A. A Continuum Model for Void Nucleation by Inclusion Debonding. *J. Appl. Mech.* **1987**, *54*, 525. [CrossRef]
60. Francfort, G.A.; Marigo, J.J. Revisiting brittle fracture as an energy minimization problem. *J. Mech. Phys. Solids* **1998**, *46*, 1319–1342. [CrossRef]
61. Bourdin, B.; Francfort, G.A.; Marigo, J.J. The variational approach to fracture. *J. Elast.* **2008**, *91*, 5–148. [CrossRef]
62. Wardetzky, M. Convergence of the cotangent formula: An overview. *Discret. Differ. Geom.* **2008**, *38*, 275–286.
63. Trabelsi, K. Nonlinear thin plate models for a family of Ogden materials. *Comptes Rendus Math.* **2003**, *337*, 819–824. [CrossRef]
64. Moerman, K.M.; Simms, C.K.; Nagel, T. Control of Tension-Compression Asymmetry in Ogden Hyperelasticity with Application to Soft Tissue Modelling. *J. Mech. Behav. Biomed. Mater.* **2016**, *56*, 218–228. [CrossRef]
65. Yu, D.; Du, D.; Yang, H.; Tu, Y. Parallel computing simulation of electrical excitation and conduction in the 3D human heart. In Proceedings of the 2014 36th Annual International Conference of the IEEE Engineering in Medicine and Biology Society, EMBC 2014, Chicago, IL, USA, 26–30 August 2014; pp. 4315–4324.
66. Thuy, P.; Fatiesa, S.; Erica, S.; Di, W.; Wei, S. Quantification andcomparison of the mechanical properties of four human cardiac valves. *Acta Mater.* **2017**, *54*, 345–355.
67. Tang, S.; Zhang, G.; Guo, T.F.; Guo, X.; Liu, W.K. Phase field modeling of fracture in nonlinearly elastic solids via energy decomposition. *Comput. Methods Appl. Mech. Eng.* **2019**, *347*, 477–494. [CrossRef]
68. Zhou, Z.; Li, Y.; Weihin, W.; Guo, T.; Tang, S.; Luo, J. Transition of Surface-Interface Creasing in Bilayer Hydrogels. *Soft Matter* **2017**, *13*, 6011–6020. [CrossRef]
69. Jordan, B.; Gorji, M.B.; Mohr, D. Neural network model describing the temperature- and rate-dependent stress-strain response of polypropylene. *Int. J. Plast.* **2020**, *135*, 102811. [CrossRef]
70. Najjar, Y.M.; Huang, C. Simulating the stress-strain behavior of Georgia kaolin via recurrent neuronet approach. *Comput. Geotech.* **2007**, *34*, 346–361. [CrossRef]
71. Mahdi, E.S.; El Kadi, H. Crushing behavior of laterally compressed composite elliptical tubes: Experiments and predictions using artificial neural networks. *Compos. Struct.* **2008**, *83*, 399–412. [CrossRef]
72. Bratskaya, S.; Skatova, A.; Privar, Y.; Boroda, A.; Kantemirova, E.; Maiorova, M.; Pestov, A. Stimuli-Responsive Dual Cross-Linked N-Carboxyethylchitosan Hydrogels with Tunable Dissolution Rate. *Gels* **2021**, *7*, 188. [CrossRef]
73. Danko, M.; Kronekova, Z.; Krupa, I.; Tkac, P.; Kasak, P. Exchange Counterion in Polycationic Hydrogels: Tunability of Hydrophobicity, Water State, and Floating Capability for a Floating pH Device. *Gels* **2021**, *7*, 109. [CrossRef]
74. Jafari, H.; Dadashzadeh, A.; Moghassemi, S.; Zahedi, P.; Amorim, C.A.; Shavandi, A. Ovarian Cell Encapsulation in an Enzymatically Crosslinked Silk-Based Hydrogel with Tunable Mechanical Properties. *Gels* **2021**, *7*, 138. [CrossRef]
75. Kuntawong, P.; Kongintr, U.; Promptmas, C. M021 3D printed-lactate amperometric biosensor for real-time noninvasive health monitoring in human sweat. *Clinica Chimica Acta* **2022**, *530*, S11. [CrossRef]

76. Zhu, P.; Peng, H.; Rwei, A.Y. Flexible, wearable biosensors for digital health. *Med. Nov. Technol. Devices* **2022**, *14*, 100118. [CrossRef]
77. Laliberte, K.E.; Scott, P.; Khan, N.I.; Mahmud, M.S.; Song, F. A wearable graphene transistor-based biosensor for monitoring IL-6 biomarker. *Microelectron. Eng.* **2022**, *262*, 111835. [CrossRef]
78. Komkova, M.A.; Eliseev, A.A.; Poyarkov, A.A.; Daboss, E.V.; Evdokimov, P.V.; Eliseev, A.A.; Karyakin, A.A. Simultaneous monitoring of sweat lactate content and sweat secretion rate by wearable remote biosensors. *Biosens. Bioelectron.* **2022**, *202*, 113970. [CrossRef] [PubMed]
79. Lu, H.; Zhang, N.; Ma, M. Electroconductive hydrogels for biomedical applications. *WIREs Nanomed. Nanobiotechnol.* **2019**, *11*, 1568–1583. [CrossRef]
80. Tpgimages. Pulse Sign. Available online: https://m.quanjing.com/imgbuy/qj7118890985.html (accessed on 7 June 2022).
81. Del Corso, G.; Verzicco, R.; Viola, F. A fast computational model for the electrophysiology of the whole human heart. *J. Comput. Phys.* **2022**, *457*, 111084. [CrossRef]
82. Budday, S.; Ovaert, T.C.; Holzapfel, G.A.; Steinmann, P.; Kuhl, E. Fifty shades of brain: A review on the mechanical testing and modeling of brain tissue. *Arch. Comput. Methods Eng.* **2020**, *27*, 1187–1230. [CrossRef]
83. Deng, Y.; Hussain, I.; Kang, M.; Li, K.; Yao, F.; Liu, S.; Fu, G. Self-recoverable and mechanical-reinforced hydrogel based on hydrophobic interaction with self-healable and conductive properties. *Chem. Eng. J.* **2018**, *353*, 900–910. [CrossRef]
84. Huang, S.; Shuyi, S.; Gan, H.; Linjun, W.; Lin, C.; Danyuan, X.; Zhou, H.; Lin, X.; Qin, Y. Facile fabrication and characterization of highly stretchable lignin-based hydroxyethyl cellulose self-healing hydrogel. *Carbohydr. Polym.* **2019**, *223*, 115080. [CrossRef]
85. Jia, Z.; Zeng, Y.; Pengfei, T.; Gan, D.; Xing, W.; Hou, Y.; Wang, K.; Xie, C.; Lu, X. Conductive, Tough, Transparent, and Self-Healing Hydrogels Based on Catechol-Metal Ion Dual Self-Catalysis. *Chem. Mater.* **2019**, *31*, 5625–5632. [CrossRef]
86. Chen, W.; Bu, Y.; Li, D.; Liu, Y.; Chen, G.; Wan, X.; Li, N. Development of High-Strength, Tough, and Self-Healing Carboxymethyl Guar Gum-Based Hydrogels for Human Motion Detection. *J. Mater. Chem. C* **2020**, *8*, 900–908. [CrossRef]
87. Liu, Y.J.; Cao, W.; Ma, M.G.; Wan, P. Ultrasensitive Wearable Soft Strain Sensors of Conductive, Self-healing, and Elastic Hydrogels with Synergistic "Soft and Hard" Hybrid Networks. *ACS Appl. Mater. Interfaces* **2017**, *9*, 25559–25570. [CrossRef] [PubMed]
88. Sun, Y.; Lu, S.; Li, Q.; Ren, Y.; Ding, Y.; Wu, H.; He, X.; Shang, Y. High strength zwitterionic nano-micelle hydrogels with superior self-healing, adhesive and ion conductive properties. *Eur. Polym. J.* **2020**, *133*, 109761. [CrossRef]
89. Zhang, Z.; Zhiliang, G.; Wang, Y.; Guo, L.; Yin, C.; Zhang, X.; Hao, J.; Zhang, G.; Chen, L. Eco-Friendly, Self-Healing Hydrogels for Adhesive and Elastic Strain Sensors, Circuit Repairing, and Flexible Electronic Devices. *Macromolecules* **2019**, *52*, 2531–2541. [CrossRef]
90. Zhao, M.; Tang, Z.; Zhang, X.; Li, Z.; Xiao, H.; Zhang, M.; Liu, K.; Ni, Y.; Huang, L.; Chen, L.; et al. A self-healing, stretchable, and conductive Poly(N-vinylpyrrolidone)/gallic acid composite hydrogel formed via hydrogen bonding for wearable electronic sensors. *Compos. Sci. Technol.* **2020**, *198*, 108294. [CrossRef]
91. An, R.; Zhang, X.; Han, L.; Wang, X.; Zhang, Y.; Shi, L.; Ran, R. Healing, flexible, high thermal sensitive dual-network ionic conductive hydrogels for 3D linear temperature sensor. *Mater. Sci. Eng. C* **2019**, *107*, 110310. [CrossRef] [PubMed]
92. Pang, J.; Wang, L.; Xu, Y.; Wu, M.; Wang, M.; Liu, Y.; Yu, S.; Li, L. Skin-inspired cellulose conductive hydrogels with integrated self-healing, strain, and thermal sensitive performance. *Carbohydr. Polym.* **2020**, *240*, 116360. [CrossRef] [PubMed]
93. Zhou, H.; Zheng, S.; Qu, C.; Wang, D.; Liu, C.; Wang, Y.; Fan, X.; Xiao, W.; Li, H.; Zhao, D.; et al. Simple and environmentally friendly approach for preparing high-performance polyimide precursor hydrogel with fully aromatic structures for strain sensor. *Eur. Polym. J.* **2019**, *114*, 346–352. [CrossRef]
94. Wu, J.; Wu, Z.; Xu, H.; Wu, Q.; Liu, C.; Yang, B.R.; Gui, X.; Xie, X.; Tao, K.; Shen, Y.; et al. An intrinsically stretchable humidity sensor based on anti-drying, self-healing and transparent organohydrogels. *Mater. Horizons* **2019**, *6*, 595–603. [CrossRef]
95. Liu, S.; Qiu, Y.; Yu, W.; Zhang, H. Highly Stretchable and Self-Healing Strain Sensor Based on Gellan Gum Hybrid Hydrogel for Human Motion Monitoring. *ACS Appl. Polym. Mater.* **2020**, *2*, 1325–1334. [CrossRef]
96. Tong, R.; Chen, G.; Pan, D.; Qi, H.; Li, R.; Tian, J.; Lu, F.; He, M. Highly Stretchable and Compressible Cellulose Ionic Hydrogels for Flexible Strain Sensors. *Biomacromolecules* **2019**, *20*, 2096–2104. [CrossRef]
97. Zou, P.; Yao, J.; Cui, Y.N.; Zhao, T.; Che, J.; Yang, M.; Li, Z.; Gao, C. Advances in Cellulose-Based Hydrogels for Biomedical Engineering: A Review Summary. *Gels* **2022**, *8*, 364. [CrossRef] [PubMed]
98. Ansari, M.J.; Rajendran, R.R.; Mohanto, S.; Agarwal, U.; Panda, K.; Dhotre, K.; Manne, R.; Deepak, A.; Zafar, A.; Yasir, M.; et al. Poly(N-isopropylacrylamide)-Based Hydrogels for Biomedical Applications: A Review of the State-of-the-Art. *Gels* **2022**, *8*, 454. [CrossRef] [PubMed]
99. Yang, Y.; Wang, X.; Yang, F.; Wang, L.; Wu, D. Highly elastic and ultratough hybrid ionic–covalent hydrogels with tunable structures and mechanics. *Adv. Mater.* **2018**, *30*, 1707071. [CrossRef] [PubMed]
100. Peppas, N.A.; Van Blarcom, D.S. Hydrogel-based biosensors and sensing devices for drug delivery. *J. Control. Release* **2016**, *240*, 142–150. [CrossRef] [PubMed]
101. Culver, H.R.; Clegg, J.R.; Peppas, N.A. Analyte-Responsive Hydrogels: Intelligent Materials for Biosensing and Drug Delivery. *Accounts Chem. Res.* **2017**, *50*, 170–178. [CrossRef]
102. PantherMedia. Blood Corpuscle. Available online: https://m.quanjing.com/imgbuy/qj6714931654.html (accessed on 20 June 2022).

103. Dimatteo, R.; Darling, N.; Segura, T. In Situ forming injectable hydrogels for drug delivery and wound repair. *Adv. Drug Deliv. Rev.* **2018**, *127*, 167–184. [CrossRef]
104. Li, Y.; Yang, H.; Lee, D. Advances in biodegradable and injectable hydrogels for biomedical applications. *J. Control. Release* **2021**, *330*, 151–160. [CrossRef]
105. Wang, Q.; Qu, Y.; Zhang, Z.; Huang, H.; Xu, Y.; Shen, F.; Wang, L.; Sun, L. Injectable DNA Hydrogel-Based Local Drug Delivery and Immunotherapy. *Gels* **2022**, *8*, 400. [CrossRef]
106. Xie, Y.; Guan, Q.; Guo, J.; Chen, Y.; Yin, Y.; Han, X. Hydrogels for Exosome Delivery in Biomedical Applications. *Gels* **2022**, *8*, 328. [CrossRef]
107. Yang, Y.; Guo, H.; Du, Z.; Hong, W.; Lu, T.; Wang, T. Rate-dependent fracture of hydrogels due to water migration. *J. Mech. Phys. Solids* **2022**, *167*, 105007. [CrossRef]
108. Li, C.; Wang, Z.; Wang, Y.; He, Q.; Long, R.; Cai, S. Effects of network structures on the fracture of hydrogel. *Extrem. Mech. Lett.* **2021**, *49*, 101495. [CrossRef]
109. Desai, R.M.; Koshy, S.T.; Hilderbrand, S.A.; Mooney, D.J.; Joshi, N.S. Versatile click alginate hydrogels crosslinked via tetrazine-norbornene chemistry. *Biomaterials* **2015**, *50*, 30–37. [CrossRef] [PubMed]
110. Liu, Y.; Luo, Y.N.; Zhang, P.; Yang, W.F.; Zhang, C.Y.; Yin, Y.L. The Preparation of Novel P(OEGMA-co-MEO2MA) Microgels-Based Thermosensitive Hydrogel and Its Application in Three-Dimensional Cell Scaffold. *Gels* **2022**, *8*, 313. [CrossRef]
111. Jiao, W.; Li, X.; Shan, J.; Wang, X. Study of Several Alginate-Based Hydrogels for In Vitro 3D Cell Cultures. *Gels* **2022**, *8*, 147. [CrossRef]
112. Nguyen, H.Q.D.; Kao, C.Y.; Chiang, C.P.; Hung, Y.H.; Lo, C.M. Investigating the Immunomodulatory Potential of Dental Pulp Stem Cell Cultured on Decellularized Bladder Hydrogel towards Macrophage Response In Vitro. *Gels* **2022**, *8*, 187. [CrossRef]
113. Zhang, J.; Wang, Y.; Wei, Q.; Wang, Y.; Lei, M.; Li, M.; Li, D.; Zhang, L.; Wu, Y. Self-Healing Mechanism and Conductivity of the Hydrogel Flexible Sensors: A Review. *Gels* **2021**, *7*, 216. [CrossRef]
114. Yang, Y.; Zhou, M.; Peng, J.; Wang, X.; Liu, Y.; Wang, W.; Wu, D. Robust, anti-freezing and conductive bonding of chitosan-based double-network hydrogels for stable-performance flexible electronic. *Carbohydr. Polym.* **2022**, *276*, 118753. [CrossRef] [PubMed]
115. Fan, H.; Wang, L.; Feng, X.; Bu, Y.; Wu, D.; Jin, Z. Supramolecular Hydrogel Formation Based on Tannic Acid. *Macromolecules* **2017**, *50*, 666–676. [CrossRef]
116. Chen, S.; Dong, Y.; Ma, S.; Ren, J.; Yang, X.; Wang, Y.; Lu, S. Superstretching MXene Composite Hydrogel as a Bidirectional Stress Response Thixotropic Sensor. *ACS Appl. Mater. Interfaces* **2021**, *13*, 13629–13636. [CrossRef] [PubMed]

gels

MDPI

Article

Evaluation of the Accessibility of Molecules in Hydrogels Using a Scale of Spin Probes

Iulia Matei [1], Ana-Maria Ariciu [2], Elena Irina Popescu [1], Sorin Mocanu [1], Alexandru Vincentiu Florian Neculae [1], Florenta Savonea [1] and Gabriela Ionita [1,*]

[1] "Ilie Murgulescu" Institute of Physical Chemistry of the Romanian Academy, 202 Splaiul Independentei, 060021 Bucharest, Romania; iuliamatei@icf.ro (I.M.); elena.irina.popescu@cn-caragiale.ro (E.I.P.); smocanu@icf.ro (S.M.); aniculae@icf.ro (A.V.F.N.); f.savonea@icf.ro (F.S.)

[2] IPG Health Consulting, King Edward Street, Macclesfield SK19 1AQ, UK; ana-maria.ariciu@mccannhealth.com

* Correspondence: gabi2ionita@yahoo.com or ige@icf.ro

Abstract: In this work, we explored by means of electron paramagnetic resonance (EPR) spectroscopy the accessibility of a series of spin probes, covering a scale of molecular weights in the range of 200–60,000 Da, in a variety of hydrogels: covalent network, ionotropic, interpenetrating polymer network (IPN) and semi-IPN. The covalent gel network consists of polyethylene or polypropylene chains linked *via* isocyanate groups with cyclodextrin, and the ionotropic gel is generated by alginate in the presence of Ca^{2+} ions, whereas semi-IPN and IPN gel networks are generated in a solution of alginate and chitosan by adding crosslinking agents, Ca^{2+} for alginate and glutaraldehyde for chitosan. It was observed that the size of the diffusing species determines the ability of the gel to uptake them. Low molecular weight compounds can diffuse into the gel, but when the size of the probes increases, the gel cannot uptake them. Spin-labelled Pluronic F127 cannot be encapsulated by any covalent gel, whereas spin-labelled albumin can diffuse in alginate gels and in most of the IPN networks. The EPR spectra also evidenced the specific interactions of spin probes inside hydrogels. The results suggest that EPR spectroscopy can be an alternate method to evaluate the mesh size of gel systems and to provide information on local interactions inside gels.

Keywords: polymeric gels; diffusion; EPR spectroscopy; nitroxides

1. Introduction

Hydrogels represent a class of materials that are relevant for diverse applications in the field of cosmetics, in food and medicinal industries, due to the particularity of their solid-like network to sequester a large quantity of water—a biologically relevant liquid. The hydrogel "solid network" can be the result of a chemical or of a physico-chemical assembly of gelators. Most hydrogels result from physically-driven assembly or chemical crosslinking processes of natural or/and synthetic polymers. The main difference between chemical and physical hydrogels is that once the chemical bonds of the chemical network are broken, the gel cannot self-heal, whereas physical hydrogels can be recovered by changing the physical conditions. These materials are able to encapsulate and to release compounds in the water pools delimited by the solid network, and these properties are exploited in a variety of their applications [1–9].

The properties of encapsulation and release of compounds entrapped in the hydrogel network are related to the mesh size or correlation length, ξ, which is defined as the average distance between consecutive cross-links [10,11]. The estimation of this parameter can be made by various methods such as solute exclusion, mercury porosimetry, nitrogen adsorption/capillary condensation, microscopy and pulsed field gradient-NMR (PFG-NMR) [12–14]. To estimate the mesh size, in some cases, it is necessary to obtain the corresponding xerogel, which means that the determined mesh size can be smaller than that in hydrogel. This parameter provides direct information on the size of the particles that can be entrapped in the hydrogel.

In this study, we provide an alternate method to evaluate the ability of a hydrogel to encapsulate molecules. The electron paramagnetic resonance (EPR) spectroscopy, with its spin probe method, has been used recently by Vesković et al. to determine the water content of hydrogels based on changes in the spectral parameters of PROXYL and TEMPO spin probes [15]. In our study, we use EPR spectroscopy to monitor the diffusion in hydrogels of various spin probes with molecular weights in the range 200–66,000 Da. The spin probes used are depicted in Figure 1 and are of low molecular weight (TEMPO, 4-amino-TEMPO, 4-carboxy-TEMPO, ~200 Da), medium molecular weight (spin-labelled Pluronics L62NO/~2500 Da, P123NO/~5800 Da and F127NO/~12,600 Da; spin-labelled polyethylene glycol, PEG8000-TEMPO/~8000 Da) and high molecular weight (spin-labelled albumin, BSA-TEMPO/~66,000 Da). We analyzed the diffusion of these probes in hydrogels that have polysaccharides and β-cyclodextrins as building blocks of their solid network.

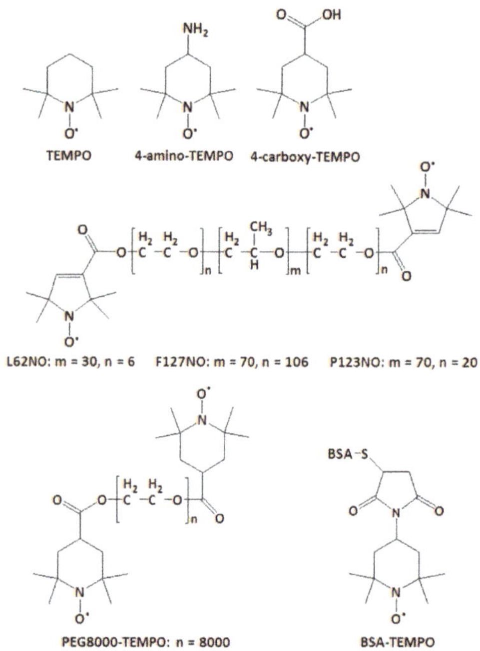

Figure 1. Chemical structures of the spin probes tested for diffusion in hydrogels.

The three types of hydrogels investigated in this study are shown in Figure 2. The first one refers to covalent hydrogel networks formed by reaction of isocyanato-end-capped polyethylene glycol (PEG) or polypropylene glycol (PPG) with β-cyclodextrin (β-CD) using poly glycol chains of different length and different poly glycol:β-CD initial ratios, as described previously [16]. The following covalent hydrogels containing cyclodextrins were prepared: PEG900/β-CD (4:1), PEG900/β-CD (10:1), PEG900/β-CD (14:1), PEG2000/β-CD (10:1) and PPG2000/β-CD (10:1). A covalent gel in which β-CD was replaced by pentaerythritol in the nodes of the network was also prepared, in order to demonstrate that host–guest complexation with β-CD cavities in the gel influences the dynamics of the spin probe. Alginate gel and IPN or semi-IPN gel networks were formed by adding crosslinking agents to alginate/chitosan solutions (Ca^{2+} for alginate and glutaraldehyde for chitosan).

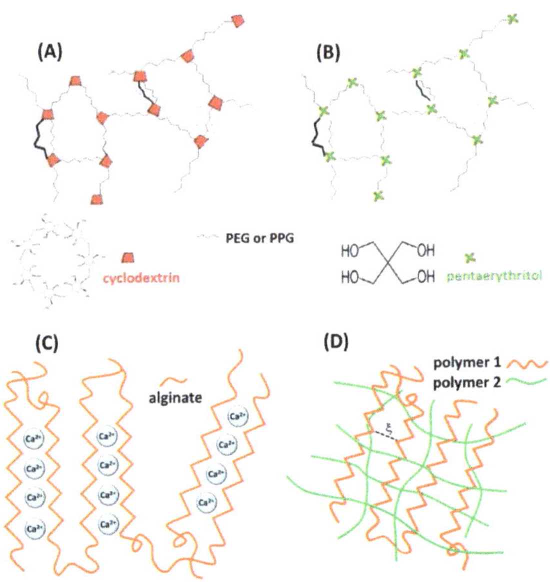

Figure 2. Schematic representation of hydrogel networks: (**A**) covalent gel PEG/β-CD or PPG/β-CD, (**B**) covalent gel pentaerythritol/β-CD, (**C**) ionotropic alginate gel, (**D**) interpenetrating gel network (IPN).

2. Results and Discussion

For all hydrogels, in order to evidence the diffusion of each spin probe, a certain quantity of hydrogel (e.g., 500 mg) was immersed in 1 mL aqueous solution of spin probe (5×10^{-4} M), and the system was kept overnight for equilibration. After this, the EPR signal was checked both in solution (termed supernatant) and in gel. In the aqueous solution, nitroxide spin probes have isotropic or quasi-isotropic motion (Supplementary Materials, Figure S1). Compared with solution, in the gel systems considered in this study, the encapsulation of nitroxides is often accompanied by specific interactions of the probes inside the gel, and their distribution in the non-uniform compartments of the gel. From this point of view, EPR spectroscopy is a suitable method to investigate such systems as the EPR parameters of the spin probes (hyperfine coupling constant, a_N, and rotational correlation time, τ) that are sensitive to changes of dynamics and to the polarity of the microenvironment [17]. The interactions of spin probes inside the gel can be translated into changes in dynamics from an isotropic motion to a highly restricted one, which are reflected in the shape of the EPR spectra [18].

Covalent hydrogels are good examples to highlight the non-homogeneity in gel systems, as they consist of solvent pools in which the spin probe can move freely, and cyclodextrin cavities that can accommodate spin probes fully or partially. In addition, spin probes can interfere with the fiber network of the gel. The diffusion of TEMPO, 4-amino-TEMPO and 4-carboxy-TEMPO in all covalent hydrogels considered in this study shows an unchanged spin probe dynamic as compared to solution. It is known that TEMPO-type radicals form complexes with cyclodextrins in solution, as proven by the slight variation in dynamics and the decrease in the a_N value, although the binding constants are relatively low, in the range 10^2–10^3 M^{-1} [19–21]. Differently, encapsulation of adamantyl-TEMPO [16] or of dual molecular probes bearing a pyrene or dansyl fluorophore and a TEMPO paramagnetic moiety [22,23] determines changes in the EPR spectra of these probes, proving the non-homogeneity of these covalent hydrogels; the presence of two spectral components was observed, corresponding to an equilibrium between the free probe located in the water pools of the gel, and the probe complexed with β-CDs from the network nodes. In the case of the low molecular weight TEMPO radicals used in this study, the fact that their dynamic is not changed after encapsulation, revealing only one component with EPR parameters similar to the solution, indicates that these probes are not complexed by β-CD cavities, but are located in the water pools delimited by the solid network of the gels. This might be a surprising result compared to those reported in the case of adamantyl-TEMPO [16] or dual molecular probes [22,23]. However, we have to take into account that the binding constants for adamantyl derivatives and dansyl probes are 2–3 orders of magnitude higher than those of TEMPO derivatives, and this can explain the differences.

In the case of polymeric spin probes (L62NO, P123NO, F127NO and PEG8000-TEMPO), a different behavior is observed as the molecular weight of the probe increases. Thus, L62NO and P123NO diffuse in all covalent gels, whereas F127NO and PEG8000-TEMPO are only encapsulated by some of the gels. Table S1 summarizes the ability of covalent gels to selectively uptake polymeric spin probes. It can be observed that, in the case of polymeric probes with molecular weight around 10,000 Da, the accessibility in these gels is difficult or does not occur. In the previous EPR study regarding the influence of the ratio between the number of PEG chains and β-CD units on the gel properties, it was observed that, as this ratio increases, the gel network becomes less rigid [16]. Therefore, for PEG900/β-CD (14:1), the diffusion of probes with molecular weight around 10,000 Da was not observed. This has been proven by the fact that only the supernatant has an EPR signal (Figure S2).

Increasing the gel fiber length by increasing the number of PEG units from 900 to 2000 made possible the diffusion of the polymeric probes F127NO and PEG8000-TEMPO in the PEG2000/β-CD gel (Table S1). Replacing PEG2000 with PPG2000, which in fact reduces the fiber length, led once again to the impossibility of diffusion for F127NO and PEG8000-TEMPO. It was also observed that the PPG2000/β-CD hydrogel can uptake completely spin probes with low molecular weight, whereas in the case of polymeric spin probes, a distribution between gel and supernatant ensued.

In the case of the L62NO spin probe, partial diffusion takes place in the PEG900/β-CD (10:1) hydrogel (Figure 3). The same behavior was observed for the P123NO spin probe (Figure S3). For P123NO, the spectra in water, supernatant and gel are presented. The EPR signals are similar, which means that the formation of micelles of these block copolymers does not take place inside the gel.

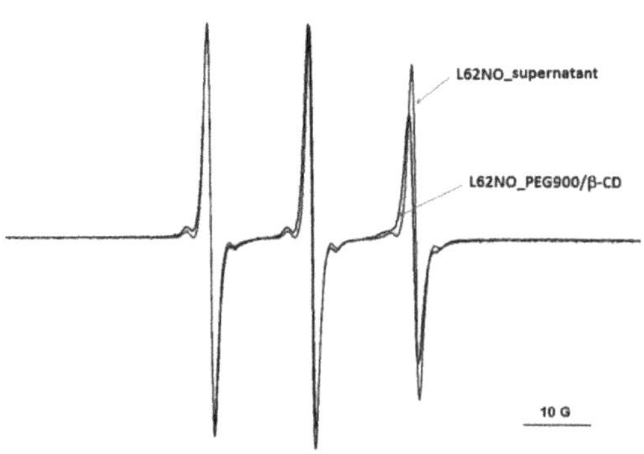

Figure 3. The EPR spectra of L62NO in PEG900/β-CD (10:1) hydrogel and in supernatant, after equilibration.

The PEG/β-CD covalent hydrogels have different mesh sizes, which depends on the initial ratio of reactants. Thus, in the case of the PEG900/β-CD (4:1) hydrogel, it was observed that the probes with molecular weight in the range 200–12,600 Da diffuse into the gel (Figure 4). Moreover, in the case of polymeric probes encapsulated in covalent hydrogels, the EPR spectra exhibit a two-component feature, with one component presenting a restricted motion and another component corresponding to a faster dynamic.

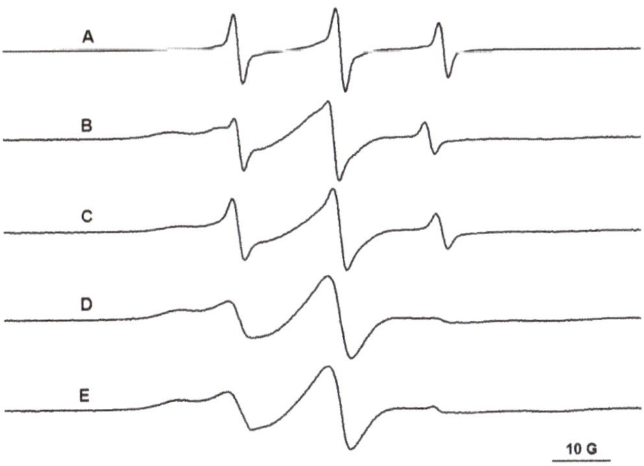

Figure 4. The EPR spectra of TEMPO (**A**), L62NO (**B**), P123NO (**C**), PEG8000-TEMPO (**D**) and F127NO (**E**) encapsulated in PEG900/β-CD (4:1) hydrogel.

The fact that the spin probe motion inside the hydrogel depends on the length of the chain bridging adjacent cyclodextrin units is suggested in Figure 5 for the case of L62NO. In this figure are shown the EPR spectra in PEG900/β-CD (4:1) and PPG2000/β-CD (10:1), and in the corresponding supernatants after equilibration. It can be noted that the immobilized component contribution is less evident in the PPG2000/β-CD hydrogel. The EPR spectra of L62NO encapsulated in PEG900/β-CD (10:1) and PEG2000/β-CD (10:1) are similar to the spectrum of L62NO in water. These results are in accordance with those previously reported, which show that the rigidity and, to some extent, the ability of these systems to encapsulate molecules depend on the PEG/β-CD ratio [16].

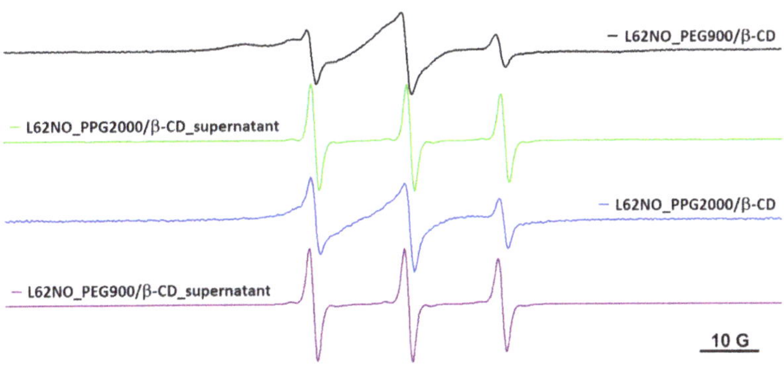

Figure 5. The EPR spectra of L62NO in PEG900/β-CD (4:1), PPG2000/β-CD (10:1) and corresponding supernatant solutions, after equilibration.

The immobilized component of polymeric spin probes in PEG/β-CD hydrogels can be assigned to the complexation of these probes by the cyclodextrins placed in the nodes of the gel network. To prove this, we monitored the diffusion in a hydrogel in which cyclodextrins were replaced with pentaerythritol units. In this case, the immobilized component is not observed (Figure 6).

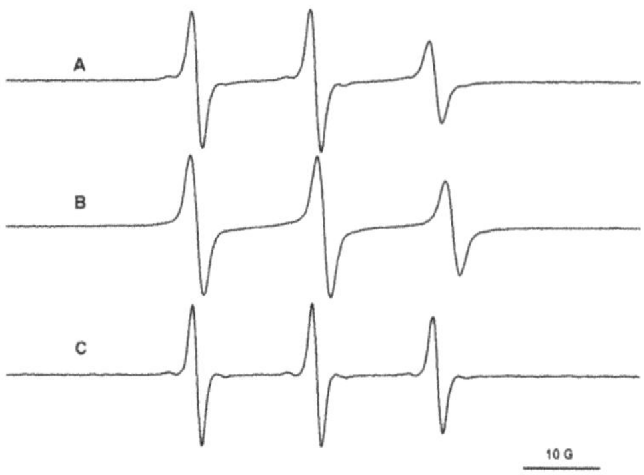

Figure 6. The EPR spectra of L62NO (**A**), P123NO (**B**) and F127NO (**C**) in PEG900/pentaerythritol hydrogel.

In what concerns spin-labelled BSA, this probe was not encapsulated by any covalent hydrogel. This has already been shown in the case of PEG900/β-CD (10:1), in a study reporting the ability of this hydrogel to strip sodium dodecyl sulphate (SDS) molecules from their complex with human serum albumin (HSA) [24]. In that study, it was proved that although SDS micelles have a similar diameter to HSA [25,26], SDS micelles are removed from the surface of HSA by diffusion of SDS into the gel at the molecular level, followed by reorganization into micelles inside the water pools of the hydrogel. Moreover, another study regarding the formation of gold nanoparticles inside the PEG900/β-CD hydrogel showed that nanoparticles with the size of 2–5 nm cannot be released from this gel [27]. This recommended the PEG900/β-CD hydrogel/gold nanoparticles hybrid material for its catalytic ability to neutralize 4-nitrophenol [28].

As it was mentioned in Introduction, other polysaccharide-based hydrogels were explored as well: the ionotropic gel resulted by complexation of alginate with Ca^{2+} ions (denoted ALG_Ca), the semi-IPN networks formed either by adding Ca^{2+} to a solution of alginate and chitosan (ALG_CHIT_Ca) or by adding glutaraldehyde (GA) to a solution of alginate and chitosan (ALG_CHIT_GA) and the IPN gels formed by adding Ca^{2+} first and GA second to a solution of alginate and chitosan (ALG_CHIT_Ca_GA) or by adding glutaraldehyde first and Ca^{2+} second to a solution of alginate and chitosan (ALG_CHIT_GA_Ca). Diffusion experiments in these hydrogels revealed that all spin probes are encapsulated by polysaccharide gels, with no changes occurring in the dynamics irrespective of the molecular weight of the spin probe. Thus, all spin probes obtained by spin labelling block copolymers F127, L62 and P123 diffused into alginate gel and into semi-IPN and IPN alginate/chitosan hydrogels, as exemplified in Figure 7 for the case of F127NO. The EPR parameters of the spin probes in these gels are given in Table S2.

A different behavior was noted for the spin-labelled albumin probe, BSA-TEMPO. The EPR measurements demonstrated that the semi-IPN and IPN systems resulted by adding crosslinking agents have a different capacity to entrap BSA-TEMPO, depending on the order in which the crosslinking agents are added. As it can be observed from Figure 8, BSA-TEMPO diffuses in the alginate gel (ALG_Ca), showing a free motion that indicates no interaction between BSA-TEMPO and the gel fibers. This can be explained by taking into consideration the fact that the BSA surface is negatively charged in our working conditions [29], and that the alginate gel also presents free carboxylate groups. These groups belong mainly to the D-mannuronate residues, as it is known that Ca^{2+} ions bind preferentially to the L-guluronate residues of alginate [30]. The literature data show that the alginate gel can accommodate large macromolecules such as proteins [31] and DNA [32], and even cells [33]. However, in the case of semi-IPN networks, the diffusion experiments show different behaviors. When only Ca^{2+} is added (ALG_CHIT_Ca), the structure of the semi-IPN network is largely that of the alginate gel. The EPR spectrum reveals a two-component feature, with one component similar to the one in alginate gel and another revealing a more restricted motion. The latter component arises from an interaction between BSA-TEMPO and the amino groups present in the structure of chitosan. In the case of the semi-IPN network obtained by adding only GA to the alginate/chitosan solution (ALG_CHIT_GA), BSA-TEMPO diffuses in the gel and shows a highly restricted motion, which can be explained by a strong interaction of BSA-TEMPO with the free amino groups in the chitosan gel fibers.

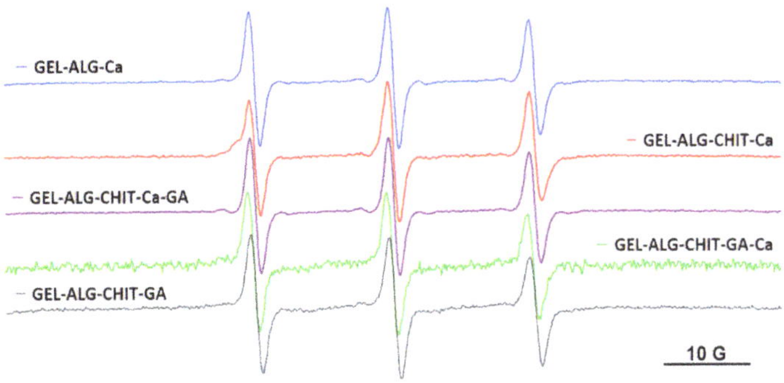

Figure 7. The EPR spectra of F127NO in alginate, semi-IPN and IPN alginate/chitosan hydrogels.

The mesh size of IPN hydrogels, and thus the ability of the hydrogels to encapsulate large molecules such as spin-labelled BSA, depends on the order in which the two crosslinkers are added. As it can be observed in Figure 8, BSA-TEMPO diffuses and has a restricted motion in the IPN formed by adding Ca^{2+} first and then GA (ALG_CHIT_Ca_GA), whereas in the case of the IPN formed by adding first GA and then Ca^{2+} (ALG_CHIT_GA_Ca), diffusion hardly occurs and the EPR signal becomes very weak. The spectral features can be rationalized as follows. In the ALG_CHIT_Ca_GA system, the less dense alginate gel network is formed first, followed by grafting of the chitosan network. The interaction between BSA-TEMPO and the remaining free amino groups in chitosan can be observed. In the case of the ALG_CHIT_GA_Ca system, the addition of Ca^{2+} to the ALG_CHIT_GA semi-IPN determines an increase in the density of the gel fibers that has as effect the decrease in the mesh size of the IPN network. As a consequence, the capacity of BSA-TEMPO to diffuse in this IPN is significantly reduced. Thus, we can conclude that, for the IPN hydrogel resulting from the successive addition of GA and Ca^{2+} to the alginate/chitosan solution, the mesh size is smaller than the size of BSA-TEMPO.

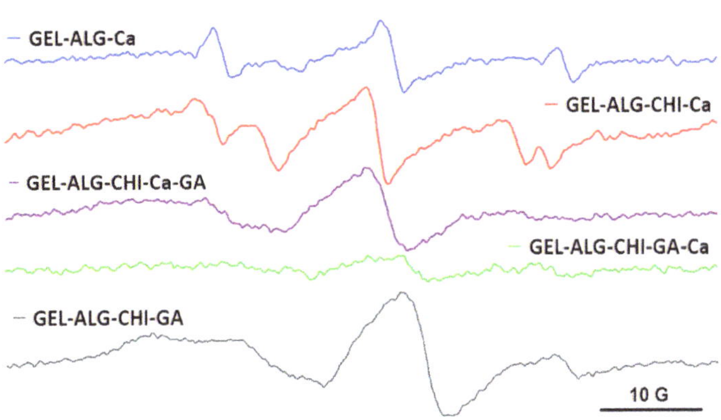

Figure 8. The EPR spectra of BSA-TEMPO in alginate, semi-IPN and IPN hydrogels.

3. Conclusions

To summarize, we have shown that by using EPR spectroscopy, it was possible to obtain information on the encapsulation properties of a series of polymeric gels obtained by different methods, and to correlate in some cases with the mesh size of these systems. We observed that, in the case of the covalent gels containing cyclodextrin, the encapsulation depends on the length of the polymer linkers and also involves complexation by cyclodextrins. In the case of small mesh sizes, the spin probes cannot penetrate the gel network and no EPR signal appears in the gel samples. In the case of large meshes with strong interactions, an EPR spectrum appears, indicating a restricted or strongly restricted motion. Experiments have shown that polymer gels have a mesh size of less than 10 nm (BSA size) compared to alginate and to IPN hydrogels with chitosan. The IPN and semi-IPN networks of alginate and chitosan have the mesh size greater than the size of BSA, and also exhibit interaction with this protein. The EPR spectroscopy is a suitable method for obtaining information on interactions inside hydrogel networks involving encapsulated molecules, and can also provide information on the mesh size.

4. Materials and Methods

Polyethylene glycols PEG900, PEG2000, PPG2000 and β-CD were supplied by Sigma-Aldrich (St. Louis, MO, USA) and Alfa Aesar (Ward Hill, MA, USA). Hexamethylene diisocyanate (HDMI), dibutyltin dilaurate, pentaerythritol (2,2-bis(hydroxymethyl)-1,3-propanediol), dimethylformamide (DMF), TEMPO, 4-amino-TEMPO and 4-carboxy-TEMPO were purchased from Aldrich. Sodium alginate (very low viscosity) and chitosan (low molecular weight) were from Alfa Aesar and Aldrich, respectively.

The covalent gels formed by reaction of PEG900, PEG2000 or PPG2000 with β-CD are referred to as PEG900/β-CD, PEG2000/β-CD and PPG2000/β-CD. The gels were prepared according to the procedure reported by Cesteros et al. [34,35]. Before reacting, the glycols and cyclodextrin were dehydrated. PEGs and PPG were functionalized by reaction with HDMI in DMF, and after that the β-CD solution was added dropwise to the reaction mixture, as reported earlier [27]. The gelation was completed in a few days. For PEG900/β-CD hydrogels, the PEG900 to β-CD ratio was 10:1, 14:1 or 4:1. In the case of PEG2000/β-CD and PPG2000/β-CD, the ratio was 10:1.

The polymeric spin probes L62NO, F127NO, P123NO and PEG8000-TEMPO were obtained following the procedures indicated in the literature [36]. Spin-labelled BSA, BSA-TEMPO, was obtained as described by Feix et al. [37].

For all diffusion experiments, 25 μL of spin probe stock solution (2×10^{-2} M in ethanol) was evaporated, then the spin probe was redissolved in 1 ml water in which 0.5 g of hydrogel was left overnight to equilibrate.

The EPR spectra were recorded on an X-band spectrometer FA100 from JEOL (Tokyo, Japan) at room temperature using the following settings: frequency modulation of 100 kHz, microwave power of 0.998 mW, sweep time of 480 s, modulation amplitude of 1 G, time constant of 0.3 s and a magnetic field scan range of 100 G.

Supplementary Materials: The following supporting information can be downloaded at: https://www.mdpi.com/article/10.3390/gels8070428/s1, Figure S1: The EPR spectra of the nitroxide-type spin probes used for diffusion experiments in hydrogels; Table S1: Encapsulation of polymeric spin probes in covalent hydrogels; Figure S2: The EPR spectra of F127NO in PEG900/β-CD (1:10) hydrogel and in supernatant, after equilibration; Figure S3: The EPR spectra of P123NO in water, in PEG900/β-CD (10:1) hydrogel and in supernatant, after equilibration; Table S2: Hyperfine coupling constants (a_N) and rotational correlation times (τ) of spin probes in alginate, semi-IPN and IPN hydrogels.

Author Contributions: Conceptualization, G.I.; methodology, G.I.; validation, S.M.; formal analysis, E.I.P., A.V.F.N. and F.S.; investigation, E.I.P., A.V.F.N., S.M. and A.-M.A.; writing—original draft preparation, G.I. and I.M.; writing—review and editing, G.I. and I.M.; supervision, G.I. All authors have read and agreed to the published version of the manuscript.

Funding: Part of the experiments were performed in the frame of a grant from the Romanian National Authority for Scientific Research, CNCS-UEFISCDI, project number PN-III-P4-IDPCE-2016-0734, and carried out within the research direction "Non-covalent interactions investigated by using dual molecular probes" of the "Ilie Murgulescu" Institute of Physical Chemistry at the Romanian Academy.

Institutional Review Board Statement: Not applicable.

Informed Consent Statement: Not applicable.

Conflicts of Interest: The authors declare no conflict of interest. The funders had no role in the design of the study; in the collection, analyses or interpretation of data; in the writing of the manuscript, or in the decision to publish the results.

References

1. Appel, E.A.; del Bario, J.; Loh, X.J.; Scherman, O.A. Supramolecular polymeric hydrogels. *Chem. Soc. Rev.* **2012**, *41*, 6195–6214. [CrossRef] [PubMed]
2. Buwalda, S.J.; Boere, K.W.; Dijkstra, P.J.; Feijen, J.; Vermonden, T.; Hennink, W.E. Hydrogels in a historical perspective: From simple networks to smart materials. *J. Control. Release* **2014**, *190*, 254–273. [CrossRef] [PubMed]
3. Du, X.; Zhou, J.; Shi, J.; Xu, B. Supramolecular hydrogelators and hydrogels: From soft matter to molecular biomaterials. *Chem. Rev.* **2015**, *115*, 13165–13307. [CrossRef] [PubMed]
4. Peppas, N.A.; Huang, Y.; Torres-Lugo, M.; Ward, J.H.; Zhang, J. Physicochemical foundations and structural design of hydrogels in medicine and biology. *Annu. Rev. Biomed. Eng.* **2000**, *2*, 9–29. [CrossRef]
5. Tan, S.; Ladewig, K.; Fu, Q.; Blencowe, A.; Qiao, G.G. Cyclodextrin-based supramolecular assemblies and hydrogels: Recent advances and future perspectives. *Macromol. Rapid Commun.* **2014**, *35*, 1166–1184. [CrossRef]
6. Goodrich, C.P.; Brenner, M.P.; Ribbeck, K. Enhanced diffusion by binding to the crosslinks of a polymer gel. *Nat. Commun.* **2018**, *9*, 4348. [CrossRef]
7. Luo, X.; Zhu, L.; Wang, Y.C.; Li, J.; Nie, J.; Wang, Z.L. A flexible multifunctional triboelectric nanogenerator based on MXene/PVA hydrogel. *Adv. Funct. Mater.* **2021**, *31*, 2104928. [CrossRef]
8. Li, S.N.; Yu, Z.R.; Guo, B.F.; Guo, K.Y.; Li, Y.; Gong, L.X.; Zhao, L.; Bae, J.; Tang, L.C. Environmentally stable, mechanically flexible, self-adhesive, and electrically conductive $Ti_3C_2T_X$ MXene hydrogels for wide-temperature strain sensing. *Nano Energy* **2021**, *90*, 106502. [CrossRef]
9. Li, S.N.; Li, B.; Yu, Z.R.; Gong, L.X.; Xia, Q.Q.; Feng, Y.; Jia, D.; Zhou, Y.; Tang, L.C. Chitosan in-situ grafted magnetite nanoparticles toward mechanically robust and electrically conductive ionic-covalent nanocomposite hydrogels with sensitive strain-responsive resistance. *Compos. Sci. Technol.* **2020**, *195*, 108173. [CrossRef]
10. Kirchhof, S.; Abrami, M.; Messmann, V.; Hammer, N.; Goepferich, A.M.; Grassi, M.; Brandl, F.P. Diels-Alder hydrogels for controlled antibody release: Correlation between mesh size and release rate. *Mol. Pharm.* **2015**, *12*, 3358–3368. [CrossRef]
11. Ionita, G. Characterisation and tailoring the properties of hydrogels using spectroscopic methods. In *Emerging Concepts in Analysis and Applications of Hydrogels*; Majee, S.B., Ed.; IntechOpen: London, UK, 2016. [CrossRef]
12. Bag, D.S.; Alam, S. Chiral chemical absorption property of a cross-linked poly(N-isopropyl acrylamide-co-sodium acrylate) thermoresponsive smart gel. *Chirality* **2012**, *24*, 506–511. [CrossRef] [PubMed]
13. Roy, S.; Javid, N.; Sefcik, J.; Halling, P.J.; Ulijn, R.V. Salt-induced control of supramolecular order in biocatalytic hydrogelation. *Langmuir* **2012**, *28*, 16664–16670. [CrossRef] [PubMed]
14. Ma, M.; Kuang, Y.; Gao, Y.; Zhang, Y.; Gao, P.; Xu, B. Aromatic-aromatic interactions induce the self-assembly of pentapeptidic derivatives in water to form nanofibers and supramolecular hydrogels. *J. Am. Chem. Soc.* **2010**, *132*, 2719–2728. [CrossRef]
15. Vesković, A.; Nakarada, Đ.; Popović Bijelić, A. A novel methodology for hydrogel water content determination by EPR: The basis for real-time monitoring of controlled drug release and hydrogel swelling and degradation. *Polym. Test.* **2021**, *98*, 107187. [CrossRef]
16. Ionita, G.; Ariciu, A.M.; Turcu, O.; Chechik, V. Properties of polyethylene glycol/cyclodextrin hydrogels revealed by spin probe and spin labelling methods. *Soft Matter* **2014**, *10*, 1778–1783. [CrossRef]
17. Hinderberg, D. EPR spectroscopy in polymer science. In *EPR Spectroscopy*; Topics in Current Chemistry; Drescher, M., Jeschke, G., Eds.; Springer: Berlin/Heidelberg, Germany, 2012; Volume 321, pp. 67–90. [CrossRef]
18. Eaton, G.R.; Eaton, S.S.; Salikhov, K.M. *Foundations of Modern EPR*; World Scientific: Singapore, 1998.
19. Ionita, G.; Caragheorgheopol, A.; Caldararu, H.; Jones, L.; Chechik, V. Inclusion complexes of cyclodextrins with nitroxide-based spin probes in aqueous solutions. *Org. Biomol. Chem.* **2009**, *7*, 598–602. [CrossRef] [PubMed]
20. Ionita, G.; Chechik, V. Mobility of spin probes in viscous cyclodextrin solutions. *Phys. Chem. Chem. Phys.* **2010**, *12*, 6956–6960. [CrossRef] [PubMed]
21. Ionita, G.; Mocanu, S.; Matei, I. Conformational preferences of TEMPO type radicals in complexes with cyclodextrins revealed by a combination of EPR spectroscopy, induced circular dichroism and molecular modeling. *Phys. Chem. Chem. Phys.* **2020**, *22*, 12154–12165. [CrossRef]

22. Mocanu, S.; Matei, I.; Ionescu, S.; Tecuceanu, V.; Marinescu, G.; Ionita, P.; Culita, D.; Leonties, A.; Ionita, G. Complexation of β-cyclodextrin with dual molecular probes bearing fluorescent and paramagnetic moieties linked by short polyether chains. *Phys. Chem. Chem. Phys.* **2017**, *19*, 27839–27847. [CrossRef]
23. Mocanu, S.; Matei, I.; Leonties, A.; Tecuceanu, V.; Hanganu, A.; Minea, Z.; Stancu, A.; Popescu, E.I.; Ionita, G. New flexible molecular probes bearing dansyl and TEMPO moieties for host–guest interactions in solution and gels. *New J. Chem.* **2019**, *43*, 11233–11240. [CrossRef]
24. Matei, I.; Ariciu, A.M.; Neacsu, M.V.; Collauto, A.; Salifoglou, A.; Ionita, G. Cationic spin probe reporting on thermal denaturation and complexation–decomplexation of BSA with SDS. Potential applications in protein purification processes. *J. Phys. Chem. B* **2014**, *118*, 11238–11252. [CrossRef]
25. Bruce, C.D.; Berkowitz, M.L.; Perera, L.; Forbes, M.D. Molecular dynamics simulation of sodium dodecyl sulfate micelle in water: Micellar structural characteristics and counterion distribution. *J. Phys. Chem. B* **2002**, *106*, 3788–3793. [CrossRef]
26. Tojo, A.; Kinugasa, S. Mechanisms of glomerular albumin filtration and tubular reabsorption. *Int. J. Nephrol.* **2012**, *2012*, 481520. [CrossRef] [PubMed]
27. Ionita, G.; Marinescu, G.; Ilie, C.; Anghel, D.F.; Smith, D.K.; Chechik, V. Sorption of metal ions by poly(ethylene glycol)/β-CD hydrogels leads to gel-embedded metal nanoparticles. *Langmuir* **2013**, *29*, 9173–9178. [CrossRef] [PubMed]
28. Matei, I.; Bem, M.; Leonties, A.R.; Radutiu, C.; Popescu, E.I.; Mocanu, S.; Savonea, F.; Baratoiu, R.; Ionita, G. Processes mediated by gold nanoparticles encapsulated in polymeric gels evidenced by EPR spectroscopy. *Rev. Roum. Chim.* **2021**, *66*, 295–302. [CrossRef]
29. Peters, T., Jr. Serum albumin. *Adv. Protein Chem.* **1985**, *37*, 161–245. [CrossRef]
30. Grant, G.T.; Morris, E.R.; Rees, D.A.; Smith, P.J.C.; Thom, D. Biological interactions between polysaccharides and divalent cations: The egg-box model. *FEBS Lett.* **1973**, *32*, 195–198. [CrossRef]
31. Zhang, Z.; Zhang, R.; Zou, L.; McClements, D.J. Protein encapsulation in alginate hydrogel beads: Effect of pH on microgel stability, protein retention and protein release. *Food Hydrocoll.* **2016**, *58*, 308–315. [CrossRef]
32. Machado, A.H.E.; Lundberg, D.; Ribeiro, A.J.; Veiga, F.J.; Miguel, M.G.; Lindman, B.; Olsson, U. Encapsulation of DNA in macroscopic and nanosized calcium alginate gel particles. *Langmuir* **2013**, *29*, 15926–15935. [CrossRef]
33. Ashimova, A.; Yegorov, S.; Negmetzhanov, B.; Hortelano, G. Cell encapsulation within alginate microcapsules: Immunological challenges and outlook. *Front. Bioeng. Biotechnol.* **2019**, *7*, 380. [CrossRef]
34. Cesteros, L.C.; Ramirez, C.A.; Pecina, A.; Katime, I. Poly(ethylene glycol-β-cyclodextrin) gels: Synthesis and properties. *J. Appl. Polym. Sci.* **2006**, *102*, 1162–1166. [CrossRef]
35. Cesteros, L.C.; Ramirez, C.A.; Pecina, A.; Katime, I. Synthesis and properties of hydrophilic networks based on poly(ethylene glycol) and β-cyclodextrin. *Macromol. Chem. Phys.* **2007**, *208*, 1764–1772. [CrossRef]
36. Caragheorgheopol, A.; Caldararu, H.; Dragutan, I.; Joela, H.; Brown, W. Micellization and micellar structure of a poly(ethylene oxide)/poly(propylene oxide)/poly(ethylene oxide) triblock copolymer in water solution, as studied by the spin probe technique. *Langmuir* **1997**, *13*, 6912–6921. [CrossRef]
37. Feix, J.B.; Bachowski, G.J.; Girotti, A.W. Photodynamic action of Merocyanine 540 on erythrocyte membranes: Structural perturbation of lipid and protein constituents. *Biochim. Biophys. Acta* **1991**, *1075*, 28–35. [CrossRef]

gels

MDPI

Article

A Highly Mechanical, Conductive, and Cryophylactic Double Network Hydrogel for Flexible and Low-Temperature Tolerant Strain Sensors

Quan Diao [1,*,†], Hongyan Liu [2,†] and Yanyu Yang [2,*]

1 College of Materials & Chemical Engineering, Zhongyuan University of Technology, Zhengzhou 450007, China
2 College of Materials Science and Engineering, Zhengzhou University, Zhengzhou 450001, China; liuhongyan@ipe.ac.cn
* Correspondence: dq_chem@126.com (Q.D.); yyyang@zzu.edu.cn (Y.Y.)
† These authors contributed equally to this work.

Abstract: Due to their stretchability, conductivity, and good biocompatibility, hydrogels have been recognized as potential materials for flexible sensors. However, it is still challenging for hydrogels to meet the conductivity, mechanical strength, and freeze-resistant requirements in practice. In this study, a chitosan-poly (acrylic acid-co-acrylamide) double network (DN) hydrogel was prepared by immersing the chitosan-poly (acrylic acid-co-acrylamide) composite hydrogel into $Fe_2(SO_4)_3$ solution. Due to the formation of an energy-dissipative chitosan physical network, the DN hydrogel possessed excellent tensile and compression properties. Moreover, the incorporation of the inorganic salt endowed the DN hydrogel with excellent conductivity and freeze-resistance. The strain sensor prepared using this DN hydrogel displayed remarkable sensitivity and reliability in detecting stretching and bending deformations. In addition, this DN hydrogel sensor also worked well at a lower temperature (−20 °C). The highly mechanical, conductive, and freeze-resistant DN hydrogel revealed a promising application in the field of wearable devices.

Keywords: hydrogel; double network; mechanical property; freeze-resistant; strain sensor

1. Introduction

With the increasing demands of pressure and deformation detection in artificial intelligence devices, strain sensors have attracted massive attention in recent years [1–4]. A strain sensor can detect a subtle tensile and bending deformation by outputting electrical signals such as current, resistance, and capacitance changes [5–7]. For better application in human body detection, strain sensors based on soft and biocompatible materials need to be developed. Unlike ordinary polymers such as plastic and rubber, hydrogels, which consist of a 3D-polymeric network swelling in water, have the advantages of softness and biocompatibility. Therefore, the hydrogel can be used in tissue engineering, flexible electronics, and soft robots [8–11]. In addition, the conductivity, mechanical strength, and freeze-resistant properties are the essential factors for strain sensors [12–15].

Due to the crystallization of internal water molecules, the hydrogel's performance becomes poor in sub-zero temperatures. One effective way to improve the low temperature performance of hydrogels is incorporating organic solvents, such as glycerol [16], ethylene glycol [17], and dimethyl sulfoxide (DMSO) [18]. For instance, Rong et al. reported that the strain sensor made by H_2O/ ethylene glycol organohydrogel could work in sub-zero temperatures, such as −40 °C [17]. The incorporation of ethylene glycol was beneficial for generating hydrogen bonds with water molecules and impeding the formation of ice crystals. Li et al. developed the DMSO/H_2O system, which contained polyvinyl alcohol, cellulose nanofibers, and $AlCl_3$. The DMSO in the system formed hydrogen bonds with water molecules and effectively extended the operating temperature (−50 °C or 50 °C) [18]. However, some organic fluids, such as DMSO and ethylene glycol, are not as ecofriendly as water [18,19]. Moreover, ionic liquids also have been used to prepared conductive gels with a broad range in their operation temperature [20,21]. Notably, the introduction of inorganic salts is the most economical method for enhancing the freeze-resistant property of hydrogels. Suo et al. proposed a freeze-resistant alginate-Ca^{2+}/polyacrylamide double network (DN) hydrogel that could maintain stretchability, toughness, and conductivity at temperatures far below the freezing point of water [22].

Currently, many strain sensors based on hydrogels have been devised. Most hydrogels are given electrical conductivity with conductive fillers (carbon materials [23], inorganic salts [24], metal nanoparticles [25], etc.). Among these fillers, inorganic salts can disperse in the hydrogels homogenously without phase separation [26]. The water in the hydrogel dissolves salts, and the network structure provides migration channels for the dissociated ions. Additionally, the ions can combine with the hydrogel network through hydrogen bonds or coordination bonds, which can lower the freezing point of the hydrogel. Meanwhile, the dissociated ions can be utilized to build a newly formed network with additives in hydrogels. The generated network provides the hydrogel with stronger mechanical properties [27]. Inspired by this, the highly mechanical DN hydrogels were candidates for flexible strain sensors [28–31].

In this study, we fabricated a DN hydrogel with excellent stretchability, conductivity, and cryophylactic properties. As shown in Figure 1, this chitosan-poly (acrylic acid-co-acrylamide) (CS/P (AA-co-AM)) double network hydrogel comprised of chitosan with an ionically cross-linked network. The DN hydrogel was prepared by soaking CS/P (AA-co-AM) composites into an $Fe_2(SO_4)_3$ solution. The prepared DN hydrogel exhibited outstanding mechanical properties: remarkable stretchability ($\varepsilon \approx 400\%$) and compressibility ($\varepsilon = 98\%$), and excellent anti-fatigue performance. The strain transducer made by this DN hydrogel showed prominent sensing performance for tensile deformation: the response (defined as $\frac{R-R_0}{R_0}$) was 4.92 at a strain of 500%. Moreover, the low-temperature property of the prepared transducer was also good: the response was 1.15 (at the tensile of 180%) at −20 °C. Generally, the study offers a potential material for monitoring the human body's movements.

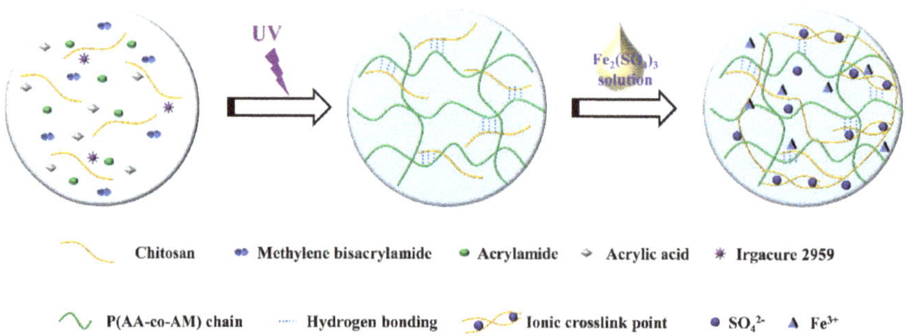

Figure 1. The schematic diagram of the preparation of CS/P(AA-co-AM) DN hydrogel.

2. Results and Discussion

2.1. Mechanical Properties of Hydrogels

As shown in Figure 2A, the DN hydrogel withstood rigorous compression and recovered quickly without any deformation. In addition, the hydrogel could bear several stretching behaviors, such as direct stretching, twisting–stretching, and crossing–stretching (Figure 2B–D). The hydrogel also withstood a load of 1 kg without cracking (Figure 2E). The tensile strength and toughness of the DN hydrogel were clearly superior to that of composite hydrogel (Figure S1). The prominent mechanical performance of the DN hydrogel was due to the synergic effect of the internal dual network: one was the poly(hydroxyethyl acrylamide) network derived from the photo-initiated crosslinking of AA and AM, while the other was the ionic network of CS and Fe^{3+}. The SEM images (Figure 2F,G) exhibited the comparison of micromorphology. The DN hydrogel showed denser microstructure compared with the composite hydrogel due to the formation of the CS ionic network.

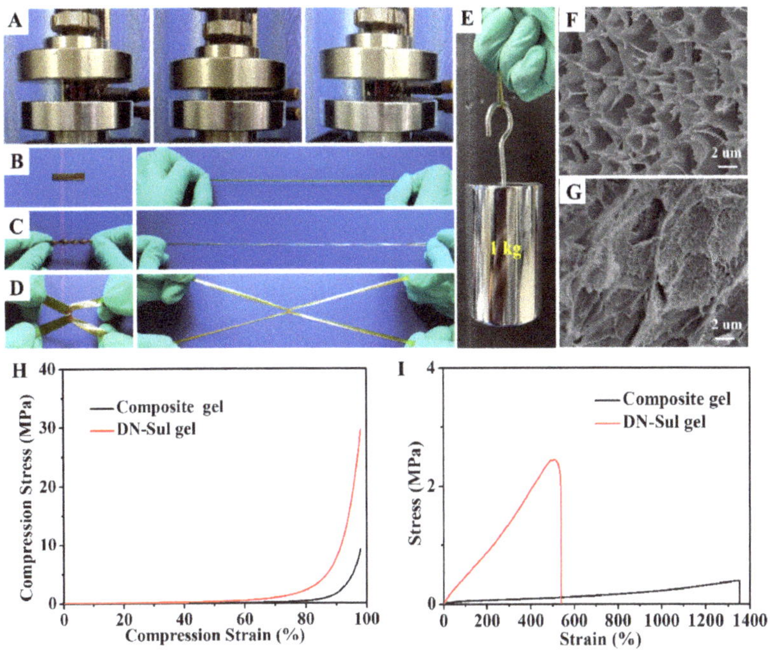

Figure 2. The mechanical properties of the DN hydrogel. (**A**) Compression, (**B**) stretching, (**C**) twisting–stretching, and (**D**) crossing–stretching. (**E**) Load-bearing ability (loading weight: 1 kg). SEM images of (**F**) the composite hydrogel and (**G**) the DN hydrogel. (**H**) Compressive and (**I**) tensile curves of the composite and DN hydrogels.

The detailed mechanical properties were measured by compression and tensile tests. At 98% compressive strain, the compression stress of DN hydrogel reached 29.59 MPa, while the value of the composite hydrogel was 9.23 MPa (Figure 2H). Remarkably, the DN hydrogel remained intact during the whole compressive process, demonstrating its superior compressibility. Meanwhile, the DN hydrogel also exhibited an improved tensile strength of 2.43 MPa, which was six times that of the original composite hydrogel (Figure 2I). Figure S2A shows that the elastic modulus of the DN hydrogel reached 0.5 MPa, much larger than that of the composite hydrogel (0.03 MPa). The toughness was calculated by the area integration of the stress–strain curves (Figure S2B). The toughness of the DN hydrogel (6.79 MJ/m^3) was triple that of the composite hydrogel (2.18 MJ/m^3). In addition, the fracture strain of the DN hydrogel was 537%, as shown in Figure 2I. As reported, when the physicochemical DN hydrogel was deformed, its physical bonds broke and dissipated energy to protect the chemical bonds [32,33]. At the first stage of elongation, the dynamic physical interactions (CO_2LFe^{III} and hydrogen bond) together with the chitosan physical network were destroyed for dissipating energy. Once the tensile strain exceeded 537%, both the physical and covalent cross-links were ruptured.

Figure 3A shows the energy-dissipation capacity of the hydrogels measured by a stretching–restoring test. The area of the loading–unloading curve of the composite hydrogel was very small, and the loading curve and unloading curve almost coincided. The hysteresis loop area of the ion–covalent DN hydrogel was larger, and the dissipation energy was 1.96 MJ/m^3 by integration (Figure 3B). This was attributed to the fact that the chitosan's physical cross-linking network and ionic coordination (CO_2LFe^{III}) provided an effective energy-dissipation mechanism for the DN hydrogel. The energy-dissipation capacity of the DN hydrogel with different tensile strains (150–400%) is shown in Figure 3C. It was found that the hysteresis loop area was aggrandized with the hydrogel's deformation increase. The dissipation energy increased from 0.47 MJ/m^3 at the tensile strain of 150% to 3.56 MJ/m^3 at a 400% tensile strain, while the dissipation coefficient increased from 50% to 70.5% (Figure 3D).

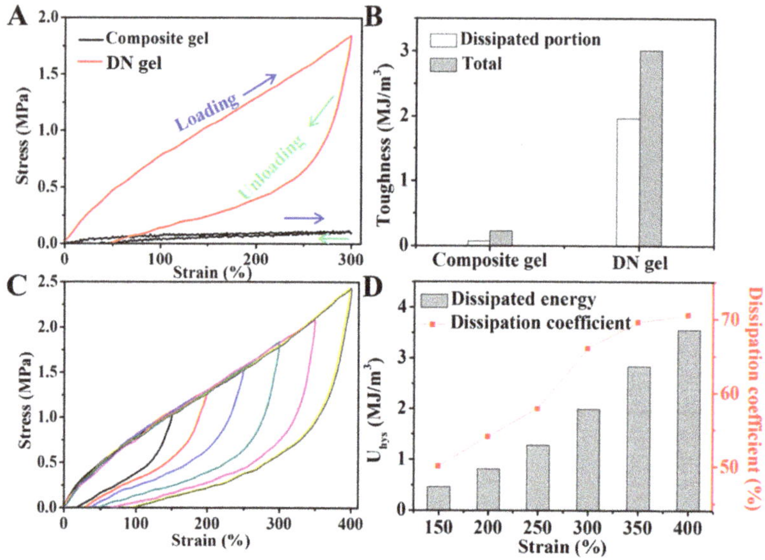

Figure 3. (**A**) Loading–unloading curves of composite and DN hydrogels at a strain of 300% and (**B**) their corresponding calculated dissipated energy and energy-dissipation coefficient. (**C**) Loading–unloading curves of DN hydrogel at different strains (150%, 200%, 250%, 300%, 350%, and 400%) and (**D**) corresponding calculated dissipated energy and energy-dissipation coefficient.

Figure 4A,B show the DN hydrogel's compression self-recovery ability. During the test, six DN hydrogel specimens were compressed at a 50% strain and then left to rest for different amounts of time (0, 1, 3, 5, 10, and 20 min) before being recompressed. Compared with the original sample, the dissipated energy of the 0 min specimen remained 59.5%, with remarkable elastic modulus (77.7%) and maximum stress (94.5%). After resting for 20 min, the elastic modulus reverted to 95%, with the maximum stress restored to 98.4%, very close to the initial state. In addition, the tensile self-recovery ability was also tested as the specimens were stretched to 200% deformation (Figure 4C,D). After 20 min, the dissipated energy of the stretched sample recovered to 75.4%. Meanwhile, the elastic coefficient and tensile strength basically recovered, and the recovery efficiency was 91.5% and 98.5%, respectively. These tests demonstrated the eminent self-restorability of the DN hydrogel after compression and tensile deformation.

In order to further study the stability and anti-softening properties, DN hydrogel was encapsulated with 3 M adhesive tape to avoid the loss and evaporation of water during the testing process. The hydrogel sample was then subjected to a continuous loading and unloading cycle test 1000 times (ε = 100%). As shown in Figure 5A, the stress and dissipation ring area tested in the first cycle were larger than those of the subsequent ones, and the maximum stress was 0.83 MPa. It was found from the stretching curve of the 2nd to 1000th cycles that the mechanical properties of DN hydrogel were relatively stable and essentially unchanged. As shown in Figure 5B, according to the maximum stress and dissipation energy of 1000 tensile cycles, the values of the 2nd to 1000th tensile cycles were basically the same, which strongly demonstrated the stability of the mechanical properties of DN hydrogel. As shown in Figure 5C, the 1000th tensile cycle curve was locally amplified during the 500th to 540th cycles. The maximum tensile strength basically remained at the same level, indicating that the DN hydrogel had not only ultra-high mechanical strength but also stability.

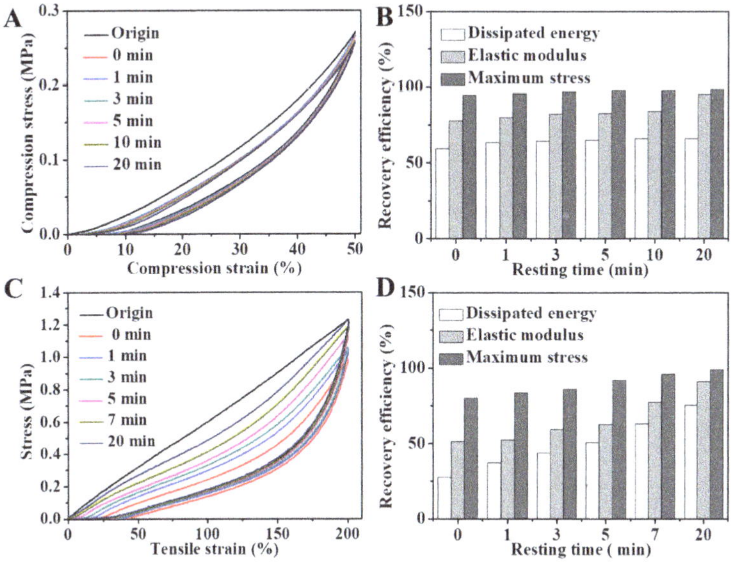

Figure 4. (**A**) Self-recovery behavior and (**B**) self-recovery efficiency upon compression. (**C**) Self-recovery behavior and (**D**) self-recovery efficiency upon elongation.

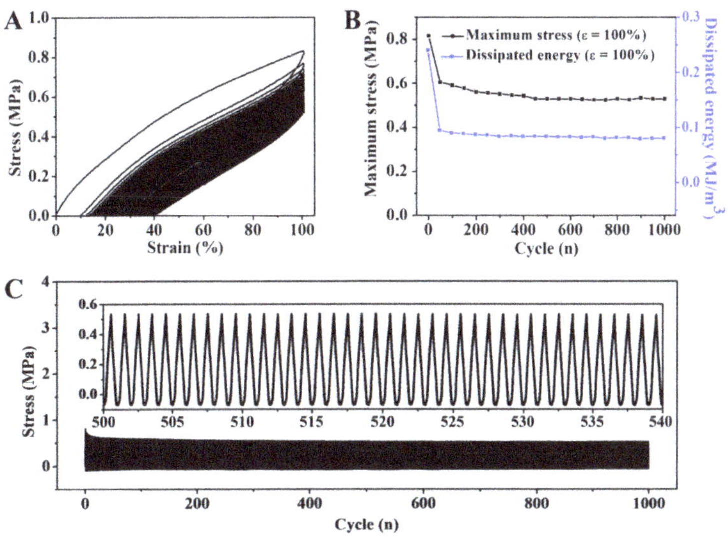

Figure 5. (**A**) 1000 continuous elongation–relaxation cycles of the DN hydrogel. (**B**) The evolution of maximal stress and dissipated energy during 1000 cycles. (**C**) Real-time stress curve during a continuous 1000 cycles at a strain of 100%. The built-in figure was the stress curve during the 500th to 540th cycles.

2.2. Modulation on Mechanical Performance of DN Hydrogel

Due to the ionic coordination between CS and sulfate solution, the increase of CS content directly affected the CS ionically crosslinked degree and changed the rigidity of the physical network. Thus, the CS content could be used to regulate the mechanical properties of the DN hydrogel. As shown in Figure 6A,B, when the molar ratio of acrylic acid to acrylamide was 15%, the concentration of ferric sulfate was 2 M, and the soaking time was 50 min, the hydrogel's rigidity was increased, elastic modulus enlarged, ductility reduced, and toughness enlarged as the CS content increased (the detailed data of the mechanical properties of the DN hydrogels with different CS contents are listed in Table 1). According to the influence of the chitosan content on the mechanical properties and considering the tensile strength and ductility of the hydrogel, the final fixed CS content was 7.5%. In addition, the molar ratio of the soft and tough chemical network can also affect the mechanical properties of hydrogels. Therefore, by adjusting the molar ratio of acrylic acid to acrylamide, the mechanical properties of hydrogels could be regulated. As shown in Figure 6C,D, when the molar ratio of acrylic acid to acrylamide increased, the tensile strength of DN hydrogel gradually increased, the ductility decreased, and the elastic modulus increased (the detailed data of the mechanical properties of the DN hydrogels with different molar ratios of AA to AM are listed in Table 2). Taking into account the toughness, the hydrogel prepared with 7.5% CS content, and 15% acrylic acid was the optimal sample.

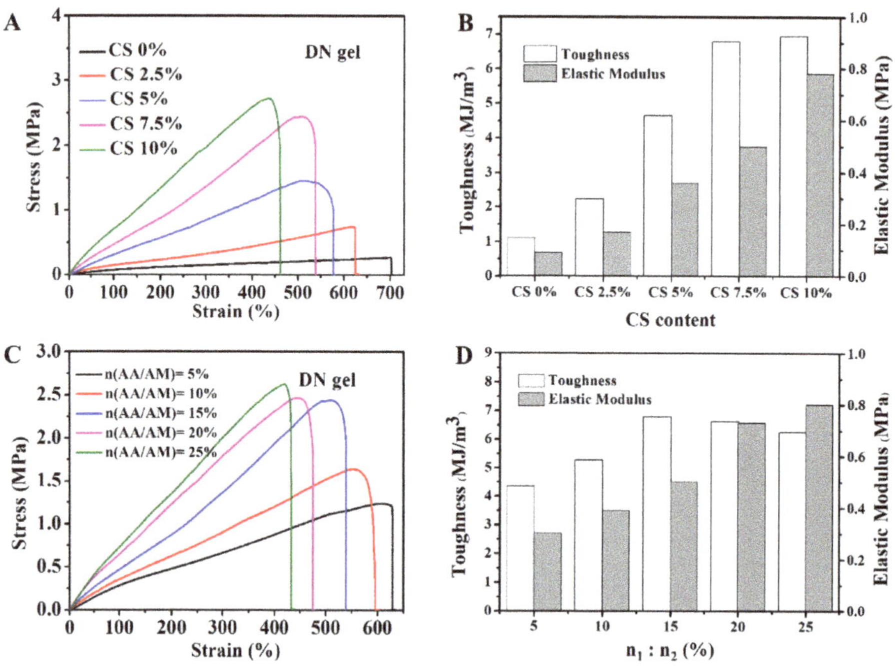

Figure 6. (**A**) The stress–strain curves of the DN hydrogels with various CS content and (**B**) their corresponding toughness and elastic modulus. (**C**) The stress–strain curves of the DN hydrogels with various molar ratios of AA to AM (n_1:n_2) and (**D**) their corresponding toughness and elastic modulus.

Table 1. The detailed mechanical properties of the DN hydrogels with different CS contents.

CS Content	Elastic Modulus (MPa)	Fracture Strength (MPa)	Fracture Strain (%)	Toughness (MJ/m^3)
0%	0.09	0.26	702	1.11
2.5%	0.17	0.73	623	2.24
5%	0.36	1.46	576	4.65
7.5%	0.50	2.44	539	6.80
10%	0.78	2.71	460	6.95

Table 2. The detailed mechanical properties of the DN hydrogels with different molar ratios of AA to AM.

$n_{(AA)}$:$n_{(AM)}$	Elastic Modulus (MPa)	Fracture Strength (MPa)	Fracture Strain (%)	Toughness (MJ/m^3)
5%	0.29	1.24	629	4.35
10%	0.39	1.64	594	5.29
15%	0.50	2.44	539	6.79
20%	0.73	2.47	474	6.64
25%	0.80	2.63	433	6.26

2.3. Mechanical Flexibility at Low Temperatures

Because the water inside hydrogels tends to crystallize below 0 °C, the mechanical properties of the hydrogel will be damaged. Incorporating salt in the hydrogel is an effective way to depress the ice point [22,34,35]. Compared with the composite hydrogel, the DN hydrogel still maintained good mechanical properties: after being stored at −20 °C for 24 h, the DN hydrogel could be stretched and bended, while the composite hydrogel became stiff and easy to crack (Figure S3). In addition, to study the cryogenic mechanical properties of the composite and DN hydrogel in more detail, the tensile tests were made at low temperature. When the temperature went down to −20 °C, the breaking elongation of the composite hydrogel decreased to 28% (Figure S4A). However, even at −20 °C, the DN hydrogel still maintained good mechanical properties, with 250% breaking elongation (Figure 7A). As the temperature dropped below −20 °C, the gel converted into a mixture of ice crystals and polymer networks, leading to reduced flexibility and ductility, together with enhanced rigidity [36]. The detailed mechanical properties parameters are illustrated in Figures 7B–D and S4B–D. As the temperature decreased, the tensile strength and elastic modulus of the hydrogels were aggrandized, whereas the breaking elongation decreased rapidly. The Fe^{3+} and SO_4^{2-} were expected to prevent water molecules from gathering and crystallizing, resulting in the outstanding freeze-resistant property.

2.4. Sensing Performance of Hydrogel Flexible Sensor

In addition to the outstanding mechanical properties, the ions in the DN hydrogel also provided good ionic conductivity, which led the hydrogel to transmit electric signals through ionic migration. The conductivity of the hydrogel varied with the type of deformation, such as stretching, compression, and twisting. Therefore, the hydrogel can be used in flexible sensors. As shown in Figure 8A, the hydrogel exhibited a resistant-type behavior to the tensile deformation; as the resistance increased, the tensile strain continued to grow. The hydrogel sensor exhibited good sensitivity to a small strain (10–50%) and large strain (50–400%), as shown in Figure 8B,C. As the previous literature reported, the gauge factor (GF) of the sensor can be calculated from the curve of the relative resistance variation ($\frac{R-R_0}{R_0}$) and strain [37]. In detail, the relationship between $\frac{R-R_0}{R_0}$ and strain (ε) can be turned into a formula: $\frac{R-R_0}{R_0} = 0.0065\varepsilon^2 + 1.57\varepsilon$. Therefore, the corresponding derivative data (GF) followed a linear trend (GF = 0.013ε + 1.57) as shown in Figure S5. The sensor also showed a stable response to 60 stretching cycles, which demonstrated the eminent repeatability (Figure 8D).

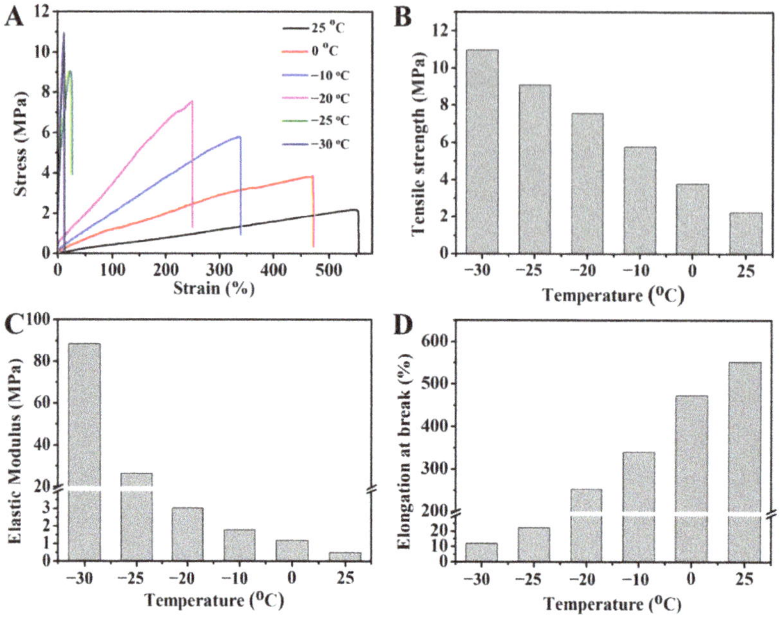

Figure 7. (**A**) The stress–strain curves and the detailed mechanical property parameters of the DN hydrogel at different temperatures: (**B**) tensile strength, (**C**) elastic modulus, and (**D**) fracture strain.

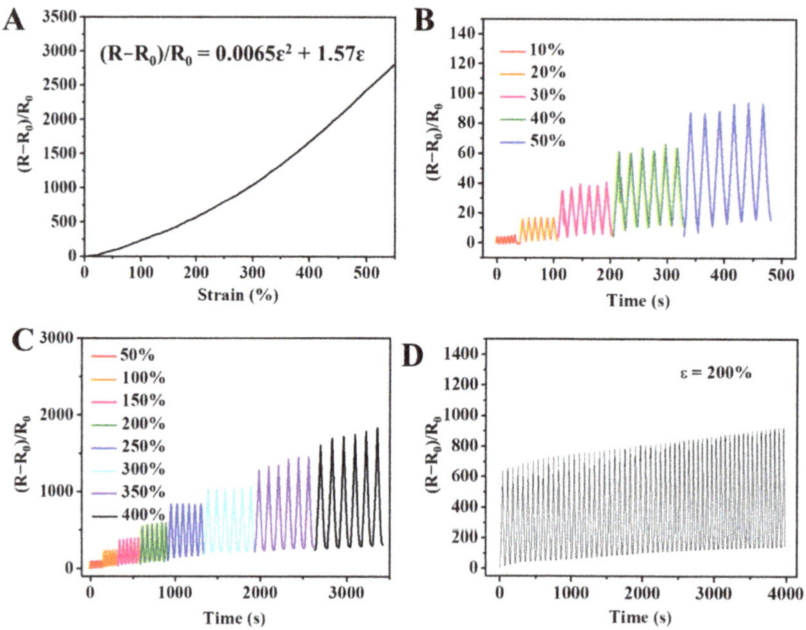

Figure 8. The performance of the DN hydrogel sensor: (**A**) relationship between resistance variation and tensile strain; (**B**,**C**) relative resistance-variation curves of multiple tensile strains; (**D**) real-time resistance-variation curves within 60 cycles (ε = 200%).

2.5. Human Motion Detection

Additionally, the hydrogel sensor was validated to detect human motions, such as the flexing of a finger, an arm, and a leg (Figure 9). The bending movement caused the relative resistance to increase, while the straightening movement restored the sensor to the initial resistance. Compared with the motions of fingers and arms, the deformation degree of knee-bending was larger, resulting in a larger resistance change than others. Meanwhile, the hydrogel sensor responded very well to repeated joint-bending movements, indicating that the sensor had excellent reliability in human-motion monitoring.

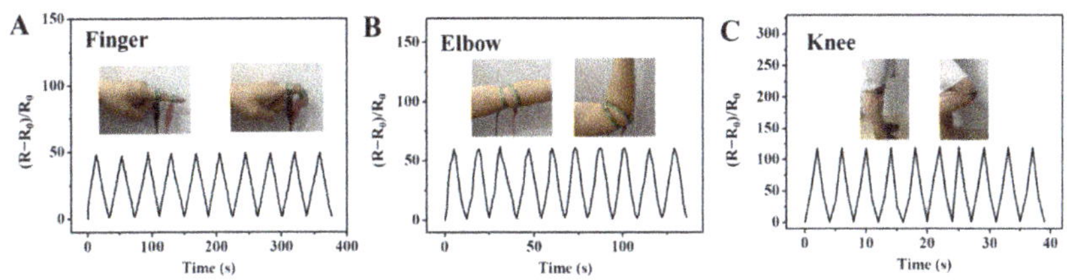

Figure 9. Monitoring of human body movements: (**A**) finger bending, (**B**) elbow bending, and (**C**) knee bending.

2.6. Low-Temperature Strain Responsiveness

As demonstrated, the DN hydrogel showed outstanding mechanical properties at sub-zero temperatures. Furthermore, the cryogenic sensitivity of the hydrogel was also investigated. An LED bulb connected to a circuit containing a piece of hydrogel was used to study the conductivity (Figure S6). As shown in Figure 10A, the conductivity of the hydrogel varied inversely with the temperature. When the temperature decreased, the migration of ions was impeded due to the limited movement of the polymer chains and the crystallization of water molecules. Figure 10B shows the response–strain curves of the gel sensors at 25 °C and −20 °C. Compared with the response value at normal temperatures, the response value at −20 °C was reduced to a certain extent. However, the gel sensor could still be utilized at low temperatures, as the response to 150% strain rose to 160%. Figure 10C illustrates the continuous response to the tensile strain ranging from 10% to 150% at −20 °C. The repeated curves at each strain demonstrated the excellent stability of the hydrogel sensor.

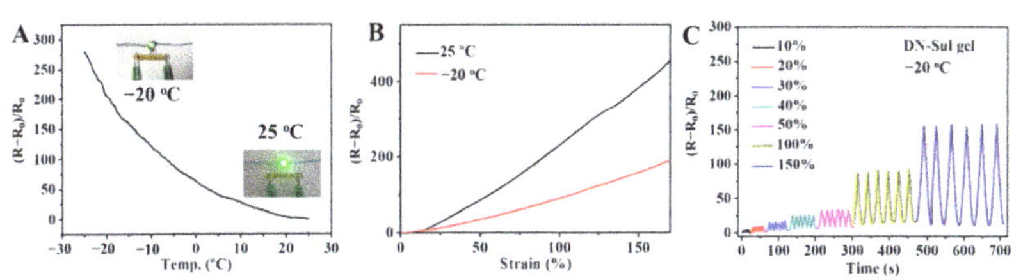

Figure 10. (**A**) The relative resistance change of the DN hydrogel with the temperature decreasing from 25 °C to −20 °C (the resistance at 25 °C was defined as R_0). Inset pictures showed the bulb light's variation in the circuit connected by the DN hydrogel at 25 °C and −20 °C. (**B**) The performance comparison of the DN hydrogel sensor at 25 °C and −20 °C. (**C**) Real-time relative resistance-change curves on multiple tensile strains at −20 °C.

3. Conclusions

In conclusion, we prepared a DN hydrogel by immersing a CS/P(AA-co-AM) composite hydrogel into an $Fe_2(SO_4)_3$ solution. Inorganic salt ions combined with the CS to form the energy-dissipative ionic network that enhanced the tensile property and conductivity simultaneously. Since the effective energy-dissipation mechanism was derived from the generated CS ionic network, the DN hydrogel exhibited prominent mechanical properties: tensile strength of 2.43 MPa, elastic modulus of 0.5 MPa, and toughness of 6.79 MJ/m^3, as well as an outstanding self-recovery ability and excellent durability. Additionally, the introduction of ions provided the DN hydrogel with good freeze resistance and conductivity. Impressively, the strain sensor assembled from the DN hydrogel exhibited strain sensitivity and cycling stability. Even at low temperatures, the DN hydrogel still possessed good strain-sensing performance with slightly reduced sensitivity. The sensor also demonstrated a superior ability to detect various human body movements in real time. This study provides a new strategy for constructing functional hydrogel materials for stretchable, flexible sensors.

4. Materials and Methods

4.1. Materials

Acrylamide (AM, 98%), acrylic acid (AA, 99.5%), and methylene bisacrylamide (MBA, 98%) were bought from Alfa Aesar. Short-chain chitosan (CS, $Mw \approx 10$ kDa, degree of deacetylation > 90%) and iron sulfate hydrate were obtained from Aladdin Industrial Corp., China. The pure water was homemade by a water purification machine.

4.2. Fabrication of CS/P (AA-co-AM) DN Hydrogel

As shown in Figure 1, the DN hydrogel was fabricated by following two steps. First, the short-chain CS (0.75 g, 7.5 wt%), AM (2.13 g), AA (0.31 mL), MBA (0.16 mL), and Irgacure (54 mg) were added to pure water (10 mL) with continuous stirring. The homogeneous solution was then transferred into a glass mold and became CS/P (AA-co-AM) hydrogel under UV light (150 W) irradiation. Second, to get the DN hydrogel, the prepared CS/P (AA-co-AM) hydrogel was immersed in 2 M $Fe_2(SO_4)_3$ solution for 30 min. A serial of DN hydrogels were prepared with different CS contents (0 wt%, 2.5 wt%, 5 wt%, 7.5 wt%, and 10 wt%).

4.3. Fabrication of the Hydrogel-Based Flexible Sensor

The flexible sensor based on the DN hydrogel was assembled using two copper tapes attached at the DN hydrogel's edges (Figure 11). The hydrogel-based flexible sensor was then sealed by a 3 M tape to lock the water in the hydrogel.

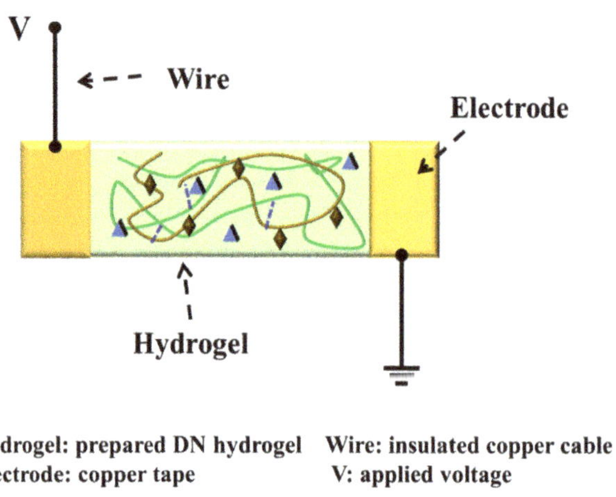

Figure 11. The schematic diagram of the strain sensor assembled using DN hydrogel and Cu electrodes.

4.4. Characterization

The microtopography of the DN hydrogel was observed with a field-emission scanning electron microscope (FESEM, FEI Quanta 250 FEG). The hydrogel's mechanical properties were measured by a universal tensile tester (MTS CMT4104) with hydrogel strips (25 mm × 3 mm × 1 mm) for an elongation test and hydrogel cylinders (diameter × height: 10 mm × 12 mm) for a compression test. According to the previous literature, the rates of the elongation test and compression test were 50 mm/min and 5 mm/min, respectively, while the load cells were 500 N and 10 kN, respectively [27,32]. The mechanical properties at low temperatures were measured with a universal tensile tester (MTS CMT4104) attached to a cryogenic box (MTS GDX200).

Supplementary Materials: The following supporting information can be downloaded at: https://www.mdpi.com/article/10.3390/gels8070424/s1, Figure S1: Mechanical properties of the composite hydrogel. (A) Compression. (B) Various tensile deformations: (B_1) stretching, (B_2) twisting–stretching, (B_3 and B_4) knotted–stretching, and (B_5) crossing–stretching; Figure S2: (A) Elastic modulus. (B) Toughness of the composite and DN hydrogels calculated from the tensile curves; Figure S3: Pictures of the composite and DN hydrogels at −20 °C; Figure S4: (A) Stress–strain curves and detailed mechanical properties' parameters of the composite hydrogel at different temperatures. (B) Tensile strength, (C) elastic modulus, and (D) elongation; Figure S5: GF variation of the DN hydrogel with strain; Figure S6: Circuit diagram of the lamp-lighting experiment.

Author Contributions: Conceptualization, Y.Y. and Q.D.; methodology, H.L.; data curation, H.L.; writing—original draft preparation, Q.D.; writing—review and editing, Q.D. and Y.Y.; project administration, Y.Y.; funding acquisition, Q.D. and Y.Y. All authors have read and agreed to the published version of the manuscript.

Funding: This research was funded by the National Natural Science Foundation of China (Nos. 52173144, 61703446, and 51803188), the Excellent Youth Fund Project of the Henan Natural Science Foundation (No. 222300420079), and the Key Scientific and Technological Project of Henan Province (Nos. 212102210635 and 222102210264).

Institutional Review Board Statement: Not applicable.

Informed Consent Statement: Not applicable.

Data Availability Statement: Not applicable.

Conflicts of Interest: The authors declare no conflict of interest.

References

1. Gao, Y.; Peng, J.; Zhou, M.; Yang, Y.; Wang, X.; Wang, J.; Cao, Y.; Wang, W.; Wu, D. A multi-model, large range and anti-freezing sensor based on a multi-crosslinked poly(vinyl alcohol) hydrogel for human-motion monitoring. *J. Mater. Chem. B* **2020**, *8*, 11010–11020. [CrossRef] [PubMed]
2. Cai, G.; Wang, J.; Qian, K.; Chen, J.; Li, S.; Lee, P.S. Extremely stretchable strain sensors based on conductive self-healing dynamic cross-links hydrogels for human-motion detection. *Adv. Sci.* **2017**, *4*, 1600190. [CrossRef] [PubMed]
3. Zhang, J.; Wan, L.; Gao, Y.; Fang, X.; Lu, T.; Pan, L.; Xuan, F. Highly stretchable and self-healable mxene/polyvinyl alcohol hydrogel electrode for wearable capacitive electronic skin. *Adv. Electron. Mater.* **2019**, *5*, 1900285. [CrossRef]
4. Wang, X.; Chen, G.; Tian, J.; Wan, X. Chitin/Ca solvent-based conductive and stretchable organohydrogel with anti-freezing and anti-drying. *Int. J. Biol. Macromol.* **2022**, *207*, 484–492. [CrossRef] [PubMed]
5. Zhang, J.-W.; Zhang, Y.; Li, Y.-Y.; Wang, P. Textile-based flexible pressure sensors: A review. *Polym. Rev.* **2022**, *62*, 65–94. [CrossRef]
6. Zhang, L.; Wang, J.; Wang, S.; Wang, L.; Wu, M. Neuron-inspired multifunctional conductive hydrogels for flexible wearable sensors. *J. Mater. Chem. C* **2022**, *10*, 4327–4335. [CrossRef]
7. Liu, C.; Wang, X.; Zhang, H.J.; You, X.; Yue, O. Self-healable, high-strength hydrogel electrode for flexible sensors and supercapacitors. *ACS Appl. Mater. Interfaces* **2021**, *13*, 36240–36252. [CrossRef]
8. Park, K.; Choi, H.; Kang, K.; Shin, M.; Son, D. Soft stretchable conductive carboxymethylcellulose hydrogels for wearable sensors. *Gels* **2022**, *8*, 92. [CrossRef]
9. Zhang, J.; Wang, Y.; Wei, Q.; Wang, Y.; Lei, M.; Li, M.; Li, D.; Zhang, L.; Wu, Y. Self-healing mechanism and conductivity of the hydrogel flexible sensors: A review. *Gels* **2021**, *7*, 216. [CrossRef]
10. Mao, Z.; Kuroki, M.; Otsuka, Y.; Maeda, S. Contraction waves in self-oscillating polymer gels. *Extrem. Mech. Lett.* **2020**, *39*, 100830. [CrossRef]
11. Maeda, S.; Hara, Y.; Yoshida, R.; Hashimoto, S. Peristaltic motion of polymer gels. *Angew. Chem. Int. Ed.* **2008**, *47*, 6690–6693. [CrossRef] [PubMed]
12. Rukmanikrishnan, B.; Lee, J. Anti-freezing and thermally self-healing polymer composite comprising polyvinyl alcohol, polyethylene oxide, and sodium carboxymethyl cellulose. *Eur. Polym. J.* **2021**, *154*, 110565. [CrossRef]

13. Chen, S.; Huang, J.; Zhou, Z.; Chen, Q.; Hong, M.; Yang, S.; Heqing, F. Highly elastic anti-fatigue and anti-freezing conductive double network hydrogel for human body sensors. *Ind. Eng. Chem. Res.* **2021**, *60*, 6162–6172. [CrossRef]
14. Ding, Q.; Wu, Z.; Tao, K.; Wei, Y.; Wang, W.; Yang, B.R.; Xie, X.; Wu, J. Environment tolerant, adaptable and stretchable organohydrogels: Preparation, optimization, and applications. *Mater. Horiz.* **2022**, *9*, 1356–1386. [CrossRef] [PubMed]
15. Li, G.; Li, C.; Li, G.; Yu, D.; Song, Z.; Wang, H.; Liu, X.; Liu, H.; Liu, W. Development of conductive hydrogels for fabricating flexible strain sensors. *Small* **2022**, *18*, e2101518. [CrossRef]
16. Wang, J.; Dai, T.; Zhou, Y.; Mohamed, A.; Yuan, G.; Jia, H. Adhesive and high-sensitivity modified $Ti_3C_2T_X$ (MXene)-based organohydrogels with wide work temperature range for wearable sensors. *J. Colloid Interface Sci.* **2022**, *613*, 94–102. [CrossRef]
17. Rong, Q.; Lei, W.; Chen, L.; Yin, Y.; Zhou, J.; Liu, M. Anti-freezing, conductive self-healing organohydrogels with stable strain-sensitivity at subzero temperatures. *Angew. Chem. Int. Ed.* **2017**, *56*, 14159–14163. [CrossRef]
18. Li, M.; Chen, D.; Sun, X.; Xu, Z.; Yang, Y.; Song, Y.; Jiang, F. An environmentally tolerant, highly stable, cellulose nanofiber-reinforced, conductive hydrogel multifunctional sensor. *Carbohydr. Polym.* **2022**, *284*, 119199. [CrossRef]
19. Wu, J.; Wu, Z.; Huang, W.; Yang, X.; Liang, Y.; Tao, K.; Yang, B.R.; Shi, W.; Xie, X. Stretchable, stable, and room-temperature gas sensors based on self-healing and transparent organohydrogels. *ACS Appl. Mater. Interfaces* **2020**, *12*, 52070–52081. [CrossRef]
20. Liu, Y.; Wang, W.; Gu, K.; Yao, J.; Shao, Z.; Chen, X. Poly(vinyl alcohol) hydrogels with integrated toughness, conductivity, and freezing tolerance based on ionic liquid/water binary solvent systems. *ACS Appl. Mater. Interfaces* **2021**, *13*, 29008–29020. [CrossRef]
21. Wang, J.; Li, Q.; Li, K.; Sun, X.; Wang, Y.; Zhuang, T.; Yan, J.; Wang, H. Ultra-high electrical conductivity in filler-free polymeric hydrogels toward thermoelectrics and electromagnetic interference shielding. *Adv. Mater.* **2022**, *34*, e2109904. [CrossRef] [PubMed]
22. Morelle, X.P.; Illeperuma, W.R.; Tian, K.; Bai, R.; Suo, Z.; Vlassak, J.J. Highly stretchable and tough hydrogels below water freezing temperature. *Adv. Mater.* **2018**, *30*, e1801541. [CrossRef] [PubMed]
23. Liu, X.; Zhang, Q.; Gao, G. Solvent-resistant and nonswellable hydrogel conductor toward mechanical perception in diverse liquid media. *ACS Nano* **2020**, *14*, 13709–13717. [CrossRef] [PubMed]
24. Jiang, C.; Zhu, T.; Liu, H.; Yang, G.; He, Z.; Wang, M.; Ji, M.; Cong, G.; Yu, J.; Zhu, C.; et al. A one-step aqueous route to prepare polyacrylonitrile-based hydrogels with excellent ionic conductivity and extreme low temperature tolerance. *J. Mater. Chem. A* **2020**, *8*, 22090–22099. [CrossRef]
25. Liang, C.; Liu, Y.; Ruan, Y.; Qiu, H.; Song, P.; Kong, J.; Zhang, H.; Gu, J. Multifunctional sponges with flexible motion sensing and outstanding thermal insulation for superior electromagnetic interference shielding. *Compos. Part A Appl. Sci. Manuf.* **2020**, *139*, 106143. [CrossRef]
26. Wang, Z.; Cong, Y.; Fu, J. Stretchable and tough conductive hydrogels for flexible pressure and strain sensors. *J. Mater. Chem. B* **2020**, *8*, 3437–3459. [CrossRef]
27. Liu, H.; Wang, X.; Cao, Y.; Yang, Y.; Yang, Y.; Gao, Y.; Ma, Z.; Wang, J.; Wang, W.; Wu, D. Freezing-tolerant, highly sensitive strain and pressure sensors assembled from ionic conductive hydrogels with dynamic cross-links. *ACS Appl. Mater. Interfaces* **2020**, *12*, 25334–25344. [CrossRef]
28. Zhang, J.; Chen, L.; Shen, B.; Mo, J.; Tang, F.; Feng, J. Highly stretchable and self-healing double network hydrogel based on polysaccharide and polyzwitterion for wearable electric skin. *Polymer* **2020**, *194*, 122381. [CrossRef]
29. Zhang, J.; Chen, L.; Shen, B.; Wang, Y.; Peng, P.; Tang, F.; Feng, J. Highly transparent, self-healing, injectable and self-adhesive chitosan/polyzwitterion-based double network hydrogel for potential 3D printing wearable strain sensor. *Mater. Sci. Eng. C* **2020**, *117*, 111298. [CrossRef]
30. Xia, M.; Pan, S.; Li, H.; Yi, X.; Zhan, Y.; Sun, Z.; Jiang, X.; Zhang, Y. Hybrid double-network hydrogel for highly stretchable, excellent sensitive, stabilized, and transparent strain sensors. *J. Biomater. Sci. Polym. Ed.* **2021**, *32*, 1548–1563. [CrossRef]
31. Li, H.; Zheng, H.; Tan, Y.J.; Tor, S.B.; Zhou, K. Development of an ultrastretchable double-network hydrogel for flexible strain sensors. *ACS Appl. Mater. Interfaces* **2021**, *13*, 12814–12823. [CrossRef] [PubMed]
32. Yang, Y.; Yang, Y.; Cao, Y.; Wang, X.; Chen, Y.; Liu, H.; Gao, Y.; Wang, J.; Liu, C.; Wang, W.; et al. Anti-freezing, resilient and tough hydrogels for sensitive and large-range strain and pressure sensors. *Chem. Eng. J.* **2021**, *403*, 126431. [CrossRef]
33. Yang, Y.; Wu, D. Energy-dissipative and soften resistant hydrogels based on chitosan physical network: From construction to application. *Chin. J. Chem.* **2022**, *40*, 2118–2134. [CrossRef]
34. Zhang, X.F.; Ma, X.; Hou, T.; Guo, K.; Yin, J.; Wang, Z.; Shu, L.; He, M.; Yao, J. Inorganic salts induce thermally reversible and anti-freezing cellulose hydrogels. *Angew. Chem. Int. Ed.* **2019**, *58*, 7366–7370. [CrossRef]
35. Wu, D.; Wang, X.; Yang, Y. Chitosan-based high-mechanical double-network hydrogels: Construction, modulation and applications. *Acta Chim. Sin.* **2021**, *79*, 1–9.
36. Meijer, H.E.H.; Govaert, L.E. Mechanical performance of polymer systems: The relation between structure and properties. *Prog. Polym. Sci.* **2005**, *30*, 915–938. [CrossRef]
37. Zhao, L.; Ren, Z.; Liu, X.; Ling, Q.; Li, Z.; Gu, H. A Multifunctional, Self-healing, self-adhesive, and conductive sodium alginate/poly(vinyl alcohol) composite hydrogel as a flexible strain sensor. *ACS Appl. Mater. Interfaces* **2021**, *13*, 11344–11355. [CrossRef]

gels

Article

Synthesis and Hydrogelation of Star-Shaped Graft Copolypetides with Asymmetric Topology

Thi Ha My Phan [1,†], Yu-Hsun Yang [1,†], Yi-Jen Tsai [1], Fang-Yu Chung [1], Tooru Ooya [2,3], Shiho Kawasaki [2] and Jeng-Shiung Jan [1,4,*]

1 Department of Chemical Engineering, National Cheng Kung University, Tainan 70101, Taiwan; n36087113@gs.ncku.edu.tw (T.H.M.P.); e34074093@gs.ncku.edu.tw (Y.-H.Y.); n36094704@gs.ncku.edu.tw (Y.-J.T.); f54081036@gs.ncku.edu.tw (F.-Y.C.)

2 Graduate School of Engineering, Kobe University, Kobe 657-8501, Japan; ooya@tiger.kobe-u.ac.jp (T.O.); shihoshi@stu.kobe-u.ac.jp (S.K.)

3 Center of Advanced Medical Engineering Research & Development (CAMED), Kobe University, Kobe 657-8501, Japan

4 Hierarchical Green-Energy Materials (Hi-GEM) Research Center, National Cheng Kung University, Tainan 70101, Taiwan

* Correspondence: jsjan@mail.ncku.edu.tw

† These authors contributed equally to this work.

Abstract: To study the self-assembly and hydrogel formation of the star-shaped graft copolypeptides with asymmetric topology, star-shaped poly($_L$-lysine) with various arm numbers were synthesized by using asymmetric polyglycerol dendrimers (PGDs) as the initiators and 1,1,3,3-tetramethylguanidine (TMG) as an activator for OH groups, followed by deprotection and grafting with indole or phenyl group on the side chain. The packing of the grafting moiety via non-covalent interactions not only facilitated the polypeptide segments to adopt more ordered conformations but also triggered the spontaneous hydrogelation. The hydrogelation ability was found to be correlated with polypeptide composition and topology. The star-shaped polypeptides with asymmetric topology exhibited poorer hydrogelation ability than those with symmetric topology due to the less efficient packing of the grafted moiety. The star-shaped polypeptides grafted with indole group on the side chain exhibited better hydrogelation ability than those grafted with phenyl group with the same arm number. This report demonstrated that the grafted moiety and polypeptide topology possessed the potential ability to modulate the polypeptide hydrogelation and hydrogel characteristics.

Keywords: hydrogel; star-shaped polymer; polypeptide; topology

1. Introduction

Hydrogels are a type of biomaterials possessing the ability to contain a large amount of water inside their three-dimensional network, which makes their special properties such as living cells mimetic, biodegradable, function and conformation adjustable [1–5]. Moreover, with tunable mechanical properties, hydrogels have drawn significant attention to investigate their applications in biomedical fields such as wound dressing, drug delivery systems, and tissue engineering [6–8]. Hydrogel was defined first time by Bemmelen in 1894, and its first application was reported in biomedical fields by Wichterle and Lim in 1960 [9–11]. Amongst all types of hydrogels, peptide-based hydrogels have self-assembly ability for hydrogel formation through hydrogen bonding, aromatic (cation-π, π-π), and hydrophobic/hydrophilic interactions [12–15]. Additionally, peptide segments can adopt secondary conformations (random coil, β-sheet/β-turn) inside their gel network to enhance the mechanical properties of hydrogels [16–20]. As a type of synthetic hydrogel, polypeptides can modify the characteristics of hydrogels by varying the functional groups on the side chain or the chain composition on the backbone [21–23].

Traditionally, star-shaped polypeptides were synthesized by using dendrimers bearing many amino groups with symmetric topology as the initiators [24,25]. In 2016, the Heise group reported the synthesis of 8-armed block copolypeptides using a polypropylene imine dendrimer as the core of the polymers, and the hydrogelation of those samples was good with critical gelation concentrations (CGCs) lower than 2.0 wt% [25]. Recently, the polyol initiators were used with the promotion of 1,1,3,3-tetramethylguanidine (TMG) as an activator for hydroxyl (OH) groups to favor the ring-opening polymerization (ROP) of *N*-carboxyanhydride (NCA) reaction [21,22,26,27]. Most of the previous studies investigated the hydrogelation of the star-shaped block/graft polypeptides with symmetric structures [22,23]. In 2018 and 2019, our group reported that the star-shaped graft copolypeptides using the symmetric polyol initiators exhibited good hydrogelation ability in deionized water [23,28].

In this study, we synthesized the star-shaped polypeptides grafted with indole and phenyl groups, using polyglycerol dendrimers of generations 1, 2, and 3 (PGDs G1, G2, and G3) [27,29] as the asymmetric initiators to investigate the effect of grafted moiety, arm number, and polypeptide topology on the hydrogelation of the star-shaped graft copolypeptides. One previous paper has used PGD G2 as an initiator for the synthesis of the block copolypeptides and studied the hydrogelation of these polypeptides [21]. The PGD G1 and G3 have never been used as initiators for the synthesis of polypeptides. Herein, this study reported the synthesis and hydrogelation of asymmetric, star-shaped graft copolypeptides with 6, 12, and 24 arms using PGD G1, G2, and G3 as initiators, respectively. Moreover, indole and phenyl groups were employed as the grafting moieties for the polypeptides. The as-prepared polypeptides were characterized by a variety of analytical instruments to confirm the successful synthesis. Furthermore, these samples were analyzed to characterize their polypeptide chain conformation, assembled nano-/micro-structure in the hydrogel network, and hydrogel mechanical strength. Recovery behavior was also conducted to demonstrate their injectable ability.

2. Results and Discussion

2.1. Star-Shaped Graft Copolypeptide Synthesis and Characterization

As shown in Scheme 1, the 6-armed poly(Z-L-lysine), 12-armed poly(Z-L-lysine), and 24-armed poly(Z-L-lysine) (6s-PZLL, 12s-PZLL, 24s-PZLL) homopolypeptides were synthesized by ROP of ZLL NCA with PGD G1, G2, and G3 as an asymmetric initiator, respectively, using TMG to activate the OH groups. The homopolypeptides were dissolved in trifluoroacetic acid-d_1 (TFA-d_1) and *N*, *N*-dimethylformamide (DMF) for proton nuclear magnetic resonance (^{1}H NMR) and gel permeation chromatography–light scattering (GPC-LS) analyses, respectively, and the results for homopolypeptide characterization were summarized in Table S1, Figure 1 and Figure S1. Based on ^{1}H NMR data (Figure S1 and Figure 1a), the degree of polymerization (DP) was calculated from the ratio of integral areas of d protons on the phenyl group of ZLL block to b protons on the initiator. The integrated values of other protons on ZLL could be used to cross-check the results. Moreover, the calculation of DPs and number-averaged molecular weights (M_n) derived from the GPC-LS analysis gave good agreement with those from ^{1}H NMR analysis (Table S1). As shown in Figure S1, the proton b on the initiators at around 4.0–4.1 ppm demonstrated the successful synthesis of the star-shaped homopolypeptides by the activating ability of TMG on OH groups of the PGD initiators. This result is consistent with those from ^{1}H–^{13}C two-dimensional Heteronuclear Single Quantum Correlation (^{1}H–^{13}C 2D HSQC) analysis in a previous study, which reported that the proton b on the initiator was shifted from 3.82 ppm to 4.1 ppm after blocked with PLL [21]. After removing the Z groups on ZLL segments by using HBr, the star-shaped poly(L-lysine) (s-PLL) samples were dialyzed against deionized water (DIW) and obtained by freeze-drying. The s-PLL samples dissolved in deuterated water (D_2O) were analyzed by ^{1}H NMR. For an example of 12s-PLL in Figure 1b′, it is obvious that the Z groups were successfully removed with the remaining percentage lower than 5.0%, as reported in the literature [21,22,27].

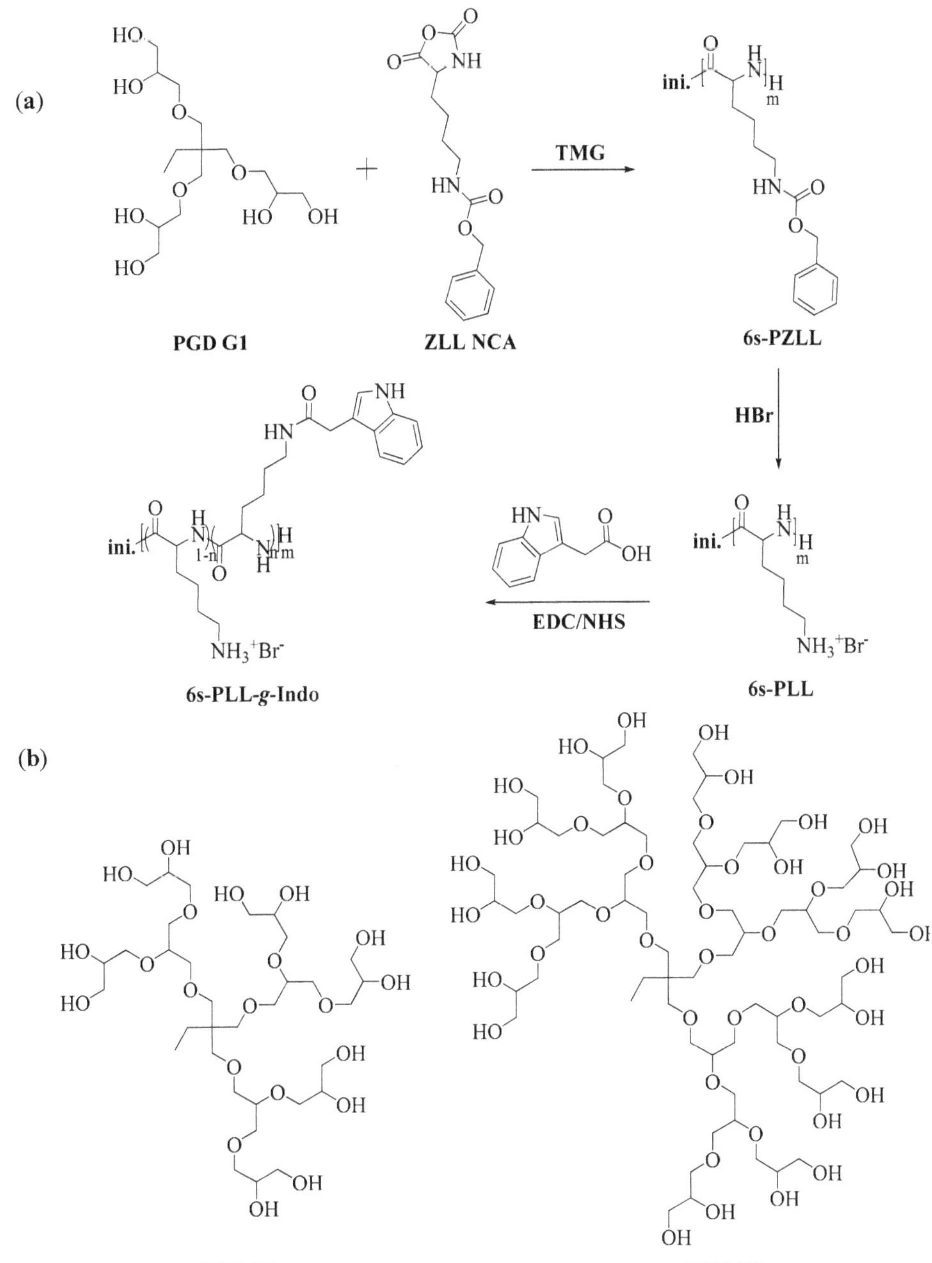

Scheme 1. (**a**) The representative procedure of synthesis of 6-armed poly(L-lysine)-*graft*-indole (6s-PLL-*g*-Indo) using polyglycerol dendrimer of generation 1 (PGD G1) as the 6-armed initiator. (**b**) The structures of PGDs of generations 2 and 3 (G2 and G3) as 12-armed and 24-armed initiators, respectively. Abbreviations: Z-L-lysine *N*-carboxyanhydride (ZLL NCA), 1,1,3,3-tetramethylguanidine (TMG), 6-armed poly(Z-L-lysine) (6s-PZLL), hydrogen bromide (HBr), 6-armed poly(L-lysine) (6s-PLL), 1-ethyl-3-(-3-dimethylaminopropyl) carbodiimide/*N*-hydroxysuccinimide (EDC/NHS).

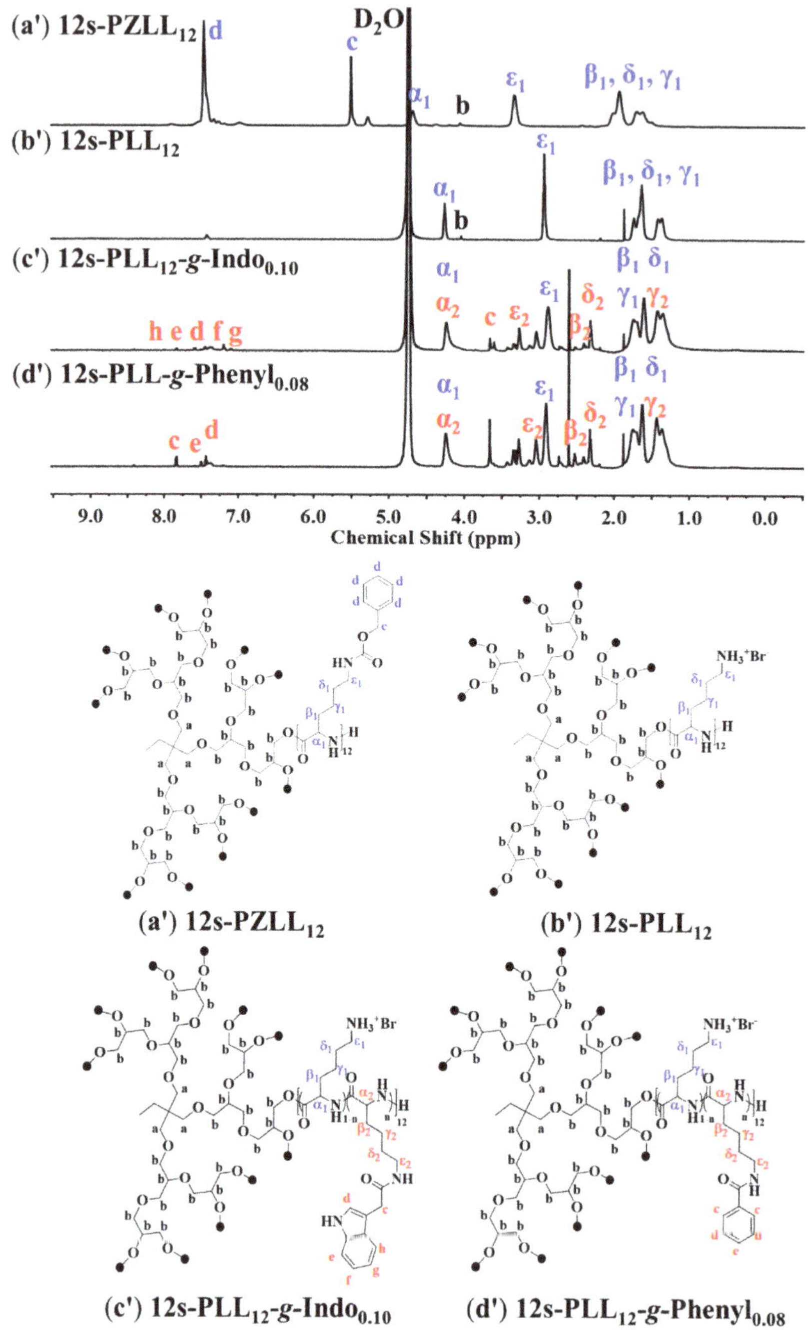

Figure 1. ^{1}H NMR spectra of (**a′**) 12-armed poly(Z-L-lysine)$_{12}$ (12s-PZLL$_{12}$) in TFA-d_1, (**b′**) 12-armed poly(L-lysine)$_{12}$ (12s-PLL$_{12}$) in D_2O, (**c′**) 12-armed poly(L-lysine)$_{12}$-*graft*-indole$_{0.10}$ (12s-PLL$_{12}$-*g*-Indo$_{0.10}$) in D_2O, and (**d′**) 12-armed poly(L-lysine)$_{12}$-*graft*-phenyl$_{0.08}$ (12s-PLL$_{12}$-*g*-Phenyl$_{0.08}$) in D_2O. The symbols were used to represent the different protons.

The s-PLL polypeptides were conjugated with indolyl and phenyl groups by reacting with 3-indoleacetic acid and benzoic acid, respectively, using 1-ethyl-3-(-3-dimethylaminopropyl) carbodiimide/*N*-hydroxysuccinimide (EDC/NHS) as activators (an example of 6s-PLL-*g*-Indo as shown in Scheme 1a). After dialysis and lyophilization, the samples were dissolved in D_2O for ^1H NMR analysis. The integrated ratio of the protons on the graft moiety to the α_1 proton on the PLL block was calculated to determine the grafting ratio and grafting efficiency (Figure 1c′,d′). In Table 1, the grafting efficiency was in the range 26–43%, which is much lower than those in a reported study (75–83%) for 6-armed graft copolypeptides using dipentaerythritol as an initiator [23]. It could be possibly explained by the steric hindrance of the higher number of arms and asymmetric topology.

Table 1. The grafting ratio, grafting efficiency, critical gelation concentration (CGC) and the percentages of various secondary conformations of star-shaped graft copolypeptides. The secondary structure data was fitted from the CD analysis using the BeStSel website.

Polypeptide	Grafting Ratio (%)	Grafting Efficiency (%)	CGC (wt%)	Secondary Structure	
				Random Coil (%)	β-Sheet/β-Turn (%)
6s-PLL$_{32}$-*g*-Indo$_{0.08}$	7.9	26	- *	41.9	37.2/16.1
6s-PLL$_{32}$-*g*-Phenyl$_{0.08}$	7.9	26	- *	42.8	37.3/16.2
12s-PLL$_{12}$-*g*-Indo$_{0.10}$	9.8	33	5.85	43.1	36.7/16.8
12s-PLL$_{12}$-*g*-Phenyl$_{0.08}$	8.1	27	6.3	44.5	35.5/16.6
24s-PLL$_{12}$-*g*-Indo$_{0.11}$	11.2	37	2.9	42.6	39.1/15.4
24s-PLL$_{12}$-*g*-Phenyl$_{0.13}$	12.8	43	3.6	43.3	36.7/15.9

* The polypeptide sample did not form hydrogel at 10.0 wt%.

2.2. Critical Gelation Concentrations (CGCs) and Hydrogelation of Star-Shaped Graft Copolypeptides

The star-shaped graft copolypeptides were homogeneously dissolved in DIW by vortex mixing and sonication before the vial inversion approach to determine whether their CGCs were lower than 10.0 wt%. It is obvious that the 12-armed and 24-armed graft copolypeptides have good hydrogelation ability, with the CGC values ranging from 2.9 to 6.3 wt%, whilst the 6-armed ones cannot form hydrogel at a concentration lower than 10.0 wt% (Table 1). Moreover, the 24-armed poly(L-lysine)$_{12}$-*graft*-indole$_{0.11}$ (24s-PLL$_{12}$-*g*-Indo$_{0.11}$) (2.9 wt%) and 24-armed poly(L-lysine)$_{12}$-*graft*-phenyl$_{0.13}$ (24s-PLL$_{12}$-*g*-Phenyl$_{0.13}$) (3.6 wt%) exhibited lower CGCs than the 12s-PLL$_{12}$-*g*-Indo$_{0.10}$ (5.85 wt%) and 12s-PLL$_{12}$-*g*-Phenyl$_{0.08}$ (6.3 wt%). It revealed that the star-shaped polypeptides with comparable grafting ratio exhibited better hydrogelation ability upon the increase in the arm number. Moreover, the star-shaped polypeptides grafted with indole group exhibited better hydrogelation ability than those grafted with phenyl group. The results are consistent with those reported in the previous studies [21–23,27]. It can be explained that higher number of arms gave more branching chains in the polypeptide structure. We previously reported that 6-armed poly(L-lysine)$_{30}$-*graft*-indole$_{0.30}$ (6s-PLL$_{30}$-*g*-Indo$_{0.30}$) synthesized by using dipentaerythritol as the initiator possessed an excellent hydrogelation with the CGC of 0.75 wt%, whilst this study showed that 6s-PLL$_{32}$-*g*-Indo$_{0.08}$ using PGD G1 with asymmetric topology did not form hydrogel below 10.0 wt% [23]. It demonstrated that polymer topology has a strong effect on the hydrogelation of the star-shaped polypeptides and the symmetric topology could render the star-shaped polypeptides exhibiting a better gel-forming behavior via efficient packing. With the same arm number, the star-shaped polypeptides conjugated with indole group showed better hydrogelation ability than the ones conjugated with phenyl group. The hydrogen bonding, hydrophobic, and aromatic (including cation-π, π-π) interactions of the indole group could establish efficient facilitation for the hydrogelation of the star-shaped polypeptides. Although the grafted phenyl group cannot efficiently facilitate the formation of hydrogels as good as the grafted indole group, the hydrophobic and aromatic interactions of the phenyl group still could assist the hydrogelation of the corresponding polypeptides with low CGCs (6.3 wt% for 12s-PLL_{12}-*g*-$Phenyl_{0.08}$ and 3.6 wt% for 24s-PLL_{12}-*g*-$Phenyl_{0.13}$). Using PGD G2 as a polyol initiator, a previous study reported that 12-armed-poly(L-lysine)$_{16}$-*block*-poly(L-phenylalanine)$_5$ (12s-PLL_{16}-*b*-PLF_5) exhibited a CGC of 3.0 wt% [21], which is lower than that of 12s-PLL_{12}-*g*-$Phenyl_{0.08}$ (6.3 wt%). It suggested that the star-shaped block polypeptides have a better ability to form hydrogel than the star-shaped graft counterparts. The hydrophobic segment on the star-shaped block polypeptides could exhibit higher efficiency in packing in an aqueous environment than the hydrophobic group on the star-shaped graft counterparts. Amongst all samples, the 24s-PLL_{12}-*g*-$Indo_{0.11}$ exhibited the best hydrogelation ability, evidenced by its lowest CGC of 2.9 wt%. The results demonstrate that the functional groups and symmetry have a critical effect on the hydrogel formation of the star-shaped graft polypeptides.

2.3. Secondary Structure of Star-Shaped Graft Copolypeptides

Circular dichroism (CD) measurement was employed to determine the adoption of the secondary conformation for the polypeptide samples under neutral environments (Figure S2). It was obvious that a negative dominant peak was shown at 190–200 nm in the CD spectra of all polypeptide samples, which indicated the existence of random coil conformation. Moreover, a curved peak existing at a wavelength range of 215–225 nm was representative of β-motif conformation. The CD results were fitted to compute the percentage of polypeptide secondary conformation by using BeStSel website as shown in Table 1. The major conformations adopted by the star-shaped graft copolypeptides were random coil and β-motifs (β-sheet and β-turn). It was renowned that the random coil conformation was originated from the PLL segment at a neutral environment [21,22,29,30]. Upon grafting the functional group, the PLL segment would undergo coil-to-sheet conformational transition, due to the multiple interactions between the functional group. The results are in a good agreement with those reported in the previous studies [31–33]. As shown in Table 1, the percentage of β-sheet/β-turn conformations (50.0–60.0%) was much higher than the molar ratio of the functional groups grafted onto the star-shaped polypeptides. It demonstrated that the confinement of the PLL segment exerted by the packing of the grafted moiety would facilitate the polypeptide chains to adopt more ordered conformation (that is, β-sheet/β-turn) as well as hydrogelation [21,22,27]. It can be seen that the polypeptides grafted with indole group adopted slightly higher β-motif conformation than these grafted with phenyl group with the same arm number. The results exhibited that β-motif structures could enhance the hydrogelation in the network of the star-shaped graft polypeptides.

2.4. Self-Assembled Nanostructure and Gel Morphology of Star-Shaped Graft Copolypeptides

The macrostructures of the star-shaped graft copolypeptides were observed by field emission scanning electron microscopy (FE-SEM) analysis. The polypeptides were dissolved in DIW for gel formation and lyophilized to obtain the samples which possessed the native morphology for the SEM analysis [34,35]. All polypeptide samples after lyophilization showed three-dimensional (3D) networks with porous structures (Figure 2). Moreover, 24-armed samples exhibited larger pore sizes than 12-armed ones, which could facilitate the 24-armed samples to accommodate more water in the networks with lower CGCs [21,22].

X-ray diffraction (XRD) analysis was operated on lyophilized polypeptide samples with a range of 2θ = 5°–40° to characterize the molecular packing in the polypeptide hydrogel networks. As shown in Figure 3, all lyophilized polypeptide samples exhibited two major peaks at 2θ = 10°–12.5° and 20°–22.5°, which indicated the existence of the aromatic stacking by the grafted moieties and the amorphous state, respectively. Moreover, XRD data was used to compute the corresponding d spacing of the aromatic stacking ranged between 0.70 and 0.77 mm. [36,37].

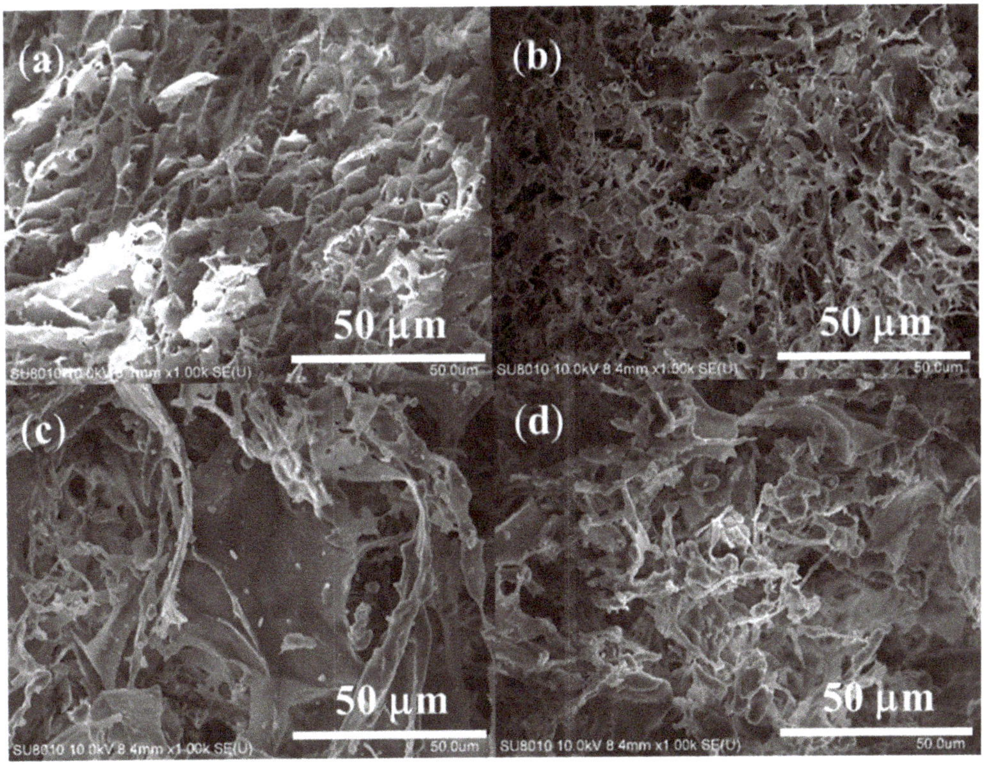

Figure 2. FE-SEM images of lyophilized (**a**)12s-PLL$_{12}$-*g*-Indo$_{0.10}$, (**b**) 12s-PLL$_{12}$-*g*-Phenyl$_{0.08}$, (**c**) 24s-PLL$_{12}$-*g*-Indo$_{0.11}$ and (**d**) 24s-PLL$_{12}$-*g*-Phenyl$_{0.13}$ samples. The polypeptide concentrations of 12-armed and 24-armed polypeptides were 8.0 wt% and 5.0 wt%, respectively.

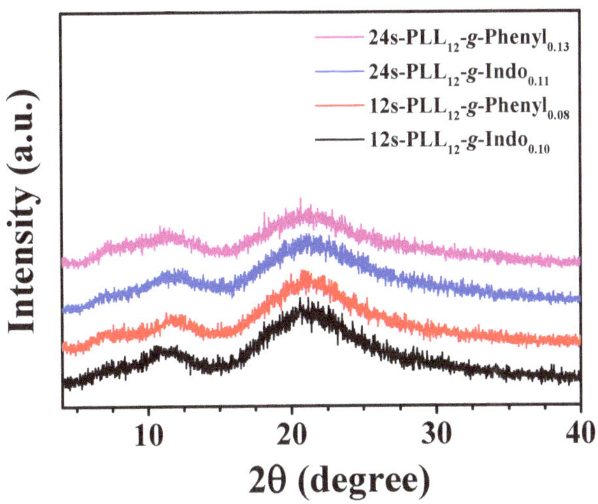

Figure 3. XRD patterns of 12-armed and 24-armed graft polypeptides. All the samples were lyophilized before analyzing. The concentration of 12-armed and 24-armed polypeptide hydrogels were 8.0 wt% and 5.0 wt%, respectively.

The self-assembled nanostructures of the polypeptide hydrogels (8.0 wt%) were characterized by small-angle X-ray scattering (SAXS) analysis (Figure 4). For all hydrogel samples, the scattering intensity (I) exhibited a relation of $I(q) \propto q^{-n}$ with n = 3.2–3.8, indicating the network forming 3D assembled nanostructures with ill-defined morphology [38,39]. The SAXS data were fitted by SasView software to compute the radius of gyration (R_g) of nano-assemblies formed by the packing of the grafted moieties. As shown in Table S2, it is obvious that the 24-armed polypeptide hydrogels exhibited higher R_g than the 12-armed ones. The result showed that the polypeptide samples with the higher arm number had a tendency to form larger nano-assemblies in the hydrogel network.

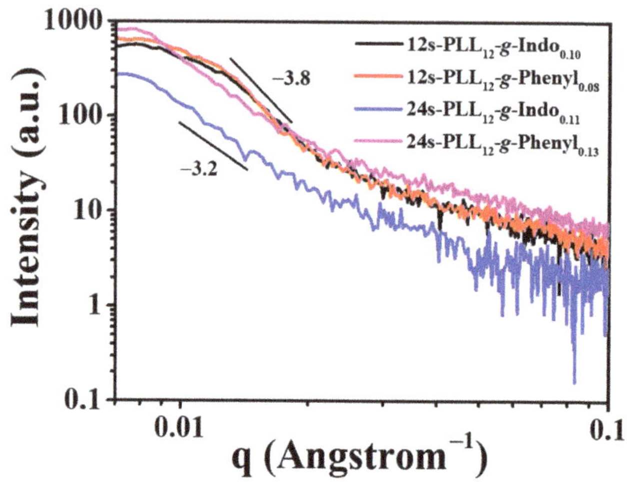

Figure 4. SAXS spectra of 12-armed and 24-armed polypeptide hydrogel samples with the polypeptide concentration of 8.0 wt%.

2.5. Mechanical Strength and Recovery Behavior of Star-Shaped Graft Copolypeptides

The mechanical strengths of the polypeptide hydrogels (concentration: 8.0 wt%) were characterized by employing rheological measurements at room temperature (RT) (Figure 5). The strength properties of the samples were observed under two experimental conditions: changing of strain and angular frequency. The tendency of storage modulus (G′) and loss modulus (G″) following the experiments were considered to estimate the mechanical properties of the polypeptide hydrogels. It was obvious that all hydrogel samples showed the G′ values higher than G″ regardless at various frequency (Figure 5a), showing no occurrence of sol-to-gel transition upon varying frequency. The 24-armed polypeptide hydrogels exhibited a higher mechanical strength than the 12-armed ones. As reported in the previous studies, the mechanical properties enhanced with the increase in polypeptide arm number [21–23,27]. The hydrogels formed by the polypeptides grafted with indole group exhibited higher mechanical properties than those formed by the polypeptides grafted with phenyl group with the same arm number. Previous studies have shown that the hydrogels formed by the polypeptides adopting more β-motifs exhibited higher mechanical properties [21,22,27,40,41]. Amongst all the hydrogel samples, 24s-PLL$_{12}$-g-Indo$_{0.11}$ adopted the highest percentage of β-motif conformation possessed the highest mechanical strength. As shown in Figure 5b, all the polypeptide hydrogel samples possessed shear-thinning behavior, evidenced by the decrease in G′ and G″ upon the increase in strain. It could be seen a crossover point of G′ and G″ at the strain lower than 100.0% for each sample, showing the strain limitation to deform the polypeptide hydrogel network. Moreover, the 24-armed polypeptide hydrogels exhibited crossover points at larger strain amplitude than the 12-armed ones, which indicated that the rigidity of the 24-armed polypeptide hydrogels were higher than that of 12-armed ones. The results showed that the mechanical strengths of polypeptide hydrogels were strongly dependent on grafted moiety and polypeptide topology.

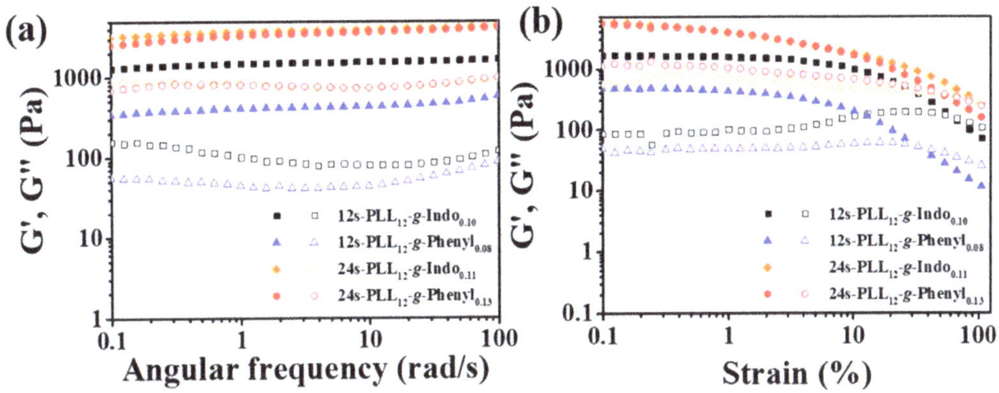

Figure 5. Storage modulus (G′: closed symbol) and loss modulus (G″: open symbol) of (**a**) 12-armed and 24-armed graft polypeptide hydrogels under different angular frequencies at fixed strain (0.1%) and (**b**) various strains at an angular frequency of 1.0 rad/s. The polypeptide concentration of all samples was 8.0 wt%.

To characterize the recovery behavior of the polypeptide hydrogels, the rheological measurements were operated at RT with the following conditions: fixed-frequency = 1 rad/s, various strain sweep values (1.0% for 100 s, 100.0% for 300 s, and 1.0% for 600 s). As shown in Figure S3, the G′ of hydrogel samples declined sharply as the strain increased to 100.0%, which is an indication of the breaking down of the gel structures, and the G′ value recovered to the original state after returning the strain to the initial value. The results showed that the polypeptide hydrogels exhibited good recovery ability, demonstrating the potential to be used as injectable materials [23,28].

3. Conclusions

In this study, we have reported the successful synthesis of the star-shaped poly(L-lysine) grafted with indole and phenyl groups using PGDs as the asymmetric initiators with various arm numbers. The 12-armed and 24-armed graft copolypeptides exhibited good hydrogelation ability with CGCs from 2.9 wt% to 6.3 wt% whilst the 6-armed counterparts could not undergo hydrogelation at a polypeptide concentration lower than 10.0 wt%. The polypeptide topology (arm number and symmetry) and various non-covalent interactions exerted by the grafted moiety played a critical role in the hydrogelation of the polypeptides. The packing of the grafted moiety resulted in the confinement of the PLL segment, facilitating the polypeptide chains to adopt more ordered conformation as well as hydrogelation. This study has given a potential insight for the synthesis and mechanical characteristic modification of the star-shaped polypeptide hydrogels by changing the polypeptide composition and network topology.

4. Materials and Methods

4.1. Materials

All chemicals were purchased and used directly without purification unless otherwise noted. ZLL NCA was synthesized by using the reported procedure [21–23,42]. PGD G1 (M_n = 365), G2(M_n = 800), and G3 (M_n = 1690) were synthesized according to our previous reports [43]. Anhydrous hexane (ECHO, Tainan, Taiwan) and tetrahydrofuran (THF, J.T.Baker, Phillipsburg, NJ, USA) were used after the removal of water by stirring with calcium hydride (90–95%, Alfa Aesar, Lancashire, England) and sodium (99.95%, in mineral oil, Aldrich, Auvergne Rhone Alpes, France) overnight, respectively. DMF (Macron, Croydon, PA, USA) was dried by activated 4A molecular sieves (UniRegion Bio-Tech, Hsinchu, Taiwan).

4.2. Synthesis and Characterization of Star-Shaped Poly(Z-L-lysine) Polypeptides (s-PZLL)

Following the reported procedure, s-PLL homopolypeptides were synthesized by using ROP as shown in Scheme 1 [21,22,27]. The reaction was promoted by using TMG as an activator for the OH groups on the initiators for ROP of ZLL NCA. In this study, PGDs G1, G2, and G3 were used as 6-armed, 12-armed, and 24-armed initiators for the synthesis of star-shaped polypeptides with an asymmetric topology, respectively. The synthesis of 6-armed poly(Z-L-lysine) (6s-PZLL) is given below as a detailed example.

Under an inert atmosphere, 11.65 mg of PGD G1 was dissolved in anhydrous DMF (1.634 mL) contained in a 60 mL round bottom flask with the addition of TMG (7.42 μL, Aldrich, Beijing, China) as a promoter. The solution was stirred and heated for 30 min until the mixture was becoming homogeneous. Then, 1.2 g of ZLL NCA was totally dissolved in anhydrous DMF (3.92 mL) in another bottle before dropping into the initiator solution. The flask with the reacting solution was sealed and stirred at RT for 48 h. Then, the solution was poured into a cellulose dialysis tube (MWCO: 6000–8000 g mol^{-1}, Spectrum Laboratories, Rancho Dominguez, CA, USA) and dialyzed against methanol and DIW for 48 and 24 h, respectively. Finally, the dialyzed solution was lyophilized for 72 h to collect the white product (yield: 80–90%).

Each s-PZLL homopolypeptide sample was separately dissolved in TFA-d_1 and DMF for ^{1}H NMR and GPC-LS analyses, respectively, for polypeptide characterization. In this study, BRUKER ADVANCE III HD NMR (600 MHz, Bruker Corporation, Karlsruhe, Germany) system was employed to conduct ^{1}H NMR at RT. For GPC-LS, a system with three Viscotek detectors and two Shodex columns was used to operate the analysis at 55 °C with 0.8 mL/min of flow rate. Moreover, 0.1 M LiBr in DMF solution and polystyrene (molecular weight: 25,000 g/mol) were requested as an eluent and a calculation standard, respectively, for this operation. After dissolving in DMF, the solutions used for GPC-LS analysis were passed through a filter (polytetrafluoroethylene, 0.45 mm, 13 mm, Finetech, Taichung, Taiwan) to remove the impurities.

4.3. Deprotection of Z-Group on Polypeptides

To synthesize star-shaped poly(L-lysine) homopolypeptides (s-PLL), the Z groups were removed by using hydrogen bromide (HBr, 33.0 wt% in acetic acid, Acros, Yehud, Israel) (Scheme 1a) [21,22,27]. In detail, each 1.0 g of s-PZLL was dissolved in 50 mL of trifluoroacetic acid (TFA, Alfa Aesar, Lancashire, England) before HBr was dropped into the mixture slowly (molar ratio of HBr:Z group = 5:1). The reaction occurred at RT during stirring for 1 h. After completion of reaction, the solution was poured into diethyl ether for precipitation and the solvent was removed by centrifugation. The product was washed with diethyl ether twice and dried completely under vacuum overnight. The dried solid was dissolved in DIW and dialyzed against DIW for 72 h using a cellulose membrane with MWCO of 6000–8000 g mol^{-1}. Lyophilization was employed for the removal of water to obtain the final product with a 90.0–97.0% of yield. The polypeptide sample after deprotection was dissolved in D_2O for ^{1}H NMR-600NMR analysis.

4.4. Synthesis and Characterization of Star-Shaped Poly(L-lysine) Graft Copolypeptides

For functionalization of star-shaped polypeptide, the amino group on s-PLL was reacted with benzoic acid and 3-indoleacetic acid using EDC/NHS as activators (Scheme 1a) [23,28]. The synthesis of 6-armed poly(L-lysine)-*graft*-indole (6s-PLL-*g*-Indo) is shown as a reference.

In the sequence, EDC (0.56 mL, Alfa Aesar, Lancashire, England) and NHS (0.368 g, Alfa Aesar, Lancashire, England) were added to the mixture of 3-indoleacetic acid (83.96 mg) dissolved in anhydrous methanol (2.1 mL, AENCORE, Box Hill, Australia) under a nitrogen atmosphere. The solution was sealed and stirred at RT for 24 h for EDC/NHS to activate the carboxyl groups on 3-indoleacetic acid (Sigma-Aldrich, St. Louis, MO, USA). The 6s-PLL polypeptide (425 mg) was dissolved in anhydrous methanol (4.25 mL) before being added to the activated solution. The final solution was stirred for 48 h, dialyzed against DI water for 48 h with a dialysis membrane (MWCO of 6000–8000 g mol^{-1}), and lyophilized for 72 h to obtain a spongy product (yield: 90.0–95.0%). The graft copolypeptide was dissolved in D_2O for ^1H NMR-600NMR analysis.

4.5. Preparation of Star-Shaped Graft Copolypeptide Hydrogel Samples and Determination of Critical Gelation Concentrations (CGCs)

Star-shaped graft copolypeptide samples were dissolved in DIW with various concentrations to analyze their hydrogelation ability with the assistance of vortex and sonication to obtain homogeneous mixtures. To determine the CGCs of the samples, the inverting vial method was employed to observe whether the polypeptide mixtures will flow down after the containers were reversed [44–46]. The sample exhibit hydrogelation ability if it is stable at the vial bottom after 60 s with a concentration lower than 10.0 wt%. CGC is the lowest concentration of polypeptide at which it could undergo hydrogelation.

4.6. Determination of Secondary Structure Adoption

CD analysis was employed to characterize the polypeptide secondary conformation. The polypeptide samples dissolved in DIW at a concentration of 0.1 mg/mL were analyzed by CD measurement using a JASCO J-815 spectrometer (JASCO Corporation, Tokyo, Japan) at RT. The percentages of secondary conformations were calculated from CD data by using BeStSel fitting website.

4.7. Characterization of Graft Copolypeptide Hydrogel Self-Assembled Nanostructures

The self-assembled nanostructures of hydrogel samples were determined by using XRD, HR-SEM, and SAXS analyses. The 12-armed and 24-armed polypeptide hydrogels were prepared at concentrations of 8.0 and 5.0 wt%, respectively, and lyophilized for XRD and SEM measurements. Ultima IV-9407F701 Model of X-ray spectrometer (Rigaku Corporation, Tokyo, Japan) equipped with Cu Kα radiation (250 mA, 50 kV, 0.154 nm) at 4°/min of speed was used to determine XRD data of freeze-dried samples from 2θ = 5°–40°. The lyophilized samples were sputter-coated with platinum (Pt) before the operation of SEM imaging by using A Hitachi SU8010 microscope (Hitachi High-Technologies Corporation, Tokyo, Japan). For SAXS analysis, the functioned copolypeptide hydrogel samples were prepared at a concentration of 8.0 wt%, contained in quartz capillary tubes (1 mm diameter), and analyzed by a NANOSTAR U SYSTEM (Bruker AXS Gmbh, Karlsruhe, Germany) under 650 μA of current and 45 kV of voltage at RT.

4.8. Characterization of Mechanical Strength and Recovery Behavior

The mechanical strength and recovery behavior of hydrogel samples were characterized by the storage modulus (G′) and loss modulus (G″) values of rheological measurements. The hydrogel samples were prepared at a concentration of 8.0 wt% for all conditions of rheological measurements at RT. The mechanical properties of samples were studied under various strain sweep (frequency = 1 rad/s, strain = 0.01–100.0%) and frequency (strain = 1.0%, frequency = 0.1–100 rad/s) conditions. To observe the recovery ability of the hydrogel samples, the rheological system was operated at a fixed frequency of 1 rad/s and changing strain sweep under three different periods: 1.0% for the first 100 s, 100.0% for the next 300 s to break the hydrogel structure down, and 1.0% for the last 600 s. A rheometer HR-2 system (TA Instruments/Waters Corporation, New Castle, DE, USA) was employed to measure the rheological properties in this study, with a parallel aluminum plate of 25 mm diameter.

Supplementary Materials: The following supporting information can be downloaded at: https://www.mdpi.com/article/10.3390/gels8060366/s1, Table S1: Characterization of star-shaped poly (Z-L-lysine) homopolypeptides (s-PZLL). The degree of polymerizations (DP), number-averaged molecular weights (M_n), and ratio of weight-averaged molecular weight to number-averaged molecular weight (M_w/M_n) were calculated from proton nuclear magnetic resonance (^{1}H NMR) and gel permeation chromatography-light scattering (GPC-LS) analyses; Table S2: Radius of gyration (R_g) of nano-assemblies in star-shaped graft copolypeptides. The concentration of all samples was 8.0 wt%. The data was fitted from Small-angle X-ray scattering (SAXS) analysis using SasView software (version 5.0.4); Figure S1: ^{1}H NMR of (a') 6s-PZLL$_{32}$, (b') 12s-PZLL$_{12}$, and (c') 24s-PZLL$_{12}$ homopolypeptides in trifluoroacetic acid-d_1 (TFA-d_1); Figure S2: Circular dichroism (CD) spectra of graft polypeptides at 0.1 mg/mL; Figure S3: Recovery behavior of 12-amed and 24-armed polypeptide hydrogels. The concentration of hydrogel samples was 8.0 wt%. The rheological measurement was operated at 1 rad/s of frequency, room temperature and various strain at three continuous periods: 1.0% for 100 s, 100.0% for 300 s, and 1.0% for 600 s.

Author Contributions: Conceptualization, J.-S.J.; methodology, T.H.M.P. and Y.-H.Y.; validation, T.H.M.P. and Y.-H.Y.; formal analysis T.H.M.P., Y.-H.Y., Y.-J.T., F.-Y.C. and S.K.; investigation, T.H.M.P., Y.-H.Y., Y.-J.T. and S.K.; data curation, T.H.M.P. and Y.-H.Y.; writing—original draft preparation, T.H.M.P. and Y.-H.Y.; writing—review and editing, J.-S.J.; visualization, T.O.; supervision, T.O. and J.-S.J.; project administration, J.-S.J.; funding acquisition, T.O. and J.-S.J. All authors have read and agreed to the published version of the manuscript.

Funding: The authors acknowledge the funding support by the Ministry of Science and Technology, Taiwan (MOST 107-2923-M-006-002-MY3, 108-2221-E-006-034-MY3 and 110-2221-E-006-002-MY3). This work was financially supported by the Hierarchical Green-Energy Materials (Hi-GEM) Research Center, from The Featured Areas Research Center Program within the framework of the Higher Education Sprout Project by the Ministry of Education (MOE) and the Ministry of Science and Technology (MOST 111-2634-F-006-008) in Taiwan, and partially supported by Kobe University grant for promoting international joint research (to TO).

Data Availability Statement: Not applicable.

Acknowledgments: We thank Bi-Yun Lin and Kun-Hsu Lee (Instrument Center, National Cheng Kung University) for their help in performing the NMR and SAXS experiments, respectively.

Conflicts of Interest: The authors declare no conflict of interest. The funders had no role in the design of the study, in the collection, analyses, or interpretation of data, in the writing of the manuscript, or in the decision to publish the results.

References

1. Fu, J. Hydrogel properties and applications. *J. Mater. Chem. B* **2019**, *7*, 1523–1525. [CrossRef] [PubMed]
2. Oyen, M. Mechanical characterisation of hydrogel materials. *Int. Mater. Rev.* **2014**, *59*, 44–59. [CrossRef]
3. Vedadghavami, A.; Minooei, F.; Mohammadi, M.H.; Khetani, S.; Kolahchi, A.R.; Mashayekhan, S.; Sanati-Nezhad, A. Manufacturing of hydrogel biomaterials with controlled mechanical properties for tissue engineering applications. *Acta Biomater.* **2017**, *62*, 42–63. [CrossRef] [PubMed]
4. Chyzy, A.; Plonska-Brzezinska, M.E. Hydrogel properties and their impact on regenerative medicine and tissue engineering. *Molecules* **2020**, *25*, 5795. [CrossRef]
5. Saini, K. Preparation method, Properties and Crosslinking of hydrogel: A review. *PharmaTutor* **2017**, *5*, 27–36.
6. Darge, H.F.; Andrgie, A.T.; Tsai, H.-C.; Lai, J.-Y. Polysaccharide and polypeptide based injectable thermo-sensitive hydrogels for local biomedical applications. *Int. J. Biol. Macromol.* **2019**, *133*, 545–563. [CrossRef]
7. Hamidi, M.; Azadi, A.; Rafiei, P. Hydrogel nanoparticles in drug delivery. *Adv. Drug Del. Rev.* **2008**, *60*, 1638–1649. [CrossRef]
8. Deligkaris, K.; Tadele, T.S.; Olthuis, W.; van den Berg, A. Hydrogel-based devices for biomedical applications. *Sens. Actuators B Chem.* **2010**, *147*, 765–774. [CrossRef]
9. Thakur, S.; Thakur, V.K.; Arotiba, O.A. History, classification, properties and application of hydrogels: An overview. *Hydrogels* **2018**, 29–50. [CrossRef]
10. Van Bemmelen, J. Das hydrogel und das krystallinische hydrat des kupferoxyds. *Z. Anorg. Chem.* **1894**, *5*, 466–483. [CrossRef]
11. Wichterle, O.; Lim, D. Hydrophilic gels for biological use. *Nature* **1960**, *185*, 117–118. [CrossRef]
12. Zhang, L.; Furst, E.M.; Kiick, K.L. Manipulation of hydrogel assembly and growth factor delivery via the use of peptide–polysaccharide interactions. *J. Control. Release* **2006**, *114*, 130–142. [CrossRef] [PubMed]

13. Xing, R.; Li, S.; Zhang, N.; Shen, G.; Mohwald, H.; Yan, X. Self-assembled injectable peptide hydrogels capable of triggering antitumor immune response. *Biomacromolecules* **2017**, *18*, 3514–3523. [CrossRef]
14. Li, S.; Xing, R.; Chang, R.; Zou, Q.; Yan, X. Nanodrugs based on peptide-modulated self-assembly: Design, delivery and tumor therapy. *Curr. Opin. Colloid Interface Sci.* **2018**, *35*, 17–25. [CrossRef]
15. Kiick, K.L. Peptide-and protein-mediated assembly of heparinized hydrogels. *Soft Matter* **2008**, *4*, 29–37. [CrossRef]
16. Bonduelle, C. Secondary structures of synthetic polypeptide polymers. *Polym. Chem.* **2018**, *9*, 1517–1529. [CrossRef]
17. Chen, C.; Lan, J.; Li, Y.; Liang, D.; Ni, X.; Liu, Q. Secondary structure-governed polypeptide cross-linked polymeric hydrogels. *Chem. Mater.* **2019**, *32*, 1153–1161. [CrossRef]
18. Fan, J.; Li, R.; Wang, H.; He, X.; Nguyen, T.P.; Letteri, R.A.; Zou, J.; Wooley, K.L. Multi-responsive polypeptide hydrogels derived from N-carboxyanhydride terpolymerizations for delivery of nonsteroidal anti-inflammatory drugs. *Org. Biomol. Chem.* **2017**, *15*, 5145–5154. [CrossRef]
19. Zhao, S.; Zhu, H.; Chen, Z.; Shuai, S.; Zhang, N.; Liu, Y.; Rao, Z.; Li, Y.; Zhao, C.; Zhou, K. Preparation and properties of a temperature-and pH-responsive polypeptide hydrogel. *Mater. Res. Express* **2019**, *6*, 085711. [CrossRef]
20. Lin, J.-Y.; Lai, P.-L.; Lin, Y.-K.; Peng, S.; Lee, L.-Y.; Chen, C.-N.; Chu, I.-M. A poloxamer-polypeptide thermosensitive hydrogel as a cell scaffold and sustained release depot. *Polym. Chem.* **2016**, *7*, 2976–2985. [CrossRef]
21. Huang, C.-C.; Phan, T.H.M.; Ooya, T.; Kawasaki, S.; Lin, B.-Y.; Jan, J.-S. Effect of tethered sheet-like motif and asymmetric topology on hydrogelation of star-shaped block copolypeptides. *Polymer* **2022**, *250*, 124864. [CrossRef]
22. Phan, T.H.M.; Huang, C.-C.; Tsai, Y.-J.; Hu, J.-J.; Jan, J.-S. Polypeptide Composition and Topology Affect Hydrogelation of Star-Shaped Poly (L-lysine)-Based Amphiphilic Copolypeptides. *Gels* **2021**, *7*, 131. [CrossRef] [PubMed]
23. Shen, X.-Y.; Tang, C.-C.; Jan, J.-S. Synthesis and hydrogelation of star-shaped poly (L-lysine) polypeptides modified with different functional groups. *Polymer* **2018**, *151*, 108–116. [CrossRef]
24. Shen, Y.; Zhang, S.; Wan, Y.; Fu, W.; Li, Z. Hydrogels assembled from star-shaped polypeptides with a dendrimer as the core. *Soft Matter* **2015**, *11*, 2945–2951. [CrossRef]
25. Murphy, R.; Borase, T.; Payne, C.; O'Dwyer, J.; Cryan, S.-A.; Heise, A. Hydrogels from amphiphilic star block copolypeptides. *RSC Adv.* **2016**, *6*, 23370–23376. [CrossRef]
26. Lu, D.; Li, Y.; Wang, X.; Zhang, Y.; Guo, H.; Sun, S.; Wang, X.; Zhang, Y.; Lei, Z. All-in-one hyperbranched polypeptides for surgical adhesives and interventional embolization of tumors. *J. Mater. Chem. B* **2018**, *6*, 7511–7520. [CrossRef]
27. Tang, C.-C.; Zhang, S.-H.; Phan, T.H.M.; Tseng, Y.-C.; Jan, J.-S. Block length and topology affect self-assembly and gelation of poly (L-lysine)-block-poly (S-benzyl-l-cysteine) block copolypeptides. *Polymer* **2021**, *228*, 123891. [CrossRef]
28. Chen, Y.-F.; Lai, Y.-D.; Chang, C.-H.; Tsai, Y.-C.; Tang, C.-C.; Jan, J.-S. Star-shaped polypeptides exhibit potent antibacterial activities. *Nanoscale* **2019**, *11*, 11696–11708. [CrossRef]
29. Wang, Y.; Chang, Y.C. Synthesis and conformational transition of surface-tethered polypeptide: Poly (L-lysine). *Macromolecules* **2003**, *36*, 6511–6518. [CrossRef]
30. St. Pierre, S.; Ingwall, R.; Verlander, M.; Goodman, M. Conformational studies of sequential polypeptides containing lysine and tyrosine. *Biopolym. Orig. Res. Biomol.* **1978**, *17*, 1837–1848. [CrossRef]
31. Stanojlovic, V.; Müller, A.; Moazzam, A.; Hinterholzer, A.; Ożga, K.; Berlicki, Ł.; Schubert, M.; Cabrele, C. A Conformationally Stable Acyclic β-Hairpin Scaffold Tolerating the Incorporation of Poorly β-Sheet-Prone Amino Acids. *ChemBioChem* **2022**, *23*, e202100604. [CrossRef] [PubMed]
32. Sengupta, A.; Mahalakshmi, R.; Shamala, N.; Balaram, P. Aromatic interactions in tryptophan-containing peptides: Crystal structures of model tryptophan peptides and phenylalanine analogs. *J. Pept. Res.* **2005**, *65*, 113–129. [CrossRef] [PubMed]
33. Mahalakshmi, R.; Sengupta, A.; Raghothama, S.; Shamala, N.; Balaram, P. Tryptophan rich peptides: Influence of indole rings on backbone conformation. *Pept. Sci. Orig. Res. Biomol.* **2007**, *88*, 36–54. [CrossRef]
34. Sun, Y.; Wollenberg, A.L.; O'Shea, T.M.; Cui, Y.; Zhou, Z.H.; Sofroniew, M.V.; Deming, T.J. Conformation-directed formation of self-healing diblock copolypeptide hydrogels via polyion complexation. *J. Am. Chem. Soc.* **2017**, *139*, 15114–15121. [CrossRef]
35. Huang, J.; Hastings, C.L.; Duffy, G.P.; Kelly, H.M.; Raeburn, J.; Adams, D.J.; Heise, A. Supramolecular hydrogels with reverse thermal gelation properties from (oligo) tyrosine containing block copolymers. *Biomacromolecules* **2013**, *14*, 200–206. [CrossRef]
36. Liu, Y.; Zhan, G.; Zhong, X.; Yu, Y.; Gan, W. Effect of pi–pi stacking on the self-assembly of azomethine-type rod–coil liquid crystals. *Liq. Cryst.* **2011**, *38*, 995–1006. [CrossRef]
37. Das, A. Studies on complex π-π and T-stacking features of imidazole and phenyl/p-halophenyl units in series of 5-amino-1-(phenyl/p-halophenyl) imidazole-4-carboxamides and their carbonitrile derivatives: Role of halogens in tuning of conformation. *J. Mol. Struct.* **2017**, *1147*, 520–540. [CrossRef]
38. Chu, B. *Laser Light Scattering: Basic Principles and Practice*; Courier Corporation: Chelmsford, MA, USA, 2007.
39. Tsai, Y.-L.; Tseng, Y.-C.; Chen, Y.-M.; Wen, T.-C.; Jan, J.-S. Zwitterionic polypeptides bearing carboxybetaine and sulfobetaine: Synthesis, self-assembly, and their interactions with proteins. *Polym. Chem.* **2018**, *9*, 1178–1189. [CrossRef]
40. Beychok, S.; Fasman, G.D. Circular dichroism of poly-L-tyrosine. *Biochemistry* **1964**, *3*, 1675–1678. [CrossRef]
41. Fasman, G.D.; Bodenheimer, E.; Lindblow, C. Optical rotatory dispersion studies of Poly-L-tyrosine and copolymers of L-glutamic acid and L-tyrosine. Significance of the tyrosyl Cotton effects with respect to protein conformation. *Biochemistry* **1964**, *3*, 1665–1674. [CrossRef]

42. Hirschmann, R.; Schwam, H.; Strachan, R.; Schoenewaldt, E.; Barkemeyer, H.; Miller, S.; Conn, J.B.; Garsky, V.; Veber, D.F.; Denkewalter, R.G. Controlled synthesis of peptides in aqueous medium. VIII. Preparation and use of novel. alpha.-amino acid N-carboxyanhydrides. *J. Am. Chem. Soc.* **1971**, *93*, 2746–2754. [CrossRef] [PubMed]
43. Ooya, T.; Ogawa, T.; Takeuchi, T. Temperature-induced recovery of a bioactive enzyme using polyglycerol dendrimers: Correlation between bound water and protein interaction. *J. Biomater. Sci. Polym. Ed.* **2018**, *29*, 701–715. [CrossRef] [PubMed]
44. Hou, S.-S.; Hsu, Y.-Y.; Lin, J.-H.; Jan, J.-S. Alkyl-poly (L-threonine)/cyclodextrin supramolecular hydrogels with different molecular assemblies and gel properties. *ACS Macro Lett.* **2016**, *5*, 1201–1205. [CrossRef] [PubMed]
45. Jommanee, N.; Chanthad, C.; Manokruang, K. Preparation of injectable hydrogels from temperature and pH responsive grafted chitosan with tuned gelation temperature suitable for tumor acidic environment. *Carbohydr. Polym.* **2018**, *198*, 486–494. [CrossRef] [PubMed]
46. Kitagawa, M.; Maeda, T.; Hotta, A. PEG-based nanocomposite hydrogel: Thermo-responsive sol-gel transition and degradation behavior controlled by the LA/GA ratio of PLGA-PEG-PLGA. *Polym. Degrad. Stab.* **2018**, *147*, 222–228. [CrossRef]

gels

Article

Hydrogel Containing Solid Lipid Nanoparticles Loaded with Argan Oil and Simvastatin: Preparation, In Vitro and Ex Vivo Assessment

Muhammad Farhan Ali Khan [1], Asim Ur.Rehman [1], Haidar Howari [2], Aiyeshah Alhodaib [3,*], Faiz Ullah [4], Zia ul Mustafa [5], Abdelhamid Elaissari [6] and Naveed Ahmed [1,*]

1 Department of Pharmacy, Faculty of Biological Sciences, Quaid-i-Azam University, Islamabad 45320, Pakistan; farhanali@bs.qau.edu.pk (M.F.A.K.); arehman@qau.edu.pk (A.U.)
2 Department of Physics, Deanship of Educational Services, Qassim University, Buraydah 51452, Saudi Arabia; haidarh1@yahoo.com
3 Department of Physics, College of Science, Qassim University, Buraydah 51452, Saudi Arabia
4 Department of Chemistry, Quaid-i-Azam University, Islamabad 45320, Pakistan; faizullah@chem.qau.edu.pk
5 Punjab University College of Pharmacy, University of the Punjab, Lahore 54000, Pakistan; zia.ucp@gmail.com
6 Univ Lyon, University Claude Bernard Lyon-1, CNRS, ISA-UMR 5280, 69622 Villeurbanne, France; abdelhamid.elaissari@univ-lyon1.fr
* Correspondence: ahdieb@qu.edu.sa (A.A.); natanoli@qau.edu.pk (N.A.); Tel.: +92-32-1246-6312 (N.A.)

Abstract: Transdermal hydrogels have the potential to improve therapeutic outcomes via enhancing bioavailability and reducing toxicity associated with oral delivery. The goal of the present study was to formulate and optimise argan oil loaded transdermal hydrogel containing lipid nanoparticles. The high pressure homogenization (HPH) method was utilised to fabricate Simvastatin loaded solid lipid nanoparticles (SIM-SLNs) with precirol ATO 5 as a lipid core and Poloxamer 407 (P407) to stabilise the core. The optimised nanoformulation was characterised for its particle diameter, zeta potential, surface morphology, entrapment efficiency, crystallinity and molecular interaction. Furthermore, transdermal hydrogel was characterised for physical appearance, rheology, pH, bio adhesion, extrudability, spreadability and safety profile. In vitro and ex vivo assays were executed to gauge the potential of SLNs and argan oil for transdermal delivery. The mean particle size, zeta potential and polydispersity index (PDI) of the optimised nanoparticles were 205 nm, −16.6 mV and 0.127, respectively. Crystallinity studies and Fourier transform infrared (FTIR) analysis revealed no molecular interaction. The in vitro release model explains anomalous non-Fickian release of drug from matrix system. Ex vivo skin penetration studies conducted through a fluorescence microscope confirmed penetration of the formulation across the stratum corneum. Hydrogel plays a crucial role in controlling the burst release and imparting the effect of argan oil as hypolipidemic agent and permeation enhancer.

Keywords: hyperlipidaemia; simvastatin; improved therapeutic efficiency; transdermal hydrogel

1. Introduction

Hyperlipidaemia is a well-known risk factor for cardiovascular mortality and morbidity. It mainly causes atherosclerosis, which in turn leads to coronary heart disease [1]. It is usually characterised by elevated levels of low-density lipoproteins and triglycerides. Escalated interest in argan oil, obtained from *Argania spinose*, is due to its antioxidant properties. Individuals consuming argan oil are less prone to heart attacks as compared to non-users [2]. It is also a hepatoprotective and choleretic agent that prevents atherosclerosis and hypercholesterolemia [3]. Simvastatin (SIM) is an anti-hyperlipidaemic drug having low bioavailability (<5%), poor aqueous solubility (30 μg/mL) and a high partition co-efficient (log P~4.5) [4]. It is a competitive inhibitor of 3-hydroxy-3-methylglutaryl coenzyme A (HMG-CoA) reductase and is mainly metabolised in the gut and liver by the cytochrome 3A system, which presents a major obstacle in the development of oral dosage forms. This route of administration also causes gastric lesions such as ulcers and erosions. Intramuscular, subcutaneous and intravenous routes of administration can cause discomfort to the patient as well as irritation at the site of injection and local soft tissue damage [5].

To circumvent issues of poor bioavailability and Gastrointestinal tract (GIT) irritation, conventional approaches are not enough [6]. Hence, there is a dire need to develop a novel dosage form that can combat the problems of poor water solubility and excessive metabolism [7]. A transdermal delivery system can reduce various side effects associated with oral dosage forms [8]. Incorporating SLNs into the transdermal hydrogel offers several benefits: lack of bio-toxicity, sustained drug release, better physical stability, high drug incorporation, improved bioavailability and reduced side effects [9,10]. Based on the lipophilic nature of SIM, lipid nanoparticles loaded into transdermal hydrogel containing argan oil might be a promising strategy to overcome the problem of low oral bioavailability and to provide synergistic hypolipidemic activity.

In recent decades, nanoparticulate carrier systems have been investigated to modulate pharmacokinetics, physicochemical properties, biocompatibility, therapeutic efficiency and bioavailability of poorly aqueous soluble drugs [11]. SLNs have gained augmented interest due to their superior characteristics compared to their colloidal counterparts [12,13]. Lipophilic drugs can be incorporated more effectively with higher encapsulation efficiency, providing a wide range of applications [14]. Encapsulation of drug molecules within this lipid matrix can prevent degradation and also impart prolonged release properties [15]. Moreover, macro molecules such as oligonucleotides, proteins, DNA and RNA can also be delivered through this carrier system [16]. Drugs incorporated into lipid nanoparticles exhibit better permeation as compared to conventional dosage forms [17]. Glycerides have created considerable interest in the development of controlled release dosage forms. Precirol ATO 5, also known as glyceryl palmitostearate, imparts prolonged release properties [18]. P-407 is a non-ionic surfactant which is used for sustained drug delivery. According to the FDA, it enhances stabilization; thus, nanoparticles can be optimised using this surfactant [19].

The objective of current study is to evaluate characteristics of SIM-SLNs based transdermal hydrogel. To fabricate SIM-SLNs, a HPH technique was used, with a lipid core stabilised by a surfactant mixture, and they were characterised for in vitro release behaviour and physicochemical properties after loading into hydrogel. Finally, a transdermal drug delivery system was developed to achieve sustained release and enhanced permeation using argan oil as a permeation enhancer. Following transdermal administration of SIM-SLNs to rat models, various pharmacokinetic parameters were also calculated.

2. Materials and Methods

2.1. Materials

Precirol ATO 5 was a generous gift from Gattefosse France. SIM and P407 (Sigma-Aldrich, Steinheim, Germany), Ethanol and Carbopol 934 were acquired from BDH Laboratory Supplies. Argan oil was purchased from Laboratories Omega Pharma, Chatillon, France. All other chemicals used in the current study were of analytical grade and were used without further processing.

2.2. Methods

Preparation of SIM-SLNs

The SLNs were prepared by the HPH technique with slight modifications [20]. Briefly, 50 mg of precirol ATO 5 was completely dissolved in 2 mL of ethanol by heating at 60 °C in a water bath. The resultant organic phase (clear appearance) was quickly mixed with an aqueous phase (20 mL) containing surfactant (0.5%) preheated to the same temperature. This mixture was homogenised through a homogenizer (D-91126 Heidolph instruments, Schwabach, Germany) at 24,500 rpm for 5 min at 60 °C. Then, the obtained SLNs were cooled in an ice water bath to quickly crystallise the lipid. Drug-loaded SLNs were fabricated through the same method by adding SIM to the organic phase. For optimization of formulation (SIM-SLNs), a trial-and-error method was used, varying the concentration of different components such as Precirol and P-407. An optimised formulation (H) was selected on the basis of smaller particle size and higher encapsulation efficiency. The resultant trials for optimization are tabulated in Table 1.

Table 1. Formulation of SLNs using HPH technique.

Formulation Code	Lipid Phase		Aqueous Phase (20 mL Water)
	Lipid (mg)	Solvent (mL)	P407 (%)
A	50	2	0.25
B	50	2	0.5
C	50	2	0.75
D	50	2	1
E	50	2	1.5
F	30	2	0.5
G	40	2	0.5
H	**50**	**2**	**0.5**
I	60	2	0.5
J	70	2	0.5

2.3. *Physicochemical Characterization of Nanoparticles*

2.3.1. Particle Size, Polydispersity Index, Zeta Potential and Entrapment Efficiency

Photon correlation spectroscopy and electrophoretic light scattering techniques were used to determine particle size, polydispersity index (PDI) and zeta potential of the samples through zetasizer Nano ZS 90 (Malvern Instruments; Worcestershire, UK), equipped with software (version 6.34) and a He-Ne laser at a wavelength of 635 nm and static scattering angle of 90 degrees. Briefly, 10 µL of the sample was mixed with 1 mL of deionised water and vortexed for 2 min, followed by analysis with a zetasizer. Each result displayed was measured in triplicate.

Spectrophotometric analysis was used to determine the amount of SIM incorporated in SIM-SLNs. For determination of entrapment efficiency, formulation (1 mL) was taken in a falcon tube and volume was made up to 10 mL with ethanol. After that, the mixture was centrifuged for 1 h at 13,500 rpm in a centrifuge (Hermle Labortechnik, Wehingen, Germany). Supernatant (100 µL) was taken for further processing and diluted with 10 mL of ethanol. For determination of drug content, a UV-visible spectrophotometer (Dynamica, Halo DB-20, Livingston, UK) was utilised at λ_{max} of 239 nm.

2.3.2. Scanning Electron Microscopy (SEM)

SIM-SLNs were evaluated for surface morphology using a Scanning electron microscopy (Tescan vega-3, model imu vp-sem, New York, NY, USA). For this purpose, nanoparticles were lyophilised with the help of a lyophiliser (Alpha 1-2 LD plus, Christ, Osterloh, Germany) to obtain dried particles. High resolution SEM was used to analyse the surface properties, structure and size of dried particles [21].

2.3.3. Fourier Transform Infrared Spectroscopy (FTIR)

FTIR analysis was carried out to investigate any potential interaction among the individual components of the formulation. The FTIR spectra of SIM, Precirol, SIM-SLNs, and P407 were acquitted between 4000 cm^{-1} and 450 cm^{-1} using an FTIR spectrophotometer (L16000A, Perkin-Emler USA). Each 10 mg sample was triturated with 100 mg of KBr (potassium bromide), and all measurements were executed in triplicate at room temperature [22].

2.3.4. Powder X-ray Diffractometry (PXRD)

The X-ray diffraction patterns of lyophilised SIM-SLNs and their individual solid components (SIM, Precirol, and Poloxamer) were studied by a powder X-ray diffractometer (Rigaku Inc.,Tokyo, Japan) using a copper anode as a source of radiation. Scanning of samples was performed at room temperature (20 kV, 5 mA). After that, absolute intensity was recorded in the range of 5–80 against theta 2 and a scanning rate of 1.2°/min [23].

2.4. Formulation of SIM-SLNs Loaded Transdermal Hydrogel

Carbopol 934 was utilised for the preparation of the transdermal hydrogel (Scheme 1). Briefly, a 1% carbopol solution was prepared by dissolving it in distilled water with stirring at 300 rpm, followed by SIM-SLN addition to the solution. Glutaraldehyde (0.5%) was added as a cross linker during stirring. At the end, triethanolamine (0.5%) as a gelling agent was added to neutralise the pH of the medium. The hydrogel having a penetration enhancer was prepared in the same manner by adding argan oil (1 mL) before incorporation of triethanolamine [24].

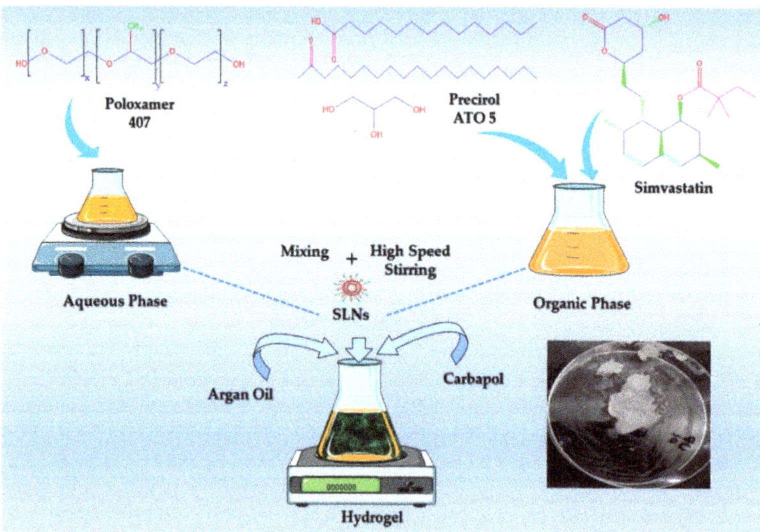

Scheme 1. Illustration of hydrogel formation.

2.5. Characterization of Transdermal Hydrogel

2.5.1. Physical Appearance and pH Measurement

The pH of the transdermal hydrogel was measured by utilising a pH metre (PHS25CW Bante instrument, Chicago, IL, USA). Physical appearance includes smoothness, transparency, homogeneity and colour. The SIM-SLNs hydrogel was also inspected for these characteristics [25].

2.5.2. Rheological Study and Spreadability of Transdermal Hydrogel

Brookfield viscometer (DV-I Prime Brookfield engineering laboratories, Inc., 11 Commerce Boulevard, Middleboro, MA, USA) with S63 spindle was used to measure the viscosity of a hydrogel at various shear rates. For the purpose, 2 g of hydrogel was taken in a cone and viscosity was measured at 0, 0.5, 1, 2.5, 4, 5, 10, 20, 50 and 100 rpm [26].

The glass slide method was utilised for the measurement of hydrogel spread ability. A glass slide was marked with a circle of 1 cm having 0.5 g of gel placed in the centre [27]. Another glass slide was placed on it in a manner that hydrogel was sandwiched between two glass slides. After that, a 500 g weight was placed on the glass slides, and the increment in diameter was measured. Using Equation (1), results were obtained with respect to applied mass and spreading area [28].

$$Spreadability\ index\ (Si) = d^2 \times \frac{\pi}{4} \quad (1)$$

2.5.3. Swelling index, Bio-adhesion and Extrudability

In order to measure the swelling index of the hydrogel, the gravimetric method was used. For this purpose, transdermal hydrogel was lyophilised and then placed in a phosphate buffer at pH 5.5. Swollen hydrogel was taken out of the medium and weighed at different time intervals (0.5, 1, 1.5, 2, 2.5, 3, 3.5 and 4 h) until weight became constant. Percentage swelling was measured using Equation (2).

$$Swelling\ ratio = \frac{w2 - w1}{w1} \times 100 \quad (2)$$

A disintegration apparatus was used to determine the bio adhesion time. An apparatus beaker (500 mL) was filled with phosphate buffer of pH 5.5 at 32 °C. Rat skin was attached with a glass slide and then was placed on the disintegration apparatus with the help of adhesive tape. In the next step, a transdermal hydrogel was applied to the rat's skin. After immersion into a phosphate buffer solution, the whole assembly was allowed to move up and down. The time it took to remove the applied hydrogel was recorded accordingly [29].

For determination of extrudability, an observation test was performed to assess expulsion of hydrogel from a tube. A closed aluminium collapsible tube was filled with 10 g of hydrogel, and a clamp was applied for prevention of any roll back. The cap was removed after placing a 500 g weight on the tube. The amount of hydrogel extruded was collected, weighed and the percentage was calculated [30].

2.6. *In Vitro Drug Release Studies*

A dialysis membrane was used to perform in vitro release studies in phosphate buffer at pH 7.4 and 5.5. The formulation was placed in a dialysis sac to assess drug release at various time intervals. After dipping a dialysis sac in a beaker, this medium was placed on a shaking water bath at 37 °C. All studies were performed in triplicate [31]. Sampling time was 0.25, 0.5, 1, 3, 6, 12, 24 and 48 h. At each time interval, 1 mL of sample was withdrawn. Phosphate buffer of the respective pH was used to replace the amount of sample withdrawn to maintain the sink conditions. A UV-visible spectrophotometer was utilised to quantify these samples at a wavelength of 239 nm. To calculate drug release, a graph was plotted between percent drug release versus time using MS Excel. First order, zero order, Higuchi, Hixon–Crowel, and Korsmeyer–Peppas kinetic models were applied for evaluation of the release profile of SIM-SLNs and SIM-SLNs hydrogel.

2.7. *Permeation Studies*

Permeation studies were performed using rat skin. A healthy animal was sacrificed, and the skin from the abdominal portion was separated. After that, a surgical blade was used to remove all the hairs from the rat's skin. Afterwards, the skin was placed in warm water to remove adipose tissue and stored in a freezer at −30 °C. To check permeation, it was placed in a buffer solution for half an hour before the experiment. Rat skin was utilised to check the permeability of SIM-SLNs through skin. The first one was a hydrogel having argan oil as a permeation enhancer, and the second was SIM-SLNs. A skin layer was placed on the donor compartment, and the formulation was applied to the skin. The receiver compartment was filled with phosphate buffer solution (pH 7.4), and samples were withdrawn with long needle syringes. The sampling interval was 0.25, 0.5, 1, 3, 6, 12 and 24 h. A sample (1 mL) was withdrawn from the receiver compartment after each interval, and sink conditions were maintained. The absorbance values of all samples were recorded using a UV-visible spectrophotometer at 239 nm. The study was performed in triplicate, and Microsoft Excel was used to interpret the results.

2.8. Ex Vivo Penetration Analysis through Fluorescent Microscope

A fluorescent microscope was utilised to study the distribution of nanoparticles in rat skin. Rhodamine-loaded nanoparticles were formulated by injecting the dye (2 mg) into the organic phase. Nanoparticles were centrifuged at 13,500 rpm for 1 h to separate un-entrapped dye. Entrapped Rhodamine-loaded nanoparticles were collected from the pellet and incorporated into the hydrogel. Rat skin was placed on the Franz diffusion cell for 5 h with the application of a hydrogel formulation. After that, the transactional part of rat skin was obtained with the help of a cryostat microtome. Images were visualised with the help of a fluorescent microscope by keeping a blank skin as a control [32].

2.9. Evaluation of Skin Structure after Treatment with Hydrogel

In order to evaluate the effect of the permeation enhancer, skin structure after application of the hydrogel was investigated with Fourier transform infrared spectroscopy (FTIR) because it is supposed that permeation enhancers may induce structural changes in the stratum corneum. Rat skin was placed on the Franz diffusion cell and treated with hydrogel for 5 h. After that, the skin was washed and subjected to FTIR analysis, while untreated rat skin was used as a control [33].

2.10. Stability Studies

According to the ICH guidelines, stability studies were carried out for transdermal hydrogel. Accelerated studies were performed for a period of six months (1, 3 and 6 months) at a temperature of 25 °C and a relative humidity of 60% [34]. Hydrogel was evaluated on the basis of changes in pH and physical appearance.

3. Results and Discussion

3.1. Preparation and Physicochemical Properties of SIM-SLNs

SIM-SLNs were successfully prepared by HPH technique with good reproducibly. Precirol ATO 5 was used as a solid lipid core of SIM-SLNs, stabilised by surfactant. Precirol is a pharmaceutically acceptable, biocompatible and biodegradable lipid having high incorporation efficiency and stable lipid core with sustained drug release. P-407, a non-ionic surfactant, was used to interact and form a miscible surfactant shell with Precirol core. The surfactant provides steric stabilization to improve the stability of SIM-SLNs by reducing surface interactions between the particles [35,36]. After analysing all stable formulations, optimised formulation (H) was selected on the basis of smaller particle size and maximum entrapment efficiency (Table 2). An increase in particle size was observed when a large amount of Precirol ATO 5 was used. In one-way ANOVA analysis, it is evident that particle size (F = 13.204, p = 0.001) and PDI (F = 8.608, p = 0.004) were significantly associated with lipid concentration as particle size and PDI increase by increasing lipid concentration (Table 2). This illustrates the fact that the size of SLNs depends upon concentration of lipids because lipids have the tendency to coalesce at high concentrations [37]. Increase in particle size can be due to density difference between external and internal phases, or it may occur due to a reduced diffusion rate of solute molecules in the outer phase. It was observed that increasing concentration of surfactant causes decrease in particle size. It might be due to reduction in surface tension between the organic and aqueous phases [38].

Table 2. Response parameters of formulation: entrapment efficiency, particle size and PDI.

Formulation Code	Entrapment Efficiency	Particle Size (nm)	PDI	Observation
A	65	230	0.400	Non uniform [a] and unstable [b]
B	79	225	0.452	Uniform and stable [c]
C	75	180	0.327	Uniform [d] and stable
D	68	175	0.252	Uniform and stable
E	63	166	0.174	Uniform and stable
F	74	177	0.268	Uniform and stable
G	77	182	0.372	Uniform and stable
H	**82**	**205**	**0.127**	**Uniform and stable**
I	79	241	0.407	Non uniform and stable
J	71	551	0.793	Non uniform and unstable

[a] Globular appearance to naked eye. [b] Did not remain uniform for more than 90 days at 25 °C. [c] Remained uniform for more than 90 days. [d] Non globular appearance.

Blank and drug loaded nanoparticles exhibited particle sizes of 173 and 205 nm with PDI values of 0.177 and 0.127 expressing narrow size distribution (Figure 1). The phenomenon of size increase is commonly observed after drug incorporation. The particle size plays a crucial role in the uptake of nanoparticles after transdermal administration. Nanoparticles smaller than 300 nm are desirable for effective transdermal transport into the systemic circulation [39].

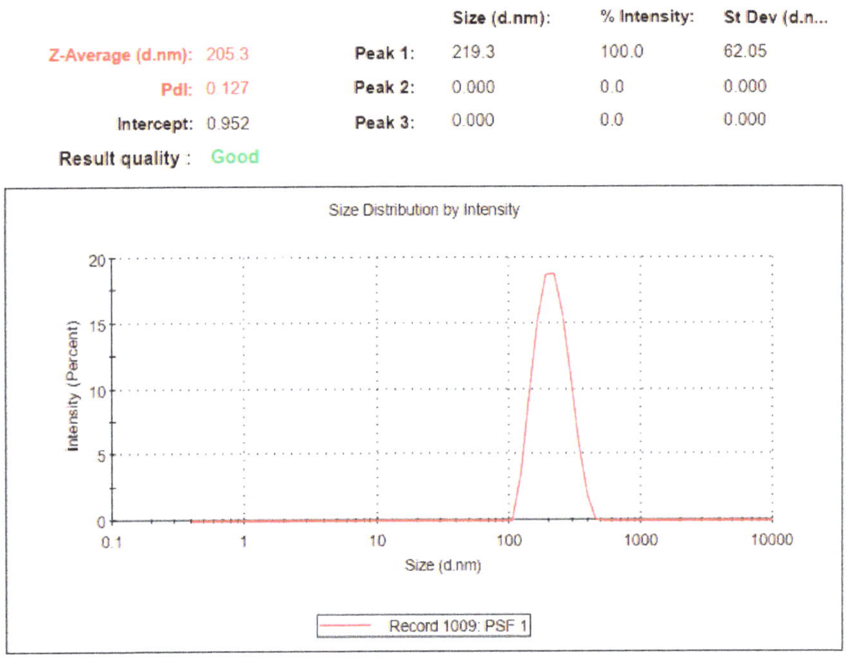

Figure 1. Particle size distribution of optimised drug loaded formulation of SLNs.

PDI value below 0.3 indicates homogenous distribution; hence, it is evident that SIM-SLNs were uniform in size [40]. The zeta potential of the nanoparticulate formulation is a key indicator for stability influencing physical stability of colloidal dispersions. Due to electrostatic repulsion between similarly charged particles, higher zeta potential prevents aggregation thus conferring stability to colloidal dispersion. SIM-SLNs exhibited a zeta potential of about −16.6 mV which ensures electrostatic stabilization [41]. SLNs prepared by the HPH method displayed a smaller particle size (205 nm), low Polydispersity index (0.127) and zeta potential (−16.6 mV), which warrants physical stability of formulation [42]. These results also augmented in linear regression models that indicates lipid concentration is a significant determinant of particle size and PDI, and any increase in lipid concentration results into increase in particle size ($\beta = 7.953$, $p < 0.05$) and PDI ($\beta = 0.011$, $p < 0.05$). While increase in Poloxamer concentration is negatively related to particle size and PDI, as increase in surfactant concentration leads to decrease in particle size ($\beta = -57.675$, $p > 0.05$) and PDI ($\beta = -0.127$, $p > 0.05$).

Encapsulation efficiency is a parameter of interest to deliver drug in higher dose more precisely at site of action. Entrapment efficiency is bound to increase with increasing concentration of lipid from 74% to 82% as shown in Table 2. The lipid acts a solubilizing agent for highly lipophilic drugs. Use of solid lipids cause enormous defects in crystal lattice resulting higher imperfections, thus, enhancing the space to entrap drug molecules [43]. As compared to solvent emulsification method, use of the HPH technique leads to higher entrapment efficiency and enhanced percentage yield [44]. One-way ANOVA analysis revealed that entrapment efficiency was increased by increase in lipid concentration (Table 2). Entrapment efficiency is bound to increase with the increasing concentration of lipid because it acts as a solubilizing agent for highly lipophilic drugs. Linear regression analysis also indicates any increase in lipid concentration results into increase in EE ($\beta = 0.437$, $p < 0.05$). Results also indicate that poloxamer concentration had a significant association with entrapment efficiency; by increasing poloxamer concentration, there is significant decrease in entrapment efficiency ($F = 3.848$, $p = 0.043$). This is also evident in linear regression as increase in surfactant concentration is negatively related to entrapment efficiency ($\beta = -6.727$, $p > 0.05$).

3.2. Particle Morphology

Scanning electron microscopy (SEM) was utilised to determine the morphology of blank and drug loaded SLNs. Figure 2 exhibits SEM images at different intensifications. Results confirm the sphericity of nanoparticles. SEM images of SIM-SLNs exhibited the uniform and spherical nature of the nanoparticles. It was noted that results are in agreement with earlier studies [45].

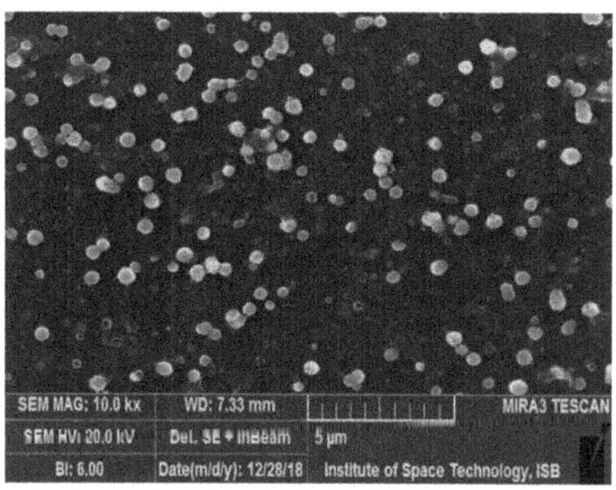

Figure 2. Scanning electron microscope (SEM) image of Precirol lipid nanoparticles with spherical morphology loaded with SIM.

3.3. FTIR Analysis

FTIR analysis was performed to investigate molecular interaction between drug and lipid matrix. The spectra of SIM exhibited characteristic peaks at 3545 cm^{-1} due to stretching vibration of the OH groups, and Peaks at 2914 cm^{-1} and 2849 cm^{-1} were due to stretching vibration of the CH group (Figure 3). Stretching vibrations of the ester group showed peaks at 1728 cm^{-1}. The same characteristic peaks were observed in the spectra of drug loaded SLNs which confirms chemical stability of SIM in SLNs formulation. FTIR analysis confirmed that there was no chemical or electrostatic interaction between ingredients of the nanoparticulate formulation. Absence of any new peak or functional group also exhibits absence of any interaction among constituents. SLNs prepared through this technique exhibit regular bands in FTIR spectra. While SLNs prepared through other techniques such as solvent evaporation show significant convolution in FTIR spectra. Other studies also reported similar conclusion from FTIR analysis [46]. SIM characteristic peaks of FTIR can be seen in the formulation indicating no change in chemical structure of drug. Absence of OH stretch can be explained by intermolecular hydrogen bonding [47].

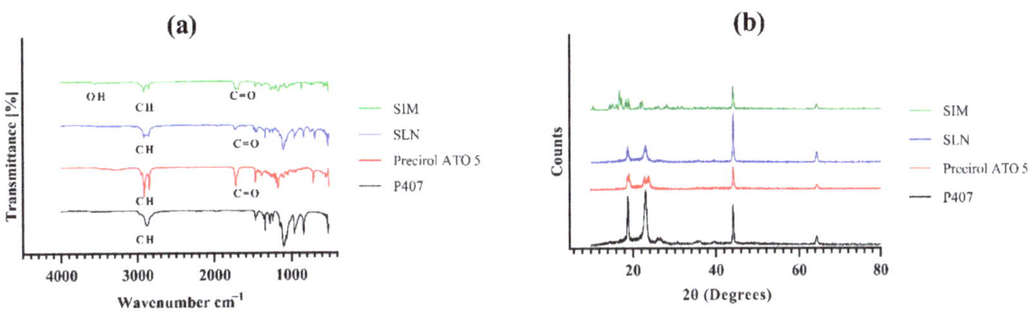

Figure 3. (**a**) FTIR and (**b**) PXRD spectrum of SIM, SLN formulation, Precirol ATO 5 and P407.

3.4. Degree of Crystallinity of SIM-SLNs

The crystalline behaviour of lyophilised SIM-SLNs was evaluated by using PXRD. X-ray diffraction analysis showed a number of sharp peaks demonstrating the crystalline nature of the drug (Figure 3). SIM exhibits sharp peaks in range of 10–30 which confirms the crystalline nature of drug, but after loading into lipid (Precirol), there were no characteristic peaks observed within this range. However, Precirol retained its crystalline form before and after drug loading. These results are in agreement with earlier study because the nature of drug changed from crystalline to amorphous while no change was observed in its lipid nature. The decrease in crystallinity might be due to absorption of the surfactant on SLNs surface. Results are in harmony with previous studies suggesting an amorphous state of SIM in the solid lipid core [48].

3.5. Physicochemical Evaluation of Transdermal Hydrogel

For the development of the transdermal hydrogel, the barrier function of the stratum corneum is of prime importance. To transport SIM-SLNs across the stratum corneum, a Carbopol based hydrogel was formulated. Fabrication of a secondary carrier was essential to control the burst release and to impart the effect of argan oil as permeation enhancer and hypolipidemic ingredient. Furthermore, transdermal hydrogel also improves the patient compliance. To deliver the drug more efficiently through the transdermal route, permeation enhancers are used. Permeation enhancers should be non-reactive, non-toxic, biocompatible, etc. For this purpose, Argan oil was utilised as permeation enhancer to augment the permeation from the lipid system [49].

3.5.1. Physical Appearance and pH of Formulation

The transdermal hydrogel was smooth in texture with no grittiness and was uniform and clear. Normal pH of skin ranges from 4–6 while pH of transdermal hydrogel was found to be 5.3 indicating slightly acidic nature. It is beneficial for transdermal formulation because pH at skin surface and within different layers remain acidic favouring intact delivery of nanoparticles across skin. pH is a significant parameter and should be within range (4–6); otherwise, it can cause skin irritation and incompatibility. Hence, formulations with pH falling in between this range are suitable for drug delivery through skin [50].

3.5.2. Spreadability and Rheological Behaviour of Transdermal Hydrogel

In order to evaluate ease of application and consistency, spreadability measurement is an important parameter. Spreadability of drug loaded hydrogel was measured by using glass slide method. The value was found to be 273.2 $\pm$ 0.5 mm^2. Ease of application and spreadability is an important property of formulation for uniform drug delivery. It also improves patient compliance. Spreadability studies confirmed ease of spreading on skin surface. These results are in accordance with previous studies as values are within an acceptable range [51].

Viscosity of transdermal hydrogel was determined at different shear rates as shown in Figure 4. It can be seen that there is an inverse relationship between viscosity of hydrogel and shear rate. Viscosity of transdermal hydrogel is going to be reduced by enhancing shear rate and vice versa. The behaviour of the hydrogel is similar to shear thinning gels, which behave like liquids under stress while converting to viscous form when the stress is removed. This kind of behaviour under stress conditions is known as shear thinning. Though the exact cause of shear thinning is not fully understood, it is widely regarded to be the effect of small structural changes within the fluid, such that microscale geometries within the fluid rearrange to facilitate shearing. For assessment of adhesive capability and retention time on skin, rheological properties are of prime importance. The thixotropic nature of the hydrogel has the benefit of easy application on skin requiring little force [52].

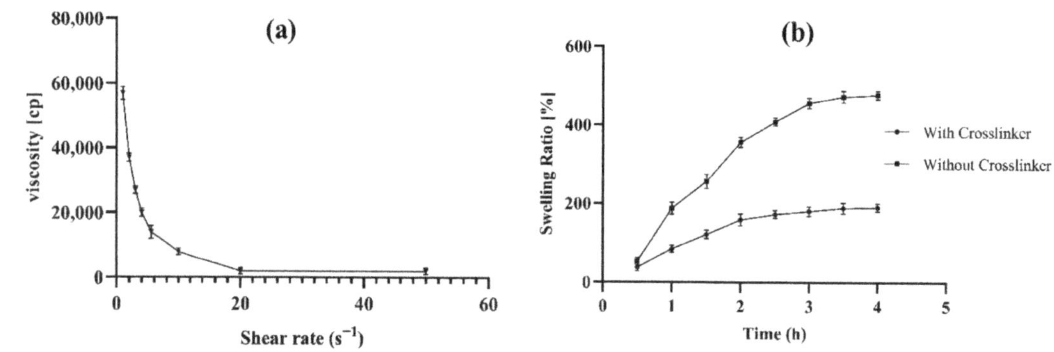

Figure 4. (**a**) Rheology of hydrogel; (**b**) swelling index of lyophilised hydrogel with and without cross linker.

3.5.3. Swelling index, Bioadhesion and Extrudability

Swelling studies of transdermal gel were performed with and without a cross linker. It was observed from swelling index results that a swelling ratio was higher in hydrogels having no cross linker. Significant reduction in swelling ratio was observed in hydrogels with cross linker (Figure 4). The reason might be that without cross linker, lipid chains create more space for diffusion of water in the gel matrix. In contrast, the cross-linked gel restricted the free movement of water, thus decreasing swelling ratio. It indicated that the cross linker imparts strength to the hydrogel and controls the burst release from hydrogel [53]. The retention time of the transdermal formulation was found to be 45 min. Slight decrease in retention time indicates presence of argan oil which reduces the viscosity. The reduction in bioadhesion might be due to the accelerated experimental conditions in the disintegration apparatus. The extrudability of Carbopol 934 based hydrogel was found to be 89% (8.9 g) which is a good property. The extrusion of hydrogel from the tube is an important parameter to enhance patient compliance and ease of application. In order to extrude gel from the tube, viscosity plays a significant role. Low viscous gels extrude more easily as compared to highly viscous gels. These results are in accordance with previous studies [30].

3.6. In-Vitro Release

A dialysis membrane method was adopted to evaluate drug release from SIM-SLNs at pH 7.4 and pH 5.5. Figure 5 illustrates that nearly 85% of all entrapped drug was released in 12 h from the SLNs formulation, while 65% of SIM was released from the transdermal hydrogel which indicates a more sustained release at pH 7.4. However, drug release was negligible at pH 5.5.

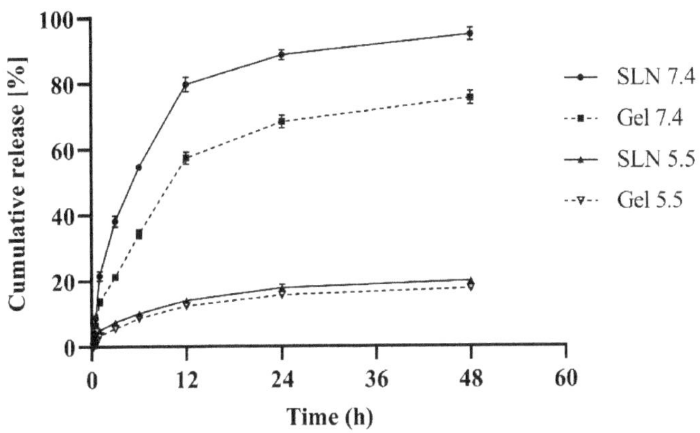

Figure 5. In vitro release profile of SLN formulation and Carbopol based hydrogel at pH 7.4 and pH 5.5.

By applying various release models such as zero order, first order, Hixon–Crowel, and Korsmeyer–Peppas, the mechanism of drug release was determined. R^2 values indicate that release from SIM-SLNs follows the Korsmeyer–Peppas model. The R^2 value was 0.991. It also illustrates that release from transdermal hydrogel follows the Korsmeyer–Peppas model. The R^2 value in this case is 0.989, and the n value was 0.651, indicating that drug release will follow anomalous non-Fickian release mechanism. Burst release from nanoparticles is a major concern for researchers all across the globe. This problem can be tackled by loading nanoparticles into secondary carriers: patches, films and gels. Keeping this in mind, Precirol based nanocarriers were successfully loaded into a Carbopol based hydrogel which is a secondary carrier. The burst release may be attributed to smaller particles with large surface area causing the fastest release at the initial stage. Moreover, burst release might be due to drug adsorbed onto or close to the SLNs surface [54]. To investigate the drug transport mechanism, the value of n in the Korsmeyer–Peppas model is important. Thus, n values between 0.5–1 correspond to anomalous non-Fickian diffusion and values below 0.5 show Fickian diffusion. Only the first 60% of the release is considered for estimation of this parameter [55]. An R^2 value near to 1 explains that release of drug from nanoparticles follows a concentration dependent release (first order kinetics). On other hand, the hydrogel follows the Korsmeyer Peppas model R^2 = (0.989). This model explains the anomalous non-Fickian release of drug from the matrix system [38].

3.7. Skin Permeability

A Franz diffusion cell was utilised for determination of skin permeability. A comparison between permeability of transdermal hydrogel with and without permeation enhancer (argan oil) was also performed (Figure 6). A statistical test (ANOVA) was applied to determine permeability of these formulations. Enhanced skin permeability was observed in the transdermal hydrogel having permeation enhancer. The steady state flux (Jss) of the nanoparticles based transdermal hydrogel was measured. Jss of the hydrogel containing argan oil was higher (615 μg/cm^2) as compared to the simple formulation (417 μg/cm^2). It was concluded statistically (0.004) that hydrogel with permeation enhancer shows better permeability as compared to other formulations. It was concluded from permeation studies by applying statistical tool analysis of variance that a significant amount of drug permeated in the buffer medium of pH 7.4 through the skin. Increased permeation might occur due to the presence of argan oil which acts as a permeation enhancer. Most probably, the increase in permeation rate was attained due to the lipid coating and permeation enhancer. The skin permeation enhancement effect of argan oil is well known. It is composed of oleic and linoleic acid which enhances penetration through the intercellular pathway by interacting with the stratum corneum. Argan oil provides a dual benefit: it acts as a permeation enhancer by interaction with the skin lipid causing reversible fluidization, and also, it acts as an anti-hypercholesterolemic agent [55].

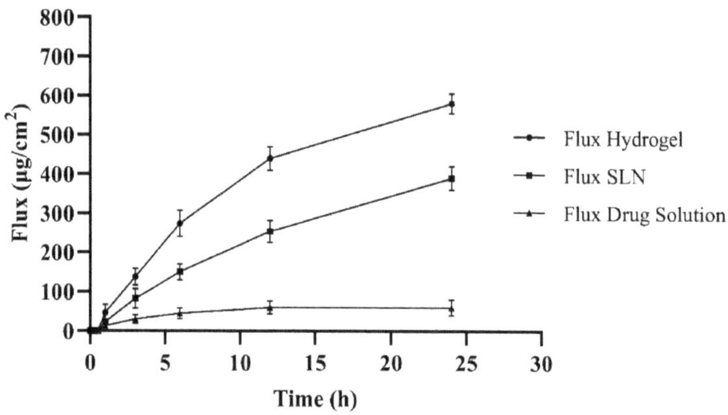

Figure 6. Ex vivo permeability studies of drug solution (control), SLNs and transdermal hydrogel at pH 7.4.

3.8. Ex Vivo Penetration Analysis through Fluorescent Microscope

Fluorescent photomicrographs of rhodamine loaded nanoparticles illustrate the deeper penetration of nanoparticles as seen in the Figure 7. Incorporation of rhodamine resulted in strong fluorescent marking in the dermis layer. These results indicate nanoparticles deliver a substantial amount of drug across the skin. A fluorescence microscope (Leica mikroskopie & Systeme GmbH., Wetziar, Germany) was used to visualize the distribution of rhodamine loaded nanoparticles across the skin following the application of the hydrogel formulation with the permeation enhancer. It diffused into the deeper layers as can be seen from the green fluorescence. These results suggest that SLNs have the capability to enhance the penetration of SIM across the skin and hence are expected to enhance the bioavailability. These results are in accordance with previous results as they have also conducted permeation studies for SLNs [47].

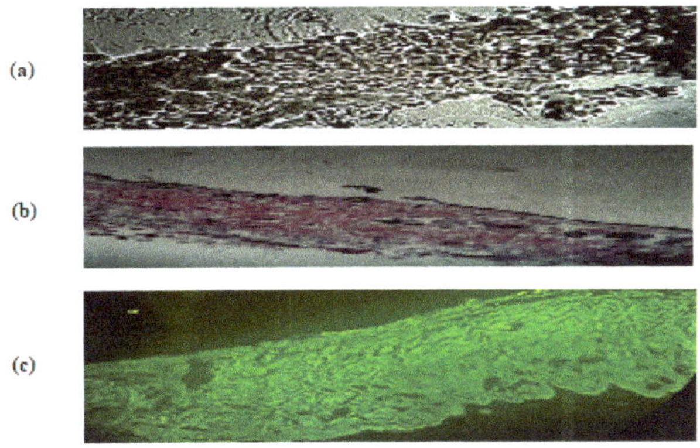

Figure 7. Fluorescence image of rat skin (transactional slice) after Rhodamine loaded nanoparticles permeation: (**a**) skin without treatment, (**b**) visualization under normal microscope and (**c**) green fluoresce exhibiting penetration across stratum corneum.

3.9. Evaluation of Skin Structure after Treatment with Hydrogel

Permeation enhancers are known to reduce barrier function of stratum corneum by destabilizing and fluidizing skin lipids, hence enhancing percutaneous absorption. In this study, changes in epidermal structures were observed through FTIR. It was detected from the vibrations that there was a minute change in control and hydrogel NPs treated skin. The incorporation of permeation enhancer caused no damage to the skin as confirmed by the FTIR spectra (Figure 8). Small convulsions were observed in the spectra which may be due to argan oil induced intercellular penetration. However, these are minor changes, and skin has the ability to recover due to its regenerative and elastic properties. The present results suggest this formulation is a safe and promising transdermal delivery system.

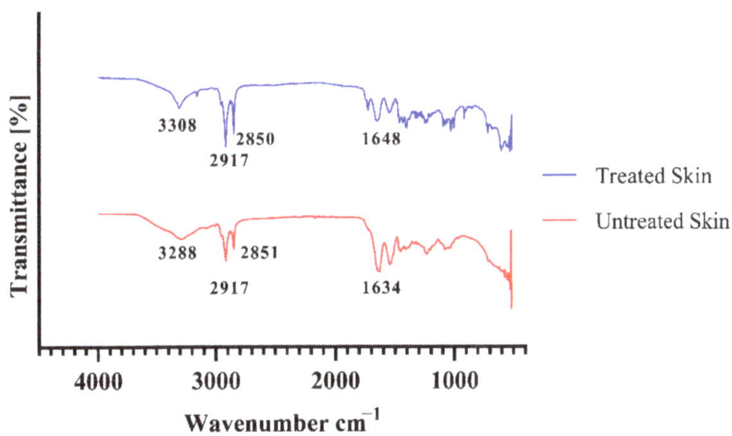

Figure 8. FTIR analysis of hydrogel treated skin and original skin without any specific treatment.

3.10. Stability Studies of Hydrogel

Physical appearance of hydrogel exhibited no change in terms of phase separation, colour and grittiness at accelerated conditions as shown in Table 3. However, a slight increase in pH was observed from 5.3 to 5.6. Stability is an important parameter to ensure the effectiveness of formulation. The results of this study exhibited no gross change; thus, it can be considered that the hydrogel has a tendency to maintain its effectiveness.

Table 3. Stability studies of hydrogel in terms of physical appearance and pH.

Time (Months)	Physical Appearance			pH
	Phase Separation	Color	Grittiness	
0	No	No change	None	5.3
3	No	No change	None	5.5
6	No	No change	None	5.6

4. Conclusions

Lipid nanoparticles have emerged as a versatile nano-platform for a number of applications in drug delivery. Hydrogels loaded with argan oil and NPs were successfully formulated with biodegradable and biocompatible ingredients for transdermal delivery. SIM-SLNs with uniform particle size distribution and spherical morphology demonstrated high encapsulation efficiency, better retention time, increased permeability and sustained in vitro drug release. Transdermal administration of nanoparticles via hydrogel might significantly enhance bioavailability. Consequently, argan oil based transdermal hydrogels containing SIM-SLNs could be promising nanocarriers to enhance therapeutic potential because argan oil acts as an antihyperlipidemic ingredient. The advantages of gels can be extended to those suffering from gastrointestinal problems or drugs having limited bioavailability. Further studies are required to confirm the synergistic effect of argan oil and SIM on the reduction of hyperlipidaemia.

Author Contributions: Conceptualization, A.U. and M.F.A.K.; formal analysis, Z.u.M. and F.U.; funding acquisition, H.H. and A.E.; methodology, M.F.A.K.; supervision, N.A.; visualization, N.A.; writing—original draft, M.F.A.K.; writing—review and editing, A.A., A.E. and N.A. All authors have read and agreed to the published version of the manuscript.

Funding: This research received no external funding.

Institutional Review Board Statement: Not applicable.

Informed Consent Statement: Not applicable.

Data Availability Statement: Not applicable.

Acknowledgments: The researchers would like to thank the Deanship of Scientific Research, Qassim University for funding the publication of this project. They are also thankful to Gattefosse, France, for providing us Precirol ATO 5 as gift through Morgan Chemicals, Pakistan.

Conflicts of Interest: All authors declare that they have no conflict of interest.

Ethical Approval: Not applicable.

References

1. Adi, D.; Abuzhalihan, J.; Wang, Y.-H.; Baituola, G.; Wu, Y.; Xie, X.; Fu, Z.Y.; Yang, Y.N.; Ma, X.; Li, X.M.; et al. IDOL gene variant is associated with hyperlipidemia in Han population in Xinjiang, China. *Sci. Rep.* **2020**, *10*, 14280. [CrossRef] [PubMed]
2. Asbahani, A.; El Miladi, K.; Badri, W.; Sala, M.; Addi, E.H.A.; Casabianca, H.; Mousadik, A.; El Hartmann, D.; Jilale, A.; Renaud, F.N.R.; et al. Essential oils: From extraction to encapsulation. *Int. J. Pharm.* **2015**, *483*, 220–243. [CrossRef] [PubMed]
3. Badreddine, A.; Zarrouk, A.; Meddeb, W.; Nury, T.; Rezig, L.; Debbabi, M.; Bessam, F.Z.; Brahmi, F.; Vejux, A.; Mejri, M.; et al. Chapter 10—Antioxidant and neuroprotective properties of Mediterranean oils: Argan oil, olive oil, and milk thistle seed oil. In *Oxidative Stress and Dietary Antioxidants in Neurological Diseases*; Martin, C.R., Preedy, V.R., Eds.; Academic Press: London, UK, 2020; pp. 143–154. Available online: https://www.sciencedirect.com/science/article/pii/B9780128177808099941 (accessed on 22 February 2022). [CrossRef]
4. Tan, W.Y.T.; Young, B.E.; Lye, D.C.; Chew, D.E.K.; Dalan, R. Statin use is associated with lower disease severity in COVID-19 infection. *Sci. Rep.* **2020**, *10*, 17458. [CrossRef]
5. Dolatabadi, J.E.N.; Hamishehkar, H.; Eskandani, M.; Valizadeh, H. Formulation, characterization and cytotoxicity studies of alendronate sodium-loaded solid lipid nanoparticles. *Colloids Surf. B Biointerfaces* **2014**, *117*, 21–28. [CrossRef]
6. Nutan, B.; Chandel, A.K.S.; Biswas, A.; Kumar, A.; Yadav, A.; Maiti, P.; Jewrajka, S.K. Gold Nanoparticle Promoted Formation and Biological Properties of Injectable Hydrogels. *Biomacromolecules* **2020**, *21*, 3782–3794. [CrossRef]

7. Yang, Z.; Peng, H.; Wang, W.; Liu, T. Crystallization behavior of poly(ε-caprolactone)/layered double hydroxide nanocomposites. *J. Appl. Polym. Sci.* **2010**, *116*, 2658–2667. [CrossRef]
8. Ahmed, T.A.; Badr-Eldin, S.M.; Ahmed, O.A.A.; Aldawsari, H. Intranasal optimized solid lipid nanoparticles loaded in situ gel for enhancing trans-mucosal delivery of simvastatin. *J. Drug Deliv. Sci. Technol.* **2018**, *48*, 499–508. [CrossRef]
9. Thoniyot, P.; Tan, M.J.; Karim, A.A.; Young, D.J.; Loh, X.J. Nanoparticle–Hydrogel Composites: Concept, Design, and Applications of These Promising, Multi-Functional Materials. *Adv. Sci.* **2015**, *2*, 1400010. [CrossRef]
10. Qindeel, M.; Ullah, M.H.; Fakhar-ud-Din; Ahmed, N.; Rehman, A. Recent trends, challenges and future outlook of transdermal drug delivery systems for rheumatoid arthritis therapy. *J. Control. Release* **2020**, *327*, 595–615. [CrossRef]
11. Xia, M.; Cheng, Y.; Meng, Z.; Jiang, X.; Chen, Z.; Theato, P.; Zhu, M. A novel nanocomposite hydrogel with precisely tunable UCST and LCST. *Macromol. Rapid Commun.* **2015**, *36*, 477–482. [CrossRef] [PubMed]
12. Duan, Y.; Dhar, A.; Patel, C.; Khimani, M.; Neogi, S.; Sharma, P.; Siva Kumar, N.; Vekariya, R.L. A brief review on solid lipid nanoparticles: Part and parcel of contemporary drug delivery systems. *RSC Adv.* **2020**, *10*, 26777–26791. [CrossRef]
13. Ma, G.; Lin, W.; Wang, Z.; Zhang, J.; Qian, H.; Xu, L.; Yuan, Z.; Chen, S. Development of polypeptide-based zwitterionic amphiphilic micelles for nanodrug delivery. *J. Mater. Chem. B* **2016**, *4*, 5256–5264. [CrossRef] [PubMed]
14. Narayanan, S.; Pavithran, M.; Viswanath, A.; Narayanan, D.; Mohan, C.C.; Manzoor, K.; Menon, D. Sequentially releasing dual-drug-loaded PLGA-casein core/shell nanomedicine: Design, synthesis, biocompatibility and pharmacokinetics. *Acta Biomater.* **2014**, *10*, 2112–2124. [CrossRef] [PubMed]
15. Martínez Rivas, C.J.; Tarhini, M.; Badri, W.; Miladi, K.; Greige-Gerges, H.; Nazari, Q.A.; Galindo Rodríguez, S.A.; Román, R.Á.; Fessi, H.; Elaissari, A. Nanoprecipitation process: From encapsulation to drug delivery. *Int. J. Pharm.* **2017**, *532*, 66–81. [CrossRef]
16. Tella, A.C.; Okoro, H.K.; Sokoya, S.O.; Adimula, V.O.; Olatunji, S.O.; Zvinowanda, C.; Ngila, J.C.; Shaibu, R.O.; Adeyemi, O.G. Synthesis, Characterization and Antifungal Activity of Fe(III) Metal–Organic Framework and its Nano-Composite. *Chem. Afr.* **2020**, *3*, 119–126. [CrossRef]
17. Mohammadi-Samani, S.; Salehi, H.; Entezar-Almahdi, E.; Masjedi, M. Preparation and characterization of sumatriptan loaded solid lipid nanoparticles for transdermal delivery. *J. Drug Deliv. Sci. Technol.* **2020**, *57*, 101719. [CrossRef]
18. Alvarez-Figueroa, M.J.; Narváez-Araya, D.; Armijo-Escalona, N.; Carrasco-Flores, E.A.; González-Aramundiz, J.V. Design of Chitosan Nanocapsules with Compritol 888 ATO® for Imiquimod Transdermal Administration. Evaluation of Their Skin Absorption by Raman Microscopy. *Pharm. Res.* **2020**, *37*, 195. [CrossRef]
19. Eltellawy, Y.A.; El-Kayal, M.; Abdel-Rahman, R.F.; Salah, S.; Shaker, D.S. Optimization of transdermal atorvastatin calcium–Loaded proniosomes: Restoring lipid profile and alleviating hepatotoxicity in poloxamer 407-induced hyperlipidemia. *Int. J. Pharm.* **2021**, *593*, 120163. [CrossRef]
20. Luo, Y.; Teng, Z.; Li, Y.; Wang, Q. Solid lipid nanoparticles for oral drug delivery: Chitosan coating improves stability, controlled delivery, mucoadhesion and cellular uptake. *Carbohydr. Polym.* **2015**, *122*, 221–229. [CrossRef]
21. Saqib, M.; Shabbir Ali Bhatti, A.; Ahmad, N.M.; Ahmed, N.; Shahnaz, G.; Lebaz, N.; Elaissari, A. Amphotericin b loaded polymeric nanoparticles for treatment of leishmania infections. *Nanomaterials* **2020**, *10*, 1152. [CrossRef] [PubMed]
22. Rana, I.; Khan, N.; Ansari, M.M.; Shah, F.A.; Din, F.U.; Sarwar, S.; Imran, M.; Qureshi, O.S.; Choi, H.I.; Lee, C.H.; et al. Solid lipid nanoparticles-mediated enhanced antidepressant activity of duloxetine in lipopolysaccharide-induced depressive model. *Colloids Surf. B Biointerfaces* **2020**, *194*, 111209. [CrossRef] [PubMed]
23. Harisa, G.I.; Alomrani, A.H.; Badran, M.M. Simvastatin-loaded nanostructured lipid carriers attenuate the atherogenic risk of erythrocytes in hyperlipidemic rats. *Eur. J. Pharm. Sci.* **2017**, *96*, 62–71. [CrossRef] [PubMed]
24. Qindeel, M.; Ahmed, N.; Sabir, F.; Khan, S.; Ur-Rehman, A. Development of novel pH-sensitive nanoparticles loaded hydrogel for transdermal drug delivery. *Drug Dev. Ind. Pharm.* **2019**, *45*, 629–641. [CrossRef] [PubMed]
25. Jana, S.; Manna, S.; Nayak, A.K.; Sen, K.K.; Basu, S.K. Carbopol gel containing chitosan-egg albumin nanoparticles for transdermal aceclofenac delivery. *Colloids Surf. B Biointerfaces* **2014**, *114*, 36–44. [CrossRef]
26. Zeb, A.; Qureshi, O.S.; Yu, C.H.; Akram, M.; Kim, H.S.; Kim, M.S.; Kang, J.H.; Majid, A.; Chang, S.Y.; Bae, O.N.; et al. Enhanced anti-rheumatic activity of methotrexate-entrapped ultradeformable liposomal gel in adjuvant-induced arthritis rat model. *Int. J. Pharm.* **2017**, *525*, 92–100. [CrossRef]
27. Siddiqui, B.; ur.Rehman, A.; Haq, I.-U.; Ahmad, N.M.; Ahmed, N. Development, optimisation, and evaluation of nanoencapsulated diacerein emulgel for potential use in osteoarthritis. *J. Microencapsul.* **2020**, *37*, 595–608. [CrossRef]
28. Seong, J.S.; Yun, M.E.; Park, S.N. Surfactant-stable and pH-sensitive liposomes coated with *N*-succinyl-chitosan and chitooligosaccharide for delivery of quercetin. *Carbohydr. Polym.* **2018**, *181*, 659–667. [CrossRef]
29. Naz, K.; Shahnaz, G.; Ahmed, N.; Qureshi, N.A.; Sarwar, H.S.; Imran, M.; Khan, G.M. Formulation and In Vitro Characterization of Thiolated Buccoadhesive Film of Fluconazole. *AAPS PharmSciTech* **2017**, *18*, 1043–1055. [CrossRef]
30. Jamadar, M.J.; Shaikh, R.H. Preparation and Evaluation of Herbal Gel Formulation. *J. Pharaceutical Res. Educ.* **2017**, *1*, 201–224.
31. Qureshi, O.S.; Kim, H.S.; Zeb, A.; Choi, J.S.; Kim, H.S.; Kwon, J.E.; Kim, M.S.; Kang, J.H.; Ryou, C.; Park, J.S.; et al. Sustained release docetaxel-incorporated lipid nanoparticles with improved pharmacokinetics for oral and parenteral administration. *J. Microencapsul.* **2017**, *34*, 250–261. [CrossRef] [PubMed]
32. Azeez, L.; Lateef, A.; Adejumo, A.L.; Adeleke, J.T.; Adetoro, R.O.; Mustapha, Z. Adsorption Behaviour of Rhodamine B on Hen Feather and Corn Starch Functionalized with Green Synthesized Silver Nanoparticles (AgNPs) Mediated with Cocoa Pods Extracts. *Chem. Afr.* **2020**, *3*, 237–250. [CrossRef]

33. Dar, M.J.; Din, F.U.; Khan, G.M. Sodium stibogluconate loaded nano-deformable liposomes for topical treatment of leishmaniasis: Macrophage as a target cell. *Drug Deliv.* **2018**, *25*, 1595–1606. [CrossRef]
34. Panchagnula, R.; Bokalial, R.; Sharma, P.; Khandavilli, S. Transdermal delivery of naloxone: Skin permeation, pharmacokinetic, irritancy and stability studies. *Int. J. Pharm.* **2005**, *293*, 213–223. [CrossRef] [PubMed]
35. Giuliano, E.; Paolino, D.; Fresta, M.; Cosco, D. Drug-Loaded Biocompatible Nanocarriers Embedded in Poloxamer 407 Hydrogels as Therapeutic Formulations. *Medicines* **2018**, *6*, 7. [CrossRef] [PubMed]
36. Ahmed, S.; Gull, A.; Aqil, M.; Danish Ansari, M.; Sultana, Y. Poloxamer-407 thickened lipid colloidal system of agomelatine for brain targeting: Characterization, brain pharmacokinetic study and behavioral study on Wistar rats. *Colloids Surf. B Biointerfaces* **2019**, *181*, 426–436. [CrossRef]
37. Schröder, A.; Sprakel, J.; Schroën, K.; Spaen, J.N.; Berton-Carabin, C.C. Coalescence stability of Pickering emulsions produced with lipid particles: A microfluidic study. *J. Food Eng.* **2018**, *234*, 63–72. [CrossRef]
38. Patil, S.G.; Jadhav, V.M.; Kadam, V.J. Formulation and Evaluation of Herbal Gel. *Indian J. Nat. Prod. Resour.* **2012**, *3*, 501–505.
39. De Mendoza, A.E.H.; Campanero, M.A.; Mollinedo, F.; Blanco-Prieto, M.J. Lipid nanomedicines for anticancer drug therapy. *J. Biomed. Nanotechnol.* **2009**, *5*, 323–343. [CrossRef]
40. Qureshi, O.S.; Zeb, A.; Akram, M.; Kim, M.S.; Kang, J.H.; Kim, H.S.; Majid, A.; Han, I.; Chang, S.Y.; Bae, O.N.; et al. Enhanced acute anti-inflammatory effects of CORM-2-loaded nanoparticles via sustained carbon monoxide delivery. *Eur. J. Pharm. Biopharm.* **2016**, *108*, 187–195. [CrossRef]
41. Dudhipala, N.; Veerabrahma, K. Improved anti-hyperlipidemic activity of Rosuvastatin Calcium via lipid nanoparticles: Pharmacokinetic and pharmacodynamic evaluation. *Eur. J. Pharm. Biopharm.* **2017**, *110*, 47–57. [CrossRef] [PubMed]
42. Jaafar-Maalej, C.; Diab, R.; Andrieu, V.; Elaissari, A.; Fessi, H. Ethanol injection method for hydrophilic and lipophilic drug-loaded liposome preparation. *J. Liposome Res.* **2010**, *20*, 228–243. [CrossRef] [PubMed]
43. Tarhini, M.; Benlyamani, I.; Hamdani, S.; Agusti, G.; Fessi, H.; Greige-Gerges, H.; Bentaher, A.; Elaissari, A. Protein-based nanoparticle preparation via nanoprecipitation method. *Materials* **2018**, *11*, 394. [CrossRef] [PubMed]
44. Md, S.; Kuldeep Singh, J.K.A.; Waqas, M.; Pandey, M.; Choudhury, H.; Habib, H.; Hussain, F.; Hussain, Z. Nanoencapsulation of betamethasone valerate using high pressure homogenization–solvent evaporation technique: Optimization of formulation and process parameters for efficient dermal targeting. *Drug Dev. Ind. Pharm.* **2019**, *45*, 323–332. [CrossRef]
45. Jiang, T.; Han, N.; Zhao, B.; Xie, Y.; Wang, S. Enhanced dissolution rate and oral bioavailability of simvastatin nanocrystal prepared by sonoprecipitation. *Drug Dev. Ind. Pharm.* **2012**, *38*, 1230–1239. [CrossRef]
46. Rauf, A.; Razzaq, S.; Tabish, T.A.; Tahseen, S.; Sandhu, M.A.; Shahnaz, G. Investigating the intracellular bactericidal effects of rifampicin loaded S-protected thiomeric chitosan nanocargoes against *Mycobacterium tuberculosis*. *J. Drug Deliv. Sci. Technol.* **2021**, *61*, 102184. [CrossRef]
47. Hasan, A.S.; Socha, M.; Lamprecht, A.; Ghazouani, F.; El Sapin, A.; Hoffman, M.; Maincent, P.; Ubrich, N. Effect of the microencapsulation of nanoparticles on the reduction of burst release. *Int. J. Pharm.* **2007**, *344*, 53–61. [CrossRef]
48. Chen, H.; Chang, X.; Du, D.; Li, J.; Xu, H.; Yang, X. Microemulsion-based hydrogel formulation of ibuprofen for topical delivery. *Int. J. Pharm.* **2006**, *315*, 52–58. [CrossRef]
49. Gangurde, A.B.; Amin, P.D. Microencapsulation by Spray Drying of Vitamin A Palmitate from Oil to Powder and Its Application in Topical Delivery System. *J. Encapsul. Adsorpt. Sci.* **2017**, *07*, 10–39. [CrossRef]
50. Li, X.; Yang, Q.; Zhao, Y.; Long, S.; Zheng, J. Dual physically crosslinked double network hydrogels with high toughness and self-healing properties. *Soft Matter* **2017**, *13*, 911–920. [CrossRef]
51. Mirzaei, B.E.; Ramazani, A.; Shafiee, M.; Danaei, M. Studies on glutaraldehyde crosslinked chitosan hydrogel properties for drug delivery systems. *Int. J. Polym. Mater. Polym. Biomater.* **2013**, *62*, 605–611. [CrossRef]
52. Puri, A.; Frempong, D.; Mishra, D.; Dogra, P. Microneedle-mediated transdermal delivery of naloxone hydrochloride for treatment of opioid overdose. *Int. J. Pharm.* **2021**, *604*, 120739. [CrossRef] [PubMed]
53. Din, F.U.; Saleem, S.; Aleem, F.; Ahmed, R.; Huda, N.U.; Ahmed, S.; Khaleeq, N.; Shah, K.U.; Ullah, I.; Zeb, A.; et al. Advanced colloidal technologies for the enhanced bioavailability of drugs. *Cogent Med.* **2018**, *5*, 1480572. [CrossRef]
54. Kawar, D.; Abdelkader, H. Hyaluronic acid gel-core liposomes (hyaluosomes) enhance skin permeation of ketoprofen. *Pharm. Dev. Technol.* **2019**, *24*, 947–953. [CrossRef]
55. Khalid, A.; Ahmed, N.; Qindeel, M.; Asad, M.I.; Khan, G.M.; Rehman, A.U. Development of novel biopolymer-based nanoparticles loaded cream for potential treatment of topical fungal infections. *Drug Dev. Ind. Pharm.* **2021**, *47*, 1090–1099. [CrossRef]

gels

Article

Functional Hydrogels with Chondroitin Sulfate Release Properties Regulate the Angiogenesis Behaviors of Endothelial Cells

Haonan Wang [1], Qian Li [1,*], Yongchao Jiang [2,*] and Xiaofeng Wang [1]

[1] School of Mechanics and Safety Engineering, Zhengzhou University, Zhengzhou 450001, China; whn18838970880@163.com (H.W.); xiaofengwang@zzu.edu.cn (X.W.)

[2] School of Materials Science and Engineering, Zhengzhou University, Zhengzhou 450001, China

* Correspondence: qianli@zzu.edu.cn (Q.L.); zzujeremy@outlook.com or ycjiang@zzu.edu.cn (Y.J.)

Abstract: Functional hydrogels with properties that mimic the structure of extracellular matrix (ECM) and regulate cell behaviors have drawn much attention in biomedical applications. Herein, gelatin-based hydrogels were designed and loaded with chondroitin sulfate (CS) to endow biological regulation on the angiogenesis behaviors of endothelial cells (ECs). Manufactured hydrogels containing various amounts of CS were characterized via methods including mechanical tests, cytocompatibility, hemolysis, and angiogenesis assays. The results showed that the prepared hydrogels exhibited excellent mechanical stability, cytocompatibility, and hemocompatibility. Additionally, the angiogenesis behaviors of ECs were obviously promoted. However, excessive loading of CS would weaken the effect due to a higher proportion of occupation on the cell membrane. In conclusion, this investigation highlights the great potential of these hydrogels in treating ischemic diseases and accelerating tissue regeneration in terms of regulating the angiogenesis process via CS release.

Keywords: hydrogel; chondroitin sulfate; endothelial cell; angiogenesis

1. Introduction

The vasculature is a key component of the tissue microenvironment due to its critical role in providing nutrients, oxygen, and signaling cues that influence biological outcomes [1,2]. Nowadays, ischemic diseases like atherosclerosis and diabetic foot have become a major cause of disability worldwide [3], because these diseases impede the normal behaviors of cells such as proliferation, migration, and interactions with pericytes [4,5]. A quick reconstruction of the vasculature is therefore needed to guarantee adequate exchange of nutrients and accelerated recovery of the abovementioned diseases [6].

Various attempts to date have focused primarily on exogenous growth factors or genetic modifications of cells to induce the angiogenesis process within tissue-engineered scaffolds [7]. However, recombinant angiogenic growth factors suffer from easy degradation and abnormal angiogenesis [8], while the gene-modified cells are difficult to implement due to uncontrollable behaviors. Thus, strategies to stimulate angiogenesis via functionalized scaffolds have been widely investigated [9,10].

It has been widely recognized that hydrogels with traits of a three-dimensional soft network and easy processibility are preferred in angiogenesis stimulation [11,12]. In a previous report by Choi et al., hydrogel scaffolds with small pores (<200 μm) favored the formation of vascular networks at high density and poor penetration depth, while those with large pores (>200 μm) presented the opposite trends [13]. In another study, a Cu-HHA/PVA@MΦ2 hydrogel scaffold was prepared by Liu et al. for synergistic improvement of impaired angiogenesis to accelerate the diabetic chronic wound healing [14]. These results indicate that the inherent properties and functionalization of hydrogel can be employed to induce the angiogenesis process.

As a complicated macromolecular sulfated polysaccharide from natural resources, as well as the major component of ECM [15], chondroitin sulfate (CS) has been approved to play multiple roles in regulating numerous activities of endothelial cells (ECs) [16,17]. For instance, a CS-coating method was employed to drive the endothelization of pericardium by Lopez-Moya et al. [18]. However, the effect of different CS concentrations was not further investigated. On the contrary, Xiong et al. fabricated a negatively charged alginate and chondroitin sulfate microsphere hydrogel (nCACSMH) with different CS loading [19]. Even though the authors indicated that nCACSMH can dramatically accelerate the angiogenesis-related genes in ECs, the angiogenesis behavior of ECs, i.e., the sprouting and formation of tubular structures, was not visibly detected, which might be attributed to the short release time of CS.

Inspired by their conclusions, gelatin-based hydrogels with CS release properties are also proposed here, and are assumed to regulate the angiogenesis behaviors of ECs via various CS loading amounts. Different from the report by Xiong et al. [19], herein, gelatin hydrogel was chemically cross-linked by genipin during preparation, with the aim of maintaining its mechanical stability, prolonging the release period of CS, and the effective stimulation of ECs. Accordingly, the objective of this work was to develop a hydrogel platform with visible modulation on the angiogenesis behaviors of ECs (shown in Figure 1). Therefore, composite hydrogels containing various CS concentrations were manufactured and characterized on physical and biological tests.

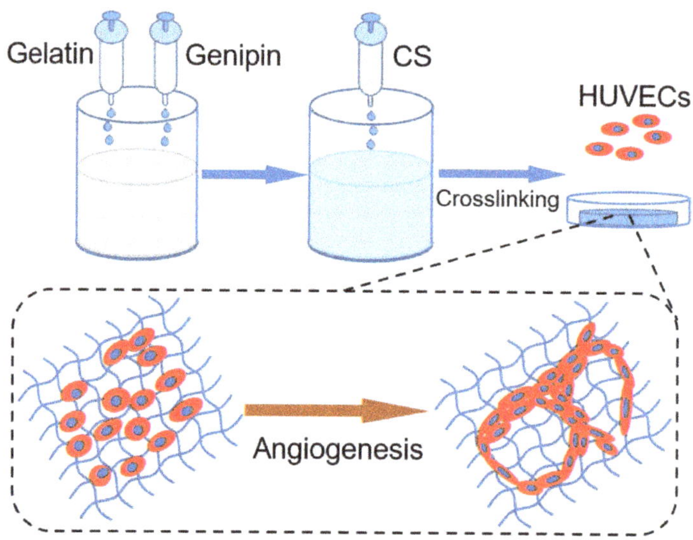

Figure 1. Schematic diagram of functional hydrogel fabrication.

2. Results and Discussion

2.1. Characterization of Hydrogels

The gelation behavior of different hydrogels was investigated, and the results are shown in Figure 2a. After 75 s at room temperature, gelatin began to gel due to the construction of hydrogen bonds within the molecular chains. Meanwhile, the gelation time increased to 84 s after CS addition, which might be ascribed to the charge repulsion between gelatin and CS molecules.

A quick reconstruction of the vasculature is necessary for adequate nutrient supply and metabolite exhaustion in biomedical fields, especially for engineered tissues [2]. To achieve this goal, the rapid and effective control of the angiogenesis process needs to be fully considered. It has been demonstrated that scaffolds with porous structures would benefit this process. From Figure 2b, it can be seen that all hydrogels show similar porous structures to the ECM. These kinds of hydrogels would be conducive to cell migration and penetration.

Due to the relatively low amount of CS, there was no obvious change in structure among the groups. However, samples in the CS-0.05 and CS 0.1 groups shrank to some degree after CS addition. This can be attributed to the Hofmeister effect between gelatin and sulfate ions, which can remove the hydration water of the protein macromolecules from the hydrogels, leading to structure folding [20]. However, the decreased pore size accelerates the formation of vascular networks according to the results of Choi et al [13].

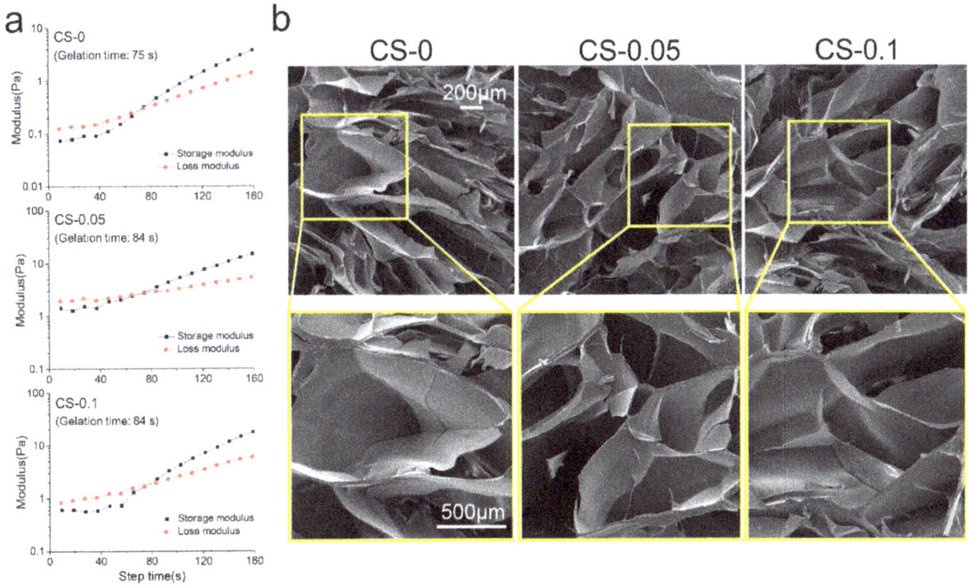

Figure 2. (**a**) The rheological behaviors, and (**b**) SEM images of different hydrogels.

In general, gelatin-based hydrogels show high swelling properties and inferior resistance to body fluids. In a previous study by Entekhabi et al., gelatin hydrogel was reported to have a swelling ratio of 500% at the same concentration as our study [21]. Thus, a genipin crosslinker was employed here. In Figure 3a, hydrogels present swelling ratios of $103.10 \pm 0.54\%$, $102.83 \pm 0.76\%$, and $104.13 \pm 0.35\%$ in the CS-0, CS-0.05 and CS-0.1 groups, respectively, indicating the development of stability after crosslinking. Due to the collagenase enzymolysis, hydrogels of three groups can biodegrade within 40 h, as can be seen from Figure 3b. However, no obvious distinction can be found among the groups, meaning that the Hofmeister effect did not affect the swelling ratios and biodegradability of the hydrogels.

2.2. Loading and Release of CS

The elemental spectra of the CS-0, CS-0.05, and CS-0.1 hydrogels were identified by XPS, and the results are displayed in Figure 4. As expected, S 2p photoelectron signals were only detected in the CS-0.05 and CS-0.1 groups. The atomic compositions of these samples were calculated on the basis of the XPS spectra, and are shown as inserts in Figure 4. According to the data, the elemental ratio of sulfur was 3.92 at% in the CS-0.05 sample, while it was 6.00 at% in the CS-0.1 group, nearly two-fold the value of the CS-0.05 sample. These results prove that CS was successfully introduced into the composite hydrogels.

The release properties of CS from the hydrogels was quantified using the Azure A staining method. Azure A is a cationic dye, and a discoloration effect can occur through electrostatic interaction between Azure A and sulfate radicals in CS [22]. Therefore, the hydrogel extract, which is the initial blue color in the CS-0 group, changes to a bluish-purple color when Azure A is added to the CS-0.05 and CS-1 groups (Figure 5a). The release profiles of CS were determined, and are in Figure 5b. The profiles of the CS-0.05 and CS-1 groups are stable and linear during the first 10 h. The CS-0.1 group shows a higher release rate than CS-0.05, which ascribes to more loading amount of CS. The CS-0.05 hydrogel reaches an equilibrium state after 10 h. While in CS-0.1 group, the equilibrium state occurred after 48 h. The distinction may be attributed to electrostatic interactions between gelatin and CS. The data above also approve that the release behavior can be tuned through different CS loading.

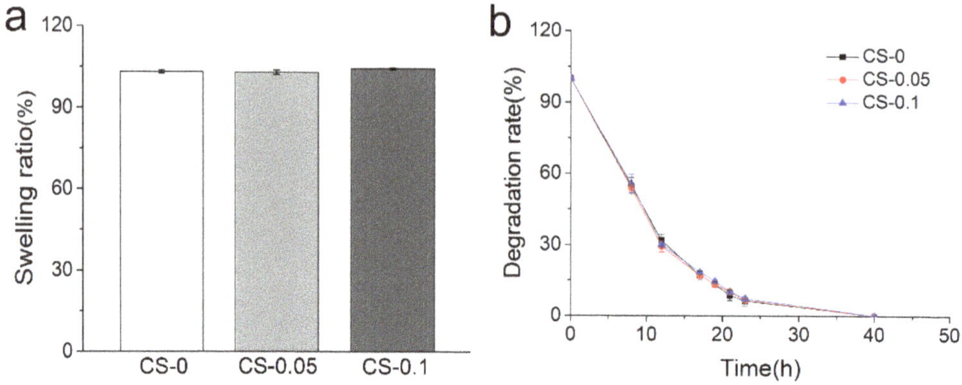

Figure 3. (**a**) Swelling ratios and (**b**) degradation rates of different hydrogels.

2.3. Mechanical Properties of Hydrogels

The mechanical properties of hydrogels were studied using a universal testing machine, and the compressive stress–strain curves and corresponding moduli of different hydrogels are presented in Figure 6a,b. As can be observed, the compressive moduli of hydrogels are around 210 kPa, which is obviously higher than that presented by the pure gelatin hydrogel in our last report [23], indicating that the combination of genipin and CS can improve the mechanical properties of gelatin hydrogels. However, with 0.1% CS addition, the compressive modulus of CS-0.1 hydrogel only increased by tens of kPa compared to that of CS-0, meaning that the crosslinking effect is the main cause of the enhanced mechanical properties.

Cyclic compression and rheological testing were also carried out to study the mechanical properties of the hydrogels. The loading–unloading curves of composite hydrogels are presented in Figure 6c. All hydrogels are recoverable when the external force is removed, which is also attributed to the reinforcing effect of genipin and CS. Among these hydrogels, the dissipation energy (the area enclosed by loading and reloading curves) decreases slightly with CS addition in Figure 6d. Due to the dehydration and folding of gelatin macromolecules caused by the Hofmeister effect, the hydrogels adsorb less energy. However, the Hofmeister effect is not strong enough to cause an obvious decrease in energy dissipation. According to the rheological curves in Figure 6e, the storage modulus increased from around 180 kPa to around 530 kPa with CS addition. It is noteworthy that these storage moduli were much higher than those in Figure 2a. In the beginning, the hydrogen bonds mainly contributed to hydrogel formation, which was weak but quick. After 5 min of reaction, the genipin crosslinking was dominant, and the degree of crosslinking increased with time increment. When the crosslinking reactions were completed after 24 h, no significant difference in mechanical properties could be found, as shown in Figure 6b,d.

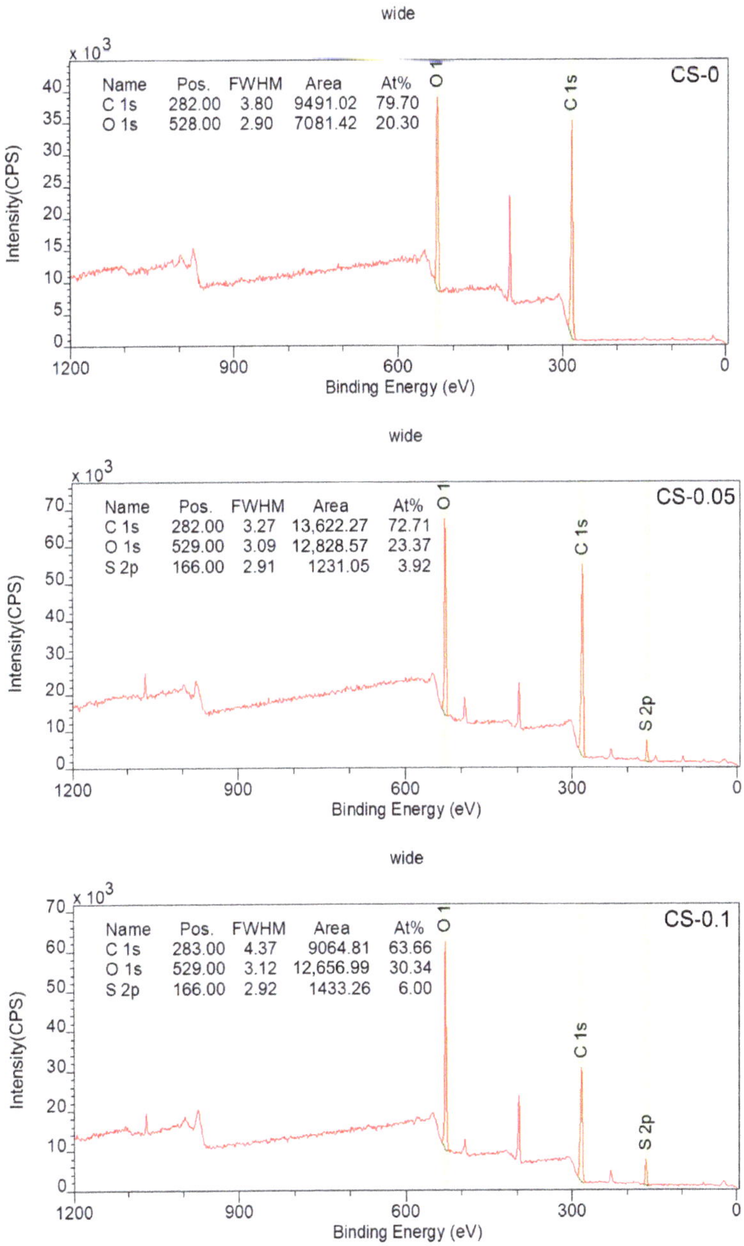

Figure 4. XPS spectra of the hydrogels.

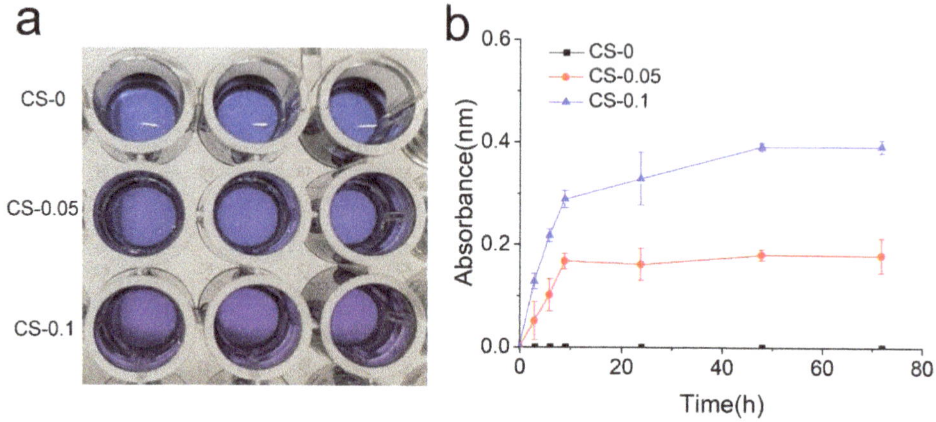

Figure 5. (**a**) Photographs of reactant, and (**b**) release profiles of CS in each group.

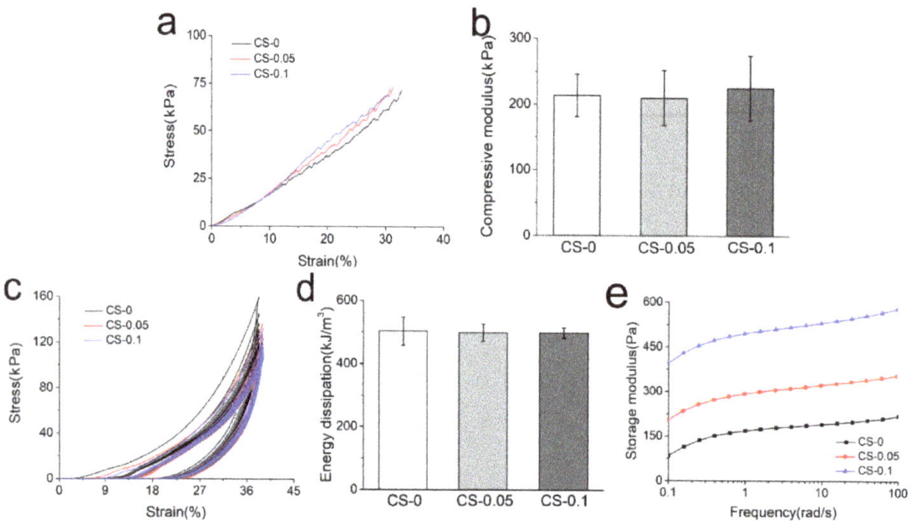

Figure 6. (**a**) The representative compressive stress–strain curves; (**b**) compressive modulus; (**c**) cyclic compressive stress–strain curves; (**d**) energy dissipation during cyclic compression on different hydrogels; and (**e**) storage modulus of each group.

2.4. Cytocompatibility and Hemocompatibility of Hydrogels

Cytocompatibility and hemocompatibility are essential criteria of hydrogels for biomedical applications. Herein, the cytocompatibility of hydrogels was evaluated via live and dead cell staining and CCK-8 proliferation assays. First, HUVECs were cultured on the surface of hydrogels for 4 days, then stained with a live and dead activity kit. Figure 7 shows the staining images under a confocal microscope. Cells in the CS-0.05 and CS-0.1 groups can adhere to the hydrogel surface and present a spindle-like morphology, which is consistent with that of the CS-0 group. Though few dead cells (labeled by the red color) can be found, the surviving rates in all groups exceed 95%. Notably, the coverage area of cells in the other two groups is larger than that of the CS-0 group, suggesting that CS addition can enhance the proliferation of HUVECs. To validate this conclusion, cell numbers from each group were measured after 1, 3 and 5 days via the CCK-8 kit, and the results are presented in Figure 7b. CS addition dramatically promotes the proliferation rate of HUVECs, leading to higher absorbance values than in the CS-0 group after 5 days. However, CS-0.05 presents the highest absorbance value, which may be ascribed to the competition among excess cells.

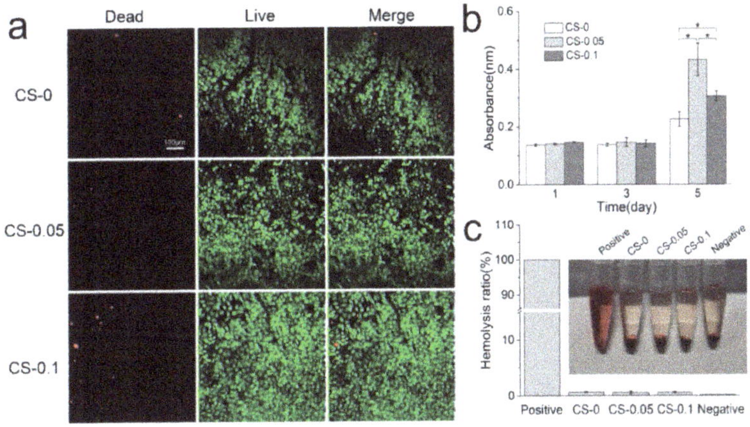

Figure 7. (**a**) Live and dead staining of HUVECs; (**b**) proliferation rates of HUVECs on different hydrogels; and (**c**) the hemolysis ratios of various hydrogels, * $p < 0.05$.

The hemocompatibility of hydrogels is identified in Figure 7c. Erythrocytes swell and burst in the positive group (DI water), while hydrogels and the negative group (PBS) show a similar canary yellow color to the normal saline. Quantitative evaluation reveals low hemolysis rates in all groups, which are below the internationally permitted level of 5% [24], indicating that these hydrogels do not cause serious hemolysis, and would be suitable for biomedical applications [25].

2.5. Modulation on Angiogenesis Behaviors

As a component of ECM, CS has been recognized to play multiple roles in regulating numerous cell activities [26]. To study the angiogenesis regulation of hydrogels on HUVECs, cells were photographed at different time points to observe dynamic changes during the sprouting assays. From Figure 8a, HUVECs are loose, without connections at the beginning of culture. After 5 h of culture, cells began to form networks with an elongated morphology, and the CS-0.05 group presents the most numerous and most dense networks. After 7 h, the tubular structures in all groups start to collapse and disappear. We further stained the cytoskeleton and nucleus at the 5 h time point, and observed cell morphology under laser confocal microscopy. Denser tubular structures and more elongated actin filaments were found in CS-0.05 group than in the other two groups. Meanwhile, in the CS-0.1 group, the angiogenesis ability of HUVECs was inhibited, with spherical and unstretched morphology at the junctions. The numbers of grids and nodes in the tubular structures shown in Figure 8b were counted via Image J software [27], and the results are presented in Figure 8c,d. As can be seen, the CS-0.05 group has the largest numbers of grids and nodes, and the CS-0.1 hydrogel has fewer grids and nodes than the CS-0 group, suggesting that a low concentration of CS might be more suitable for promoting angiogenesis and subsequent tissue regeneration.

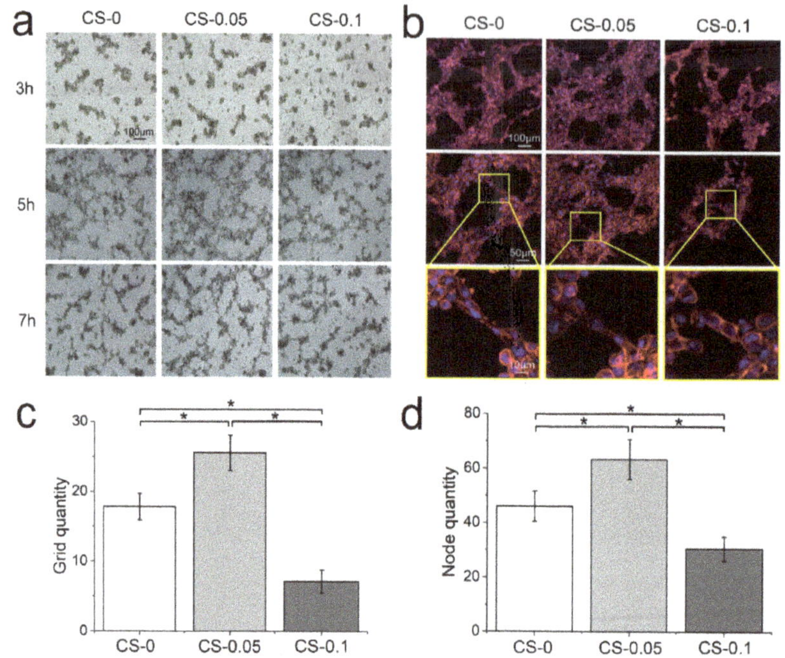

Figure 8. (**a**) Cell migration after 3 h, 5 h and 7 h of culture in sprouting assays; (**b**) fluorescent staining of nuclei and cytoskeletons after 5 h of culture; the numbers of (**c**) grids and (**d**) nodes in the tubular structure, * $p < 0.05$.

CS has been demonstrated to participate in blood vessel formation [28]. Similar to other growth factors, CS suffers from rapid dissolution and biodegradation in human fluid, which greatly restricts its targeted effect. Instead of electrostatic absorption [19], we encapsulated CS in a genipin-crosslinked gelatin hydrogel to prolong its local release time. Additionally, the release profile of CS was tuned via different loading concentrations, without any obvious influence on its physical properties. It can be concluded that CS addition endows hydrogels with angiogenesis modulation through CS release according to the biological assays. However, excessive CS will occupy a higher proportion of the cell membrane, hindering the combination of other growth factors [29,30]. Thus, the results indicate that the CS-0.05 hydrogel possesses the highest proliferation number and the best angiogenesis behaviors of ECs among these groups, showing great potential for treating ischemic diseases and achieving tissue regeneration in terms of regulating angiogenesis behaviors.

It is noteworthy that multiple growth factors (VEGF, FGF-2, and TGF-β) are necessary for steps in angiogenesis [1,17], which leads to the collapsed angiogenesis structure after 7 h, even though the CS-0.1 hydrogel is able to release CS for more than 3 days. Therefore, the synergistic effect between different growth factors and cells should be further investigated to reveal the mechanism involved in vascular reconstruction.

3. Conclusions

Herein, a kind of functional hydrogel with CS release properties was proposed and tuned using various CS amounts to modulate the angiogenesis behavior of ECs. The obtained hydrogels showed excellent mechanical stability, cytocompatibility, and hemocompatibility. Among them, CS-0.05 hydrogel was able to dramatically enhance the angiogenesis behavior, with denser tubular structures and longer stability, which has potential in treating ischemic tissue diseases and accelerating tissue regeneration. Notably, this research also indicated that an excess amount of CS could weaken the angiogenesis process due to excessive binding on the cell membrane. Further research should take the synergistic effect between different growth factors and cells into consideration to reveal the underlying mechanism of angiogenesis.

4. Materials and Methods

4.1. Materials

Gelatin, genipin, and Azure A were purchased from Aladdin Biochemical Technology Co, Ltd. (Shanghai, China). Chondroitin sulfate (CS) was obtained from Aoxing Biology (Zhejiang, China). Human umbilical vein endothelial cells (HUVECs) were obtained from Beijing Oligobio (Beijing, China). Matrigel, RPMI 1640 medium, phosphate-buffered saline (PBS), and trypsin EDTA were products of Corning Incorporated Life Science (New York, NY, USA). Triton X-100, bovine serum albumin, and glycine were products of Solarbio Life Science (Beijing, China). CF568 phalloidin and DAPI were from Biotium (Fremont, CA, USA). The CCK-8 assay kit was bought from Dojindo (Kyushu, Japan), and the live/dead viability kit was a product of Nanjing Keygen Biotech. Co, Ltd. (Nanjing, China). Deionized (DI) water was obtained from a Mill-Q water purification system (Woburn, MA, USA) and was used in all experiments [23].

4.2. Fabrication of Hydrogels

An aqueous solution method was used to produce hydrogels. Briefly, gelatin was firstly dissolved in DI water at a concentration of 10 (w/v) %. Then, genipin (0.5 w/w% to gelatin) was added to the gelatin solution with gentle stirring. Three groups of CS (0, 0.05, and 0.1 w/w% to gelatin) were dissolved in DI water, separately. Subsequently, gelatin/genipin and CS solutions with different concentrations were mixed at 45 °C to form homogeneous systems and transferred to a Teflon mold at room temperature for 24 h. The obtained hydrogels were named CS-0, CS-0.05, and CS-0.1, respectively.

A TA rheometer (DHR2 Discovery, TA Instrument, Newcastle, USA) was employed to obtain the gelation time of different hydrogels. Then, 800 μL of hydrogel precursor solution was placed between the parallel plate with a gap of 1 mm. After 5 min, the time sweep tests were then performed at 25 °C with a constant frequency of 10 rad/s and strain of 1%.

4.3. Morphology Observation

The microstructure of lyophilized hydrogels was observed using a scanning electron microscope (FEI Quanta FEG 250, Hillsboro, OR, USA) with an accelerating voltage of 20 kV. Before observing, samples were fractured in liquid nitrogen and the cross-section of samples was sputtered with gold for 60 s to increase the conductivity during imaging.

4.4. Swelling Ratio and Biodegradation Properties

Swelling ratios of hydrogels were measured by immersing samples in PBS solution (pH = 7.4) at 37 °C. At pre-determined time intervals, samples were removed from the solution, dried superficially with filter paper, and weighed. The swelling ratio (SR) was determined using the following equation:

$$\text{Swelling ratio } (\%) = [(W_t - W_i)/W_i] \times 100\% \quad (1)$$

where W_i is the initial weight of sample and W_t is the weight of hydrogel after soaking. All tests were performed in triplicate.

Likewise, the degradation properties of hydrogels were determined by immersing samples in the PBS solution containing 7.5 unit/mL collagenase. The degradation rate (DR) was calculated through equation as follows:

$$\text{Degradation rate } (\%) = \left[(W_0 - W_p)/W_0\right] \times 100\% \quad (2)$$

where W_0 is the initial mass of the sample and W_p is the residues after degradation for pre-determined intervals.

4.5. X-ray Photoelectron Spectroscopy (XPS) Analysis

The chemical compositions of hydrogels were investigated by XPS (AXIS Supra, Japan) equipment. Elements were identified in the range of 0–1200 eV with a resolution of 1 eV. The atomic concentration of each element was calculated by determining the relevant integral peak intensities.

4.6. CS Release Behavior

Azure A dye was employed to determine the release profile of CS from hydrogels [22]. Briefly, 5 g Azure A powder was dissolved in 100 mL PBS to make a stock solution. The hydrogel extract was prepared by soaking 0.6 g sample in 2 mL PBS solution. At pre-determined time intervals, 100 μL extract solution from each group was mixed with 0.5 mL stock solution, and 100 μL fresh PBS was added back. The mixture was measured at a wavelength of 540 nm via a microplate reader (Nano-Drop 2000, Thermo Scientific, Waltham, MA, USA). Serially diluted stock solutions were also measured to obtain the standard curve.

4.7. Mechanical Test

The rheological test was employed to obtain the storage modulus (G′) of hydrogels under a dynamic frequency mode. Hydrogel samples were prepared as a disc 40 mm in diameter and 1 mm in height. Samples were investigated with a strain of 1% and the frequency ranged from 0.1 to 100 rad/s.

A universal testing machine (UTM2230, SUNS Technology, Shenzhen, China) equipped with a 100 N stress sensor was used for compressive tests at room temperature. A cylindrical hydrogel specimen with a diameter of 5 mm and a height of 10 mm was placed on the fixed plate. Crosshead was set at a speed of 5 mm/min. With the same procedure, data of ten cycles were recorded in the cyclic compressive tests. At least five samples were tested in each group.

4.8. Cytocompatibility and Hemocompatibility Assays

Before cell culturing, hydrogels from different groups were tailored into a circular shape and sterilized by soaking in ethanol overnight, followed by ultraviolet light irradiation [31]. Circular samples with the same weight were plated in 24-well plates, and HUVECs with a density of 50,000 cells/cm^2 were seeded. The wells were then incubated at 37 °C in a humidified, 5% CO_2 atmosphere with the culture medium being changed every other day. After 5 days of culture, a live/dead activity kit was then applied to evaluate the cytocompatibility as described in our previous study [32].

A cell counting kit-8 (CCK-8) was employed to investigate cell proliferation behavior on different hydrogels. The CCK-8 solution (10 v/v%) was added to each well after 1, 3, and 5 days of culture, respectively. After incubating for 2 h at 5% CO_2 and 37 °C, 100 μL reaction mixture was transferred to a multi-well microplate reader and measured at the wavenumber of 450 nm.

A hemolysis test was performed in vitro using rabbit erythrocyte suspension according to the reported protocol [33]. Erythrocyte stock solution was made by diluting erythrocytes with PBS to 2 v/v%. Hydrogel extract (500 μL) from different hydrogels was added to the stock solution (500 μL) and incubated at 37 °C for 3 h. PBS and DI water were added to the stock solution and set as negative and positive controls, respectively. After incubation, the suspensions were centrifuged at 1000× g rpm for 10 min, and 200 μL supernatant was collected and determined at 540 nm by UV-vis spectrophotometer. The hemolysis rate was calculated as follows:

$$\text{Hemolysis ratio } (\%) = \left[(A_h - A_p)/(A_w - W_p)\right] \times 100\% \quad (3)$$

where A_h, A_w, and A_p are the absorbance of samples reacted with leaching medium, PBS, and DI water, respectively. The protocol of animal experiments was approved by the Animal Ethics Committee of Zhengzhou University (No. SCXK 2021-0009).

4.9. *In Vitro Sprouting Analysis*

In vitro sprouting analysis was conducted according to the reported protocol [34]. Matrigel matrix was thawed at 4 °C and added to a confocal dish (200 µL per dish) on ice. The dishes were then incubated at 37 °C for 30 min to allow the Matrigel to gel. Cells with a density of 2×10^5 cells/well were seeded. Then hydrogel extract (1 mL) from each hydrogel was added to wells. During the culture, cell morphology was captured under optical microscopy after 3, 5 and 7 h. After 5 h of culture, the cells were fixed with 4% paraformaldehyde and stained by CF568 phalloidin and DAPI and observed via laser confocal microscopy (LSM 880, Zeiss, Oberkochen, Germany). Angiogenesis behavior was evaluated by counting the number of grids and nodes in different fields via NIH ImageJ software [35]. At least 20 different areas were selected from each group.

4.10. *Statistical Analysis*

One-way analysis of variance (ANOVA) was performed to determine statistical significance. A p-value < 0.05 (denoted by an asterisk) was considered statistically significant.

Author Contributions: Investigation, methodology and writing—original draft preparation, H.W.; writing—review and editing, X.W.; supervision and writing—review and editing, Y.J.; funding acquisition, Q.L. All authors have read and agreed to the published version of the manuscript.

Funding: This work was funded by the International Science and Technology Cooperation Program of China [2015DFA30550], and the Key Science and Technology Project for Institutions of Higher Education of Henan Province [No.20B430019].

Institutional Review Board Statement: The protocols of animal experiments were approved by the Animal Ethics Committee of Zhengzhou University (No. SCXK 2021-0009).

Informed Consent Statement: Not applicable.

Data Availability Statement: Not applicable.

Acknowledgments: The authors greatly appreciate the financial support from the International Science and Technology Cooperation Program of China [2015DFA30550], and the Key Science and Technology Project for Institutions of Higher Education of Henan Province [No.20B430019].

Conflicts of Interest: The authors declare no competing financial interest.

References

1. Ngo, M.T.; Harley, B.A.C. Angiogenic biomaterials to promote therapeutic regeneration and investigate disease progression. *Biomaterials* **2020**, *255*, 120207. [CrossRef]
2. Yu, Y.; Dai, K.; Gao, Z.; Tang, W.; Shen, T.; Yuan, Y.; Wang, J.; Liu, C. Sulfated polysaccharide directs therapeutic angiogenesis via endogenous VEGF secretion of macrophages. *Sci. Adv.* **2021**, *7*, eabd8217. [CrossRef]
3. Rouwkema, J.; Khademhosseini, A. Vascularization and Angiogenesis in Tissue Engineering: Beyond Creating Static Networks. *Trends Biotechnol.* **2016**, *34*, 733–745. [CrossRef]
4. Jeong, H.-W.; Hernandez-Rodriguez, B.; Kim, J.; Kim, K.-P.; Enriquez-Gasca, R.; Yoon, J.; Adams, S.; Schoeler, H.R.; Vaquerizas, J.M.; Adams, R.H. Transcriptional regulation of endothelial cell behavior during sprouting angiogenesis. *Nat. Commun.* **2017**, *8*, e726. [CrossRef]
5. Chen, D.; Zhang, C.; Huo, H.; Ji, C.; Sun, M.; Nie, L. Injectable temperature-sensitive hydrogel with VEGF loaded microspheres for vascularization and bone regeneration of femoral head necrosis. *Mater. Lett.* **2018**, *229*, 138–141. [CrossRef]
6. Stryker, Z.I.; Rajabi, M.; Davis, P.J.; Mousa, S.A. Evaluation of Angiogenesis Assays. *Biomedicines* **2019**, *7*, 37. [CrossRef]
7. Karvinen, H.; Pasanen, E.; Rissanen, T.T.; Korpisalo, P.; Vaehaekangas, E.; Jazwa, A.; Giacca, M.; Ylae-Herttuala, S. Long-term VEGF-A expression promotes aberrant angiogenesis and fibrosis in skeletal muscle. *Gene Ther.* **2011**, *18*, 1166–1172. [CrossRef]
8. Jiang, Y.-C.; Wang, X.-F.; Xu, Y.-Y.; Qiao, Y.-H.; Guo, X.; Wang, D.-F.; Li, Q.; Turng, L.-S. Polycaprolactone nanofibers containing vascular endothelial growth factor-encapsulated gelatin particles enhance mesenchymal stem cell differentiation and angiogenesis of endothelial cells. *Biomacromolecules* **2018**, *19*, 3747–3753. [CrossRef]

9. Ghomi, E.R.; Khalili, S.; Khorasani, S.N.; Neisiany, R.E.; Ramakrishna, S. Wound dressings: Current advances and future directions. *J. Appl. Polym. Sci.* **2019**, *136*, 47738. [CrossRef]
10. Weis, S.M.; Cheresh, D.A. Tumor angiogenesis: Molecular pathways and therapeutic targets. *Nat. Med.* **2011**, *17*, 1359–1370. [CrossRef]
11. Wang, H.; Heilshorn, S.C. Adaptable Hydrogel Networks with Reversible Linkages for Tissue Engineering. *Adv. Mater.* **2015**, *27*, 3717–3736. [CrossRef] [PubMed]
12. Tsou, Y.-H.; Khoneisser, J.; Huang, P.-C.; Xu, X. Hydrogel as a bioactive material to regulate stem cell fate. *Bioact. Mater.* **2016**, *1*, 39–55. [CrossRef]
13. Choi, S.-W.; Zhang, Y.; MacEwan, M.R.; Xia, Y. Neovascularization in Biodegradable Inverse Opal Scaffolds with Uniform and Precisely Controlled Pore Sizes. *Adv. Healthc. Mater.* **2013**, *2*, 145–154. [CrossRef]
14. Liu, S.; Yu, J.; Zhang, Q.; Lu, H.; Qiu, X.; Zhou, D.; Qi, Y.; Huang, Y. Dual Cross-linked HHA Hydrogel Supplies and Regulates MΦ2 for Synergistic Improvement of Immunocompromise and Impaired Angiogenesis to Enhance Diabetic Chronic Wound Healing. *Biomacromolecules* **2020**, *21*, 3795–3806. [CrossRef]
15. Yang, J.; Shen, M.; Wen, H.; Luo, Y.; Huang, R.; Rong, L.; Xie, J. Recent advance in delivery system and tissue engineering applications of chondroitin sulfate. *Carbohydr. Polym.* **2020**, *230*, 115650. [CrossRef]
16. Zhu, W.; Ji, Y.; Wang, Y.; He, D.; Yan, Y.; Su, N.; Zhang, C.; Xing, X.-H. Structural characterization and in vitro antioxidant activities of chondroitin sulfate purified from Andrias davidianus cartilage. *Carbohydr. Polym.* **2018**, *196*, 398–404. [CrossRef]
17. de Souza Lins Borba, F.K.; Felix, G.L.Q.; Costa, E.V.L.; Silva, L.; Dias, P.F.; de Albuquerque Nogueira, R. Fractal analysis of extra-embryonic vessels of chick embryos under the effect of glucosamine and chondroitin sulfates. *Microvasc. Res.* **2016**, *105*, 114–118. [CrossRef]
18. Lopez-Moya, M.; Melgar-Lesmes, P.; Kolandaivelu, K.; de la Torre Hernandez, J.M.; Edelman, E.R.; Balcells, M. Optimizing Glutaraldehyde-Fixed Tissue Heart Valves with Chondroitin Sulfate Hydrogel for Endothelialization and Shielding against Deterioration. *Biomacromolecules* **2018**, *19*, 1234–1244. [CrossRef]
19. Xiong, X.; Xiao, W.; Zhou, S.; Cui, R.; Xu, H.H.K.; Qu, S. Enhanced proliferation and angiogenic phenotype of endothelial cells via negatively-charged alginate and chondroitin sulfate microsphere hydrogels. *Biomed. Mater.* **2021**, *16*, 025012. [CrossRef]
20. Fox, J.M.; Kang, K.; Sherman, W.; Heroux, A.; Sastry, G.M.; Baghbanzadeh, M.; Lockett, M.R.; Whitesides, G.M. Interactions between Hofmeister Anions and the Binding Pocket of a Protein. *JACS* **2015**, *137*, 3859–3866. [CrossRef]
21. Entekhabi, E.; Haghbin Nazarpak, M.; Sedighi, M.; Kazemzadeh, A. Predicting degradation rate of genipin cross-linked gelatin scaffolds with machine learning. *Mater. Sci. Eng. C* **2020**, *107*, 110362. [CrossRef] [PubMed]
22. Gao, G.Z.; Jiao, Q.C.; Ding, Y.L.; Chen, L. Study on quantitative assay of chondroitin sulfate with a spectrophotometric method of azure A. *Spectrosc. Spectr. Anal.* **2003**, *23*, 600–602.
23. Li, G.; Jiang, Y.; Li, M.; Zhang, W.; Li, Q.; Tang, K. Investigation on the tunable effect of oxidized konjac glucomannan with different molecular weight on gelatin-based composite hydrogels. *Int. J. Biol. Macromol.* **2021**, *168*, 233–241. [CrossRef] [PubMed]
24. Li, C.; Mu, C.; Lin, W.; Ngai, T. Gelatin Effects on the Physicochemical and Hemocompatible Properties of Gelatin/PAAm/Laponite Nanocomposite Hydrogels. *ACS Appl. Mater. Interfaces* **2015**, *7*, 18732–18741. [CrossRef] [PubMed]
25. Lei, J.; Li, X.; Wang, S.; Yuan, L.; Ge, L.; Li, D.; Mu, C. Facile Fabrication of Biocompatible Gelatin-Based Self-Healing Hydrogels. *ACS Appl. Polym. Mater.* **2019**, *1*, 1350–1358. [CrossRef]
26. Calamia, V.; Lourido, L.; Fernandez-Puente, P.; Mateos, J.; Rocha, B.; Montell, E.; Verges, J.; Ruiz-Romero, C.; Blanco, F.J. Secretome analysis of chondroitin sulfate-treated chondrocytes reveals anti-angiogenic, anti-inflammatory and anti-catabolic properties. *Arthritis Res. Ther.* **2012**, *14*, R202. [CrossRef]
27. Le Jan, S.; Hayashi, M.; Kasza, Z.; Eriksson, I.; Bishop, J.R.; Weibrecht, I.; Heldin, J.; Holmborn, K.; Jakobsson, L.; Soderberg, O.; et al. Functional overlap between chondroitin and heparan sulfate proteoglycans during VEGF-induced sprouting angiogenesis. *Arterioscler. Thromb. Vasc. Biol.* **2012**, *32*, 1255–1263. [CrossRef]
28. Del Gaudio, C.; Baiguera, S.; Boieri, M.; Mazzanti, B.; Ribatti, D.; Bianco, A.; Macchiarini, P. Induction of angiogenesis using VEGF releasing genipin-crosslinked electrospun gelatin mats. *Biomaterials* **2013**, *34*, 7754–7765. [CrossRef]
29. Xu, S.; Yonese, M. Two dimensional and three dimensional interactions between bovine serum albumin and chondroitin sulfate. *Polym. J.* **2007**, *39*, 298–303. [CrossRef]
30. Zhou, F.; Zhang, X.; Cai, D.; Li, J.; Mu, Q.; Zhang, W.; Zhu, S.; Jiang, Y.; Shen, W.; Zhang, S.; et al. Silk fibroin-chondroitin sulfate scaffold with immuno-inhibition property for articular cartilage repair. *Acta Biomater.* **2017**, *63*, 64–75. [CrossRef]
31. Jiang, Y.; Li, G.; Liu, J.; Li, M.; Li, Q.; Tang, K. Gelatin/oxidized konjac glucomannan composite hydrogels with high resistance to large deformation for tissue engineering applications. *ACS Appl. Bio Mater.* **2021**, *4*, 1536–1543. [CrossRef] [PubMed]
32. Li, X.; Wang, X.; Yao, D.; Jiang, J.; Guo, X.; Gao, Y.; Li, Q.; Shen, C. Effects of aligned and random fibers with different diameter on cell behaviors. *Colloids Surf. B* **2018**, *171*, 461–467. [CrossRef] [PubMed]
33. Wang, L.; Zhang, X.; Yang, K.; Fu, Y.V.; Xu, T.; Li, S.; Zhang, D.; Wang, L.-N.; Lee, C.-S. A Novel Double-Crosslinking-Double-Network Design for Injectable Hydrogels with Enhanced Tissue Adhesion and Antibacterial Capability for Wound Treatment. *Adv. Funct. Mater.* **2020**, *30*, 1904156. [CrossRef]

34. Cao, L.; Wang, J.; Hou, J.; Xing, W.; Liu, C. Vascularization and bone regeneration in a critical sized defect using 2-N,6-O-sulfated chitosan nanoparticles incorporating BMP-2. *Biomaterials* **2014**, *35*, 684–698. [CrossRef] [PubMed]
35. Moon, J.J.; Saik, J.E.; Poche, R.A.; Leslie-Barbick, J.E.; Lee, S.-H.; Smith, A.A.; Dickinson, M.E.; West, J.L. Biomimetic hydrogels with pro-angiogenic properties. *Biomaterials* **2010**, *31*, 3840–3847. [CrossRef] [PubMed]

gels

Article

High-Energy Photon Attenuation Properties of Lead-Free and Self-Healing Poly (Vinyl Alcohol) (PVA) Hydrogels: Numerical Determination and Simulation

Theerasarn Pianpanit [1] and Kiadtisak Saenboonruang [1,2,3,*]

[1] Department of Applied Radiation and Isotopes, Faculty of Science, Kasetsart University, Bangkok 10900, Thailand; fscitap@ku.ac.th
[2] Specialized Center of Rubber and Polymer Materials in Agriculture and Industry (RPM), Faculty of Science, Kasetsart University, Bangkok 10900, Thailand
[3] Special Research Unit of Radiation Technology for Advanced Materials, Faculty of Science, Kasetsart University, Bangkok 10900, Thailand
* Correspondence: kiadtisak.s@ku.th; Tel.: +66-2-562-5555 (ext. 646219)

Abstract: This work numerically determined high-energy photon shielding properties of self-healing poly(vinyl alcohol) (PVA) hydrogels containing lead-free, heavy-metal compounds, namely, bismuth oxide (Bi_2O_3), tungsten oxide (WO_3), and barium sulfate ($BaSO_4$), through XCOM software packages. In order to understand the dependencies of the shielding properties of the hydrogels on filler contents and photon energies, the filler contents added to the hydrogels were varied from 0–40 wt.% and the photon energies were varied from 0.001–5 MeV. The results, which were verified for their reliability and correctness with those obtained from PHITS (Particle and Heavy Ion Transport code System), indicated that overall shielding performances, which included the mass attenuation coefficients (μ_m), the linear attenuation coefficient (μ), the half-value layer (HVL), and the lead equivalence (Pb_{eq}), of the hydrogels improved with increasing filler contents but generally decreased with increasing photon energies. Among the three compounds investigated in this work, Bi_2O_3/PVA hydrogels exhibited the highest photon attenuation capabilities, determined at the same filler content and photon energy, mainly due to its highest atomic number of Bi and the highest density of Bi_2O_3 in comparison with other elements and compounds. Furthermore, due to possible reduction in self-healing and mechanical properties of the hydrogels with excessive filler contents, the least content of fillers providing a 10-mm sample with the required Pb_{eq} value of 0.5 mmPb was investigated. The determination revealed that only the hydrogel containing at least 36 wt.% of Bi_2O_3 exhibited the Pb_{eq} values greater than 0.5 mmPb for all photon energies of 0.05, 0.08, and 0.1 MeV (common X-ray energies in general nuclear facilities). The overall outcomes of the work promisingly implied the potential of PVA hydrogels to be used as novel and potent X-ray and gamma shielding materials with the additional self-healing and nonlead properties.

Keywords: PVA; Bi_2O_3; WO_3; $BaSO_4$; X-ray; gamma; shielding; XCOM; PHITS; simulation

1. Introduction

High-energy photons, especially X-rays and gamma rays, have been increasingly utilized in various applications, including material characterization [1–3], medical diagnostic and therapy [4,5], quality control in industrial products [6], plant mutation breeding [7], and national security [8]. Despite their great benefits, excessive exposures to X-rays and gamma rays pose serious biological effects on human populations, which could lead to a reduction in the immune system, the initiation of brain cancers, and the increase in mutation rates [9–11]. To cope with possible radiation illnesses and other side effects, a concept for radiation safety, namely, "As Low As Reasonably Achievable" or "ALARA", consisting of the management of (1) exposure time, (2) distance between radiation sources and users, and (3) utilization of sufficient and appropriate shielding equipment, must be strictly practiced in all nuclear and radiation facilities [12].

Specifically for users who are required to work in a proximity to radiation sources for extended time periods, utilization of effective shielding equipment has become a necessity to prevent the users and the public from excessive exposure to radiations. In the case of X-rays and gamma rays, shielding equipment mostly relies on the use of heavy metals, especially lead (Pb) and lead oxide (PbO) [13], mainly due to their relatively higher interaction probabilities between the incident photons and Pb atoms, which significantly improved photon attenuation capabilities of the materials/composites as well as their economical availability [14]. Nonetheless, the toxicity of Pb has raised serious safety concerns as excessive exposure to Pb potentially leads to an increase in blood pressure, slow nerve conduction, fatigue, drowsiness, fertility disorders, encephalopathy, and death [15]. Furthermore, the spread of Pb in water resources and forests could negatively affect animals and plants. As a result, significant efforts to replace the toxic Pb with safer compounds have been emphasized and discussed in recent years. Among several potential candidates, bismuth oxide (Bi_2O_3), tungsten oxide (WO_3), and barium sulfate ($BaSO_4$) have shown promising possibilities to serve for such purposes due to the high atomic numbers (Z) of Bi, W, and Ba (Z = 83, 74, and 56, respectively) as well as their high densities (ρ = 8.9, 7.16, and 4.5 g/cm^3, respectively), resulting in substantial enhancement of photon attenuation after being added to the main matrix [16]. For instance, Poltabtim et al. reported that natural rubber (NR) composites containing Bi_2O_3 numerically exhibited comparable X-ray and gamma shielding properties as those containing Pb; for instance, the μ_m values of Pb/NR and Bi_2O_3/NR composites were 0.094 and 0.092 cm^2/g, respectively, determined at the filler content of 80 phr and the photon energy of 0.662 MeV) [17]. Maleksadeh et al. also confirmed the usability of WO_3 and $BaSO_4$ as Pb alternatives, evidenced by just a slight decrease in the attenuation capabilities of silicone-based composites containing WO_3 and $BaSO_4$ in comparison with those containing PbO (determined at the same filler content and photon energy) [18]. These reports clearly implied that Bi_2O_3, WO_3, and $BaSO_4$ could be utilized as safe and effective alternatives to Pb and Pb compounds in the production of high-energy photon equipment.

Additionally, the types of the main matrix used for the production of the shielding materials are one of key factors that could define other important characteristics and properties. For instance, materials based on polyethylene (PE) and polyvinyl chloride (PVC) such as Bi_2O_3/UHMWPE and Bi_2O_3/wood/PVC composites exhibited exceptional strength and rigidity, which were suitable for use as construction parts or movable equipment in nuclear facilities [19,20]. On the other hand, materials based on natural rubber (NR) and synthetic rubber (SR) such as Bi_2O_3/NR, WO_3/NR, and Bi_2O_3/EPDM composites could be used as personal protective equipment (PPE) and covers of transporting casks due to their exceptional flexibility and elongation at break [21–23]. Nonetheless, the mentioned materials had some limitations due to the lack of self-healing capabilities, which may result in increased costs and procedures needed for repairment, replacement, and waste management of the damaged products. To alleviate such drawbacks, materials based on the autonomously self-healing poly(vinyl alcohol) (PVA) hydrogels have been recently developed to be used as radiation shielding materials [24,25]. An example of some PVA-based shielding materials is nano-Bi_2O_3/PVA hydrogels, which could attenuate gamma rays having the energies of 1.17 MeV and 1.33 MeV by 35% and 30%, respectively, while providing the percentage of recoverable strength (%Recovery) of 88.6% after being brought together for just 1 min [26]. Furthermore, these developed Bi_2O_3/PVA hydrogels were also lead-free, which additionally improve the health safety of users and the public with respect to the toxic lead. Another use of self-healing PVA hydrogels is Sm_2O_3/PVA and Gd_2O_3/PVA composites for neutron attenuation, for which both 1-cm-thick PVA hydrogels containing either 10.5 wt.% of Sm_2O_3 or Gd_2O_3 could reduce the initial intensity of incident thermal neutrons by almost 70%, while providing the %Recovery up to 70% after being brought into contact for 6 h [27]. These self-healing abilities of PVA hydrogels were due to the capabilities of PVA polymer chains to diffuse across fractured surfaces and initiate hydrogen bonds between two polymers (without any external stimuli after the damage) [26]. These two examples clearly indicated the great potentials of utilizing PVA hydrogels as autonomously self-healing and effective radiation shielding materials.

Despite the promising self-healable capabilities of PVA hydrogels, the addition of fillers, especially radiation protective fillers that tend to improve shielding properties of the composites with their increasing contents, may result in the decrease of %Recovery and, thus, the self-healing capabilities. This limitation was evidenced by the reports of Tiamduangtawan et al., which indicated that the increase in nano-Bi_2O_3 contents from 0 to 20 and 40 wt.% lowered the %Recovery of the hydrogels from 96.3% to 95.9% and 88.6%, respectively, while the increases in the Sm_2O_3 and Gd_2O_3 contents from 0 to 10.5 wt.% decreased the %Recovery of the self-healed hydrogels from 85% in pristine samples to ~65% and ~60%, respectively [26,27]. These negative relationships between filler contents and %Recovery were mainly due to increases in the cross-linking network of the PVA chains from physical interactions between the hydroxyl groups of PVA and the fillers, leading to the increase in overall crystallinity of the hydrogels that subsequently obstructed the chain diffusion and the initiation of self-healing mechanisms [24,26]. As a result, due to the competing roles of the fillers in enhancing photon attenuation capabilities and obstructing self-healing properties of the hydrogels, the least filler contents that provide the materials with sufficient photon shielding, while preserving the self-healing capabilities of the hydrogels, must be thoroughly determined.

As aforementioned, this work aimed to numerically determine high-energy photon shielding properties, which included the mass attenuation coefficient (μ_m), the linear attenuation coefficient (μ), and the half-value layer (HVL), of PVA hydrogels containing either Bi_2O_3, WO_3, or $BaSO_4$ using an online software package, namely, XCOM [17,28,29]. In order to verify results from XCOM for further investigation, the results were compared with those obtained from a Monte Carlo particle transport simulation code, namely, PHITS (Particle and Heavy Ion Transport code System) [30,31]. The contents of the fillers and the photon energies used for the determination were varied from 0–40 wt.% and 0.001–5 MeV, respectively. Furthermore, the least contents, which could be regarded as the recommended contents for each filler at the photon energies of 0.05, 0.08, and 0.1 MeV, were determined by comparing the values of lead equivalence (Pb_{eq}) of the 1-cm-thick PVA hydrogels with the required Pb_{eq} value of 0.5 mmPb. The outcomes of this work should not only reveal the numerical effectiveness of PVA hydrogels in photon attenuation but also promote the additional advantages of PVA hydrogels, especially the self-healing capability, in the applications of radiation protection.

2. Determination of High-Energy Photon Shielding Properties Using XCOM and PHITS

2.1. Determination of Mass Attenuaiton Coefficients (μ_m)

The values of μ_m (the fraction of photons attenuated by a homogeneous material per unit mass) for PVA hydrogels containing varying Bi_2O_3, WO_3, and $BaSO_4$ contents of 0–40 wt.% were numerically determined using the XCOM software. Based on our previous experimental report to produce self-healing hydrogels, the content of PVA was fixed at 20 wt.% and the content of water was varied as 80-X wt.%, where X is the content of Bi_2O_3, WO_3, and $BaSO_4$. The XCOM software was developed and provided by the National Institute of Standards and Technology (NIST) (Gaithersburg, MD, USA), for which the photon cross-section database used for the determination of μ_m was the NIST standard reference database 8 (XGAM), released in 2010. In order to understand the dependency of μ_m behaviors on the energies of incident photons, the values of μ_m were determined at varying photon energies from 0.001–5 MeV, with the inclusion of coherent scattering [32].

In order to increase the usability of the work, simple mathematical relationships between μ_m and filler content at the photon energies of 0.1, 0.5, 1, and 5 MeV were also determined using the form shown in Equation (1):

$$\mu_m = Ax + B \tag{1}$$

where μ_m is the mass attenuation coefficient, x is the filler content, and A (B) is the mathematical constant, determined using the Microsoft Excel software package. These determined mathematical relationships would allow us to reliably predict the value of μ_m at any filler content. It should be noted that this relationship in Equation (1) was selected based on our previous work [29], which showed a strong linear relationship between the values of μ_m and filler contents.

2.2. Verification of the Results from XCOM with PHITS

In order to reliably progress to further steps, the verification of the μ_m values obtained from XCOM was conducted by comparing the values with those obtained from PHITS [33] at the filler contents of 10, 20, 30, and 40 wt.% and the photon energies of 0.1, 0.5, 1, and 5 MeV. To obtain the μ_m values from PHITS, the photon beam with a diameter of 1 mm was pointed directly at the center of the sample, which had the surface area of 20 cm × 20 cm and the thickness of either 1 mm (for 0.1-MeV photons) or 1 cm (for 0.5-, 1-, and 5-MeV photons). More details of the PHITS setup can be found elsewhere [29]. The percentage differences between the values of μ_m obtained from XCOM and PHITS for all filler contents and photon energies of interest were also determined, following Equation (2):

$$\text{Difference } (\%) = \frac{\left|\mu_{m,XCOM} - \mu_{m,PHITS}\right|}{\mu_{m,XCOM}} \times 100\% \tag{2}$$

where Difference (%) is the percentage difference between the values of μ_m obtained from XCOM and PHITS, $\mu_{m,XCOM}$ is the mass attenuation coefficient obtained from XCOM, and $\mu_{m,PHITS}$ is the mass attenuation coefficient obtained from PHITS. It is notable to point out that the considerably larger surface area of the sample in comparison with the diameter of the photon beam was to minimize the underestimated values of μ_m due to buildup effects [34].

2.3. Determination of Linear Attenuation Coefficient (μ) and Half-Value Layer (HVL)

The values of μ (the fraction of photons attenuated by a material per unit length) and HVL (the required thickness of a material that could attenuate the intensity of incident photons by 50%) were calculated from the obtained values of μ_m in Section 2.1, following Equations (3) and (4), respectively:

$$\mu = \mu_m \times \rho \tag{3}$$

$$\text{HVL} = \frac{\ln(2)}{\mu} \tag{4}$$

where ρ is the density of the PVA hydrogels, calculated from Equation (5):

$$\rho = \frac{100}{\frac{C_{PVA}}{\rho_{PVA}} + \frac{C_{water}}{\rho_{water}} + \frac{C_F}{\rho_F}} \tag{5}$$

where ρ_{PVA}, ρ_{water}, and ρ_F are the densities of PVA, water, and filler, respectively, and C_{PVA}, C_{water}, and C_F are the contents of PVA, water, and filler, respectively. The values of ρ for individual compounds used for the determination of hydrogel densities are shown in Table 1.

Table 1. Individual densities of PVA, Bi_2O_3, WO_3, $BaSO_4$, and water used for the determination of densities for PVA hydrogels [16,26,35].

Matrix/Compound	Chemical Formula	Density (g/cm^3)
Poly(vinyl alcohol)	$[CH_2CH(OH)]_n$	1.23
Bismuth oxide	Bi_2O_3	8.90
Tungsten oxide	WO_3	7.16
Barium sulfate	$BaSO_4$	4.50
Water	H_2O	1.00

Similar to μ_m, mathematical relationships between μ (HVL) and filler content at the photon energies of 0.1, 0.5, 1, and 5 MeV were also determined, for which the forms for μ and HVL determination follow Equations (6) and (7), respectively:

$$\mu = Ax^2 + Bx + C \tag{6}$$

$$HVL = Ax^5 + Bx^4 + Cx^3 + Dx^2 + Ex + F \tag{7}$$

where A, B, C, D, E, and F are mathematical constants determined using a trendline function available in in the Microsoft Excel software package and x is the filler content. It should be noted that the forms of Equations (6) and (7) were selected due to their values of R^2 (the correlation variance) being closest to 1 in comparison with those from other mathematical relationships.

2.4. Determination of Lead Equivalence (Pb_{eq})

The Pb_{eq} values of the Bi_2O_3/PVA, WO_3/PVA, and $BaSO_4$/PVA hydrogels at the photon energies of 0.05, 0.08, and 0.1 MeV were determined using Equation (8):

$$\mu_{Pb} Pb_{eq} = \mu_{PVA} x \tag{8}$$

where μ_{Pb} is the mass attenuation coefficient of Pb; μ_{PVA} is the mass attenuation coefficient of Bi_2O_3/PVA, WO_3/PVA, or $BaSO_4$/PVA hydrogels; Pb_{eq} is the Pb equivalence (in mmPb); and x is the thickness of the sample, which was fixed at 10 mm for this work. The μ_{Pb} values used for the calculation of Pb_{eq} at the photon energies of 0.05, 0.08, and 0.1 MeV were 91.28, 27.46, and 62.98 cm^2/g, respectively [26]. It is noteworthy that the selected photon energies for the determination of Pb_{eq} were based on the common X-ray energy ranges in nuclear facilities, for which the developed PVA hydrogels could be utilized as movable panels or PPE.

3. Results and Discussion

3.1. Mass Attenuation Coefficient (μ_m)

The behaviors of μ_m for PVA hydrogels containing varying contents of Bi_2O_3, WO_3, and $BaSO_4$ at photon energies of 0.001–5 MeV are shown in Figures 1 and 2. Figure 1, which compares the μ_m values of PVA hydrogels having the same filler type but varying filler contents of 10, 20, 30, and 40 wt.%, indicates that μ_m values generally increased with increasing filler contents but decreased with increasing photon energies. The effects of filler contents on the enhancement of photon attenuation capabilities were mainly due to their relatively higher atomic numbers (Z) of Bi, W, and Ba (Z = 83, 74, and 56, respectively) in the fillers compared with those of H, C, and O (Z = 1, 6, and 8, respectively) in the PVA and water as well as their much higher densities of Bi_2O_3, WO_3, and $BaSO_4$ in comparison with those of PVA and water (Table 1), resulting in enhanced interaction probabilities between the incident photons and Bi_2O_3, WO_3, or $BaSO_4$ fillers in PVA hydrogels. It is notable that the effects of fillers on the enhancement of photon shielding properties were more pronounced at lower photon energies (0.001–0.5 MeV) than those at higher energies (0.5–5 MeV). This was because the interactions of photons with the hydrogels at lower energies were dominated by the process of photoelectric absorption, for which its interaction probability, i.e., its photoelectric cross section (σ_{pe}), greatly depends on photon energies and atomic numbers of the material, as shown in Equation (9):

$$\sigma_{pe} \propto \frac{Z^n}{(h\nu)^3} \tag{9}$$

where σ_{pe} is the photoelectric cross section (a nuclear quantity representing the interaction probabilities of an element or a material with incident radiation through the photoelectric absorption), Z is the atomic number of the element contained in the material, h is a Planck's constant, and ν is the frequency of the incident photon, which is directly proportional to the photon energy ($E = h\nu$) [29,36]. As implied in Equation (9), σ_{pe} of a material and subsequently the ability to attenuate photons would be substantially enhanced when more heavy elements (Bi, W, and Ba in this case) were added to the hydrogels, which were clearly illustrated in Figure 1a,c,d.

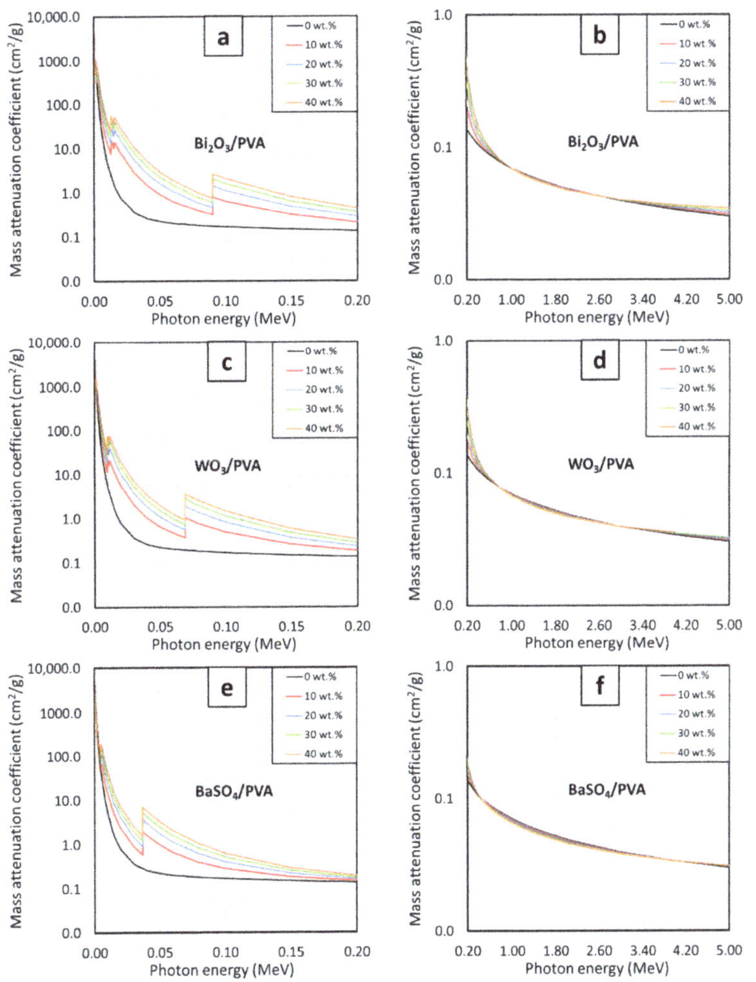

Figure 1. The μ_m values of (**a**,**b**) Bi_2O_3/PVA, (**c**,**d**) WO_3/PVA, and (**e**,**f**) $BaSO_4$/PVA hydrogels containing varying filler contents of 0, 10, 20, 30, and 40 wt.%, determined at photon energies of (**a**,**c**,**e**) 0.001–0.2 MeV and (**b**,**d**,**f**) 0.2–5 MeV using XCOM.

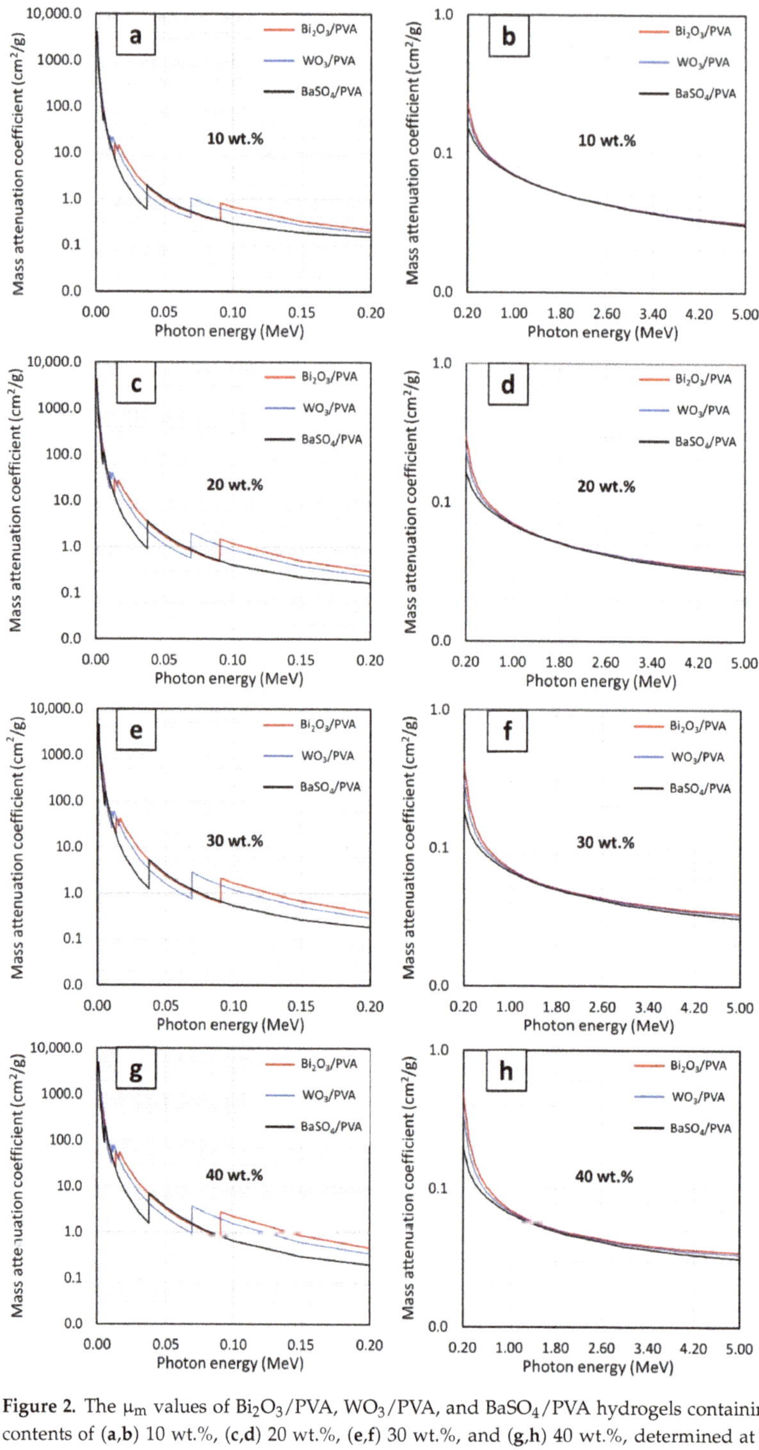

Figure 2. The μ_m values of Bi_2O_3/PVA, WO_3/PVA, and $BaSO_4$/PVA hydrogels containing filler contents of (**a**,**b**) 10 wt.%, (**c**,**d**) 20 wt.%, (**e**,**f**) 30 wt.%, and (**g**,**h**) 40 wt.%, determined at photon energies of (**a**,**c**,**e**,**g**) 0.001–0.2 MeV and (**b**,**d**,**f**,**h**) 0.2–5 MeV using XCOM.

On the other hand, Equation (9) depicts that the μ_m values would rapidly decrease with increasing photon frequencies (energies). This behavior was observed due to the inverse relationship between σ_{pe} and ν ($\sigma_{pe} \propto 1/\nu^3$) that results in sharp declines of interaction probabilities between the incident photons and the hydrogels at higher photon energies, which subsequently reduced their attenuation capabilities. It should be noted that the μ_m values for all hydrogels shown in Figure 1b,d,f were very close to each other (regardless of their filler contents), especially at the photon energies of 0.5–3 MeV. This was due to the diminishing roles of photoelectric absorption at these energies and the change in the dominant interaction mechanism from photoelectric absorption to a less effective attenuation mechanism, namely, Compton scattering. The dependency of the Compton scattering cross section (σ_{comp}) on the characteristics of the material could be described as shown in Equation (10):

$$\sigma_{comp} \propto \frac{1}{n_e} \tag{10}$$

where n_e is the electron density of the material [29,37].

However, as the photon energies further increased to higher than 3 MeV, increases in filler contents started to affect the μ_m values again. This was mainly due to the initiation of another important interaction of the photon, namely, the pair production, which starts at the photon energies greater than 1.022 MeV and becomes the dominant interaction mechanism at the energies around 3 MeV. The pair production probabilities, which are represented by the values of its cross section (σ_{pp}), are directly proportional to the square of atomic number (Z), as shown in Equation (11):

$$\sigma_{pp} \propto Z^2 \tag{11}$$

According to Equation (11), it implies that the μ_m values of the hydrogels are enhanced as more fillers are added to the hydrogels, resulting in more pronounced differences in μ_m values, as observed in Figure 1b,d,f (after 3 MeV) [17,38].

Figure 2a–d, which compares μ_m behaviors of PVA composites containing Bi_2O_3, WO_3, or $BaSO_4$ at the same filler contents of 10, 20, 30, and 40 wt.%, respectively, reveals that Bi_2O_3/PVA hydrogels generally exhibited the highest μ_m values (determined at the same filler content and photon energy), except at the photon energies of 0.010–0.012 MeV and 0.069–0.090 MeV, for which WO_3/PVA hydrogels had higher μ_m values than Bi_2O_3/PVA and $BaSO_4$/PVA hydrogels. The superior photon attenuation capabilities of Bi_2O_3/PVA hydrogels at most photon energies were observed due to their relatively higher atomic number and density than those of WO_3, $BaSO_4$, PVA, and water (Figure 3), leading to higher μ_m values for Bi_2O_3/PVA hydrogels. It is noteworthy that, at the photon energies of 0.010–0.012 MeV and 0.069–0.090 MeV, WO_3/PVA hydrogels had uncharacteristically higher μ_m values than other hydrogels. This was due to the effects of W's K-absorption edge (K-edge) and L-absorption edge (L-edge) that occurred at 0.069 MeV and 0.010–0.012 MeV, respectively, for which the σ_{pe} and, subsequently, the μ_m of WO_3/PVA hydrogels abruptly increased at these particular energies [39].

Figure 4, which compares the μ_m values for the Bi_2O_3/PVA, WO_3/PVA, and $BaSO_4$/PVA hydrogels with varying filler contents of 0–40 wt.% determined at the photon energies of 0.1, 0.5, 1, and 5 MeV, reveals that strong linear correlations between the filler contents and the μ_m values were observed for all investigated energies. While most of the relationships were positively correlated (Figure 4a,d), mainly due to the roles of the fillers in photon attenuation through the dominant photoelectric absorption (Figure 4a) and the pair production (Figure 4d), some of the relationships such as those of $BaSO_4$/PVA (Figure 4b,c) and WO_3/PVA (Figure 4c) were found to be negatively correlated. These negative trends were observed because, at these particular photon energies, the pristine PVA hydrogels, which contain higher light-element (C, H, and O) contents, could better interact with the photons through the dominant Compton scattering than those of PVA hydrogels containing the fillers [26]. This phenomenon could be mathematically explained using Equation (10) as

σ_{comp} is inversely proportional to the electron densities (n_e) of the composites, implying that the pristine PVA hydrogels would have higher σ_{comp} and, consequently, slightly higher photon attenuation capabilities (Figure 3b).

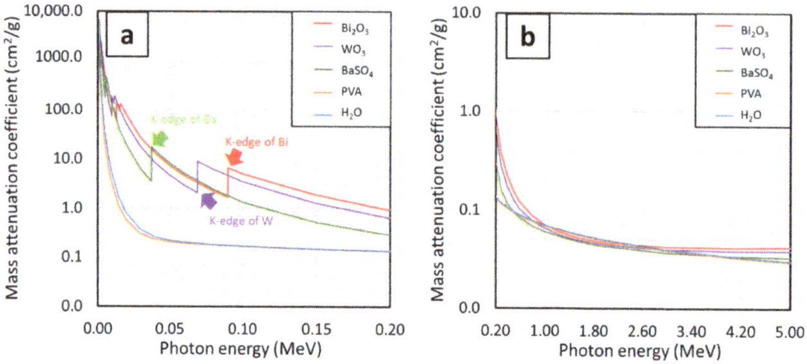

Figure 3. The μ_m values of Bi_2O_3, WO_3, $BaSO_4$, PVA, and H_2O showing K-edge and L-edge behaviors of Bi, W, and Ba at photon energies of (**a**) 0.001–0.2 MeV and (**b**) 0.2–5 MeV using XCOM.

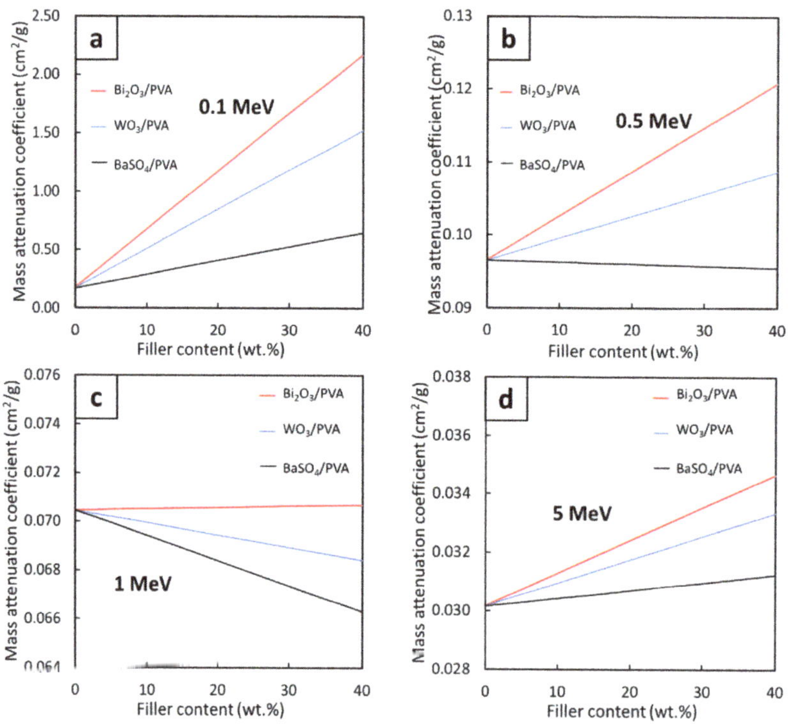

Figure 4. The μ_m values of Bi_2O_3/PVA, WO_3/PVA, and $BaSO_4$/PVA hydrogels containing varying filler contents of 0–40 wt.%, determined at photon energies of (**a**) 0.1 MeV, (**b**) 0.5 MeV, (**c**) 1 MeV, and (**d**) 5 MeV using XCOM.

In order to enable the ability to estimate the μ_m values for all filler contents at the photon energies of 0.1, 0.5, 1, and 5 MeV (Figure 4), mathematical equations with the form shown in Equation (1) were determined. The results, which are shown in Table 2, indicated that Bi_2O_3/PVA hydrogels exhibited the strongest correlations between the filler contents and the μ_m values, evidenced by the highest slopes (shown as the values of the constant A) for all photon energies. These were mainly due to the highest atomic number of Bi (Z = 83) in comparison with those of W (Z = 74) and Ba (Z = 56), which resulted in higher interaction probabilities and greater effects of the filler on the enhancement of photon shielding properties. It is notable that the constants A for all hydrogels at the photon energy of 0.1 MeV were relatively higher than those found at higher energies. This was due to a greater dependence of σ_{pe} values on the atomic number (Z) of the hydrogels (Equation (9)), for which the photoelectric absorption was the dominant photon interaction for the 0.1-MeV photons.

Table 2. Mathematical constants (A and B) for μ_m in the form μ_m = Ax + B (Equation (1)), determined from Figure 4.

Photon Energy (MeV)	Bi_2O_3		WO_3		$BaSO_4$	
	A	B	A	B	A	B
0.1	0.0499	0.1698	0.0338	0.1696	0.0119	0.1697
0.5	6.06×10^{-4}	0.0965	3.05×10^{-4}	0.0965	-2.49×10^{-5}	0.0965
1.0	5.53×10^{-6}	0.0705	-5.05×10^{-5}	0.0705	-1.04×10^{-4}	0.0705
5.0	1.13×10^{-4}	0.0301	7.99×10^{-5}	0.0301	2.73×10^{-5}	0.0301

To verify the reliability and correctness of the numerical results obtained in this work, the results from XCOM were compared with those obtained from a Monte Carlo simulation code, PHITS. The filler contents used for the comparison were 10, 20, 30, and 40 wt.% (determined at the photon energies of 0.1, 0.5, 1, and 5 MeV). The results of the comparison as well as their percentage of difference (Difference (%)) are shown in Table 3, which indicates that the results obtained from the two methods were in good agreement, with the range and the average of the Difference (%) being 0.02–1.40% and 0.53%, respectively. Hence, based on the comparison, the results from XCOM could be reliably used in later determinations of μ, HVL, and Pb_{eq}.

Table 3. Comparative μ_m values obtained from XCOM and PHITS for PVA hydrogels containing Bi_2O_3, WO_3, and $BaSO_4$ at varying photon energies of 0.1, 0.5, 1, and 5 MeV.

Filler	Photon Energy (MeV)	Content (wt.%)	μ (cm^{-1})		Difference (%)
			XCOM	PHITS	
Bi_2O_3	0.1	10	0.76547	0.76845	0.39%
		20	1.48776	1.48390	0.26%
		30	2.39405	2.39545	0.06%
		40	3.56517	3.55330	0.33%
	0.5	10	0.11741	0.11707	0.29%
		20	0.13833	0.13817	0.11%
		30	0.16473	0.16584	0.67%
		40	0.19883	0.19968	0.43%
	1.0	10	0.08069	0.08112	0.54%
		20	0.08988	0.09008	0.22%
		30	0.10142	0.10196	0.54%
		40	0.11632	0.11661	0.25%
	5.0	10	0.03578	0.03591	0.39%
		20	0.04127	0.04142	0.36%
		30	0.04815	0.04881	1.39%
		40	0.05703	0.05782	1.40%

Table 3. *Cont.*

Filler	Photon Energy (MeV)	Content (wt.%)	μ (cm^{-1})		Difference (%)
			XCOM	PHITS	
WO_3	0.1	10	0.57919	0.58535	1.06%
		20	1.06978	1.06957	0.02%
		30	1.68062	1.68674	0.36%
		40	2.46092	2.46264	0.07%
	0.5	10	0.11357	0.11371	0.13%
		20	0.12979	0.12937	0.32%
		30	0.14989	0.15053	0.43%
		40	0.17576	0.17692	0.66%
	1.0	10	0.07980	0.08048	0.86%
		20	0.08784	0.08811	0.31%
		30	0.09786	0.09827	0.42%
		40	0.11064	0.11102	0.34%
	5.0	10	0.03530	0.03552	0.64%
		20	0.04015	0.04029	0.36%
		30	0.04619	0.04661	0.92%
		40	0.05391	0.05460	1.28%
$BaSO_4$	0.1	10	0.32651	0.32995	1.06%
		20	0.50555	0.51169	1.22%
		30	0.72278	0.73043	1.06%
		40	0.99203	0.99395	0.19%
	0.5	10	0.10879	0.10925	0.43%
		20	0.11896	0.11930	0.29%
		30	0.13131	0.13147	0.13%
		40	0.14660	0.14654	0.04%
	1.0	10	0.07846	0.07920	0.95%
		20	0.08473	0.08554	0.97%
		30	0.09235	0.09272	0.41%
		40	0.10178	0.10181	0.04%
	5.0	10	0.03438	0.03461	0.69%
		20	0.03803	0.0383	0.84%
		30	0.04245	0.04271	0.63%
		40	0.04795	0.04833	0.80%

3.2. Linear Attenuation Coefficient (μ) and Half-Value Layer (HVL)

To determine the values of μ and HVL, the densities of all PVA hydrogels containing either Bi_2O_3, WO_3, or $BaSO_4$ were theoretically calculated based on Equation (5) and individual densities (Table 1). The results, which are shown in Table 4, indicated that the densities of PVA hydrogels increased with increasing filler content, mainly due to much higher densities of the fillers in comparison with those of PVA and water. Furthermore, the results suggested that Bi_2O_3/PVA hydrogels exhibited higher densities than those of WO_3/PVA and $BaSO_4$/PVA hydrogels at the same filler content. This behavior was observed due to relatively higher densities on Bi_2O_3 (Table 1).

The linear attenuation coefficients (μ) and the half-value layer (HVL) of the PVA hydrogels, calculated using Equations (3) and (4), respectively, are shown in Figures 5 and 6, respectively. The results indicated that, similarly to the behaviors of μ_m, the values of μ and HVL generally improved with increasing filler contents, evidenced by the highest values of μ and the lowest values of HVL observed in the PVA hydrogels containing 40 wt.% of the fillers. In addition, the effects of additional filler contents on the μ and HVL values were found to be more pronounced than those observed in the case of μ_m, evidenced by larger differences in the μ values for each 10 wt.% addition of the fillers. This behavior was observed due to the high densities of the PVA hydrogels (Table 4), which subsequently amplified the μ values of the hydrogels according to Equation (3).

Table 4. Densities of PVA hydrogels containing Bi_2O_3, WO_3, and $BaSO_4$ with varying contents of 0–40 wt.% (in 5 wt.% increments), calculated using Equation (5).

Content (wt.%)	Density (g/cm^3)		
	Bi_2O_3	WO_3	$BaSO_4$
0	1.039	1.039	1.039
5	1.089	1.087	1.083
10	1.144	1.141	1.130
15	1.206	1.200	1.182
20	1.274	1.265	1.239
25	1.350	1.338	1.302
30	1.436	1.419	1.371
35	1.534	1.512	1.448
40	1.646	1.617	1.535

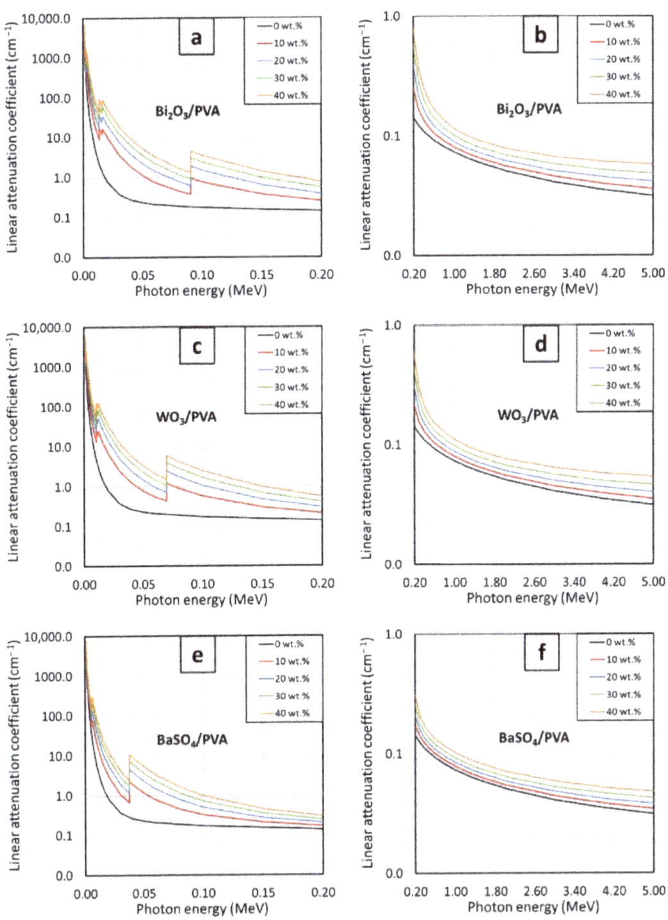

Figure 5. The μ values of (**a**,**b**) Bi_2O_3/PVA, (**c**,**d**) WO_3/PVA, and (**e**,**f**) $BaSO_4$/PVA hydrogels containing varying filler contents of 0, 10, 20, 30, and 40 wt.%, determined at photon energies of (**a**,**c**,**e**) 0.001–0.2 MeV and (**b**,**d**,**f**) 0.2–5 MeV using XCOM.

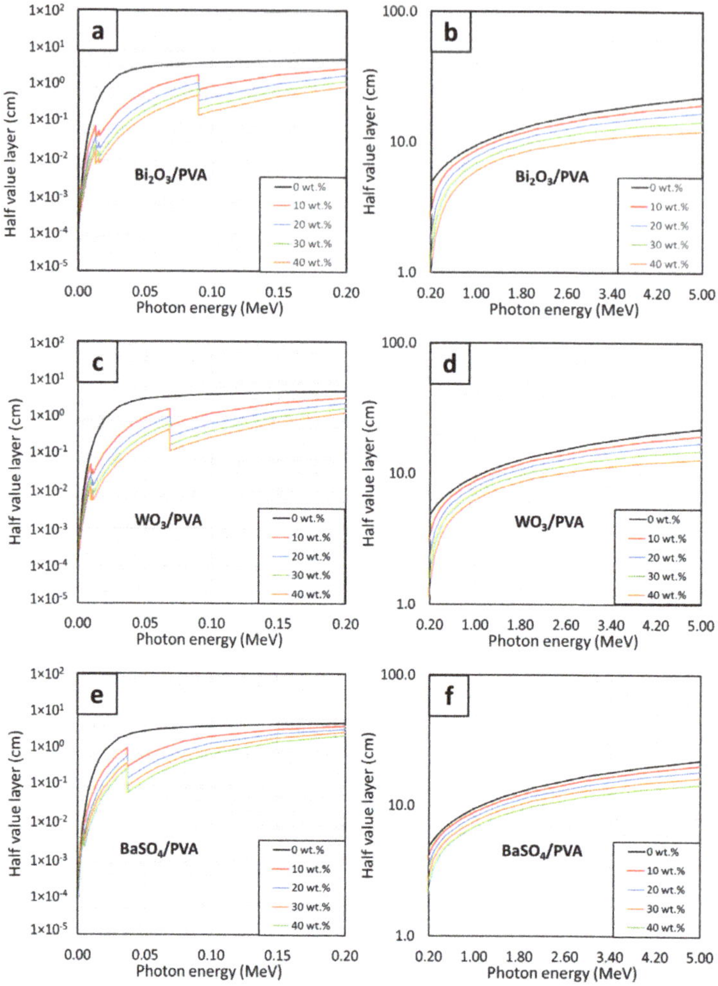

Figure 6. HVL values of (**a**,**b**) Bi_2O_3/PVA, (**c**,**d**) WO_3/PVA, and (**e**,**f**) $BaSO_4$/PVA hydrogels containing varying filler contents of 0, 10, 20, 30, and 40 wt.%, determined at photon energies of (**a**,**c**,**e**) 0.001–0.2 MeV and (**b**,**d**,**f**) 0.2–5 MeV using XCOM.

Figures 7 and 8 showed correlations between the filler contents and the μ and HVL values, respectively, at the photon energies of 0.1, 0.5, 1, and 5 MeV. Both figures suggested that additional filler contents could increase (decrease) the values of μ (HVL) with greater effects than those observed in μ_m, which was evidenced by the non-linear correlations between the filler contents and μ (HVL) values (Figure 7). Similar to μ_m, mathematical relationships between the filler contents and the μ and HVL values at the photon energies of 0.1, 0.5, 1, and 5 MeV were determined based on Equations (6) and (7), respectively. The results, which are shown in Tables 5 and 6, clearly showed that the addition of the fillers could improve photon attenuation capabilities of the hydrogels, as seen by the positive constants A in the case of μ for all photon energies and the negative constants A (E) in the case of HVL for the photon energy of 0.1 MeV (0.5, 1, and 5 MeV). Furthermore, the relatively greater effects of Bi_2O_3 on the enhancement of photon shielding properties in comparison with those of WO_3 and $BaSO_4$ were confirmed from this determination as the magnitudes of constants A in the case of Bi_2O_3/PVA hydrogels were the greatest among other hydrogels.

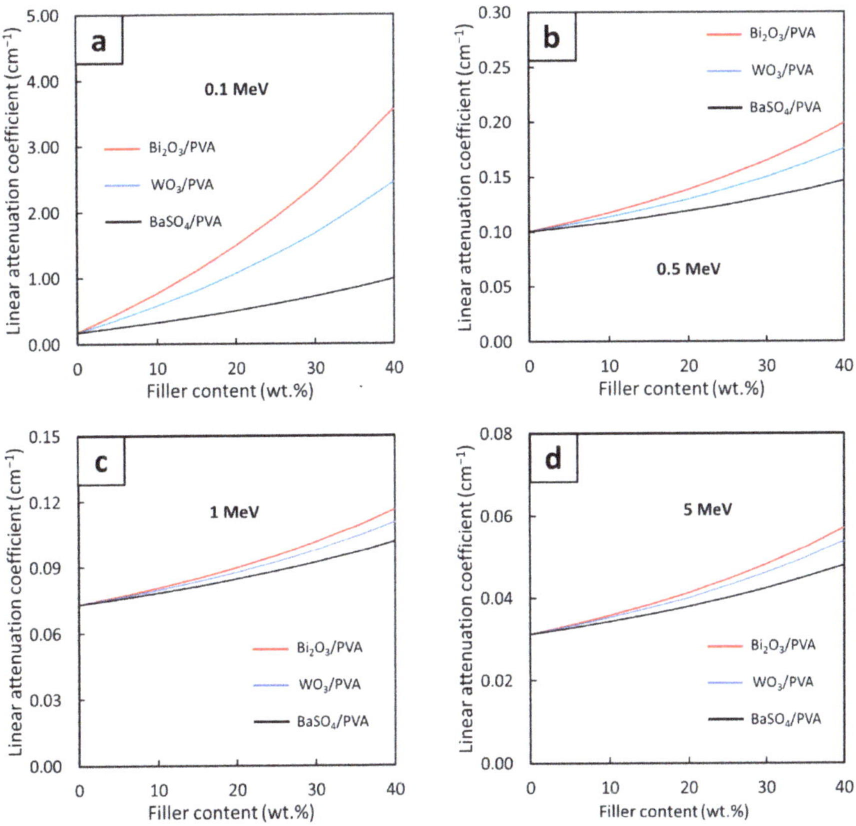

Figure 7. The μ values of Bi_2O_3/PVA, WO_3/PVA, and $BaSO_4$/PVA hydrogels containing varying filler contents of 0–40 wt.%, determined at photon energies of (**a**) 0.1 MeV, (**b**) 0.5 MeV, (**c**) 1 MeV, and (**d**) 5 MeV using XCOM.

Table 5. Mathematical constants (A, B, and C) for μ in the form $\mu = Ax^2 + Bx + C$ (Equation (6)), determined from Figure 7.

Photon Energy (MeV)	Bi_2O_3			WO_3			$BaSO_4$		
	A	B	C	A	B	C	A	B	C
0.1	9.58×10^{-4}	4.52×10^{-2}	1.98×10^{-1}	6.23×10^{-4}	3.15×10^{-2}	1.90×10^{-1}	1.97×10^{-4}	1.23×10^{-2}	1.80×10^{-1}
0.5	2.79×10^{-5}	1.31×10^{-3}	1.01×10^{-1}	2.06×10^{-5}	1.04×10^{-3}	1.01×10^{-1}	1.12×10^{-5}	7.00×10^{-3}	1.00×10^{-1}
1.0	1.22×10^{-5}	5.76×10^{-4}	7.35×10^{-2}	1.02×10^{-5}	5.17×10^{-4}	7.34×10^{-2}	6.89×10^{-6}	4.32×10^{-4}	7.33×10^{-2}
5.0	7.28×10^{-6}	3.43×10^{-4}	3.15×10^{-2}	6.17×10^{-6}	3.11×10^{-4}	3.14×10^{-2}	4.03×10^{-6}	2.51×10^{-4}	3.14×10^{-2}

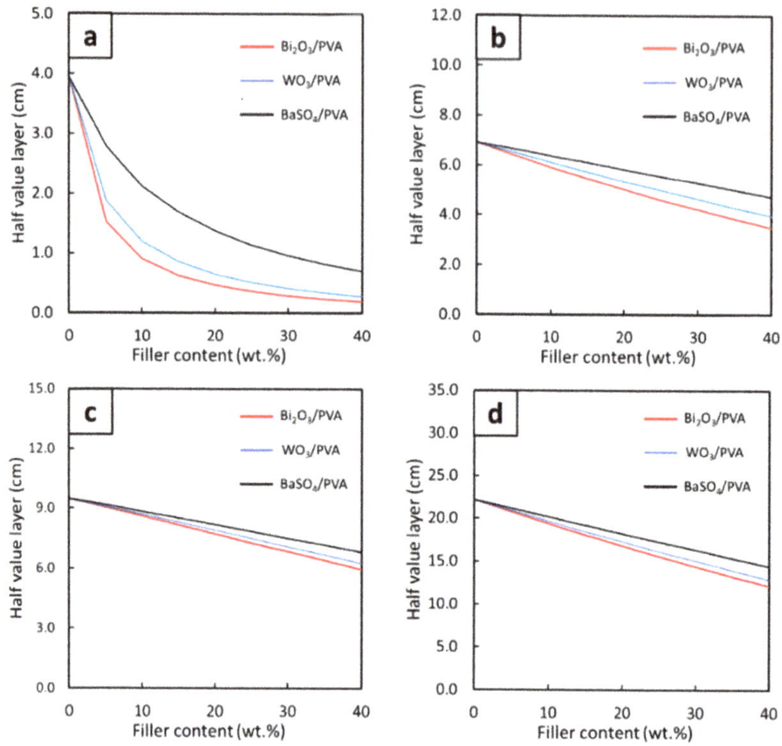

Figure 8. The HVL values of Bi_2O_3/PVA, WO_3/PVA, and $BaSO_4$/PVA hydrogels containing varying filler contents of 0–40 wt.%, determined at photon energies of (**a**) 0.1 MeV, (**b**) 0.5 MeV, (**c**) 1 MeV, and (**d**) 5 MeV using XCOM.

Table 6. Mathematical constants (A, B, C, D, and F) for HVL in the form HVL = $Ax^5 + Bx^4 + Cx^3 + Dx^2 + Ex + F$ (Equation (7)), determined from Figure 8.

Photon Energy (MeV)	Bi_2O_3					
	A	B	C	D	E	F
0.1	-5.83×10^{-7}	7.00×10^{-5}	-3.19×10^{-3}	6.92×10^{-2}	-7.46×10^{-2}	3.92
0.5	0	0	0	0	−0.0854	6.796
1.0	0	0	0	0	−0.0878	9.469
5.0	0	0	0	0	−0.2492	21.923
Photon Energy (MeV)	**WO_3**					
	A	B	C	D	E	F
0.1	-3.95×10^{-7}	4.82×10^{-5}	-2.25×10^{-3}	5.13×10^{-2}	-6.09×10^{-2}	3.93
0.5	0	0	0	0	−0.0742	6.861
1.0	0	0	0	0	−0.0802	9.484
5.0	0	0	0	0	−0.2317	21.995
Photon Energy (MeV)	**$BaSO_4$**					
	A	B	C	D	E	F
0.1	-7.63×10^{-8}	1.01×10^{-5}	-5.38×10^{-4}	1.56×10^{-2}	-2.93×10^{-1}	3.93
0.5	0	0	0	0	−0.0546	6.917
1.0	0	0	0	0	−0.0665	9.494
5.0	0	0	0	0	−0.1919	22.093

3.3. Lead Equivalence (Pb_{eq})

The Pb_{eq} values of PVA hydrogels containing Bi_2O_3, WO_3, or $BaSO_4$ with the filler contents of 10, 20, 30, and 40 wt.% (determined at the photon energies of 0.05, 0.08, and 0.1 MeV) are shown in Table 7. The results indicated that the Pb_{eq} values for all hydrogels increased with increasing filler contents, which was consistent with the behaviors of μ_m and μ. Generally, the Bi_2O_3/PVA hydrogels exhibited higher Pb_{eq} values than WO_3/PVA and $BaSO_4$/PVA hydrogels, mostly due to their higher atomic number of Bi and higher density of Bi_2O_3, for which the highest Pb_{eq} values achieved in 40 wt.% Bi_2O/PVA hydrogels were found to be 0.57, 0.61, and 0.57 mmPb at the photon energies of 0.05, 0.08, and 0.1 MeV, respectively. However, at the photon energy of 0.08 MeV, WO_3/PVA hydrogels offered higher Pb_{eq} values than those of Bi_2O_3/PVA hydrogels, determined at the same filler content. This behavior was observed due to the effects of the K-absorption edge of W, which occurred at 0.069 MeV, that resulted in uncharacteristically high interaction probabilities between the incident photons and W atoms at the specific photon energy (Figure 3), leading to the Pb_{eq} values for 40 wt.% WO_3/PVA hydrogels as high as 1.53 mmPb [39].

Table 7. Pb_{eq} values of Bi_2O_3/PVA, WO_3/PVA, and $BaSO_4$/PVA hydrogels containing varying filler contents of 10, 20, 30, and 40 wt.%, determined at photon energies of 0.05, 0.08, and 0.1 MeV using XCOM.

Photon Energy (MeV)	Filler	Lead Equivalence (mmPb)				Recommended Content (wt.%)
		10 wt.%	20 wt.%	30 wt.%	40 wt.%	
0.05	Bi_2O_3	0.12	0.24	0.38	0.57	36
	WO_3	0.08	0.16	0.25	0.36	— *
	$BaSO_4$	0.13	0.25	0.40	0.58	35
0.08	Bi_2O_3	0.16	0.28	0.42	0.61	34
	WO_3	0.33	0.64	1.03	1.53	15
	$BaSO_4$	0.17	0.28	0.43	0.60	34
0.1	Bi_2O_3	0.12	0.24	0.38	0.57	36
	WO_3	0.09	0.17	0.27	0.39	—
	$BaSO_4$	0.05	0.08	0.11	0.16	—

* — implies the recommended content for the filler at the specific photon energy was beyond the range of 0–40 wt.%.

Although the requirements for the Pb_{eq} of shielding materials varied, depending on applications and photon energies, a common Pb_{eq} requirement for shielding equipment used in general nuclear facilities is to be at least 0.5 mmPb [40]. Hence, Table 7 suggests that Bi_2O_3/PVA hydrogels with the filler content of at least 36 wt.% (interpolated from the hydrogels with 30 wt.% and 40 wt.% of Bi_2O_3) were the recommended conditions for all investigated photon energies (0.05, 0.08, and 0.1 MeV). On the other hand, the results showed that WO_3/PVA and $BaSO_4$/PVA hydrogels required the filler contents beyond 40 wt.% at some photon energies (0.05 MeV for WO_3/PVA hydrogels and 0.1 MeV for both WO_3/PVA and $BaSO_4$/PVA hydrogels) as their Pb_{eq} values were lower than the required 0.5 mmPb at the maximum filler content investigated in this work of 40 wt.%. Although 34 wt.% of Bi_2O_3 was sufficient for the photon attenuation at the energy of 0.08 MeV, the actual photon energies inside nuclear facilities were distributed as a spectrum; hence, 36 wt.% of Bi_2O_3 would be a better choice in order to cover all photon energies of concern. It should be noted that, since Pb_{eq} values depend on sample thickness, one could reduce the recommended filler contents for each photon energy by increasing sample thicknesses, which would also improve the recoverable strength of the self-healed hydrogels. Nonetheless, the choice of using thicker samples would be applicable only in applications that allowed or accepted thicker materials to serve their intended requirements.

Furthermore, based on our previous work, it could be implied that PVA hydrogels containing 36 wt.% of Bi_2O_3 would offer percentages of recoverable strength (%Recovery) of at least 75%, the initial tensile strengths of at least 219 MPa, and the elongation at break of at least 560%, for which the actual values depended on several factors such as the particle sizes of Bi_2O_3, a procedure to prepare the hydrogels, and the types of PVA granules [26]. Lastly, as the recommended contents of WO_3 and $BaSO_4$ were close to or beyond 40 wt.% and no actual experiment was conducted to confirm the producibility of such formulations, we positively believed that WO_3/PVA and $BaSO_4$/PVA hydrogels were producible as the %Recovery of 40 wt.% Bi_2O_3/PVA hydrogel was as high as 88.6%, which implied that 40 wt.% WO_3/PVA and 40 wt.% $BaSO_4$/PVA hydrogels should have similar or slightly lower mechanical properties and %Recovery in comparison with those of a 40 wt.% Bi_2O_3/PVA hydrogel. The similarity of the production and characteristics among Bi_2O_3, WO_3, and $BaSO_4$ fillers in radiation shielding materials was supported by previous works of Poltabtim et al. [16], Toyen et al. [23], and Abdolahzadeh et al. [41], who experimentally prepared Bi_2O_3-, WO_3-, or $BaSO_4$-filled EPDM, NR, and HDPE, respectively.

4. Conclusions

This work determined the numerical high-energy photon shielding properties of autonomously self-healing PVA hydrogels containing Bi_2O_3, WO_3, or $BaSO_4$ with varying filler contents of 0–40 wt.% using XCOM. The shielding properties investigated in this work consisted of μ_m, μ, HVL, and Pb_{eq} at the photon energies of 0.001–5 MeV as well as the recommended filler contents, determined at the photon energies of 0.05, 0.08, and 0.1 MeV. The results, which were in good agreement with those obtained from PHITS, revealed that the values of μ_m and μ (HVL) increased (decreased) with increasing filler contents but decreased (increased) with photon energies. In addition, the results suggested that Bi_2O_3/PVA hydrogels generally attenuated photons with higher efficiencies than those of WO_3/PVA and $BaSO_4$ hydrogels at the same filler content and photon energy. Lastly, the determination of recommended filler contents showed that Bi_2O_3/PVA hydrogels with the filler content of 36 wt.% exhibited the Pb_{eq} values higher than the required 0.5 mmPb for all investigated photon energies, implying that the PVA hydrogels with at least 36 wt.% of Bi_2O_3 were suitable for use as flexible and self-healing X-ray and gamma shielding equipment in general nuclear facilities.

Author Contributions: Conceptualization, K.S.; formal analysis, T.P. and K.S.; funding acquisition, K.S.; investigation, T.P. and K.S.; methodology, T.P. and K.S.; supervision, K.S.; validation, T.P. and K.S.; visualization, K.S.; writing—original draft, K.S.; writing—review and editing, T.P. and K.S. All authors have read and agreed to the published version of the manuscript.

Funding: This research was funded by the Kasetsart University Research and Development Institute (KURDI), Bangkok, Thailand, grant number FF(KU)25.64.

Informed Consent Statement: Not applicable.

Acknowledgments: Technical support was provided by the Department of Applied Radiation and Isotopes, Kasetsart University, Bangkok, Thailand. The Kasetsart University Research and Development Institute (KURDI) and Specialized Center of Rubber and Polymer Materials in Agriculture and Industry (RPM), Faculty of Science, Kasetsart University, provided publication support.

Conflicts of Interest: The authors declare no conflict of interest. The funders had no role in the design of the study; the collection, analyses, or interpretation of data; the writing of the manuscript; or the decision to publish the results.

References

1. Sakdinawat, A.; Attwood, S. Nanoscale X-ray imaging. *Nat. Photonics* **2010**, *4*, 840–848. [CrossRef]
2. Poltabtim, W.; Saenboonruang, K. Assessment of activity concentrations and their associated radiological health risks in commercial infant formulas in Thailand. *Chiang Mai J. Sci.* **2019**, *46*, 778–786.
3. Rittirong, A.; Saenboonruang, K. Comparative measurement of metal contents in raw and cooked rice samples prepared in different rice cookers using SR-XRF and health risk assessment. *J. Food Meas. Charact.* **2018**, *12*, 2801–2808. [CrossRef]

4. Basu, S.; Mitra, S.; Saha, N. Deep learning for screening COVID-19 using chest X-ray images. In Proceedings of the 2020 IEEE Symposium Series on Computational Intelligence (SSCI), Canberra, ACT, Australia, 1–4 December 2020; pp. 2521–2527.
5. Ni, L.; Lu, J. Interferon gamma in cancer immunotherapy. *Cancer Med.* **2018**, *7*, 4509–4516. [CrossRef] [PubMed]
6. Bhatnagar, J.; Novotny, J.; Huq, M.S. Dosimetric characteristics and quality control tests for the collimator sectors of the Leksell Gamma Knife® Perfexion^{TM}. *Med. Phys.* **2012**, *39*, 231–236. [CrossRef] [PubMed]
7. Sikder, S.; Biswas, P.; Harza, P.; Akhtar, S.; Chattopadhyay, A.M.; D'Souza, S.F. Induction of mutation in tomato (*Solanum lycoersicum* L.) by gamma irradiation and EMS. *Indian J. Genet.* **2013**, *73*, 392–399. [CrossRef]
8. Hamrashdi, H.A.; Monk, S.D.; Cheneler, D. Passive gamma-ray and neutron imaging systems for national security and nuclear non-proliferation in controlled and uncontrolled detection areas: Review of past and current status. *Sensors* **2019**, *19*, 2638. [CrossRef]
9. Manda, K.; Glasow, A.; Paape, D.; Hildebrandt, G. Effects of ionizing radiation on the immune system with special emphasis on the interaction of dendritic and T cells. *Front. Oncol.* **2012**, *2*, 102. [CrossRef]
10. Hladik, D.; Tapio, S. Effects of ionizing radiation on the mammalian brain. *Mutat. Res. Rev. Mutat. Res.* **2016**, *770*, 219–230. [CrossRef]
11. Moller, A.P.; Mousseau, T.A. Strong effects of ionizing radiation from Chernobyl on mutation rates. *Sci. Rep.* **2015**, *5*, 8363. [CrossRef]
12. McGiff, T.J.; Danforth, R.A.; Herschaft, E.E. Maintaining radiation exposures as low as reasonably achievable (ALARA) for dental personnel operating portable hand-held X-ray equipment. *Health Phys.* **2012**, *103*, S179–S185. [CrossRef] [PubMed]
13. Cinan, Z.M.; Erol, B.; Baskan, T.; Mutlu, S.; Savaskan Yilmaz, S.; Yilmaz, A.H. Gamma irradiation and the radiation shielding characteristics: For the lead oxide doped the crosslinked polystyrene-b-polyethyleneglycol block copolymers and the polystyrene-b-polyethyleneglycol-boron nitride nanocomposites. *Polymers* **2021**, *13*, 3246. [CrossRef] [PubMed]
14. Cinan, Z.M.; Baskan, T.; Erol, B.; Mutlu, S.; Misirlioglu, Y.; Yilmaz, S.S.; Yilmaz, A.H. Gamma irradiation, thermal conductivity, and phase change tests of the cement-hyperbranched poly amino-ester-block-poly cabrolactone-polyurathane plaster-lead oxide and arsenic oxide composite for development of radiation shielding material. *Int. J. Energy Res.* **2021**, *45*, 20729–20762. [CrossRef]
15. Charkiewicz, A.E.; Backstrand, J.R. Lead toxicity and pollution in Poland. *Int. J. Environ. Res. Public Health* **2020**, *17*, 4385. [CrossRef] [PubMed]
16. Poltabtim, W.; Wimolmala, E.; Saenboonruang, K. Properties of lead-free gamma-ray shielding materials from metal oxide/EPDM rubber composites. *Radiat. Phys. Chem.* **2018**, *153*, 1–9. [CrossRef]
17. Poltabtim, W.; Toyen, D.; Saenboonruang, K. Theoretical determination of high-energy photon attenuation and recommended protective filler contents for flexible and enhanced dimensionally stable wood/NR and NR composites. *Polymers* **2021**, *13*, 869. [CrossRef]
18. Malekzadeh, R.; Zali, V.S.; Jahanbakhsh, O.; Okutan, M.; Mesbahi, A. The preparation and characterization of silicone-based composites doped with $BaSO_4$, WO_3, and PbO nanoparticles for shielding applications in positron emission tomography (PET) and nuclear medicine facilities. *Nanomed. J.* **2020**, *7*, 324–334.
19. Abdalsalam, A.H.; Sayyed, M.I.; Hussein, T.A.; Sakar, E.; Mhareb, M.H.A.; Sakar, B.C.; Alim, B.; Kaky, K.M. A study of gamma attenuation property of UHMWPE/Bi_2O_3 nanocomposites. *Chem. Phys.* **2019**, *523*, 92–98. [CrossRef]
20. Poltabtim, W.; Wimolmala, E.; Markpin, T.; Sombatsompop, N.; Rosarpitak, V.; Saenboonruang, K. X-ray shielding, mechanical, physical, and water absorption properties of wood/PVC composites containing bismuth oxide. *Polymers* **2021**, *13*, 2212. [CrossRef]
21. Intom, S.; Kalkornsurapranee, E.; Johns, J.; Kaewjaeng, S.; Kothan, S.; Hongtong, W.; Chaiphaksa, W.; Kaewkhao, J. Mechanical and radiation shielding properties of flexible materials based on natural rubber/Bi_2O_3 composites. *Radiat. Phys. Chem.* **2020**, *172*, 108772. [CrossRef]
22. Thumwong, A.; Wimolmala, E.; Markpin, T.; Sombatsompop, N.; Saenboonruang, K. Enhanced X-ray shielding properties of NRL gloves with nano-Bi_2O_3 and their mechanical properties under aging conditions. *Radiat. Phys. Chem.* **2021**, *186*, 109530. [CrossRef]
23. Toyen, D.; Rittirong, A.; Poltabtim, W.; Saenboonruang, K. Flexible, lead-free, gamma-shielding materials based on natural rubber/metal oxide composites. *Iranian Polym. J.* **2018**, *27*, 33–41. [CrossRef]
24. Zhang, H.; Xia, H.; Zhao, Y. Poly(vinyl alcohol) hydrogel can autonomously self-heal. *ACS Macro Lett.* **2012**, *1*, 1233–1236. [CrossRef]
25. Mirkovic, M.; Kljajevic, L.; Dolenec, S.; Nenadovic, M.; Pavlovic, V.; Rajacic, M.; Nenadovic, S. Potential usage of hybrid polymers binders based on fly ash with the addition of PVA with satisfying mechanical and radiological properties. *Gels* **2021**, *7*, 270. [CrossRef]
26. Tiamduangtawan, P.; Kamkaew, C.; Kuntonwatchara, S.; Wimolmala, E.; Saenboonruang, K. Comparative mechanical, self-healing, and gamma attenuation properties of PVA hydrogels containing either nano- or micro-sized Bi_2O_3 for use as gamma-shielding materials. *Radiat. Phys. Chem.* **2020**, *177*, 109164. [CrossRef]
27. Tiamduangtawan, P.; Wimolmala, E.; Meesat, R.; Saenboonruang, K. Effects of Sm_2O_3 and Gd_2O_3 in poly(vinyl alcohol) hydrogels for potential use as self-healing thermal neutron shielding materials. *Radiat. Phys. Chem.* **2020**, *172*, 108818. [CrossRef]
28. XCOM. National Institute of Standards and Technology (NIST). Available online: https://physics.nist.gov/PhysRefData/Xcom/html/xcom1.html (accessed on 23 February 2022).

29. Saenboonruang, K.; Poltabtim, W.; Thumwong, A.; Pianpanit, T.; Rattanapongs, C. Rare-earth oxides as alternative high-energy photon protctiev fillers in HDPE composites: Theoretical aspects. *Polymers* **2021**, *13*, 1930. [CrossRef]
30. Ratliff, H.N.; Matsuda, N.; Abe, S.; Miura, T.; Furuta, T.; Iwamoto, Y.; Sato, T. Modernization of the DCHAIN-PHITS activation code with new features and updated data libraries. *Nucl. Instrum. Methods Phys. Res. B* **2020**, *484*, 29–41. [CrossRef]
31. Toyen, D.; Paopun, Y.; Changjan, D.; Wimolmala, E.; Mahathanabodee, S.; Pianpanit, T.; Anekratmontree, T.; Saenboonruang, K. Simulation of neutron/self-emitted gamma attenuation and effects of silane surface treatment on mechanical and wear resistance properties of Sm_2O_3/UHMWPE composites. *Polymers* **2021**, *13*, 3390. [CrossRef]
32. Livet, F.; Sutton, M. X-ray coherent scattering in metal physics. *Comptes Rendus Phys.* **2012**, *13*, 227–236. [CrossRef]
33. Sato, T.; Iwamoto, Y.; Hasimoto, S.; Ogawa, T.; Furuta, T.; Abe, S.; Kai, T.; Tsai, P.; Matsuda, N.; Iwase, H.; et al. Features of Particle and Heavy Ion Transport code System (PHITS) version 3.02. *J. Nucl. Sci. Technol.* **2018**, *55*, 684–690. [CrossRef]
34. Obaid, S.S.; Sayyed, M.I.; Gaikwad, D.K.; Pawar, P.P. Attenuation coefficients and exposure buildup factor of some rocks for gamma ray shielding applications. *Radiat. Phys. Chem.* **2018**, *148*, 86–94. [CrossRef]
35. Tasnim, A.; Sahadath, M.H.; Khan, M.N.I. Development of high-density radiation shielding materials containing $BaSO_4$ and investigation of the gamma-ray attenuation properties. *Radiat. Phys. Chem.* **2021**, *189*, 109772. [CrossRef]
36. Han, M.C.; Kim, H.S.; Pia, M.G.; Basaglia, T.; Batic, M.; Hoff, G.; Kim, C.H.; Saracco, P. Validation of cross sections for Monte Carlo simulation of the photoelectric effect. *IEEE Trans. Nucl. Sci.* **2016**, *63*, 1117–1146. [CrossRef]
37. Pratt, R.H.; Lajohn, L.A.; Florescu, V.; Suric, T.; Chatterjee, B.K.; Roy, S.C. Compton scattering revisited. *Radiat. Phys. Chem.* **2010**, *79*, 124–131. [CrossRef]
38. Ridgers, C.P.; Kirk, J.G.; Duclous, R.; Blackburn, T.G.; Brady, C.S.; Bennett, K.; Arber, T.D.; Bell, A.R. Modelling gamma-ray photon emission and pair production in high-intensity laser-matter interactions. *J. Comput. Phys.* **2014**, *260*, 273–285. [CrossRef]
39. Purans, J.; Kuzmin, A.; Parent, P.; Laffon, C. X-ray absorption study of the electronic structure of tungsten and molybdenum oxides on the O K-edge. *Electrochim. Acta* **2001**, *46*, 1973–1976. [CrossRef]
40. McCaffrey, J.P.; Shen, H.; Downton, B.; Mainegra-Hing, E. Radiation attenuation by lead and nonlead materials used I radiation shielding garments. *Med. Phys.* **2007**, *34*, 530–537. [CrossRef]
41. Abdolahzadeh, T.; Morshedian, J.; Ahmadi, S. Preparation and characterization of nano WO_3/Bi_2O_3/GO and $BaSO_4$/GO dispersed HDPE composites for x-ray shielding application. *Polyolefins J.* **2022**. [CrossRef]

gels

Article

Mechanochemical Effect on Controlled Drug Release of Konjac Glucomannan Matrix Tablets during Dry Grinding

Fuminori Okazaki [1], Yusuke Hattori [1], Tetsuo Sasaki [2] and Makoto Otsuka [1,3,*]

1 Faculty of Pharmacy, Musashino University, 1-1-20 Shinmachi, Nishi-Tokyo 202-8585, Tokyo, Japan; s1343048@stu.musashino-u.ac.jp (F.O.); yhattori@musashino-u.ac.jp (Y.H.)
2 Graduate School of Medical Photonics, Shizuoka University, 3-5-1 Johoku, Naka-ku, Hamamatsu 432-8011, Shizuoka, Japan; sasaki.tetsuo@shizuoka.ac.jp
3 Research Institute of Electronics, Shizuoka University, 3-5-1 Johoku, Naka-ku, Hamamatsu 432-8011, Shizuoka, Japan
* Correspondence: ohtsuka.makoto@shizuoka.ac.jp; Tel./Fax: +81-053-478-3264

Abstract: To design a controlled drug release preparation based on a safe natural material, a Konjac glucomannan (KGM) mixture containing 16.0 *w/w*% calcium hydroxide ($Ca(OH)_2$) was ground in a planetary ball mill for 0–120 min. The mechanochemical effect on the physicochemical properties of the KGM ground product was investigated by Fourier-transform infrared spectroscopy (FT-IR), powder X-ray spectroscopy, scanning electron microscopy with energy-dispersive X-ray spectroscopy, and drug release testing. The FT-IR spectra of the ground KGM indicated that the deacetylation reaction of KGM was accelerated in the $Ca(OH)_2$-containing sols by mechanochemical energy, and the degree of deacetylation of KGM was dependent on the grinding time. The time required for tablet disintegration of the KGM matrix tablets containing theophylline increased as the grinding time increased; therefore, drug release was sustained. The Higuchi plots of the matrix tablets obtained from KGM ground for 60–120 min exhibited good linearity because they maintained their gel matrix tablet shape during the release test. However, KGM tablets ground for 0–30 min exhibited nonlinear curves, which were caused by tablet disintegration. This suggests that drug release from the KGM matrix tablet can be freely controlled by the degree of mechanochemical treatment.

Keywords: konjac glucomannan; solid-state mechanochemical reaction; swelling and disintegration property; Fourier-transform infrared spectral change; polymer matrix tablets; sustained drug release kinetics

1. Introduction

Sustained-release tablets assist in dosing compliance [1] and improve patient quality of life by reducing dosing frequency. Various types of synthetic hydrophilic polymers that can form a gel matrix are used as bases for sustained-release tablets [2]. Since hydrated gel layers form on the surface of the gel matrix tablets, the dissolution and erosion rates of the gel layer depend on the properties and contents of the polymer excipients used, and drug release can be controlled for 8–24 h [3–6].

Naturally abundant polysaccharides have attracted considerable attention as a base for sustained-release preparations because they are non-toxic, safe, hydrophilic, and biodegradable. Glucomannan (GM) is a non-ionic polysaccharide extracted from the tubers of Konjac, which is a high-molecular-weight compound that has also been used in food [7,8]. Konjac glucomannan (KGM) fine powder obtained from natural products is partially acetylated GM, which can be gelatinized by heating in an alkaline solution, followed by deacetylation [9]. According to the FDA, GM is recognized as a safe biomedical material and is used worldwide as a functional healthcare material for diabetes and steatosis [10]. KGM is not hydrolyzed by salivary and pancreatic amylase owing to the presence of the 1,4 bonds of β-D-mannose and glucose units; therefore, KGM can enter the colon as ingested and is fermented by colonic bacteria [10]. KGM is one of the most viscous dietary fibers because it is water-soluble and highly absorbent. KGM forms a strong, elastic, and heat-stable gel matrix when heated in a weakly alkaline solution. Because of the effective properties of KGM gel powder (such as hydrophilicity, thickening, stability, and the ability to form emulsions, suspensions, gels, and films [9]), KGM can control drug release and has great potential as a pharmaceutical additive. Therefore, KGM was evaluated for its ability to control drug release in gel matrix tablets containing various drugs, and its application as a drug delivery system was investigated [11,12]. The gel matrix tablets were compression-prepared via a conventional wet granulation process [13–15]. However, since the controlled-release function of these matrix tablets depended on the amount of the prescribed ingredient, it was difficult to adjust and change the release rates of the tablets because the prescription of the formulations had to be changed. Studies were conducted on the development of pharmaceutical additives with a controlled drug release function [16,17]. Therefore, to utilize natural resources as safe medical additives without complicated process operations, this study focused on mechanochemical treatment technology, which controls the crystallinity of natural product-derived additives. Mechanochemical reactions allow the synthesis of compounds, phases, and microstructures that are essentially different from those obtained by conventional chemical reactions [18]. Using the mechanochemical reactions of organic compounds generated by grinding energy, drugs can be bonded with biopolymers [19], amorphous solids can be formed to improve the solubility of drugs with poor water solubility [20], and drugs can form co-crystals with specific compounds to become chemically stable [21,22]. In previous studies [17,23], the amorphization of starch and its gel formation in solution were controlled by ball mill grinding, and the drug release rates from the gel matrix tablets were controlled using drug release kinetic tests and X-ray computed tomography imaging.

Therefore, in this study, to control drug release from the device by a simple manufacturing process and avoid the formation of a KGM complex with other polymer compounds in wet granulation, as previously reported [11–15], KGM and alkaline substances were mechanochemically treated in a dry-state-grinding process using a planetary ball mill. The KGM crystallinity and the dispersion uniformity of alkaline fine particles depend on their mechanochemical energy. The ground KGM preparations obtained by mechanochemical reactions and their compressed tablets were examined for their chemical and pharmaceutical properties related to controlling drug release.

2. Results

2.1. Change in the Physicochemical Properties of the Ground KGM and $Ca(OH)_2$ Mixed Powder

Figure 1 shows the XRD profiles of the KGMH mixtures containing 16 *w/w*% $Ca(OH)_2$ after various grinding times. The XRD profile of $Ca(OH)_2$ showed specific diffraction peaks at 18°, 29°, and 34°, whereas the profile of KGM showed a halo pattern without any diffraction peaks. Therefore, the profile of KGMH0 (a physical mixture) only showed specific peaks due to $Ca(OH)_2$, and the profiles of the ground KGMH mixtures (KGMH120) did not change significantly up to 120 min of grinding.

Figure 2 shows SEM/EDX images of the KGMH0 and KGMH120 powders. In the SEM image of KGMH0, many particles (5–150 μm in diameter) were observed, and they had a wide particle size distribution. However, only smaller particles (5–20 μm in diameter) were observed in the EDX image. In contrast, in the SEM image of KGMH120, a large number of particles (5–40 μm in diameter) were observed, and the EDX and SEM images of the particles were almost identical.

Figure 3 shows FT-IR spectra of the KGMH mixtures after various grinding times. The IR spectra of the ground KGMH mixtures exhibited absorption peaks at 3631, 3288, 2878, 2366, 1724, 1654, and 1427 cm^{-1}. The absorption intensity of the peaks at 3621 and 1724 cm^{-1} decreased with increasing grinding time. In contrast, the absorption intensity of the peaks at 3288, 2366, and 1427 cm^{-1} increased with increasing grinding time. These results suggest that the molecular interaction of KGM changed significantly after grinding.

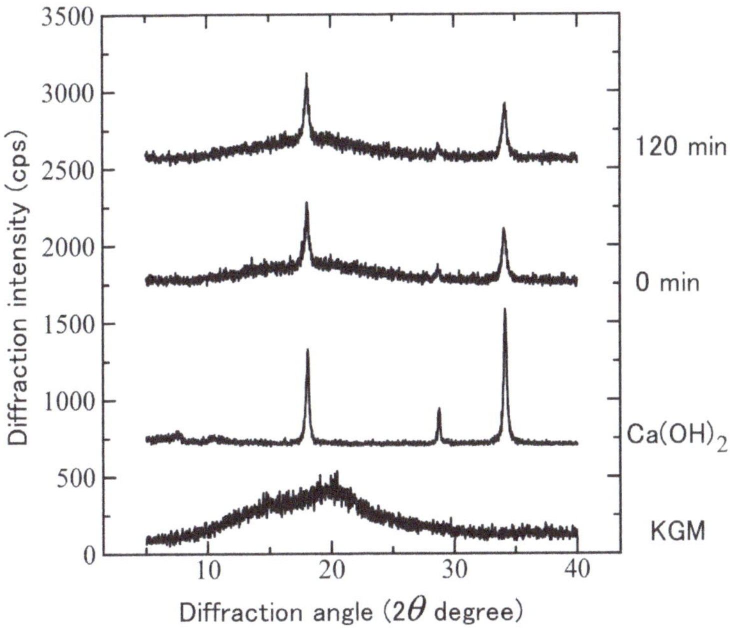

Figure 1. Powder X-ray diffraction profiles of the KGM mixtures containing 16.0 *w/w*% $Ca(OH)_2$ after various grinding times.

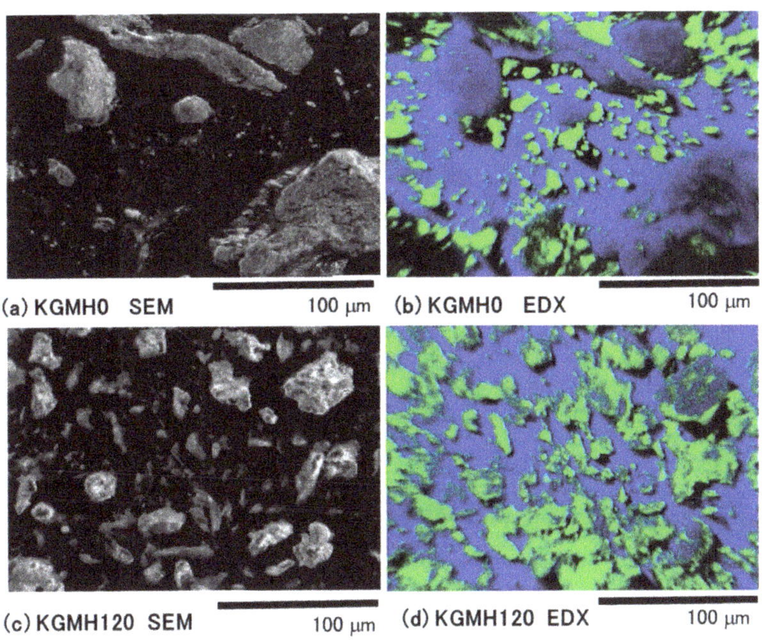

Figure 2. Scanning electron microscopy/energy-dispersive X-ray spectroscopy images of the KGMH0 and KGMH120 particles, respectively. SEM, scanning electron microscopy (**a**,**c**); EDX, energy-dispersive X-ray spectroscopy Ca (green) in EDX (**b**,**d**).

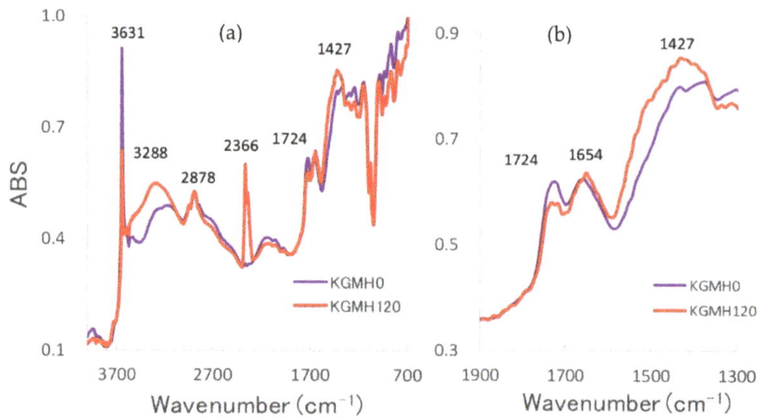

Figure 3. Fourier-transform infrared spectra of the KGMH mixtures ground for various grinding times. (**a**) Spectral range: 4000–700 cm^{-1}; (**b**) spectral range: 1900–1300 cm^{-1}.

2.2. Effect of Grinding on the Drug Release Profiles of the KGM Matrix Tablets

Figure 4 shows the drug release profiles of the polymer matrix tablets prepared from KGMH mixtures ground for 0–120 min. The KGMH0 tablets disintegrated at approximately 30 min and released over 75% of the drug within 150 min. However, based on the drug release profiles of ground KGMH tablets, the time required for tablet disintegration increased as the grinding time increased, and drug release was suppressed. Drug release from the KGMH tablets was delayed with increasing grinding time because a typical gel layer was visibly formed on the outside of the KGMH60 and KGMH120 tablets.

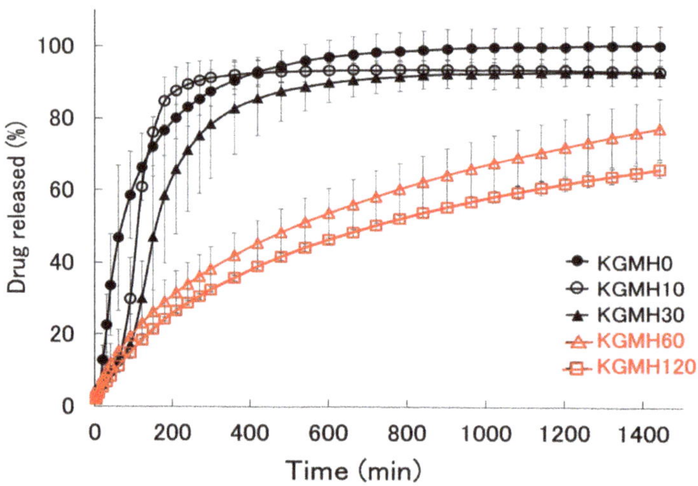

Figure 4. Drug release profiles of the polymer matrix tablets prepared from the KGMH mixtures ground for 0–120 min.

Figure 5 shows photographs of the KGMH tablets after 24 h during the drug release tests. After 24 h, the KGMH0, KGMH10, and KGMH30 tablets had completely disintegrated, but the KGMH60 and KGMH120 tablets retained a tablet shape. This result indicates that it was possible to control drug release of the KGMH polymeric matrix tablets by mechanochemical treatment in a ball mill. Therefore, it was hypothesized that the drug release profiles from the KGMH60 and KGMH120 matrix tablets followed the Higuchi Equation (1) [24]:

$$Q = \sqrt{\frac{D\varepsilon}{\tau}(2A - \varepsilon C_s)C_s t} = K_H t^{\frac{1}{2}} \quad (1)$$

where Q is the amount of drug released per unit surface area, A is the drug amount per unit volume in the device, D is the diffusion coefficient, ε is the device porosity, τ is the tortuosity, C_s is the drug solubility, and K_H is the Higuchi drug release rate constant.

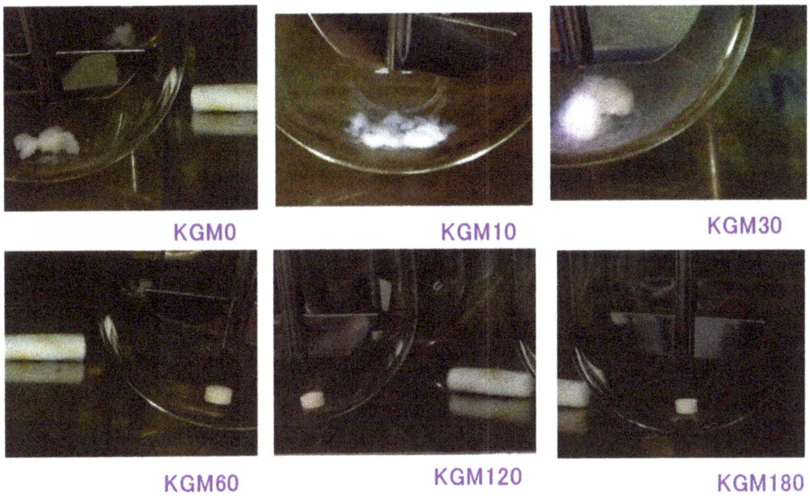

Figure 5. Photographs of the KGMH tablets after 24 h during the drug release tests.

Figure 6 shows Higuchi-type plots of the amount of drug released from the KGMH matrix tablets against the square root of time. The plots of the matrix tablets composed of KGMH ground for 60 min or longer exhibited good linearity (coefficient of determination above 0.99) and followed the Higuchi Equation (1). In contrast, the curves of the KGMH tablets ground for 30 min or less were nonlinear owing to tablet disintegration and did not follow the same drug release mechanism. Therefore, the drug release rate constants (K_H) of the KGMH60 and KGMH120 matrix tablets, which were calculated based on Equation (1) using the least-squares method, were 0.246 and 0.215 mg/min$^{1/2}$, respectively. Since the Higuchi release theory could not be applied to all tablet samples (as shown in Figure 6), a different method was applied to quantitatively evaluate the amount of drug released in a limited time. Figure 7 shows the effect of grinding time on the drug release amount (RA4) of the KGMH matrix tablets after the 240 min dissolution test. The RA4 values of the KGMH matrix tablets decreased significantly with increasing grinding time, and the RA4 values of the KGMH0, KGMH10, and KGMH30 matrix tablets were significantly higher than those of the KGMH60 and KGMH120 matrix tablets.

To investigate the effect of the amount of $Ca(OH)_2$, the KGM mixtures containing 5.2 and 10.5 *w/w*% $Ca(OH)_2$ were ground for 120 min, respectively, and labeled KGML120 and KGMM120. Figure 8 shows the effect of $Ca(OH)_2$ on the drug release profiles of the KGML120, KGMM120, and KGMH120 matrix tablets. While the Higuchi plot of the KGMH120 matrix tablets was a straight line, the plots of the KGMM120 and KGML120 matrix tablets were nonlinear curves, and their drug release did not follow the Higuchi theory. The release profiles of the matrix tablets were suppressed, reflecting the visually observed delayed tablet disintegration times.

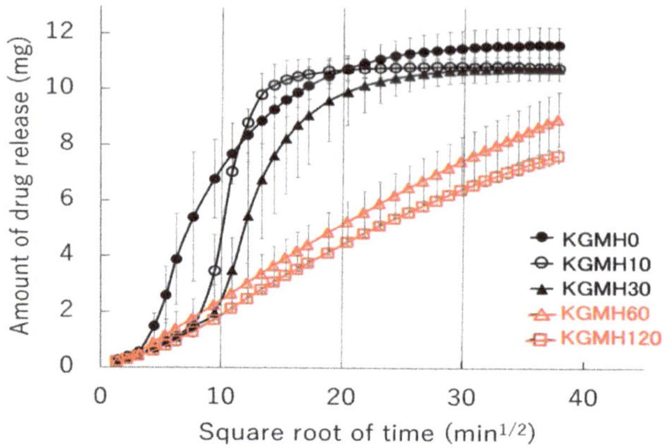

Figure 6. Higuchi-type plots of the amount of drug released from the KGMH matrix tablets against the square root of time.

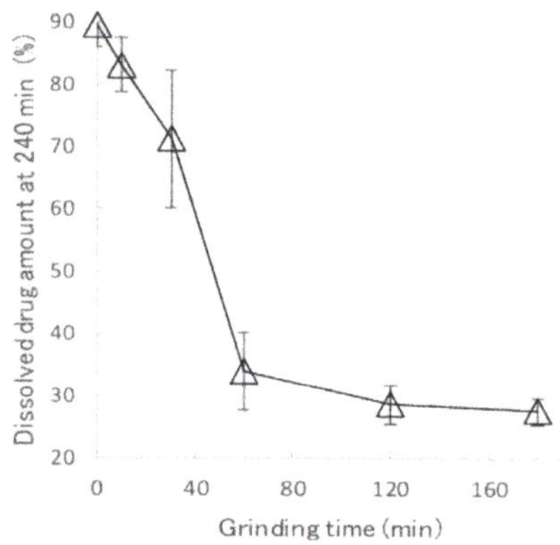

Figure 7. Effect of grinding time on the drug release amount (RA4) of the KGMH matrix tablets after 240 min dissolution test.

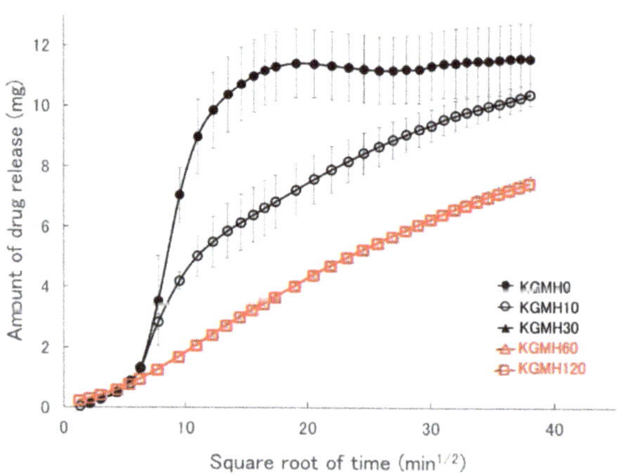

Figure 8. Effect of calcium hydroxide content on the drug release profiles of the KGML120, KGMM120, and KGMH120 matrix tablets.

Figure 9 shows the effect of $Ca(OH)_2$ on the RA4 value of KGM matrix tablets. The RA4 value of KGMH matrix tablets decreased significantly with increasing amounts of $Ca(OH)_2$, with values of 101.5%, 59.1%, and 28.6% for KGML120, KGMH120, and KGMH120, respectively.

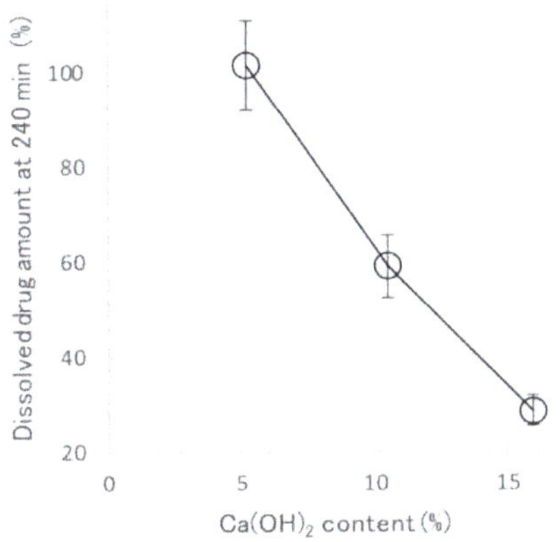

Figure 9. Effect of calcium hydroxide content on the RA4 of the KGM matrix tablets.

Figure 10 shows the effect of grinding on the water penetration rate of KGMH0 and KGMH120 matrix tablets. The KGMH0 tablets immediately absorbed water at the start of the release test. In contrast, the KGMH120 tablet slowly absorbed water, forming a gel layer on the surface at the beginning of the release test, but rapidly absorbed water after 100 min.

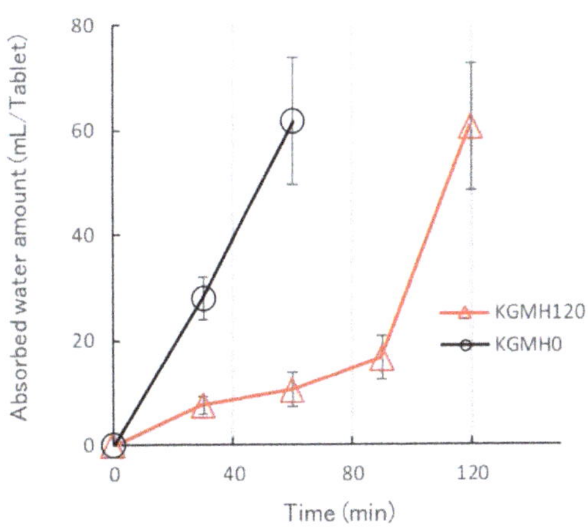

Figure 10. Effect of grinding on the water penetration rate of the KGMH0 and KGMH120 matrix tablets.

3. Discussion

3.1. Mechanochemical Reaction of KGM with $Ca(OH)_2$ during Grinding

Since the KGM fine powder is a natural product, it cannot form a gel in water owing to partial acetylation of the GM. However, when KGM is partially deacetylated by heating in an alkaline solution, it can form a strong gel in water, depending on the degree of deacetylation, as previously reported [10]. When KGM is used as a base for sustained-release preparation for a drug delivery system, gelatinization treatment by heating in an alkaline solution is required, in which the chemical stability of the active pharmaceutical ingredients can be compromised. Therefore, in this study, we aimed to develop a dry solid-state gelatinization treatment for KGM that does not decompose active pharmaceutical ingredients.

After many trial studies, we found that KGM formulations with 5–16 *w/w*% $Ca(OH)_2$ under dry grinding conditions were optimal for gelatinized matrix tablets for sustained drug release. The XRD profiles of the KGMH mixtures before and after the mechanochemical treatment showed a halo pattern with additional diffraction peaks due to $Ca(OH)_2$, indicating that the XRD profile of KGMH120 was not significantly different before and after the treatment (Figure 1). The results suggested that most of the $Ca(OH)_2$ did not change its basic crystal structure during the grinding process under the experimental conditions. In addition, the morphologies of the KGMH120 and KGMH0 particles observed by SEM/EDX (Figure 2) indicated that fine $Ca(OH)_2$ particles (less than a few micrometers in diameter) were uniformly dispersed in the KGMH120 composite particles (5–40 μm in diameter) after mechanochemical treatment.

Furthermore, in the FT-IR spectra, the peaks of the ground KGMH mixtures (Figure 3) were assigned based on a previous study. The peaks at 3631 and 3288 cm^{-1} were attributed to the free and hydrogen-bonded stretching vibrations of the OH group [25]. The peaks at 2878 and 1427 cm^{-1} were due to the stretching and angular vibrations of CH in the alkene [25] and $CO_3{}^{2-}$ of $CaCO_3$ [26], respectively. The peak at 1724 cm^{-1} was attributed to the stretching vibration of C=O in the esters' acetyl groups [27]. The peaks at 2366 and 1654 cm^{-1} were attributed to the CO stretching vibration in adsorbed carbon dioxide gas [28] and the stretching vibration of OH in water [25], respectively.

The mechanochemical reaction between KGM and $Ca(OH)_2$ can be explained using the following Equation (2):

$$\begin{array}{l} \mathrm{KGM} + n\mathrm{Ca(OH)_2} + n\mathrm{H_2O} + n\mathrm{CO_2} \\ \quad | \\ (\mathrm{CH_2OCOCH_3})_{2n} \\ \\ \quad \Delta E \\ ======> \quad \mathrm{KGM} + 2n\mathrm{CH_3COOH} + n\mathrm{CaCO_3} \\ \qquad\qquad\qquad\quad | \\ \qquad\qquad\qquad (\mathrm{CH_2OH})_{2n} \end{array} \tag{2}$$

where ΔE is mechanochemical energy during grinding.

First, when $Ca(OH)_2$ particles were ground in a ball mill, the reduction in particle size increased the surface area by more than 100 times (Figure 2c); therefore, carbon dioxide gas was adsorbed on the surface from the air in the environment (as shown by the increasing intensity of the absorbance peak at 2366 cm^{-1} in Figure 3). Fine $Ca(OH)_2$ particles were uniformly distributed on KGM (Figure 2d), which may have formed alkaline sols (containing OH- ions) on the surface of the KGM powder with a small amount of water (peak at 1654 cm^{-1} in Figure 3). Second, the KGM deacetylation reaction was catalyzed and accelerated by $Ca(OH)_2$-containing sols, as shown by the decrease in the peak at 1724 cm^{-1} due to solid-state grinding. Ca^{2+} ions reacted with carbon dioxide to produce $CaCO_3$ (as shown by the increasing intensity of the absorbance peak at 1427 cm^{-1} in Figure 3). Third, when the grinding time for the partially deacetylated KGM [9] was increased, the absorbance of the peak at 3621 cm^{-1} decreased while the absorbance of the peak at 3288 cm^{-1} increased (Figure 3). This indicates that free OH groups were easily transformed into intermolecular hydrogen bond OH groups by mechanochemical energy. This phenomenon indicates that the steric hindrance caused by the KGM acetyl groups was reduced during partial deacetylation by the mechanochemical reaction. On the other hand, the solid-state intermolecular hydrogen bonds between the particle surfaces of the partially deacetylated KGM increased. Therefore, the ground KGMH mixtures facilitated the formation of a strong gel in an aqueous solution and suppressed tablet disintegration.

3.2. Effect of Grinding on the Drug Release Kinetics of KGM Matrix Tablets and Hydrogel Formation Mechanism

To optimize the pharmaceutical formulation of KGM as a sustained-release drug preparation, the effect of $Ca(OH)_2$ content on the drug release profile of the KGM matrix tablets was investigated. When KGM was heat-treated with 5 *w/w*% $Ca(OH)_2$ in a solution at over 90 °C, it formed a sufficiently strong gel [29,30]. Therefore, the standard formulation of KGM for gel formation was labeled as KGML, which contained approximately 5 *w/w*% $Ca(OH)_2$ [29]. KGML120 and KGMM120 tablets obtained after grinding for 120 min began to disintegrate after approximately 30 min, as shown in the drug release profile (Figure 8). However, because the KGMH120 tablet containing 16 *w/w*% $Ca(OH)_2$ formed a strong gel during the drug release test, the dissolution kinetics (Figure 8) were based on the Higuchi drug release theory [24]. Therefore, using the dry grinding method, 16 *w/w*% $Ca(OH)_2$ was required to obtain a sustained-release drug preparation that followed Higuchi's theory. To clarify the intermolecular formation mechanism of the KGM matrix tablets during the drug release test, the relationship between the grinding time and sustained drug release kinetics of the KGM matrix tablets was investigated. The drug release profiles of the KGMH matrix tablets showed that the drug release was delayed (Figure 6) as the mechanochemical reaction progressed (shown in the FT-IR spectra in Figure 3) with increasing grinding time.

As described in the previous section, KGM reacted with $Ca(OH)_2$ sol in the solid-state through mechanochemical energy, and the KGM acetyl groups partially decomposed with increasing grinding time (Figure 3). Partially deacetylated KGM obtained by mechanical treatment can form a strong hydrogel [31]. After tablet compression, the deacetylated KGMH120 particles in the matrix tablet may form crosslinks between the interparticle surfaces through solid-state intermolecular interactions. When water was introduced, the particles slowly absorbed water from the tablet surface, forming a hydrogel between and inside the particles (Figure 10). Therefore, the disintegration of the tablet was suppressed (Figure 5), and the entire tablet became a stable gel matrix, which controlled drug release as a sustained-release tablet. In contrast, the KGMH0 matrix tablet might not form crosslinks between the interparticle surfaces; therefore, when water was introduced, the individual particles rapidly absorbed water (Figure 10), and the tablet disintegrated from the pressure of the expanded particles (Figure 5). This result also suggests that the disintegration time of matrix tablets can be changed to freely control the drug release rate by varying the degree of grinding.

4. Conclusions

Optimal formulations of KGM containing 16 *w/w*% $Ca(OH)_2$ were prepared to develop a sustained drug release gel matrix tablet using a dry solid-state method without degrading the active ingredient of the drug, which occurs during wet granulation in the conventional method. Fine $Ca(OH)_2$ particles were uniformly dispersed in KGM by mechanochemical treatment, and the deacetylation reaction of KGM with fine $Ca(OH)_2$ particles in an alkaline sol may be accelerated by mechanochemical energy during solid grinding. It can be concluded that the compressed matrix tablets containing partially deacetylated KGM promoted the formation of strong gels, suppressed tablet disintegration, and controlled drug release following the Higuchi theory during the drug release test. These results indicate that the mechanochemical effect of KGM with $Ca(OH)_2$ can freely control the drug release mechanism and its rate by changing the water penetration rate and disintegration time of KGM matrix tablets.

5. Materials and Methods

5.1. Materials

Theophylline anhydrate (TH) bulk powder was obtained from Shizuoka Caffeine Co., Ltd. (Lot No. F22224, Shizuoka, Japan). Water-soluble dietary fiber (KGM) was obtained from Shimizu Chemical Co., Ltd. (Leorex®® RS, Mihara, Japan) as a food additive. Leorex®® RS (KGM) is a polysaccharide with β-1 binding between D-glucose and D-mannose (ratio of approximately 1:1.6). Its molecular weight was approximately 1 million or more, its degree of polymerization was approximately 6200, and its molecular length was R0 = 1.3 Å. The viscosity of the 1.0 *w/v*% aqueous solution was 36,900 mPa·s at 25 °C, as previously reported [29]. Calcium hydroxide ($Ca(OH)_2$) was obtained from Wako Pure Chemical Industries, Ltd. (Tokyo, Japan).

5.2. Grinding Treatment

The KGM formulation samples (KGML, KGMM, and KGMH) contained 5.2, 10.5, and 16.0 *w/w*% $Ca(OH)_2$, respectively. Ten grams of the weakly mixed KGM and $Ca(OH)_2$ powder samples were placed in a 250 mL agate ball mill pot containing 10 agate balls (∅15 mm × 10 mm) and were ground for 120 min at 200 rpm with a planetary ball mill crusher (P-6; Fritsch), respectively. The ground samples were labeled KGML120, KGMM120, and KGMH120. KGMH samples containing 16.0 *w/w*% $Ca(OH)_2$ were ground for 0, 10, 30, 60, and 120 min and labeled KGMH0, KGMH10, KGMH30, KGMH60, and KGMH120, respectively. The temperature at the outside of the ball mill under the almost same experimental conditions was about 5 °C higher than room temperature, as reported previously [32].

5.3. Scanning Electron Microscopy with Energy-Dispersive X-ray Spectroscopy

The morphologies of the surface and the interior of the adsorbent particles were evaluated by scanning electron microscopy (SEM) (JSM-7600F, Field Emission SEM, JEOL, Ltd., Tokyo, Japan) with an accelerating voltage of 10 kV and 250× to 1000× magnification. Elemental analysis was performed using energy-dispersive X-ray spectroscopy (EDX) (JED-2300, JEOL, Ltd., Tokyo, Japan) at an accelerating voltage of 8.0 kV.

5.4. Powder X-ray Diffraction (XRD)

Powder X-ray diffraction (XRD) profiles of the ground samples were obtained using a RINT 2100 Ultima III (Rigaku Co. Ltd., Tokyo, Japan). The measurements were performed at 2°/min using an X-ray tube with Cu Kα radiation at a voltage of 40 kV and a current of 40 mA.

5.5. Fourier-Transform Infrared (FT-IR) Spectroscopy

Fifty milligrams of the sample powder was softly ground and dispersed in 950 mg of KBr powder in an agate mortar with a pestle. The FT-IR spectra of the sample powders were measured with 200 scans of powder diffuse reflectance on an FT-IR spectrophotometer (FT-IR/4100, Jasco Co., Ltd., Tokyo, Japan) and corrected using the Kubelka–Munk equation.

5.6. Tablet Preparation

Ground KGM powder (9.5 g) containing various amounts of $Ca(OH)_2$ was placed with TH powder (0.5 g) in a plastic bag (1.0 L capacity) and mixed well by hand for 5 min. Then, 200 mg of the mixed KGM powder with 5 *w/w*% TH was placed in an 8 mm diameter die and pressed at 10.0 kN for 10 s with a flat-surface punch at a compression speed of 10 mm/min at about 25 °C using a compression/tension tester (Techno Graph TG-50kN, NMB Co., Ltd., Tokyo, Japan).

5.7. Drug Release Kinetics Measured by the Dissolution Test

Dissolution tests of the test tablets were performed in 900 mL of water in a 1000 mL capacity round-bottomed flask at 37.0 ± 0.5 °C with stirring at 100 rpm using a puddle in an automatic dissolution test apparatus (Type DT-610, JASCO Co., Ltd., Tokyo, Japan; Japanese Pharmacopeia 18). Dissolved drug concentrations were automatically measured by ultraviolet (UV) absorption at 271 nm at predetermined time intervals in quartz flow cells with a length of 10 mm using a UV–visible spectrophotometer [33]. All measurements were repeated thrice.

5.8. Water Penetration into Tablets

The rate of water penetration into the tablets was measured using a custom-modified glass pipette, as reported by Takayama et al. [34]. After placing the tablets on the glass filter of the modified glass pipette filled with water, the amount of water absorbed was measured at predetermined times.

Author Contributions: Conceptualization, M.O.; methodology, F.O. and Y.H.; software, F.O. and Y.H.; validation, F.O.; formal analysis, F.O. and Y.H.; investigation, F.O. and Y.H.; resources, M.O.; data curation, T.S. and M.O.; writing—original draft preparation, F.O. and M.O.; writing—review and editing, T.S.; visualization, M.O.; supervision, T.S. and M.O.; project administration, M.O.; funding acquisition, T.S. All authors have read and agreed to the published version of the manuscript.

Funding: This research was supported, in part, by a grant from the Cooperative Research Project Program of the Research Institute of Electronics, Shizuoka University.

Data Availability Statement: Not applicable.

Acknowledgments: The authors acknowledge Shimizu Chemical Co. Ltd. (Mihara, Japan) for providing KGM and Leorex® RS as the sample materials.

Conflicts of Interest: The authors declare that they have no conflict of interest.

References

1. Huang, X.; Brazel, C.S. On the importance and mechanisms of burst release in matrix-controlled drug delivery systems. *J. Control. Release* **2001**, *73*, 121–136. [CrossRef]
2. Colombo, P.; Bettini, R.; Santi, P.; Peppas, N.A. Swellable matrices for controlled drug delivery: Gel-layer behaviour, mechanisms and optimal performance. *Pharm. Sci. Technol. Today* **2000**, *3*, 198–204. [CrossRef]
3. Kiortsis, S.; Kachrimanis, K.; Broussali, T.; Malamataris, S. Drug release from tableted wet granulations comprising cellulosic (HPMC or HPC) and hydrophobic component. *Eur. J. Pharm. Biopharm.* **2005**, *59*, 73–83. [CrossRef]
4. Kumar, M.N.V.R.; Kumar§, N. Polymeric Controlled Drug-Delivery Systems: Perspective Issues and Opportunities. *Drug Dev. Ind. Pharm.* **2001**, *27*, 1–30. [CrossRef]
5. Levina, M.; Rajabi-Siahboomi, A.R. The Influence of Excipients on Drug Release from Hydroxypropyl Methylcellulose Matrices. *J. Pharm. Sci.* **2004**, *93*, 2746–2754. [CrossRef] [PubMed]
6. Melia, C.D. Hydrophilic matrix sustained release systems based on polysaccharide carriers. *Crit. Rev. Ther. Drug Carr. Syst.* **1991**, *8*, 395–421.
7. Bown, D. *Aroids: Plants of the Arum Family*; Timber Press: Portland, OR, USA, 2010.
8. Talbott, S.M. *A Guide to Understanding Dietary Supplements*, 1st ed.; Haworth Press: Philadelphia, PA, USA, 2003. [CrossRef]
9. Yang, D.; Yuan, Y.; Wang, L.; Wang, X.; Mu, R.; Pang, J.; Xiao, J.; Zheng, Y. A Review on Konjac Glucomannan Gels: Microstructure and Application. *Int. J. Mol. Sci.* **2017**, *18*, 2250. [CrossRef] [PubMed]
10. Keithley, J.; Swanson, B. Glucomannan and obesity: A critical review. *Altern. Ther. Health Med.* **2005**, *11*, 30–34.
11. Shevkar, B.; Ahirrao, S.; Bhavsar, G.; Patel, A.; Rajurkar, V.; Amale, P. Konjac Glucomannan Matrix Tablet for Extended Release of Diclofenac Sodium. *Int. J. Adv. Pharm. Sci.* **2014**, *5*, 2098–2108.
12. Alvarez-Manceñido, F.; Landin, M.; Martínez-Pacheco, R. Konjac glucomannan/xanthan gum enzyme sensitive binary mixtures for colonic drug delivery. *Eur. J. Pharm. Biopharm.* **2008**, *69*, 573–581. [CrossRef]
13. Annable, P.; Williams, P.A.; Nishinari, K. Interaction in Xanthan-Glucomannan Mixtures and the Influence of Electrolyte. *Macromolecules* **1994**, *27*, 4204–4211. [CrossRef]
14. Goycoolea, F.M.; Richardson, R.K.; Morris, E.R.; Gidley, M.J. Stoichiometry and Conformation of Xanthan in Synergistic Gelation with Locust Bean Gum or Konjac Glucomannan: Evidence for Heterotypic Binding. *Macromolecules* **1995**, *28*, 8308–8320. [CrossRef]
15. Wang, K.; He, Z. Alginate–konjac glucomannan–chitosan beads as controlled release matrix. *Int. J. Pharm.* **2002**, *244*, 117–126. [CrossRef]
16. Boldyrev, V. Mechanical activation of solids and its application to technology. *J. Chim. Phys.* **1986**, *83*, 821–829. [CrossRef]
17. Hattori, Y.; Takaku, T.; Otsuka, M. Mechanochemical effect on swelling and drug release of natural polymer matrix tablets by X-ray computed tomography. *Int. J. Pharm.* **2018**, *539*, 31–38. [CrossRef]
18. Takacs, L. What Is Unique About Mechanochemical Reactions? *Acta Phys. Pol. A* **2014**, *126*, 1040–1043. [CrossRef]
19. Doi, N.; Yamauchi, Y.; Ikegami, R.; Kuzuya, M.; Sasai, Y.; Kondo, S.-I. Photo-responsive polymer micelles from o-nitrobenzyl ester-based amphiphilic block copolymers synthesized by mechanochemical solid-state copolymerization. *Polym. J.* **2020**, *52*, 1375–1385. [CrossRef]
20. Kobayashi, M.; Hattori, Y.; Sasaki, T.; Nishizawa, J.-I.; Otsuka, M. Characteristic Evaluation of the Pseu-do-Polymorphism of Amorphous Atorvastatin Calcium Hydrates by Terahertz Spectroscopy. *Colloids Surf. A Physi-Cochem. Eng. Asp.* **2021**, 127551. [CrossRef]

21. Otsuka, Y.; Ito, A.; Takeuchi, M.; Sasaki, T.; Tanaka, H. Effects of Temperature on Terahertz Spectra of Caf-feine/Oxalic Acid 2:1 cocrystal and Its Solid-State Density Functional Theory. *J. Drug Deli Sci. Technol.* **2020**, *56*, 101215. [CrossRef]
22. Otsuka, Y.; Ito, A.; Takeuchi, M.; Tanaka, H. Dry Mechanochemical Synthesis of Caffeine/Oxalic Acid Cocrystals and Their Evaluation by Powder X-Ray Diffraction and Chemometrics. *J. Pharm. Sci.* **2017**, *106*, 3458–3464. [CrossRef]
23. Takaku, T.; Hattori, Y.; Sasaki, T.; Sakamoto, T.; Otsuka, M. Evaluation of swelling properties and drug release from mechanochemical pre-gelatinized glutinous rice starch matrix tablets by near infrared spectroscopy. *J. Near Infrared Spectrosc.* **2021**, *29*, 92–101. [CrossRef]
24. Higuchi, T. Mechanism of sustained-action medication. Theoretical analysis of rate of release of solid drugs dispersed in solid matrices. *J. Pharm. Sci.* **1963**, *52*, 1145–1149. [CrossRef]
25. Wu, C.; Peng, S.; Wen, C.; Wang, X.; Fan, L.; Deng, R.; Pang, J. Structural characterization and properties of konjac glucomannan/curdlan blend films. *Carbohydr. Polym.* **2012**, *89*, 497–503. [CrossRef]
26. Rey, C.; Collins, B.; Goehl, T.; Dickson, I.R.; Glimcher, M.J. The carbonate environment in bone mineral: A resolution-enhanced fourier transform infrared spectroscopy study. *Calcif. Tissue Int.* **1989**, *45*, 157–164. [CrossRef] [PubMed]
27. Merck Home Page, IR Spectrum Table & Chart. Available online: https://www.sigmaaldrich.com/JP/ja/technical-documents/technical-article/analytical-chemistry/photometry-and-reflectometry/ir-spectrum-table (accessed on 25 December 2021).
28. National Institute of Standards and Technology (NIST). Chemistry WebBook, SRD 69, Carbon Dioxide. Available online: https://webbook.nist.gov/cgi/cbook.cgi?ID=C124389&Units=SI (accessed on 25 December 2021).
29. Home Page of Shimizu Chemical Co. Ltd. General Food/Functional Food Material "Leorex Series". Available online: https://www.ipros.jp/product/detail/2000524072/ (accessed on 26 December 2021).
30. Zhang, T.; Li, Z.; Wang, Y.; Xue, Y.; Xue, C. Effects of konjac glucomannan on heat-induced changes of physicochemical and structural properties of surimi gels. *Food Res. Int.* **2016**, *83*, 152–161. [CrossRef]
31. Du, X.; Li, J.; Chen, J.; Li, B. Effect of degree of deacetylation on physicochemical and gelation properties of konjac glucomannan. *Food Res. Int.* **2012**, *46*, 270–278. [CrossRef]
32. Otsuka, M.; Matsumoto, T.; Kaneniwa, N. Effect of environmental temperature on polymorphic solid-state transformation of indomethacin during grinding. *Chem. Pharm. Bull.* **1986**, *34*, 1784–1793. [CrossRef] [PubMed]
33. Hattori, Y.; Sugata, M.; Kamata, H.; Nagata, M.; Nagato, T.; Hasegawa, K.; Otsuka, M. Real-time monitoring of the tablet-coating process by near-infrared spectroscopy—Effects of coating polymer concentrations on pharmaceutical properties of tablets. *J. Drug Deliv. Sci. Technol.* **2018**, *46*, 111–121. [CrossRef]
34. Takayama, K.; Nambu, N.; Nagai, T. Factors affecting the dissolution of indomethacin dispersed in various water-soluble polymers. *Chem. Pharm. Bull.* **1982**, *30*, 673–678. [CrossRef]

gels

MDPI

Review

Applications and Mechanisms of Stimuli-Responsive Hydrogels in Traumatic Brain Injury

Xingfan Li [1], Linyan Duan [1], Mingyue Kong [2], Xuejun Wen [3], Fangxia Guan [1,*] and Shanshan Ma [1,2,*]

[1] School of Life Sciences, Zhengzhou University, Zhengzhou 450001, China; lixingfan0929@163.com (X.L.); duanlinyan4881@163.com (L.D.)

[2] NHC Key Laboratory of Birth Defects Prevention, Henan Key Laboratory of Population Defects Prevention, Zhengzhou 450002, China; mykong2021@163.com

[3] Department of Chemical and Life Science Engineering, School of Engineering, Virginia Commonwealth University, Richmond, VA 23284, USA; xwen@vcu.edu

* Correspondence: guanfx@zzu.edu.cn (F.G.); mss@zzu.edu.cn (S.M.)

Abstract: Traumatic brain injury (TBI) is a global neurotrauma with high morbidity and mortality that seriously threatens the life quality of patients and causes heavy burdens to families, healthcare institutions, and society. Neuroinflammation and oxidative stress can further aggravate neuronal cell death, hinder functional recovery, and lead to secondary brain injury. In addition, the blood–brain barrier prevents drugs from entering the brain tissue, which is not conducive to the recovery of TBI. Due to their high water content, biodegradability, and similarity to the natural extracellular matrix (ECM), hydrogels are widely used for the delivery and release of various therapeutic agents (drugs, natural extracts, and cells, etc.) that exhibit beneficial therapeutic efficacy in tissue repair, such as TBI. Stimuli-responsive hydrogels can undergo reversible or irreversible changes in properties, structures, and functions in response to internal/external stimuli or physiological/pathological environmental stimuli, and further improve the therapeutic effects on diseases. In this paper, we reviewed the common types of stimuli-responsive hydrogels and their applications in TBI, and further analyzed the therapeutic effects of hydrogels in TBI, such as pro-neurogenesis, anti-inflammatory, anti-apoptosis, anti-oxidation, and pro-angiogenesis. Our study may provide strategies for the treatment of TBI by using stimuli-responsive hydrogels.

Keywords: stimuli-responsive hydrogels; traumatic brain injury; biomedical applications; therapeutic mechanisms

1. Introduction

Traumatic brain injury (TBI) is a major public health problem with an increasing global prevalence and high mortality that seriously threatens the life quality of patients and causes heavy burdens to families, healthcare institutions, and society [1,2]. The global incidence of TBI is about 870–1000 cases per 100,000 people, and an estimated 64–74 million individuals worldwide suffer from TBI every year [3]. According to the Glasgow Coma Score (GCS), TBI can be divided into mild injury, moderate injury, and severe injury [4]. The pathological process of TBI includes primary injury and secondary injury. Primary brain injury generally refers to the direct damage to nerve tissue caused by external forces such as skull deformation, intracranial hemorrhage, brain tissue contusion, axonal injury, etc. [5]. Secondary brain injury occurs several minutes after injury and lasts for a long time, and includes the disturbance of ion homeostasis, excitatory toxicity, oxidative stress, inflammation, and cell death [6]. Compared with primary brain injury, secondary brain injury brings more serious harm [7]. Over the past few decades, the therapeutic outcomes for TBI patients have substantially improved due to the increasing number of basic, nursing, and clinical studies related to TBI [8]. However, due to the existence of the blood–brain barrier (BBB), most compounds or drugs cannot effectively cross the BBB to reach the injury site, thereby affecting their efficacy [9]. Therefore, there is still no effective treatment for TBI till now. Research has focused on alleviating secondary brain injury, enhancing neural network reorganization, and promoting functional recovery.

The boom in biomaterials has provided new techniques and strategies to help overcome the BBB [10]. Hydrogels are widely used for the delivery and release of various therapeutic agents (drugs, natural extracts, and cells, etc.) due to their high water content, biodegradability, and similarity to the natural extracellular matrix (ECM); these exert beneficial therapeutic efficacy in many diseases, such as TBI [11]. In addition to these features, injectable hydrogels also have special advantages, such as minimal invasiveness, effective defect filling, etc. [12]. Due to the variability and adjustability of the microenvironment, hydrogels need to be responsive to biophysical and biochemical signals, which is beneficial for function repair. Therefore, researchers have developed stimuli-responsive hydrogels [13].

Stimuli-responsive hydrogels can undergo reversible or irreversible changes in properties, structures, and functions in response to internal/external stimuli (e.g., magnetic fields, light, ultrasound, mechanical force) or physiological/pathological environmental stimuli (e.g., enzymatic reactions, temperature), and further gain some specific effects (e.g., sustained release, transport, alignment) [14,15]. Furthermore, stimuli-responsive hydrogels can rapidly detect and respond to stimuli without damaging normal cells and tissues, thereby greatly improving the therapeutic efficacy and reducing side effects [16,17]. Several articles have discussed the use of bioactive matrices (stem cells, hydrogels, and their combined systems) in central nervous system diseases, such as TBI, but they focused on hydrogel-related stem cell therapy or bioactive matrices in neural repair [10,11]. In this paper, we reviewed the characteristics, synthesis mechanisms of stimuli-responsive hydrogels, and further analyzed the applications and different therapeutic effects of hydrogels in TBI, such as anti-inflammatory, anti-oxidation, anti-apoptosis, pro-neurogenesis, and pro-angiogenesis. Our study may provide strategies for the recovery of TBI by using stimuli-responsive hydrogels, which may facilitate the synthesis of more intelligent biomaterials in the future to satisfy applications in TBI.

2. Stimuli-Responsive Hydrogels

In the past decades, stimuli-responsive hydrogels have received increasing attention as "smart hydrogels" with various advantages and functions [15,17,18]. Here, we summarized the common stimuli-responsive hydrogels, and described their synthetic materials, gelation mechanisms, and properties.

2.1. Thermo-Responsive Hydrogels

Thermo-responsive hydrogels, one of the most widely studied stimuli-responsive hydrogels, usually contain different hydrophobic groups (such as methyl, ethyl, and propyl groups), which can change their physical and conformational properties with the change of temperature [19,20]. Biomaterials used in the synthesis of thermo-responsive hydrogels can be divided into natural polymers and synthetic polymers. Natural polymers include polysaccharides (cellulose, chitosan, and xyloglucan, etc.), proteins (gelatin), and so on. Synthetic polymers include poly(*N*-isopropylacrylamide) (pNIPAAm), poly (ethylene oxide)-poly (propylene oxide)-poly (ethylene oxide) copolymer (PEO-PPO-PEO), polyethylene glycol ester (PEG)/biodegradable polyester, poly (organophosphoronitrile), and dimethylamine (ethyl methacrylate), etc. [21].

The gelling principle of thermoresponsive hydrogels is that when the temperature is lower than the lower critical solution temperature (LCST), the hydrophilic groups in the system form hydrogen bonds with hydrophilic molecules in the environment, which increases the solubility of hydrogels and can be in situ injected in the form of a solution. With the increase in temperature, the hydrogen bond force gradually weakens, while the interaction between hydrophobic groups in the system gradually strengthens, and the solubility decreases (Figure 1). After reaching the LCST, a gel is formed [20,22]. Therefore, thermoresponsive hydrogels can be divided into two systems with LCST and upper critical solution temperature (UCST) according to different properties. For LCST hydrogels, the solubility is nonlinear and inversely proportional to temperature [23]. When the ambient temperature is above the LCST, the system is transformed from a solution to a gel. Due to this mechanism, the hydrogel solution becomes liquid at room temperature and becomes a gel at physiological temperature, making the hydrogel injectable [24]. UCST hydrogels have the opposite properties to LCST hydrogels. When the ambient temperature is below the UCST, the system changes from a solution to a gel. As such hydrogels often require high-temperature injection, which may lead to denaturation of the loaded drugs or proteins, they are not widely used in the biomedical field [25].

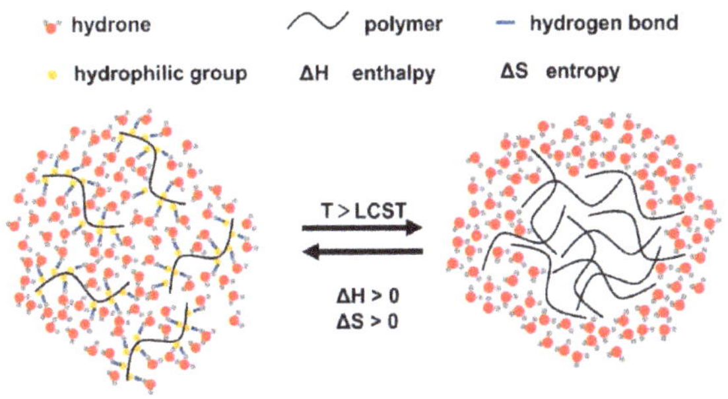

Figure 1. Schematic of thermoresponsive hydrogels (LCST hydrogels). With the increase in temperature, the hydrogen bond force between polymers and water molecules is weakened, and the hydrophobic effect is enhanced, forming the hydrogel.

Thermoresponsive hydrogels are widely used in drug release and cell encapsulation [26]. Pan et al. loaded black phosphorus nanosheets (BPNs) on a platelet-rich plasma (PRP)-chitosan thermoresponsive hydrogel. With an increase in temperature, the proton transfer of the amino group on the chitosan molecule reduced the electrostatic repulsion between chitosan molecules and increased the aggregation of hydrophobic groups, thus forming the hydrogel. Under the action of a photothermal effect, this thermoresponsive hydrogel controlled the slow release of BPNs degradation products, which is beneficial to the treatment of rheumatoid arthritis (RA) [27]. In addition, the poly[(propylenesulfide)-block-(*N*,*N*-dimethylacrylamide)-block-(*N*-isopropylacrylamide)] (PPS-b-PDMA-b-PNIPAAM) hydrogel was a thermoresponsive hydrogel. At 25 °C, it assembled into micelles that included a hydrophobic PPS core and PNIPAAM on the outer corona. When the temperature exceeded the LCST of PNIPAAM, the micelles sharply transitioned into stable, hydrated gels that demonstrated ROS-dependent drug release [28]. Collectively, thermoresponsive hydrogels are particularly promising for future applications in drug delivery. However, the translation and application of this hydrogel at the clinical level still require more in-depth research to improve their biocompatibility, safety, and experimental reproducibility.

2.2. *Photoresponsive Hydrogels*

Generally speaking, photoresponsive hydrogels mainly respond to ultraviolet (UV), visible, and near infrared (NIR) light [29]. Photoresponsive hydrogels are composed of polymer networks including photosensitive moieties with reversible crosslinking and photothermal capability [30]. After being stimulated by light, the photosensitive part captures the light signal and converts it into a chemical signals through the reactions of isomerization, cleavage and dimerization, causing changes in the structure or properties, such as shrinking swelling or crosslinking [31].

There are three gelation mechanisms for photoresponsive hydrogels. (i) The chromophores groups are added into the whole hydrogel system. These chromophores have the ability to convert light energy into heat energy, so that the hydrogen bond force in the system will weaken and then the structure or properties of the hydrogels will change; this is similar to thermoresponsive hydrogels (Figure 2A) [32]. (ii) The chromophore groups incorporated into the hydrogel matrix undergo isomerization, photocleavage, or photooxidation after photo/light stimulation, leading to changes in their physical or chemical properties, thus affecting the structure of hydrogels and even causing the hydrogels to become uncrosslinked (Figure 2B) [33]. (iii) The photosensitive molecules dissociate after being stimulated by light, resulting in a large number of ions. The ionic concentration gradient generated inside and outside the gel can enhance the osmotic pressure, causing the gel to swell (Figure 2C) [34]. For example, Wang et al. prepared a polydopamine/collagen/silk fibroin photoresponsive hydrogel. Polydopamine, which could be cross-linked to collagen or silk fibroin through catechin groups, accelerated hydrogel formation and imparted superior photothermal properties to the hydrogel. Under the irradiation of NIR light, it released thrombin, thus promoting blood coagulation, preventing angiogenesis, and effectively inhibiting the recurrence and metastasis of triple-negative breast cancer [35]. In addition, Wang et al. prepared an injectable redox and light-responsive bio-inspired MnO_2 hybrid (BMH) hydrogel through non-covalent self-assembly and MnO_2 nanosheets mediated the covalent oxidative polymerization of the catechol-functionalized chitosan. Under the action of NIR, this light-responsive BMH hydrogel inhibited melanoma growth and had long-term antibacterial properties through the controllable release of DOX and photothermal therapy [36].

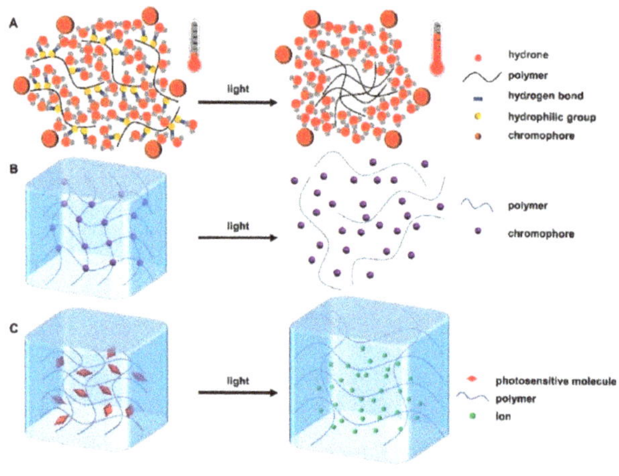

Figure 2. Three gelation mechanisms of photoresponsive hydrogels. When stimulated by light, (**A**) chromophore converts light energy into heat energy, encouraging polymers to aggregate into gels. (**B**) The chromophore structure and properties in the micelles change, causing the hydrogel to become uncrosslinked. (**C**) Photosensitive molecules dissociate, resulting in internal and external osmotic pressure, and causing hydrogel expansion.

The advantage of photoresponsive hydrogels is that the photo/light-energy stimulation is controllable and drug delivery can be precisely targeted in time and space [31]. However, there are also certain limitations. First, UV is a high-energy electromagnetic wave with carcinogenic potential and poor tissue penetration, which makes it difficult to treat deeply injured sites. The current solution is to incorporate upconverting nanoparticles (UCNPs), which can convert NIR energy into UV energy [37]. Besides, as the photosensitive groups necessary for photoresponsive hydrogels may be toxic, better biocompatible materials need to be further investigated in the future [38].

2.3. Magnetic-Responsive Hydrogels

Magnetic-responsive hydrogels are synthesized by incorporating magnetic nanoparticles into crosslinked polymers; these particles can change their structures, properties, and functions in response to magnetic fields [39]. In addition to Fe_3O_4, other common magnetic nanoparticles include $Fe_3O_4@SiO_2$, $Zn_{0.47}Mn_{0.53}Fe_2O_4$, γ-Fe_2O_3, and Co_3O_4 [40]. There are three common synthesis methods for magnetic-responsive hydrogels: co-precipitation or embedding, grafting, and blending [41].

Under the stimulation of magnetic field, the magnetic nanoparticles in the nanohydrogel can self-assemble to form an ordered structure, making the surface of the hydrogel anisotropic (Figure 3A) and thereby promoting cell proliferation, neurogenesis, signal transduction and ECM regeneration [42]. Besides, the macroscopic magnetic-responsive hydrogels perform functions such as magnetron release, magnetocaloric effect, and optical property adjustment under magnetic field stimulation (Figure 3B) [43]. Magnetic actuation does not require contact stimulation and can be operated remotely, which is suitable for complex environments [44,45]. Under the stimulation of an alternating magnetic field (AMF), the magnetic nanoparticles vibrate and the temperature increases, which accelerates the movement of drug molecules and the degradation of polymers, thus promoting drug release (Figure 3C). In addition, it is also used in combination with chemotherapy to treat tumors [46]. However, magnetic nanoparticles tend to aggregate under the action of a magnetic field, resulting in reduced pores between nanoparticles and, in turn, limiting the release of drugs [47]. In addition, under the action of a magnetic field, the magnetic-responsive hydrogel undergoes a magnetocaloric effect, which may lead to a shift in the diffraction peaks and thus the discoloration of the hydrogel (Figure 3D) [48]. For example, Zhang et al. prepared hydrogels using hyaluronic acid (HA) and PEG, and then added magnetic nanoparticles and type II collagen to the gel. This hydrogel could respond to an external magnetic field and travel to the tissue defect sites in physiological fluids under remote magnetic guidance [49]. To sum up, magnetic-responsive hydrogels have the advantages of low invasiveness, good tissue permeability, remote operation, rapid response and high controllability, but they also have some limitations, such as low mechanical strength, material toxicity, and low biocompatibility [41,43].

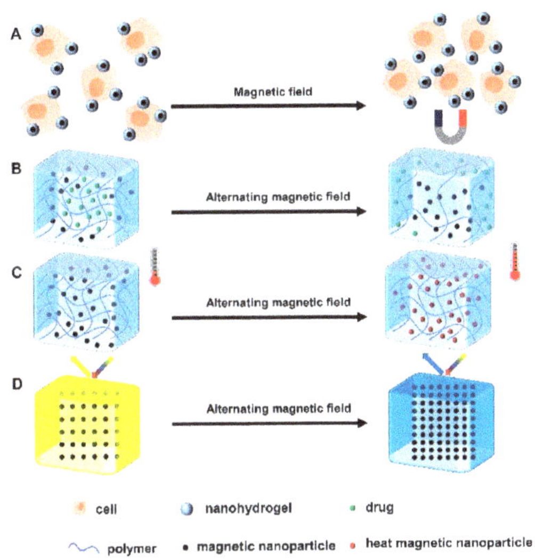

Figure 3. Four gelation mechanisms of magnetic-responsive hydrogels. (**A**) Under the influence of a magnetic field, the cells attached to the magnetic nanoparticle hydrogel clustered in a specific direction. Under the action of alternating magnetic field, (**B**) magnetic nanoparticles in hydrogels move, resulting in a decrease in crosslinking density and the release of loaded drugs. (**C**) The motion of magnetic nanoparticles generates heat, which leads to a temperature rise in the system. (**D**) Due to the magnetothermal effect, the distance between the magnetic nanoparticles decreases and the diffraction peak shifts, leading to the discoloration of the hydrogel.

2.4. Electroresponsive Hydrogels

Electroresponsive hydrogels are synthesized by adding electroactive materials into hydrogel matrix. Common electroactive materials are divided into two categories: natural materials (such as metals) and inorganic vs. organic conductive materials [such as graphene, carbon nanotubes, polypyrrole (PPy), polyaniline (PANI), and poly3,4-ethylenedioxythiophene (PEDOT)] [50,51]. Electroresponsive hydrogels can respond to electric field stimulation to modify their structures and properties and perform specific functions.

Under the stimulation of an electric field, the charges in the gel solution move directionally, and the electroactive materials undergo electrochemical reactions, resulting in structural changes to the hydrogel, thereby achieving the purpose of regulating drugs release [52,53]. In addition, electroresponsive hydrogels can promote signal transduction, thereby affecting cell growth and proliferation [54]. Moreover, electroresponsive hydrogels can also affect cellular processes, such as angiogenesis, neurogenesis, or cardiogenesis under electric field stimulation, showing good potential in biomedical engineering [52,55]. For instance, chitosan can be electro-activated by aniline–pentamer segments, and an injectable electroactive hydrogel based on pluronic–chitosan/aniline–pentamer clearly improved hippocampal-dependent learning and memory performance [56]. However, the biocompatibility, degradability, and stability of electroresponsive hydrogels still need to be improved [51].

2.5. Bioresponsive Hydrogels

Bioresponsive hydrogels can change their structures or properties (e.g., swelling, contraction, dissociation) in the presence of specific biological factors or activity and concentration changes [57]. Bioresponsive hydrogels are mainly divided into two categories: hydrogels coated with specific enzyme reaction substrates [58] and hydrogels coated with specific antigens or antibodies [59].

Enzyme-responsive hydrogels are highly selective, using a specific substrate as a crosslinking agent. In the presence of specific enzymes, the substrate undergoes enzymatic hydrolysis resulting in the cross-linking of hydrogels to change their structures or properties [22]. Enzyme-responsive hydrogels system must satisfy three conditions. First, the polymer network must contain specific enzyme substrates. Second, these substrates should be close to the active center of the enzyme. Third, the enzymatic reactions of these substrates can cause changes in the structure or properties of hydrogels [60]. Shen et al. used the polymer PLGA-PEI-MPEG (PPP) as a carrier and loaded it with activated cell-penetrating peptides (ACPP) and etanercept (ET) to construct an ET@PPP-ACPP hydrogel. ACPP and matrix metalloproteinases (MMPs) have affinities that can give hydrogels the ability to target lesions and improve the therapeutic effect in spinal cord injury (SCI) [61]. In addition, Katayama et al. synthesized polymer–peptide conjugates, NIPAM−PEP and NIPAM−PEPEP, which are graft-type copolymers, each synthesized by a methacryloyl monomer with radical copolymerization. The polymer was coated with a substrate peptide of protein kinase A (PKA) in response to PKA phosphorylation. Subsequently, the LCST of the polymer was elevated and decomposed under activation by PKA, releasing the drugs into it [62].

Antigen-responsive hydrogels covalently bind antigens and antibodies into hydrogels. Due to the non-covalent binding between antigen and antibody, the antigen-responsive hydrogels are reversibly crosslinked. The free antigens presented in the environment can compete with the antigens in the hydrogel (Figure 4), leading to the weakness of the interaction between the antigen and the antibody in the hydrogel, thereby reducing the crosslinking density of the hydrogel and causing swelling or osmosis [63]. Ye et al. prepared a bioresponsive hydrogel using antigen–antibody binding sites as crosslinking points to prepare IgG sensing gratings, which can detect human immunoglobulin G (H-IgG) in solution with high specificity and sensitivity [64]. In short, bioresponsive hydrogels generally respond to a specific substance (such as enzymes, antigens, antibodies, etc.) with high biocompatibility, and have good application prospects in the biomedical field.

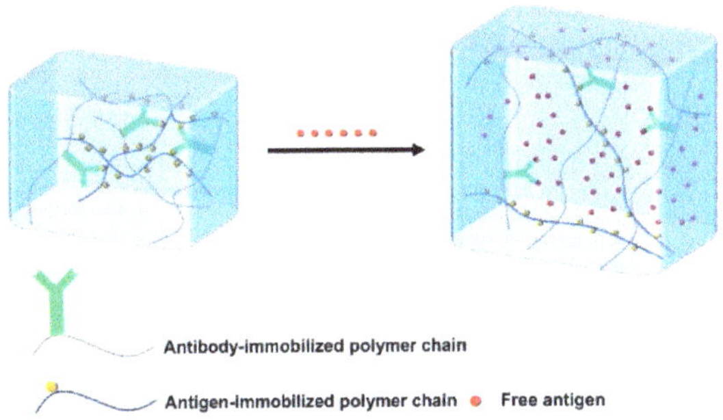

Figure 4. Schematic of the antigen-responsive hydrogels. The free antigens in the environment compete with the binding antigens in the hydrogel, resulting in the decrease of crosslinking density and swelling of the hydrogel.

3. Applications of Stimuli-Responsive Hydrogels in TBI

TBI-mediated chronic inflammation and oxidative stress in the microenvironment further induce neuronal death and lead to persistent neuronal damage, which often cause motor dysfunction, speech impairment, and intellectual disability [65,66]. Traditional drug delivery is hampered by the presence of the BBB, making it difficult to reach the target area [67]. Hydrogels have good biocompatibility and injectability, which allows them to injected in situ and adapt to the irregularity of the injury site, and have a wide range of applications in TBI [68]. Stimuli-responsive hydrogels can respond to the inflammatory and oxidative stress microenvironment at the injury site to regulate inflammation, edema, and oxidative stress levels, and promote neurogenesis and functional recovery [11].

Yao et al. synthesized a thermoresponsive chitosan–cellulose hyaluronic acid/β-glycerophosphate (CS-HEC-HA/GP) hydrogel, which was liquid below 25 °C but could rapidly transform into a hydrogel at 37 °C. Moreover, this CS-HEC-HA/GP hydrogel loaded with human umbilical cord mesenchymal stem cells (hUC-MSCs) enhanced the retention and survival of encapsulated hUC-MSCs, promoted neurogenesis, and inhibited cell apoptosis, thereby accelerating brain structural remodeling and neurological functional recovery in TBI rats [69]. Ferulic acid (FA)-loaded injectable thermoresponsive chitosan/gelatin/β-glycerophosphate (C/G/GP) hydrogel formed a gel after about 15 min at room temperature, but only 135 s at body temperature, which ensured the in situ gelation of the C/G/GP hydrogels and effectively improved the therapeutic effect of FA on TBI [70]. Hsieh et al. synthesized two thermoresponsive water-based biodegradable polyurethane hydrogels (PU1 and PU2) loaded with neural stem cells (NSCs) through 3D bioprinting technology, which could form hydrogels at 37 °C and rescued the neural function of adult zebrafish with TBI [71]. In addition, an injectable, post-trauma ROS-responsive hydrogel with embedded curcumin (Cur) could respond to ROS stimulation in the microenvironment and reduce the oxidative stress level in the TBI lesion to decrease brain edema and promote nerve regeneration after TBI [72]. Remotely activatable ECM-mimetic chitosan/collagen-based hydrogel incorporated with mesenchymal stem cell (MSC)-membrane-coated black phosphorus (BP) exerted mild photothermal effects at the implantation site under NIR illumination, which could accelerate the healing of cranial defects [73]. Moreover, an ECM-mimetic neuroprotective sulfo-functionalized peptide hydrogel could autodegrade to generate the neuroprotective hexapeptide in the presence of MMP9 enzyme, and help repair the injured neurons in TBI lesions [74]. These studies highlighted the potential of stimuli-responsive hydrogels for TBI applications due to the functional recovery or precise brain delivery. However, new approaches and further efforts are necessary to improve the biocompatibility and reduce toxicity of stimuli-responsive hydrogels, and expand their applications in preclinical studies, thus laying the foundation for their clinical applications in the future.

4. Therapeutic Mechanisms of Stimuli-Responsive Hydrogels in TBI

4.1. Pro-Neurogenesis

Neurogenesis is a multi-step process referring to the regeneration or repair of neural tissue, which plays vital roles in learning and memory function, recovery after neuronal injury, and neural plasticity [75]. Injectable hydrogels can encapsulate various cells, drugs, or cytokines, secrete a variety of growth factors, and improve the microenvironment in the lesions to promote neurogenesis and improve neurological function after TBI.

Ma et al. found that a sodium alginate/collagen/stromal cell-derived factor-1 hydrogel loaded with bone marrow-derived mesenchymal stem cells (BMSCs/SA/Col/SDF-1) promoted neurological function recovery after TBI by enhancing hippocampal neurogenesis, as evidenced by higher $EdU^+/NeuN^+$ cells and the expression of neurotrophic factors [76]. A dual-enzymatically crosslinked injectable gelatin hydrogel loaded with BMSC improved neurological function recovery of TBI rats and promoted the proliferation of endogenous neural cells by inhibiting apoptosis and secreting neurotrophic factors [77]. Zheng et al. developed an injectable hydrogel containing polydopamine/stromal cell-derived factor-1α (PDA/SDF-1α) nanoparticles using imidazole-modified gelatin methacrylate (GelMA-imid). The results showed that the GelMA-imid/SDF-1α hydrogel loaded with human amniotic mesenchymal stromal cells (hAMSCs) facilitated the regeneration of endogenous neural cells with high expression of neuronal nuclei (NeuN), neuron-specific enolase (NSE), and brain-derived neurotrophic factor (BDNF), which had great potential for the recovery from TBI [78]. Besides, Shi et al. inoculated hUC-MSCs and activated astrocytes into a specific self-assembled peptide hydrogel RADA16-BDNF scaffold (R-B-SPH scaffold). Further, the R-B-SPH scaffold could promote neuronal regeneration and neural network reconstruction after TBI [79]. After co-culture, the expression of transcription factor SOX2, neurotrophic factor BDNF, and its receptor TRKB were upregulated. In addition, SDF-1 and hypoxia-inducible factor (HIF-1) were highly expressed at the injured site, which promoted nerve regeneration and differentiation in the injured site. Furthermore, the implantation of agarose hydrogel, which can release the Fas ligand (FasL) to induce apoptosis of cytotoxic $CD8^+$ T cells in the lesion, significantly increased the expression of the nerve growth factors (NGF), BDNF and insulin-like growth factor (IGF), and promoted neural regeneration by activating the PI3K–AKT pathway in a rat model of TBI [80]. Thus, these data demonstrated that hydrogels could promote neurogenesis for repairing brain injury through encapsulated stem cells or proteins.

4.2. Anti-Inflammation

Inflammation is a dynamic response of the immune system to external stimuli or infection. However, when it is uncontrolled, cytokine storms (IL-1β, TNF-α, IL-6, etc.) occur, leading to the occurrence of various inflammatory diseases, including neurological diseases [81]. TBI can induce microglial activation and microglia polarization (M1 and M2) in the early stage, followed by macrophage infiltration. At the same time, pro-inflammatory factors induced by M1 microglia, such as interleukins (IL-1, IL-12, IL-18), gamma-interferon (IFN-γ), and tumor necrosis factor-α (TNF-α), and anti-inflammatory factors released by M2 microglia, such as interleukins (IL-4, IL-10, IL-13), α-interferon (IFN-α), etc., have competitive effects that can respectively promote or inhibit neuroinflammatory responses [5]. Therefore, neuroinflammation has a double-edged role in the functional remodeling of TBI. On the one hand, the inflammatory response can promote the response to external stimuli and promote the removal of cell debris, which is beneficial to the outcome after TBI; on the other hand, persistent chronic inflammation further accelerates neuronal death and progressive neurodegeneration [82]. Therefore, anti-inflammatory therapy is one of the strategies for the treatment of TBI, but attention should be paid to the time window of treatment and the course of the disease. Hydrogels can controllably release stem cells or anti-inflammatory drugs, improve drug availability, and have broad application prospects in the treatment of TBI [10].

Maclean et al. developed a multifunctional self-assembled peptide hydrogel, Fmoc-DIKVA, and encapsulated the anti-inflammatory agent fucoidan to treat TBI, which reduced primary glial scarring, significantly changed astrocyte morphology, and attenuated "reactive" astrocytosis at the injured site after 7 days after treatment for TBI [83]. Besides, the localized delivery of dexamethasone at a low dose via hyaluronic acid/PEG-bis-(acryloyloxyacetate) (PEG-bis-AA) composite hydrogels significantly reduced microglial activation/macrophage infiltration-mediated neuroinflammation and cell apoptosis, and promoted the recovery of motor function in TBI [84]. In addition, the transplantation of human neural stem/progenitor cells (hNS/PCs) and human adipose-derived stromal/stem cells (hADSCs)-loaded PuraMatrix (PM) hydrogels obviously inhibited the expression of TNF-α, IL-1α, and IL-6, and reduced reactive gliosis at the injury site, which is expected to reduce neuroinflammation and promote tissue remodeling and repair after TBI [85]. Moreover, SA/Col/SDF-1 neural scaffold loaded with BMSCs promoted neurological function recovery by mitigating neuroinflammation after TBI, as evidenced by less glial fibrillary acidic protein (GFAP)-positive cells and lower secretion of IL-1β and IL-6 [76]. In addition, the tyramine-modified hyaluronic acid hydrogels (HT) encapsulated with BMSC and NGF facilitated the proliferation of endogenous neural cells, probably by neurotrophic factor release and neuroinflammation regulation, and consequently improved the neurological function recovery of TBI [86]. Furthermore, the implantation of brain-derived ECM hydrogel ameliorated TBI-induced gliosis and microglial pro-inflammatory responses, thereby providing a favorable microenvironment for tissue repair and neurological recovery [87]. Taken together, injectable hydrogels hold promise as a potential therapeutic implant or cell/drug delivery vehicle for TBI tissue repair via anti-inflammatory treatment. However, there is an urgent need to further explore the complexity of the immune response after TBI and the optimal timing of treatment.

4.3. Anti-Apoptosis

Apoptosis plays an important role in neurological diseases, and inhibiting apoptosis of neural cells at the injury site is beneficial for the reconstruction of neural function [88]. The main genes involved in cell death are apoptosis inhibitors (Bcl-2) and apoptosis promoters (Bax). Studies have shown that injectable hydrogels can reduce neuronal apoptosis and promote neuronal cell survival. For example, self-assembling peptide nanofibrous hydrogel reduced acute brain injury by lowering the number of apoptotic cells, reducing the inflammatory response, as well as promoting cell survival, which may be a promising strategy in nerve repair after TBI [89]. Besides, dual-enzymatically crosslinked gelatin hydrogel significantly attenuated neuronal apoptosis, ameliorated inflammation, and facilitated the survival and proliferation of endogenous neural cells in TBI mice [90]. In addition, the transplantation of CS-HEC (chitosan, hydroxyethyl cellulose)-HA/GP thermoresponsive hydrogel loaded with hUC-MSC promoted the survival/proliferation of endogenous neurons by suppressing cell apoptosis in the brain of TBI mice [69]. Liu et al. added HA into chitosan-based self-healing hydrogels (CH). After injection into the TBI injury site, the expression of C-C-sequence chemokine ligand 2 (CCL2), toll-like receptor 2 (TLR2), IL-1β, and caspase-3 was significantly downregulated, indicating that CH hydrogel inhibited the inflammatory environment and cell apoptosis [91]. Collectively, the cell/drug-loaded hydrogels could reduce neuronal apoptosis and exert favorable effects on the neural network construction of TBI.

4.4. Anti-Oxidation

Hypoxia at the injured site after TBI can cause glutamate excitotoxicity and calcium influx, leading to the production of free radicals and oxidative stress, and then induce lipid peroxidation, DNA damage, and enzyme inactivation, which is not conducive to cell survival [92]. Therefore, functionalized hydrogels with the capacity for free radical scavenging or oxidative stress suppression is beneficial for TBI recovery and functional reconstruction.

For instance, Zhang et al. prepared an injectable hydrogel (HT/HGA hydrogel) by using antioxidant gallic acid-grafted hyaluronic acid (HGA) combined with hyaluronic acid–tyramine (HT) polymer. They found that the in situ injection of HT/HGA hydrogel with reactive oxygen species-scavenging activity significantly promoted the recovery of motor, learning, and memory functions of TBI mice by suppressing oxidative stress via the activation the of Nrf2/HO-1 pathway [93]. In addition, Qian et al. designed an injectable TM/PC hydrogel loaded with Cur that responded to a post-traumatic microenvironment (such as MMPs enzymes and ROS). This hydrogel could effectively respond to the TBI environment, and then depleted ROS and reduced brain edema to promote the regeneration and recovery of neurons [72]. Moreover, the transplantation of an injectable thermosensitive chitosan/gelatin/β-glycerol phosphate (C/G/GP) hydrogel with the controlled release of FA (a phenolic antioxidant) significantly reduced the expression of ROS, inflammation, and apoptosis-related markers at the injury site of TBI, and then ameliorated the secondary brain injury by reducing the level of oxidative stress at the injury site [70]. Furthermore, an oxi-methylcellulose-adipic acid dihydrazide (oxi-MC-ADH) hydrogel prepared as a vitamin C carrier could quickly diffuse ROS from the injured site after an acute TBI, and the subsequent 3-day sustained vitamin C release could scavenge the ROS to avoid consecutive neuronal degeneration that might lead to more serious sequelae [94]. Above all, antioxidant hydrogels show great potential in reducing secondary TBI, indicating the great value of this novel biomaterial in remodeling brain structure and function.

4.5. Pro-Angiogenesis

Vascular damage is a general pathological process in many neurodegenerative diseases as well as TBI. Angiogenesis, a highly regulated process involving the activation, migration and proliferation of vascular endothelial cells (ECs) and the formation of new blood vessels, is fundamental to brain development and repair, which contributes to the reconstruction of microenvironment for neural regeneration in the lesion area [95]. Hydrogels with angiogenic properties are ideal materials to promote angiogenesis, that is, to provide vascular supply to ischemic areas and transplanted cells. Lu et al. found that the HA-KLT (KLTWQELYQLKYKGI, a VEGF mimetic peptide,) hydrogel developed via modifying HA with a vascular endothelial growth factor (VEGF) mimetic peptide of KLT significantly improved the expression of endoglin/CD105 and enhanced the formation of blood vessels in brain lesions [96]. Besides, self-assembling peptide-based hydrogel (SAPHs) created a regenerative microenvironment for neovascularization at the injury site of TBI, which activated angiogenesis by inducing ECs adhesion and upregulating VEGF-R2, which in turn promotes axonal growth and neuronal survival [97]. In addition, a functionalized self-assembling nanopeptide hydrogel could induce angiogenesis with the expression of enhanced green fluorescent protein (EGFP) and α-smooth muscle actin (α-SMA) by ECs and pericytes, to help the reconstruction of damaged neural tissue and promote functional recovery from TBI [98]. Similarly, the combined therapy of self-assembling peptide nanofiber hydrogel (dual-functionalized with VEGF- and BDNF-mimetic peptides) and chitosan conduits synergistically contributed to the healing of injured nerves via enhancing angiogenesis and neurogenesis [99]. Thus, strategies to promote angiogenesis and neurogenesis by functionalized hydrogels have potential applications in neural tissue engineering, providing new ideas for TBI therapy.

5. Conclusions and Prospects

Since the microenvironment of TBI is not conducive to cell survival, and the BBB hinders drug transport, there is an urgent need to explore novel strategies to promote nerve regeneration. In this paper, we briefly described the characteristics and gelation mechanisms of thermoresponsive hydrogels, photoresponsive hydrogels, magnetic-responsive hydrogels, electroresponsive hydrogels, and bioresponsive hydrogels, and enumerated the applications and therapeutic mechanisms of stimuli-responsive hydrogels in TBI, including pro-neurogenesis, anti-inflammatory, anti-apoptosis, anti-oxidation, and pro-angiogenesis

(Figure 5, Table 1). Although great progress has been made in current research on stimuli-responsive hydrogels, stimuli-responsive hydrogels still have some limitations, such as thermal denaturation, UV carcinogenesis, low mechanical strength, and material toxicity. Furthermore, the high swelling properties of hydrogels may increase local tissue pressure, which not only deteriorate the mechanical property of the hydrogels but also can bring about secondary brain injury after TBI. Therefore, it is necessary to improve injection conditions to avoid additional damage to the brain, such as using syringes equipped with smaller needles to deliver small volumes at more controlled rates. In addition, the lesion volume, the location of injection, and the unique swelling properties of hydrogels formulation may all influence the acceptable range of injection volumes in the TBI lesion. Therefore, it is crucial for researchers to continue developing improved hydrogel fabrication strategies to better control the swelling of hydrogels. Taken together, future studies should preferentially focus on improving the biocompatibility of stimuli-responsive hydrogels with the body, especially deepening the study of natural materials. In addition, the optimal time for hydrogel treatment should be further explored.

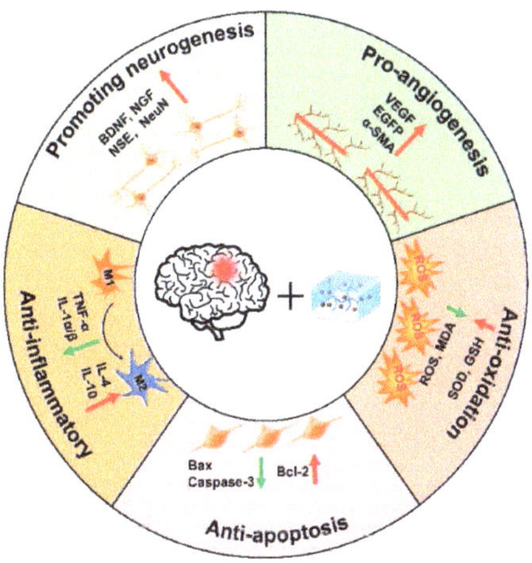

Figure 5. Therapeutic mechanisms of injectable hydrogels in TBI.

Table 1. Therapeutic mechanisms of injectable hydrogels in TBI.

Molecular Mechanisms	Hydrogels	Components	Loaded Materials	Models	Refs
Pro-neurogenesis	SA/Col hydrogel	SA, Col	BMSCs, SDF-1	SD Rats	[76]
	Gelatin hydrogel	Gelatin, GOX, HRP	BMSCs	SD Rats	[77]
	GelMA-IMID hydrogel	Gelatin, MA, IMID	hAMSCs, PDA@SDF-1α	SD Rats	[78]
	RADA16-BDNF peptide scaffolds	RADA16, RGI	CXCR4, hUC-MSCs, activated astrocytes	SD Rats	[79]
	Agarose hydrogel	Agarose	FasL	SD Rats	[80]
Anti-inflammatory	SA/Col hydrogel	SA, Col	BMSCs, SDF-1	SD Rats	[76]
	Fmoc-DIKVAV hydrogel	Fmoc	fucoidan	C57 BL/6 mice	[83]
	PEG-bis-AA/HA-DXM hydrogel	PEG, AA, HA	DXM	SD Rats	[84]
	BD™PuraMatrix™ hydrogel	PuraMatrix	hNS/PCs, hADSCs	SD Rats	[85]
	HA gelatin	HA, GOX, HRP	NGF, BMSCs	C57BL/6	[86]
	ECM hydrogel	ECM	/	C57 BL/6 mice	[87]
Anti-apoptosis	CS-HEC-HA/GP hydrogel	CS, HEC, HA, GP	hUC-MSCs	SD Rats	[69]
	GH hydrogel	gelatin, HRP, ChOx	BMSCs	C57BL/6 mice	[90]
	CH hydrogel	HA, CS	/	SD Rats, Zebrafish	[91]
Anti-oxidation	HT/HGA hydrogel	HA, Tyr	GA	C57BL/6 mice	[93]
	oxi-MC-ADH-VC hydrogel	MC	Vitamin C	SD Rats	[94]
	TM/PC hydrogel	TM, PPS120	Cur	ICR mice	[72]
	C/G/GP hydrogel	CS, gelatin, GP	FA	N2a cells	[70]
Pro-angiogenesis	HA-KLT hydrogel	HA	VEGF mimetic peptide KLT	SD Rats	[96]
	SAPH	Slanc	/	SD Rats	[97]
	SAPH	RADA16-SVVYGLR	/	Zebrafish	[98]
	SAPH	RAD	KLT, RGI	SD Rats	[99]

Author Contributions: Conceptualization, S.M.; Resources, Writing—original draft, X.L. and S.M.; Writing—review and editing, X.W. and F.G.; Resources, L.D.; Investigation, M.K.; Funding acquisition, F.G. and S.M. All authors have read and agreed to the published version of the manuscript.

Funding: This study was supported by National Natural Science Foundation of China (U2004201), Henan Overseas Expertise Introduction Center for Discipline Innovation (CXJD2021002), and Open Research Fund of National Health Commission Key Laboratory of Birth Defects Prevention and Henan Key Laboratory of Population Defects Prevention (ZD202204).

Institutional Review Board Statement: Not applicable.

Informed Consent Statement: Not applicable.

Data Availability Statement: Not applicable.

Conflicts of Interest: The authors declare that they have no conflict of interest.

Abbreviations

TBI, traumatic brain injury; BBB, blood–brain barrier; GCS, Glasgow Coma Score; pNIPAAm, poly (*N*-isopropylacrylamide); PEO-PPO-PEO, poly (ethylene oxide)-poly (propylene oxide)-poly (ethylene oxide); HA, hyaluronic acid; PEG, polyethylene glycol; LCST, lower critical solution temperature; UCST, upper critical solution temperature; BPNs, black phosphorus nanosheets; PRP, platelette-rich plasma; RA, rheumatoid arthritis; PPS-b-PDMA-b-PNIPAAM, poly [(propylenesulfide)-block-(*N*,*N*-dimethylacrylamide)-block-(*N*-isopropylacrylamide)]; UV, ultraviolet; NIR, near infrared; BMH, bio-inspired MnO_2 hybrid; UCNPs, upconverting nanoparticles; AMF, alternating magnetic field; PPy, polypyrrole; PANI, polyaniline; PEDOT, poly(3,4-ethylenedioxythiophene); PPP, PLGA-PEI-MPEG; ACPP, activated cell-penetrating peptides; ET, etanercept; MMPs, matrix metalloproteinases; spinal cord injury, SCI; PKA, protein kinase A; H-IgG, human immunoglobulin G; BP, black phosphorus; ECM, extracellular matrix; HSPs, heat shock proteins; CS-HSE-HA/GP, chitosan hydroxyethyl cellulose hyaluronic acid/β-glycerophosphate; FA, Ferulic acid; C/G/GP, chitosan/gelatin/β-glycerophosphate; PU, polyurethane; NSCs, neural stem cells; Cur, curcumin; MSC, mesenchymal stem cell; SA, sodium alginate; Col, collagen; SDF-1, stromalcell-derivedfactor-1; BMSCs, bone marrow-derived mesenchymal stem cells; PDA, polydopamine; SDF-1α, stromal-cell derived factor-1α; hAMSCs, human amniotic mesenchymal stromal cells; GelMA-imid, imidazole groups-modified gelatin methacrylate; NeuN, neuronal nuclei; NSE, neuron specific enolase; BDNF, brain derived neurotrophic factor; hUC-MSCs, human umbilical cord mesenchymal stem cells; R-B-SPH, self-assembled peptide hydrogel RADA16-BDNF; HIF-1, hypoxia-inducible factor; FasL, Fas ligand; NGF, nerve growth factor; IGF, insulin-like growth factor; TNF-α, tumor necrosis factor-α; IL, interleukin; IFN-γ, gamma-interferon; IFN-α, α-interferon; PEG-bis-AA, PEG-bis-(acryloyloxyacetate); hADSCs, human neural stem/progenitor cells; PM, PuraMatrix; GFAP, glial fibrillary acidic protein; HT, tyramine-modified hyaluronic acid hydrogels; CCL2, C-C-sequence chemokine ligand 2; TLR2, Toll-like receptor 2; CH, chitosan-based self-healing hydrogels; MDA, malondialdehyde; GSH, glutathione; oxi-MC-ADH, oxi-methylcellulose-adipic acid dihydrazide; ECs, endothelial cells; CS-HEC, chitosan, hydroxyethyl cellulose; KLT, KLTWQELYQLKYKGI; VEGF, vascular endothelial growth factor; SAPHs, self assembling peptide-based hydrogel; EGFP, enhanced green fluorescent protein; αSMA, α-smooth muscle actin.

References

1. Maas, A.I.R.; Menon, D.K.; Adelson, P.D.; Andelic, N.; Bell, M.J.; Belli, A.; Bragge, P.; Brazinova, A.; Büki, A.; Chesnut, R.M.; et al. Traumatic brain injury: Integrated approaches to improve prevention, clinical care, and research. *Lancet Neurol.* **2017**, *16*, 987–1048. [CrossRef]
2. Stocchetti, N.; Zanier, E.R. Chronic impact of traumatic brain injury on outcome and quality of life: A narrative review. *Crit. Care* **2016**, *20*, 148. [CrossRef]

3. Dewan, M.C.; Rattani, A.; Gupta, S.; Baticulon, R.E.; Hung, Y.C.; Punchak, M.; Agrawal, A.; Adeleye, A.O.; Shrime, M.G.; Rubiano, A.M.; et al. Estimating the global incidence of traumatic brain injury. *J. Neurosurg.* **2018**, *130*, 1080–1097. [CrossRef] [PubMed]
4. Haarbauer-Krupa, J.; Pugh, M.J.; Prager, E.M.; Harmon, N.; Wolfe, J.; Yaffe, K. Epidemiology of Chronic Effects of Traumatic Brain Injury. *J. Neurotrauma* **2021**, *38*, 3235–3247. [CrossRef]
5. Jassam, Y.N.; Izzy, S.; Whalen, M.; McGavern, D.B.; El Khoury, J. Neuroimmunology of Traumatic Brain Injury: Time for a Paradigm Shift. *Neuron* **2017**, *95*, 1246–1265. [CrossRef]
6. Wu, H.; Zheng, J.; Xu, S.; Fang, Y.; Wu, Y.; Zeng, J.; Shao, A.; Shi, L.; Lu, J.; Mei, S.; et al. Mer regulates microglial/macrophage M1/M2 polarization and alleviates neuroinflammation following traumatic brain injury. *J. Neuroinflamm.* **2021**, *18*, 2. [CrossRef] [PubMed]
7. Ashina, H.; Eigenbrodt, A.K.; Seifert, T.; Sinclair, A.J.; Scher, A.I.; Schytz, H.W.; Lee, M.J.; De Icco, R.; Finkel, A.G.; Ashina, M. Post-traumatic headache attributed to traumatic brain injury: Classification, clinical characteristics, and treatment. *Lancet Neurol.* **2021**, *20*, 460–469. [CrossRef]
8. Jiang, J.-Y.; Gao, G.-Y.; Feng, J.-F.; Mao, Q.; Chen, L.-G.; Yang, X.-F.; Liu, J.-F.; Wang, Y.-H.; Qiu, B.-H.; Huang, X.-J. Traumatic brain injury in China. *Lancet Neurol.* **2019**, *18*, 286–295. [CrossRef]
9. Sweeney, M.D.; Zhao, Z.; Montagne, A.; Nelson, A.R.; Zlokovic, B.V. Blood-Brain Barrier: From Physiology to Disease and Back. *Physiol. Rev.* **2019**, *99*, 21–78. [CrossRef]
10. Tan, H.X.; Borgo, M.P.D.; Aguilar, M.I.; Forsythe, J.S.; Taylor, J.M.; Crack, P.J. The use of bioactive matrices in regenerative therapies for traumatic brain injury. *Acta Biomater.* **2020**, *102*, 1–12. [CrossRef]
11. Grimaudo, M.A.; Krishnakumar, G.S.; Giusto, E.; Furlani, F.; Bassi, G.; Rossi, A.; Molinari, F.; Lista, F.; Montesi, M.; Panseri, S. Bioactive injectable hydrogels for on demand molecule/cell delivery and for tissue regeneration in the central nervous system. *Acta Biomater.* **2022**, *140*, 88–101. [CrossRef] [PubMed]
12. Narayanaswamy, R.; Torchilin, V.P. Hydrogels and Their Applications in Targeted Drug Delivery. *Molecules* **2019**, *24*, 603.
13. El-Husseiny, H.M.; Mady, E.A.; Hamabe, L.; Abugomaa, A.; Shimada, K.; Yoshida, T.; Tanaka, T.; Yokoi, A.; Elbadawy, M.; Tanaka, R. Smart/stimuli-responsive hydrogels: Cutting-edge platforms for tissue engineering and other biomedical applications. *Mater. Today Bio.* **2022**, *13*, 100186.
14. Wei, H.; Cui, J.; Lin, K.; Xie, J.; Wang, X. Recent advances in smart stimuli-responsive biomaterials for bone therapeutics and regeneration. *Bone Res.* **2022**, *10*, 17. [CrossRef] [PubMed]
15. Qiu, Y.; Park, K. Environment-sensitive hydrogels for drug delivery. *Adv. Drug Del. Rev.* **2001**, *53*, 321–339.
16. Lavrador, P.; Esteves, M.R.; Gaspar, V.M.; Mano, J.F. Stimuli-Responsive Nanocomposite Hydrogels for Biomedical Applications. *Adv. Funct. Mater.* **2020**, *31*, 2005941. [CrossRef]
17. Sood, N.; Bhardwaj, A.; Mehta, S.; Mehta, A. Stimuli-responsive hydrogels in drug delivery and tissue engineering. *Drug Deliv.* **2016**, *23*, 758–780. [CrossRef]
18. Milcovich, G.; Lettieri, S.; Antunes, F.E.; Medronho, B.; Fonseca, A.C.; Coelho, J.F.J.; Marizza, P.; Perrone, F.; Farra, R.; Dapas, B.; et al. Recent advances in smart biotechnology: Hydrogels and nanocarriers for tailored bioactive molecules depot. *Adv. Colloid Interface Sci.* **2017**, *249*, 163–180.
19. Avila-Salas, F.; Duran-Lara, E.F. An Overview, of Injectable Thermo-Responsive Hydrogens and Advances in their Biomedical Applications. *Curr. Med. Chem.* **2020**, *27*, 5773–5789. [CrossRef]
20. Zhang, K.; Xue, K.; Loh, X.J. Thermo-Responsive Hydrogels: From Recent Progress to Biomedical Applications. *Gels* **2021**, *7*, 77.
21. Klouda, L. Thermoresponsive hydrogels in biomedical applications: A seven-year update. *Eur. J. Pharm. Biopharm.* **2015**, *97*, 338–349. [CrossRef] [PubMed]
22. Koetting, M.C.; Peters, J.T.; Steichen, S.D.; Peppas, N.A. Stimulus-responsive hydrogels: Theory, modern advances, and applications. *Mater. Sci. Eng. R Rep.* **2015**, *93*, 1–49. [CrossRef] [PubMed]
23. Kasinski, A.; Zielinska-Pisklak, M.; Oledzka, E.; Sobczak, M. Smart Hydrogels—Synthetic Stimuli-Responsive Antitumor Drug Release Systems. *Int. J. Nanomed.* **2020**, *15*, 4541–4572. [CrossRef] [PubMed]
24. Taylor, M.J.; Tomlins, P.; Sahota, T.S. Thermoresponsive Gels. *Gels* **2017**, *3*, 4. [CrossRef] [PubMed]
25. Chee, P.L.; Young, D.J.; Loh, X.J. Chapter 7 Degradation Behaviour of Biodegradable Thermogels. In *Biodegradable Thermogels*; The Royal Society of Chemistry: London, UK, 2019; pp. 113–132.
26. Sponchioni, M.; Capasso Palmiero, U.; Moscatelli, D. Thermo-responsive polymers: Applications of smart materials in drug delivery and tissue engineering. *Mater. Sci. Eng. C Mater. Biol. Appl.* **2019**, *102*, 589–605. [CrossRef]
27. Pan, W.; Dai, C.; Li, Y.; Yin, Y.; Gong, L.; Machuki, J.O.; Yang, Y.; Qiu, S.; Guo, K.; Gao, F. PRP-chitosan thermoresponsive hydrogel combined with black phosphorus nanosheets as injectable biomaterial for biotherapy and phototherapy treatment of rheumatoid arthritis. *Biomaterials* **2020**, *239*, 119851. [CrossRef]
28. Gupta, M.K.; Martin, J.R.; Werfel, T.A.; Shen, T.; Page, J.M.; Duvall, C.L. Cell protective, ABC triblock polymer-based thermoresponsive hydrogels with ROS-triggered degradation and drug release. *J. Am. Chem. Soc.* **2014**, *136*, 14896–14902. [CrossRef]
29. Mahlumba, P.; Kumar, P.; du Toit, L.C.; Poka, M.S.; Ubanako, P.; Choonara, Y.E. Fabrication and Characterisation of a Photo-Responsive, Injectable Nanosystem for Sustained Delivery of Macromolecules. *Int. J. Mol. Sci.* **2021**, *22*, 3359.
30. Ji, W.; Wu, Q.; Han, X.; Zhang, W.; Wei, W.; Chen, L.; Li, L.; Huang, W. Photosensitive hydrogels: From structure, mechanisms, design to bioapplications. *Sci. China Life Sci.* **2020**, *63*, 1813–1828.
31. Li, L.; Scheiger, J.M.; Levkin, P.A. Design and Applications of Photoresponsive Hydrogels. *Adv. Mater.* **2019**, *31*, e1807333.

32. Suzuki, A. Light-induced phase-transition of poly(n-isopropylacrylamide-co-chlorophyllin) gels. *JIMSS* **1994**, *5*, 112–116. [CrossRef]
33. Kharkar, P.M.; Kiick, K.L.; Kloxin, A.M. Design of Thiol- and Light-sensitive Degradable Hydrogels using Michael-type Addition Reactions. *Polym. Chem.* **2015**, *6*, 5565–5574. [CrossRef] [PubMed]
34. Lugao, A.B.; Malmonge, S.M. Use of radiation in the production of hydrogels. *Nucl. Instrum. Methods Phys. Res. Sect. B-Beam Interact. Mater. At.* **2001**, *185*, 37–42. [CrossRef]
35. Wang, H.; Jin, Y.; Tan, Y.; Zhu, H.; Huo, W.; Niu, P.; Li, Z.; Zhang, J.; Liang, X.J.; Yang, X. Photo-responsive hydrogel facilitates nutrition deprivation by an ambidextrous approach for preventing cancer recurrence and metastasis. *Biomaterials* **2021**, *275*, 120992. [CrossRef]
36. Wang, S.; Zheng, H.; Zhou, L.; Cheng, F.; Liu, Z.; Zhang, H.; Zhang, Q. Injectable redox and light responsive MnO_2 hybrid hydrogel for simultaneous melanoma therapy and multidrug-resistant bacteria-infected wound healing. *Biomaterials* **2020**, *260*, 120314. [CrossRef]
37. Wang, X.; Wang, C.; Zhang, Q.; Cheng, Y. Near infrared light-responsive and injectable supramolecular hydrogels for on-demand drug delivery. *Chem. Commun.* **2016**, *52*, 978–981. [CrossRef]
38. Mantha, S.; Pillai, S.; Khayambashi, P.; Upadhyay, A.; Zhang, Y.; Tao, O.; Pham, H.M.; Tran, S.D. Smart Hydrogels in Tissue Engineering and Regenerative Medicine. *Materials* **2019**, *12*, 3323. [CrossRef]
39. Li, Y.; Huang, G.; Zhang, X.; Li, B.; Chen, Y.; Lu, T.; Lu, T.J.; Xu, F. Magnetic Hydrogels and Their Potential Biomedical Applications. *Adv. Funct. Mater.* **2013**, *23*, 660–672. [CrossRef]
40. Shi, W.; Huang, J.; Fang, R.; Liu, M. Imparting Functionality to the Hydrogel by Magnetic-Field-Induced Nano-assembly and Macro-response. *ACS Appl. Mater. Interfaces* **2020**, *12*, 5177–5194.
41. Zhang, J.; Huang, Q.; Du, J. Recent advances in magnetic hydrogels. *Polym. Int.* **2016**, *65*, 1365–1372. [CrossRef]
42. Santhosh, M.; Choi, J.H.; Choi, J.W. Magnetic-Assisted Cell Alignment within a Magnetic Nanoparticle-Decorated Reduced Graphene Oxide/Collagen 3D Nanocomposite Hydrogel. *Nanomaterials* **2019**, *9*, 1293. [CrossRef] [PubMed]
43. Veloso, S.R.S.; Andrade, R.G.D.; Castanheira, E.M.S. Review on the advancements of magnetic gels: Towards multifunctional magnetic liposome-hydrogel composites for biomedical applications. *Adv. Colloid Interface Sci.* **2021**, *288*, 102351. [CrossRef] [PubMed]
44. Ramanujan, R.V.; Lao, L.L. The mechanical behavior of smart magnet-hydrogel composites. *SmMaS* **2006**, *15*, 952–956. [CrossRef]
45. Nguyen, V.Q.; Ahmed, A.S.; Ramanujan, R.V. Morphing soft magnetic composites. *Adv. Mater.* **2012**, *24*, 4041–4054. [CrossRef] [PubMed]
46. Li, Z.; Li, Y.; Chen, C.; Cheng, Y. Magnetic-responsive hydrogels: From strategic design to biomedical applications. *J. Control. Release* **2021**, *335*, 541–556. [CrossRef]
47. Liu, T.Y.; Hu, S.H.; Liu, T.Y.; Liu, D.M.; Chen, S.Y. Magnetic-sensitive behavior of intelligent ferrogels for controlled release of drug. *Langmuir* **2006**, *22*, 5974–5978. [CrossRef]
48. Wang, W.T.; Fan, X.Q.; Li, F.H.; Qiu, J.J.; Umair, M.M.; Ren, W.C.; Ju, B.Z.; Zhang, S.F.; Tang, B.T. Magnetochromic Photonic Hydrogel for an Alternating Magnetic Field-Responsive Color Display. *Adv. Opt. Mater.* **2018**, *6*, 9. [CrossRef]
49. Zhang, N.; Lock, J.; Sallee, A.; Liu, H. Magnetic Nanocomposite Hydrogel for Potential Cartilage Tissue Engineering: Synthesis, Characterization, and Cytocompatibility with Bone Marrow Derived Mesenchymal Stem Cells. *ACS Appl. Mater. Interfaces* **2015**, *7*, 20987–20998. [CrossRef] [PubMed]
50. Yan, H.; Li, L.; Wang, Y.; Huang, J.; Wang, Z.; Shi, X.; Zhang, P. An electrically and magnetically responsive nanocomposite of $GdPO_4 \cdot H_2O$/P3HT/PLGA with electrical stimulation for synergistically enhancing the proliferation and differentiation of pre-osteoblasts. *NJCh* **2019**, *43*, 17315–17326. [CrossRef]
51. Kolosnjaj-Tabi, J.; Gibot, L.; Fourquaux, I.; Golzio, M.; Rols, M.P. Electric field-responsive nanoparticles and electric fields: Physical, chemical, biological mechanisms and therapeutic prospects. *Adv. Drug Deliv. Rev.* **2019**, *138*, 56–67. [CrossRef]
52. Lu, H.; Zhang, N.; Ma, M. Electroconductive hydrogels for biomedical applications. *Wiley Interdiscip. Rev. Nanomed. Nanobiotechnol.* **2019**, *11*, e1568. [CrossRef] [PubMed]
53. Qu, J.; Liang, Y.; Shi, M.; Guo, B.; Gao, Y.; Yin, Z. Biocompatible conductive hydrogels based on dextran and aniline trimer as electro-responsive drug delivery system for localized drug release. *Int. J. Biol. Macromol.* **2019**, *140*, 255–264. [CrossRef] [PubMed]
54. Qiao, K.; Guo, S.; Zheng, Y.; Xu, X.; Meng, H.; Peng, J.; Fang, Z.; Xie, Y. Effects of graphene on the structure, properties, electro-response behaviors of GO/PAA composite hydrogels and influence of electro-mechanical coupling on BMSC differentiation. *Mater. Sci. Eng. C Mater. Biol. Appl.* **2018**, *93*, 853–863. [CrossRef] [PubMed]
55. Carayon, I.; Gaubert, A.; Mousli, Y.; Philippe, B. Electro-responsive hydrogels: Macromolecular and supramolecular approaches in the biomedical field. *Biomater. Sci.* **2020**, *8*, 5589–5600. [CrossRef] [PubMed]
56. Nourbakhsh, M.; Zarrintaj, P.; Jafari, S.H.; Hosseini, S.M.; Aliakbari, S.; Pourbadie, H.G.; Naderi, N.; Zibaii, M.I.; Gholizadeh, S.S.; Ramsey, J.D.; et al. Fabricating an electroactive injectable hydrogel based on pluronic-chitosan/aniline-pentamer containing angiogenic factor for functional repair of the hippocampus ischemia rat model. *Mater. Sci. Eng. C Mater. Biol. Appl.* **2020**, *117*, 111328. [CrossRef]
57. Sgambato, A.; Cipolla, L.; Russo, L. Bioresponsive Hydrogels: Chemical Strategies and Perspectives in Tissue Engineering. *Gels* **2016**, *2*, 28. [CrossRef]
58. Ulijn, R.V. Enzyme-responsive materials: A new class of smart biomaterials. *J. Mater. Chem.* **2006**, *16*, 2217–2225. [CrossRef]

59. Zhang, R.; Bowyer, A.; Eisenthal, R.; Hubble, J. A smart membrane based on an antigen-responsive hydrogel. *Biotechnol. Bioeng.* **2007**, *97*, 976–984. [CrossRef]
60. Chandrawati, R. Enzyme-responsive polymer hydrogels for therapeutic delivery. *Exp. Biol. Med.* **2016**, *241*, 972–979. [CrossRef]
61. Shen, K.; Sun, G.; Chan, L.; He, L.; Li, X.; Yang, S.; Wang, B.; Zhang, H.; Huang, J.; Chang, M.; et al. Anti-Inflammatory Nanotherapeutics by Targeting Matrix Metalloproteinases for Immunotherapy of Spinal Cord Injury. *Small* **2021**, *17*, e2102102. [CrossRef]
62. Katayama, Y.; Sonoda, T.; Maeda, M. A polymer micelle responding to the protein kinase A signal. *Macromolecules* **2001**, *34*, 8569–8573. [CrossRef]
63. Miyata, T.; Asami, N.; Uragami, T. A reversibly antigen-responsive hydrogel. *Nature* **1999**, *399*, 766–769. [CrossRef] [PubMed]
64. Ye, G.; Yang, C.; Wang, X. Sensing diffraction gratings of antigen-responsive hydrogel for human immunoglobulin-g detection. *Macromol. Rapid Commun.* **2010**, *31*, 1332–1336. [CrossRef] [PubMed]
65. Khatri, N.; Thakur, M.; Pareek, V.; Kumar, S.; Sharma, S.; Datusalia, A.K. Oxidative Stress: Major Threat in Traumatic Brain Injury. *CNS Neurol. Disord. Drug Targets* **2018**, *17*, 689–695. [CrossRef] [PubMed]
66. Simon, D.W.; McGeachy, M.J.; Bayir, H.; Clark, R.S.; Loane, D.J.; Kochanek, P.M. The far-reaching scope of neuroinflammation after traumatic brain injury. *Nat. Rev. Neurol.* **2017**, *13*, 171–191. [CrossRef]
67. Tam, R.Y.; Fuehrmann, T.; Mitrousis, N.; Shoichet, M.S. Regenerative therapies for central nervous system diseases: A biomaterials approach. *Neuropsychopharmacology* **2014**, *39*, 169–188. [CrossRef]
68. Thomas, J.M.; Louca, I.; Bolan, F.; Sava, O.R.; Allan, S.M.; Lawrence, C.B.; Pinteaux, E. Regenerative Potential of Hydrogels for Intracerebral Hemorrhage: Lessons from Ischemic Stroke and Traumatic Brain Injury Research. *Adv. Healthc. Mater.* **2021**, *10*, e2100455. [CrossRef]
69. Yao, M.; Chen, Y.; Zhang, J.; Gao, F.; Ma, S.; Guan, F. Chitosan-based thermosensitive composite hydrogel enhances the therapeutic efficacy of human umbilical cord MSC in TBI rat model. *Mater. Today Chem.* **2019**, *14*, 100192. [CrossRef]
70. Dong, G.C.; Kuan, C.Y.; Subramaniam, S.; Zhao, J.Y.; Sivasubramaniam, S.; Chang, H.Y.; Lin, F.H. A potent inhibition of oxidative stress induced gene expression in neural cells by sustained ferulic acid release from chitosan based hydrogel. *Mater. Sci. Eng. C Mater. Biol. Appl.* **2015**, *49*, 691–699. [CrossRef]
71. Hsieh, F.Y.; Lin, H.H.; Hsu, S.H. 3D bioprinting of neural stem cell-laden thermoresponsive biodegradable polyurethane hydrogel and potential in central nervous system repair. *Biomaterials* **2015**, *71*, 48–57. [CrossRef]
72. Qian, F.; Han, Y.; Han, Z.; Zhang, D.; Zhang, L.; Zhao, G.; Li, S.; Jin, G.; Yu, R.; Liu, H. In Situ implantable, post-trauma microenvironment-responsive, ROS Depletion Hydrogels for the treatment of Traumatic brain injury. *Biomaterials* **2021**, *270*, 120675. [CrossRef] [PubMed]
73. Tan, L.; Hu, Y.; Li, M.H.; Zhang, Y.C.; Xue, C.C.; Chen, M.H.; Luo, Z.; Cai, K.Y. Remotely-activatable extracellular matrix-mimetic hydrogel promotes physiological bone mineralization for enhanced cranial defect healing. *Chem. Eng. J.* **2022**, *431*, 14. [CrossRef]
74. Adak, A.; Das, G.; Khan, J.; Mukherjee, N.; Gupta, V.; Mallesh, R.; Ghosh, S. Extracellular Matrix (ECM)-Mimicking Neuroprotective Injectable Sulfo-Functionalized Peptide Hydrogel for Repairing Brain Injury. *ACS Biomater. Sci. Eng.* **2020**, *6*, 2287–2296. [CrossRef] [PubMed]
75. Ngwenya, L.B.; Danzer, S.C. Impact of Traumatic Brain Injury on Neurogenesis. *Front. Neurosci.* **2018**, *12*, 1014. [CrossRef]
76. Ma, S.; Zhou, J.; Huang, T.; Zhang, Z.; Xing, Q.; Zhou, X.; Zhang, K.; Yao, M.; Cheng, T.; Wang, X.; et al. Sodium alginate/collagen/stromal cell-derived factor-1 neural scaffold loaded with BMSCs promotes neurological function recovery after traumatic brain injury. *Acta Biomater.* **2021**, *131*, 185–197. [CrossRef] [PubMed]
77. Yao, M.; Gao, F.; Xu, R.; Zhang, J.; Chen, Y.; Guan, F. A dual-enzymatically cross-linked injectable gelatin hydrogel loaded with BMSC improves neurological function recovery of traumatic brain injury in rats. *Biomater. Sci.* **2019**, *7*, 4088–4098. [CrossRef]
78. Zheng, Y.; Wu, G.; Chen, L.; Zhang, Y.; Luo, Y.; Zheng, Y.; Hu, F.; Forouzanfar, T.; Lin, H.; Liu, B. Neuro-regenerative imidazole-functionalized GelMA hydrogel loaded with hAMSC and SDF-1alpha promote stem cell differentiation and repair focal brain injury. *Bioact. Mater.* **2021**, *6*, 627–637. [CrossRef]
79. Shi, W.; Huang, C.J.; Xu, X.D.; Jin, G.H.; Huang, R.Q.; Huang, J.F.; Chen, Y.N.; Ju, S.Q.; Wang, Y.; Shi, Y.W.; et al. Transplantation of RADA16-BDNF peptide scaffold with human umbilical cord mesenchymal stem cells forced with CXCR4 and activated astrocytes for repair of traumatic brain injury. *Acta Biomater.* **2016**, *45*, 247–261. [CrossRef]
80. Alvarado-Velez, M.; Enam, S.F.; Mehta, N.; Lyon, J.G.; LaPlaca, M.C.; Bellamkonda, R.V. Immuno-suppressive hydrogels enhance allogeneic MSC survival after transplantation in the injured brain. *Biomaterials* **2021**, *266*, 120419. [CrossRef]
81. Dinarello, C.A. Anti-inflammatory Agents: Present and Future. *Cell* **2010**, *140*, 935–950. [CrossRef]
82. Zheng, R.Z.; Lee, K.Y.; Qi, Z.X.; Wang, Z.; Xu, Z.Y.; Wu, X.H.; Mao, Y. Neuroinflammation Following Traumatic Brain Injury: Take It Seriously or Not. *Front. Immunol.* **2022**, *13*, 855701. [CrossRef]
83. Maclean, F.L.; Wang, Y.; Walker, R.; Horne, M.K.; Williams, R.J.; Nisbet, D.R. Reducing Astrocytic Scarring after Traumatic Brain Injury with a Multifaceted Anti-Inflammatory Hydrogel System. *ACS Biomater. Sci. Eng.* **2017**, *3*, 2542–2549. [CrossRef]
84. Jeong, D.U.; Bae, S.; Macks, C.; Whitaker, J.; Lynn, M.; Webb, K.; Lee, J.S. Hydrogel-mediated local delivery of dexamethasone reduces neuroinflammation after traumatic brain injury. *Biomed. Mater.* **2021**, *16*, 035002. [CrossRef] [PubMed]
85. Jahanbazi Jahan-Abad, A.; Sahab Negah, S.; Hosseini Ravandi, H.; Ghasemi, S.; Borhani-Haghighi, M.; Stummer, W.; Gorji, A.; Khaleghi Ghadiri, M. Human Neural Stem/Progenitor Cells Derived From Epileptic Human Brain in a Self-Assembling Peptide Nanoscaffold Improve Traumatic Brain Injury in Rats. *Mol. Neurobiol.* **2018**, *55*, 9122–9138. [CrossRef] [PubMed]

86. Wang, L.; Zhang, D.; Ren, Y.; Guo, S.; Li, J.; Ma, S.; Yao, M.; Guan, F. Injectable hyaluronic acid hydrogel loaded with BMSC and NGF for traumatic brain injury treatment. *Mater. Today Bio.* **2022**, *13*, 100201. [CrossRef] [PubMed]
87. Wu, Y.; Wang, J.; Shi, Y.; Pu, H.; Leak, R.K.; Liou, A.K.F.; Badylak, S.F.; Liu, Z.; Zhang, J.; Chen, J.; et al. Implantation of Brain-Derived Extracellular Matrix Enhances Neurological Recovery after Traumatic Brain Injury. *Cell Transplant.* **2017**, *26*, 1224–1234. [CrossRef] [PubMed]
88. Radi, E.; Formichi, P.; Battisti, C.; Federico, A. Apoptosis and oxidative stress in neurodegenerative diseases. *J. Alzheimers Dis.* **2014**, *42* (Suppl. S3), S125–S152. [CrossRef] [PubMed]
89. Zhang, N.; He, L.; Wu, W. Self-assembling peptide nanofibrous hydrogel as a promising strategy in nerve repair after traumatic injury in the nervous system. *Neural. Regen. Res.* **2016**, *11*, 717–718. [PubMed]
90. Li, J.; Zhang, D.; Guo, S.; Zhao, C.; Wang, L.; Ma, S.; Guan, F.; Yao, M. Dual-enzymatically cross-linked gelatin hydrogel promotes neural differentiation and neurotrophin secretion of bone marrow-derived mesenchymal stem cells for treatment of moderate traumatic brain injury. *Int. J. Biol. Macromol.* **2021**, *187*, 200–213. [CrossRef]
91. Liu, Y.; Hsu, Y.H.; Huang, A.P.; Hsu, S.H. Semi-Interpenetrating Polymer Network of Hyaluronan and Chitosan Self-Healing Hydrogels for Central Nervous System Repair. *ACS Appl. Mater. Interfaces* **2020**, *12*, 40108–40120. [CrossRef]
92. Carvajal, F.J.; Cerpa, W. Regulation of Phosphorylated State of NMDA Receptor by STEP61 Phosphatase after Mild-Traumatic Brain Injury: Role of Oxidative Stress. *Antioxidants* **2021**, *10*, 1575. [CrossRef] [PubMed]
93. Zhang, D.; Ren, Y.; He, Y.; Chang, R.; Guo, S.; Ma, S.; Guan, F.; Yao, M. In situ forming and biocompatible hyaluronic acid hydrogel with reactive oxygen species-scavenging activity to improve traumatic brain injury repair by suppressing oxidative stress and neuroinflammation. *Mater. Today Bio.* **2022**, *15*, 100278. [CrossRef] [PubMed]
94. Kuan, C.-Y.; Lin, Y.-Y.; Chen, C.-Y.; Yang, C.-C.; Chi, C.-Y.; Li, C.-H.; Dong, G.-C.; Lin, F.-H. The preparation of oxidized methylcellulose crosslinked by adipic acid dihydrazide loaded with vitamin C for traumatic brain injury. *J. Mater. Chem. B* **2019**, *7*, 4499–4508. [CrossRef]
95. Louissaint, A.; Rao, S.; Leventhal, C.; Goldman, S.A. Coordinated interaction of Neurogenesis and angiogenesis in the adult songbird brain. *Neuron* **2002**, *34*, 945–960. [CrossRef]
96. Lu, J.; Guan, F.; Cui, F.; Sun, X.; Zhao, L.; Wang, Y.; Wang, X. Enhanced angiogenesis by the hyaluronic acid hydrogels immobilized with a VEGF mimetic peptide in a traumatic brain injury model in rats. *Regen. Biomater.* **2019**, *6*, 325–334. [CrossRef] [PubMed]
97. Ma, X.; Agas, A.; Siddiqui, Z.; Kim, K.; Iglesias-Montoro, P.; Kalluru, J.; Kumar, V.; Haorah, J. Angiogenic peptide hydrogels for treatment of traumatic brain injury. *Bioact. Mater.* **2020**, *5*, 124–132. [CrossRef] [PubMed]
98. Wang, T.W.; Chang, K.C.; Chen, L.H.; Liao, S.Y.; Yeh, C.W.; Chuang, Y.J. Effects of an injectable functionalized self-assembling nanopeptide hydrogel on angiogenesis and neurogenesis for regeneration of the central nervous system. *Nanoscale* **2017**, *9*, 16281–16292. [CrossRef] [PubMed]
99. Lu, J.; Yan, X.; Sun, X.; Shen, X.; Yin, H.; Wang, C.; Liu, Y.; Lu, C.; Fu, H.; Yang, S.; et al. Synergistic effects of dual-presenting VEGF- and BDNF-mimetic peptide epitopes from self-assembling peptide hydrogels on peripheral nerve regeneration. *Nanoscale* **2019**, *11*, 19943–19958. [CrossRef]

gels

MDPI

Review

Advances of Stimulus-Responsive Hydrogels for Bone Defects Repair in Tissue Engineering

Shuai Chang [1,2,3], Shaobo Wang [1,2,3], Zhongjun Liu [1,2,3,*] and Xing Wang [4,5,*]

1 Department of Orthopedics, Peking University Third Hospital, Beijing 100191, China; changshuaipkuth@163.com (S.C.); wangshaobopkuth@163.com (S.W.)
2 Beijing Key Laboratory of Spinal Disease Research, Peking University Third Hospital, Beijing 100191, China
3 Engineering Research Center of Bone and Joint Precision Medicine, Ministry of Education, Peking University Third Hospital, Beijing 100191, China
4 Beijing National Laboratory for Molecular Sciences, Institute of Chemistry, Chinese Academy of Sciences, Beijing 100190, China
5 Beijing National Laboratory for Molecular Sciences, Institute of Chemistry, University of Chinese Academy of Sciences, Beijing 100049, China
* Correspondence: zjliu@bjmu.edu.cn (Z.L.); wangxing@iccas.ac.cn (X.W.)

Abstract: Bone defects, as one of the most urgent problems in the orthopedic clinic, have attracted much attention from the biomedical community and society. Hydrogels have been widely used in the biomedical field for tissue engineering research because of their excellent hydrophilicity, biocompatibility, and degradability. Stimulus-responsive hydrogels, as a new type of smart biomaterial, have more advantages in sensing external physical (light, temperature, pressure, electric field, magnetic field, etc.), chemical (pH, redox reaction, ions, etc.), biochemical (glucose, enzymes, etc.) and other different stimuli. They can respond to stimuli such as the characteristics of the 3D shape and solid–liquid phase state, and exhibit special properties (injection ability, self-repair, shape memory, etc.), thus becoming an ideal material to provide cell adhesion, proliferation, and differentiation, and achieve precise bone defect repair. This review is focused on the classification, design concepts, and research progress of stimulus-responsive hydrogels based on different types of external environmental stimuli, aiming at introducing new ideas and methods for repairing complex bone defects.

Keywords: stimulus-responsive; smart hydrogels; bone defects repair; tissue engineering

1. Introduction

Bone defects, caused by trauma, bone tumor resection, infection (such as osteomyelitis), and other factors, have become one of the major issues seriously affecting patients' limb function, causing physiological and psychological damage, and always requiring a safe and effective treatment to achieve bone tissue regeneration and repair. Among previous common treatments, bone transplantation has been the gold standard for the treatment of bone defects [1]. However, bone transplantation has failed to be widely used, due to the limited bone graft source of autologous material, high treatment cost, immune rejection risk, complicate bone handling technology, as well as the risk of infection and complications [2,3]. Therefore, the traditional bone defect treatment technology is unable to effectively meet the needs of patients. Tissue engineering is a multifaceted field that uses biological substitutes to promote the regeneration of failing or damaged tissues. The basic element of tissue engineering relies on cells, scaffolds, and growth factors. Traditional strategies in tissue engineering always incorporate the use of stem cells to regenerate new tissue in damaged areas [4–8]. In this case, various scaffolds provide attachment sites for cells and a protective environment for the proliferation and differentiation of attached cells. To successfully achieve bone tissue regeneration, it is important to find a scaffold that can temporarily replace the damaged tissue and adapt to the biological environment of the host tissue with optimal porosity, thus allowing for the transport of a sufficient number of cells. In addition, scaffold must biodegrade within the host tissue at an appropriate rate. The proposed biomaterial must be able to interact with the surrounding tissue, and the patient's immune response must be minimized. Furthermore, the scaffold should provide adequate mechanical strength to the bone defect site and efficiently encapsulate and control the release of drugs or proteins [9,10].

It is well known that the process of bone defect repair is composed of an organic range of osteocytes, osteoblasts, and osteoclasts, called basic multicellular units (BMUS). BMUS act on the periosteum, trabecular surface, and cortex of bone, promoting osteogenesis absorption and bone repair. However, the adhesion, proliferation, and differentiation of BMUS should have a suitable cellular microenvironment and extracellular matrix (ECM), and, thereby, the question of how to simulate the natural ECM becomes the key to tissue engineering scaffolds [11]. The ECM is composed of a mixture of several biomolecules, including collagens and glycoproteins, arranged in distinct structures that are essentially unique to a specific tissue. The bone ECM determines the mechanical properties of the skeleton. The mineralized portion of the bone tissue imparts rigidity to the biomaterial, while the organic components of the ECM provide flexibility. The mineralized portion of the ECM is composed largely of calcium phosphate in the form of hydroxyapatite (HAP), plus an extensive type I collagen-rich organic ECM. In addition to mineralized bone ECM, other unique tissue types exist in association with bone ECM networks, such as non-mineralized marrow, the endosteum, the periosteum, and the perilucanar matrix, which can regulate the processes of osteoclast activity, osteoblast progenitor proliferation and differentiation, and osteocyte function. Until now, ECM has played a far more dynamic role in regulating cell function, tissue morphogenesis, modulating the matrix assembly, and organizing the process of matrix mineralization for bone remodeling. In addition, the ECM can bind to extracellular growth factors, cell-bound ligands and receptors, and proteases to exert its activities and functions [12–16].

Hydrogels have a hydrophilic nature with a three-dimensional structure and similar ECM components, which are easily chemically modifiable and can be further tuned to exhibit a favorable degradation profile and mechanical integrity, as well as incorporating cells, growth factors, or drugs, making them suitable scaffolds for cellular infiltration, adhesion, growth, proliferation, migration, and differentiation [17–19]. Moreover, the sustainable development of dynamic chemistries allows the use of hydrogels as nearly physiological matrices to recapitulate the dynamic interactions of native environments [20–22]. Therefore, based on the excellent hydrophilicity, adjustability, biocompatibility, and degradability for cell adhesion, proliferation, and differentiation, hydrogel scaffolds have been widely used in the biomedical field of tissue engineering research [23]. According to the differences in the source of gel materials, hydrogels are divided into synthetic polymer hydrogels and natural polymer hydrogels. On the one hand, the traditional polymer hydrogel mostly contains synthetic polymer materials, such as polyethylene glycol (PEG), polyacrylic acid (PAA), polymethyl methacrylate (PMMA), and so on. In addition, there is a complex synthesis process, large energy consumption, high cost, degradation in the rate of bone regeneration, and the elastic modulus cannot adapt to the needs of the specific microtissue environment, which has been an important reason restricting their application [24–30]. In addition, the performance of traditional hydrogel molding is individual and unchangeable in the treatment of large and complex bone defects, and it cannot accurately fill irregular defect parts. The above defects greatly limit its application breadth in bone defect repair. On the other hand, since natural materials can be used to conduct multiple structural and biological functions due to their outstanding range of macromolecular designs, which have originated from the evolution of living beings in different environments throughout millions of years, naturally derived hydrogels (chitosan, alginate, hyaluronan, collagen, agarose, etc.) are particularly appealing because of their inherent biocompatibility, biodegradability, safety, and accessible renewable resources, such as animals, plants, algae, and microorganisms around the world [31–33]. After the complex purification, fermentation, and other steps, these natural polymer hydrogels can achieve large-scale production to maintain the original biocompatibility and excellent biodegradation performance for a wide range of applications [34–36].

According to the types of stimulus sources, hydrogels are mainly divided into three categories (Figure 1): physical responsive hydrogels, chemical responsive hydrogels, and biochemical responsive hydrogels [37]. Since 1960, when Wichterle published the landmark article [38], the hydrogel has gradually developed from a simple, inert gel material to a complex, variable and controllable "smart" gel material with sensitive or responsive properties, which has been gradually introduced into the field of bone defect repair. It can respond to external physical stimuli (light, temperature, pressure, electric field, magnetic field, etc.), chemical stimuli (pH, redox response, ions, etc.), and biochemical stimuli, glucose, enzymes, etc.), and conduct shape transformation, produce injectability, and exhibit self-healing and shape memory properties to perform complex bone defect repair treatment that is minimally invasive; thus, it has become an emerging scaffold material in bone tissue engineering [39].

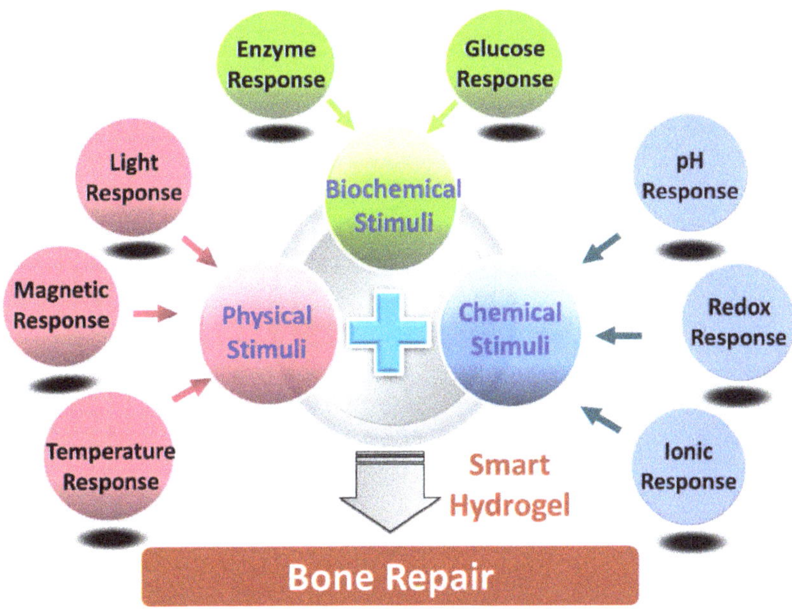

Figure 1. Schematic illustration of different types of smart hydrogels for bone repair.

Stimulus-responsive hydrogels can be prepared by the design of polymer molecular chains. Altering external environmental stimulation, such as temperature, pH, ionic strength, organic compound concentration, magnetic fields, electric fields, and light, can regulate the structure and physicochemical properties of the hydrogels, thus changing the swelling and degradation behavior of the hydrogels, which endows them with "smart" characteristics [40,41]. This allows smart hydrogels to have environmentally responsive gelation and degradation properties, so stimulus-responsive hydrogels can be widely used for bone tissue engineering. Through molecular design, the reactive hydrogel scaffolds, cells, and growth factors are injected into specific defect tissues, and, meanwhile, the stimulus-responsive hydrogels can not only positively provide mechanical and biological support for bone regeneration, but also gradually be degraded and absorbed by the tissue after implantation into the organism. With the disintegration of the hydrogel scaffold, the cells can constantly proliferate, differentiate, and secrete the ECM, which eventually enables new tissue formation and then achieves the corresponding functional bone reconstruction [42,43].

In this review, we will first introduce the classification and design concept of smart hydrogels based on different types of external environmental stimuli, and then discuss relevant examples of research progress using these stimulus-responsive hydrogels, especially focusing, but not limited to, bone repair applications.

2. Stimulus-Responsive Hydrogels

External environmental stimulation mainly includes light, temperature, pH value, redox, ionic strength, magnetic fields, electric fields, glucose, and enzymes. Therefore, the stimulus-responsive hydrogels can be divided into photo-responsive hydrogels, temperature-responsive hydrogels, pH-responsive hydrogels, redox-responsive hydrogels, magnetic-responsive hydrogels, enzyme-responsive hydrogels, and so on [44]. Different external stimulus environments require the corresponding design and fabrication of the hydrogel in the molecular structure. Once these external factors change to a critical point, stimulus-responsive hydrogels can rapidly undergo a discontinuous change or volume phase transformation. They have high biocompatibility, flexible design, high stability, and few side effects, and they can not only transport drugs for targeted release to achieve the purpose of targeted therapy in the field of clinical treatment, but also embed cells and serve as scaffolds for bone tissue repair in regenerative medicine. Table 1 summarizes the advantages and limitations of six responsive hydrogel categories.

Table 1. The advantages and limitations of stimulus-responsive hydrogel categories.

Types of Stimuli	Advantages	Limitations
Photo-responsive hydrogels	Mild reaction conditions; Low damage to human body; Spatio-temporal control of drug release without directly contacting the lesion	Ultraviolet and visible light cannot penetrate the tissue, which leads to the limited application only for in vitro system and skin-level treatments
Temperature-responsive hydrogels	injection capacity; Highly targeted and less toxic side effects; Effectively reduce the treatment cost for patients and improve their health-related quality of survival life	Low response rate; Low difference of pathological and normal tissues within the body
pH-responsive hydrogels	The pH of pathological tissues like local tissue inflammation, infection and cancer differs from that of normal tissues	The clinical prediction of the pH value in diseased sites may result in adverse tissue reactions
Redox-responsive hydrogels	Redox-responsive drug release; Relationship between metal ions and mechanical properties; Regulation of hydrogel hardness by redox reaction to promote bone regeneration	Low difference between the pathological and normal tissues limits the application
Magnetic-responsive hydrogels	Directional drug movement in a pathological state under the guidance of the environmental magnetic field can achieve the targeted therapy	The potential toxicity of magnetic nanoparticles may be harmful to live organisms
Enzyme-responsive hydrogels	Structural changes and quick degradation in response to specific enzymes promote the release of bio-factors for cell proliferation and differentiation	Weak peptides activity and low half-life limit the long-term use

2.1. Photo-Responsive Hydrogels

Light, as a remote stimulus, can present precise control in space and time [45]. Photo-responsive hydrogels include a polymer network and photochromic groups. These groups undergo light fracture, isomerization, and light dimer formation under different light conditions [46]. Photo-responsive hydrogels can be divided into two categories: covalent connection with the hydrogel (nitrobenzyl) and non-covalently connected but suspended in a network (Irgacure 2959, phenyl-2,4,6-lithium, eosin Y, etc.) of nitrobenzyl with the hydrogel [47–51]. For the first class of hydrogels, the light signal is captured by the photochromic molecules and then converted into a chemical signal by chromophores [52]. The size and mechanics of a photosensitive hydrogel can change due to the radiation of light-sensitive groups, which undergo light fracture, isomerization, and light dimer formation under different light conditions [53], such as crosslinking linearly functionalized polyacrylamide (PA) with a photosensitizer of azobenzene (AZO). For example, the photoinduced AZO-PA hydrogel prepared by Lee et al., changed its stiffness with the visible light wavelength, its degradation rate could match the bone generation time, and its biocompatibility was verified by mesenchymal stem cells. Time-resolved analysis of cell morphology showed characteristic cell spreading and increased aspect ratios in response to greater substrate stiffness. This hydrogel provided a platform to study mechanosignaling in cells responding to dynamic changes in stiffness, which is expected to play an important role in bone defects at different locations, offering a new way to study mechanotransduction signaling pathways and biological processes for bone tissue engineering [54]. For the second class of hydrogels, photoinitiators could decompose into free radicals and undergo a chemical reaction of polymerization or isomerization through the effect of light, causing changes in macromolecular chain conformation and the swelling volume. Khetan et al., reported a method for the gradual preparation of a methacrylic acid hyaluronic acid hydrogel, achieving the initial gelation of the hydrogel by introducing dithiothreitol into the methacrylic acid hyaluronic acid solution. When the initialized hydrogel was swollen by the photoinitiator of Irgacure 2959, the hydrogel could rapidly stiffen due to the free radical polymerization of the remaining methacrylate group [48]. Thus, both types of photoreactive hydrogels could regulate the physical or chemical properties of the hydrogels. The photoreaction fraction was sensitive to specific wavelengths (e.g., visible, ultraviolet, infrared light), which improved the control of the photo-responsive hydrogels [55].

2.2. Temperature-Responsive Hydrogels

Temperature-responsive hydrogels or temperature-sensitive hydrogels, based on the temperature difference from room temperature, can produce physical and chemical changes from sol to gel. Temperature-sensitive hydrogels have both hydrophilic and hydrophobic groups and phase transition properties with a temperature response at a critical solution temperature (CST), which could produce a change in affinity to the solvent. When performing the swelling–contraction state transition—that is, the sol–gel state transition—the temperature transition point is called the lowest critical dissolution temperature (LCST) or the utmost critical dissolution temperature (UCST) [56]. The most common LCST temperature-sensitive hydrogel used in bone repair is poly N-isopropyl acrylamide (PNIPAAm) and its derivatives. As a raw material for temperature-sensitive nanogels, by adding more hydrophilic acrylamide (AAm) to NIPAAm, the LCST rises from 32 °C to 37 °C, which is closer to the human body temperature. Yoshimatsu et al., synthesized poly (NIPAAm-co-AAm) copolymers and prepared their nanogels by radical polymerization [57]. The nanogels had a diameter of 50 to 450 nm and a volume phase transition temperature of 37 °C to 43 °C. Animal thermal targeting experiments combined with near-infrared fluorophore showed that gel delivery could be performed at specific locations by controlling the LCST and tissue heating processes of the nanogels. Antitumor drugs can also be loaded onto temperature-responsive nanogels. In bone defects, this nanogel can be targeted to treat cancer and bone defects. The temperature-responsive hydrogel system has been applied to release diverse growth factors and thus serve as a form of bone-regenerative medicine. The growth factors covalently bind to the injectable hydrogels and influence the growth of specific types of cells, which promotes cell proliferation, migration, recruitment, and angiogenesis, and modulates cell differentiation. Nafee et al., reported a temperature-responsive chitosan/β-GP hydrogel to effectively deliver bone resorption inhibitors and hinder the osteoclast action to treat Paget's disease and postmenopausal and glucocorticoid-induced osteoporosis. After encapsulating a BCS III bone resorption inhibitor, alendronate (ALN), the chitosan/β-GP hydrogel exhibited temperature-reversible gelation behavior and ensured controlled ALN release over 45–65 days, with a lower inflammatory response and faster proliferation and maturation of the granulation tissue. The biodegradability and biocompatibility of the system were confirmed upon analysis at 21 days after injection of the hydrogel [58].

In addition, copolymers of PEG and polycaprolactone (PCL) are also common LCST-type hydrogels. Ni et al., studied an injectable PEG-PCL-PEG hydrogel as a thermal inducing material for bone tissue engineering with reversibility recovery when PEG-PCL-PEG aqueous solution was changed from sol to gel [59]. This hydrogel had the advantages of being minimally invasive and precision matching in the treatment of bone tissue defects. Fu et al., combined collagen and HAP into a PEG-PCL-PEG copolymer to produce a hydrogel, and its biocompatibility and enhanced biomimetic microstructure endowed it with excellent performance in the treatment of bone defects [60]. Cai et al., prepared a dual network hydrogel (SHIELD) (Figure 2A), which can be used to directly inject transplanted stem cells, such as human adipose stem cells (hASCs) and human bone marrow mesenchymal stem cells (BM-MSCs) [61]. Molecular identification between peptides could form a weak network, which provided mechanical protection during injection and reduced cell damage caused by shear stress during injection. When the temperature increased to SHIELD LCST (ca. 34 °C), the enhanced hydrophobic interaction between PNIPAm polymer chains strengthened the network structure and improved the cell retention time, thus providing a microenvironment for cell proliferation and differentiation for bone repair (Figure 2B). Traditional temperature-sensitive materials include polyacrylic acid, gelatin, etc. However, their applications in bone repair are scarce because of their high solid-state temperature, which can easily cause damage to the body and is not conducive to material implantation and bone formation.

2.3. pH-Responsive Hydrogels

pH is the most intensively studied environmental stimulus in chemical stimulus-responsive hydrogels. The polymer hydrogels have many acidic or alkaline groups, which can quickly receive or release protons (protonation and deprotonation) in the environment, thus enabling an intelligent response to pH solutions. A hydrogel crosslinked by cellulose precursor macromolecules under the catalysis of phenylalanine had a pH response, and the gel time and mechanical strength could be adjusted by the content of precursor macromolecules and phenylalanine [62].

The pH-sensitive hydrogels are generally composed of the polymer backbone and ion side groups, which perform the conversion through the absorption or release of protons in response to the changes in the surrounding pH [63]. When the pH of the surrounding environment reaches pKa or pKb, the ionic strength of pH-sensitive hydrogels changes significantly, resulting in strong electrostatic repulsion and the appearance of ionic groups and volume mutation. Based on the charge properties of the charged groups, the pH-responsive hydrogels are divided into two categories, anionic and cationic hydrogels. Anionic hydrogels with negatively charged groups include carboxylic acid, sulfonic acid, etc. [64]. In high-pH conditions (pH > pKa), the charged groups expand the hydrogel. Similarly, a cationic hydrogel changes in low-pH conditions (pH > pKb). At the same time, some scholars have pointed out that in order to simulate the natural ECM, a weak polymeric electrolyte should be used to synthesize the pH-sensitive hydrogels. The ionic strength in a weakly polymerized electrolyte can be adjusted smoothly by changing the pH solution to achieve the desired mechanical properties [65]. The pH-responsive hydrogels had controllable pH by introducing weak electrolytes at their hydrophobic ends [66]. Yoshikawa et al., synthesized triblock copolymers using pH-sensitive poly(2-(diisopropylamino)ethyl methacrylate) (PDPA) and poly(2-(methacryloyloxy)ethyl phosphorylcholine) (PMPC) [67]. Simple modulation of the solution pH in a narrow physiologically relevant range produced highly adjustable hydrogels with Young's moduli of 1.4 to 40 kPa. Such hydrogels can simulate the natural environment well, even under complex stresses, and have wide applications in dealing with complex stresses caused by various bone defects. Rogina et al., demonstrated a pH-responsive chitosan–HAP hydrogel with a gelling agent of $NaHCO_3$. Tailoring of $NaHCO_3$ could achieve the quick gelation of the chitosan–HAP-based hydrogel within 4 min, exhibiting good viability for cell proliferation and differentiation as a potential cell carrier [68]. Lundberg et al., prepared a pH-dependent hydrogel by producing synthetic thiol-functionalized histamine [69]. The hardness of the hydrogels increased five-fold in the range of pH 5.0 to 8.0, and their biocompatibility in different cell lines greatly improved the mechanical properties of the hydrogel scaffolds, indicating promising applications in the repair of large-segment bone defects.

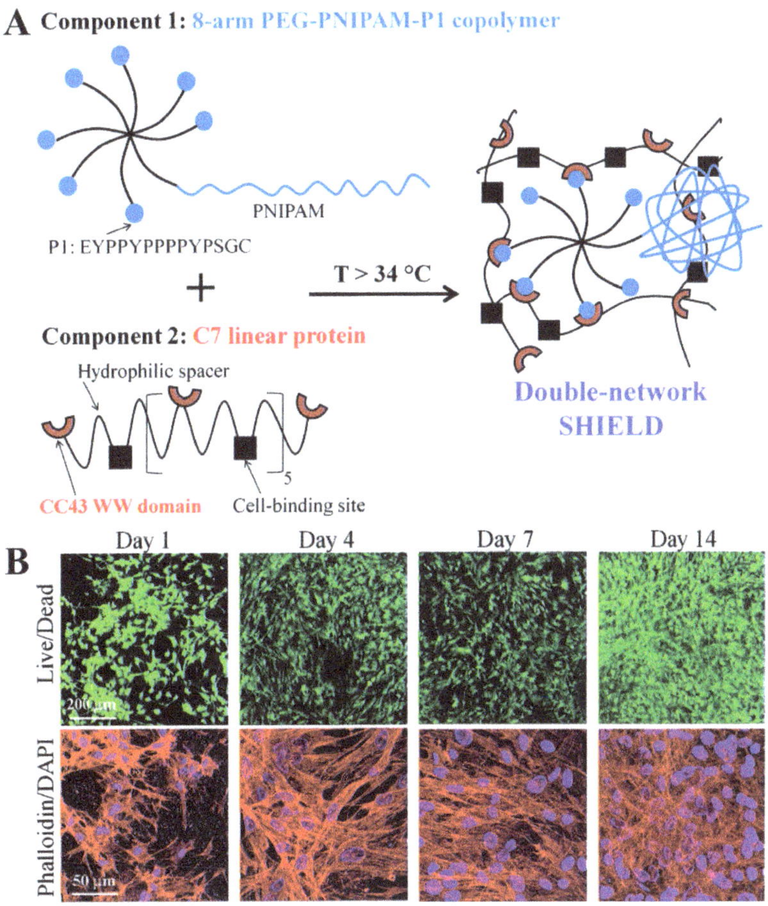

Figure 2. Retention of transplanted stem cells enhanced by thermo-responsive injectable hydrogels with in situ double network formation. (**A**) Structural design of SHIELD. (**B**) Confocal images of hASCs cultured in SHIELD. Reproduced from [61] with permission. Copyright 2015 Wiley.

2.4. Redox-Responsive Hydrogels

Redox-responsive hydrogels can react to the reduction and oxidation of their constituent molecules. The redox-responsive hydrogels undergo redox reactions through partial subunits within the polymer backbone, flooding opposing ions to balance the newly formed charges and eventually leading to the material's expansion [70]. Different metal ions confer different redox responsiveness, resulting in different mechanical properties. Iron ions are considered promising crosslinkers in biomedical applications, providing hydrogels with redox responsiveness and tunable mechanical properties by switching between two oxidation states, trivalent and divalent iron ions. This compensates for the lack of mechanical properties of simple hydrocoagulation and support. Papanikolaou et al., designed a new redox hydrogel that switched between divalent and trivalent iron ions and was reversible between soft (0.06 MPa) and hard (2.1 MPa) [71]. Many scholars have studied the relationship between metal ions and the mechanical properties of materials, and they have also confirmed that the regulation of hydrogel hardness by a redox reaction can promote bone regeneration [72–74]. However, further research is needed to investigate the relationship between metal ions and biomineralization, which may become a highly relevant topic in the field.

2.5. Magnetic Field-Responsive Hydrogels

Magnetic field-responsive hydrogels are generally composed of matrix hydrogels and magnetic components, which can remotely regulate the physical, biochemical, and mechanical properties due to their structural and functional responses to the external magnetic field. The properties of the magnetic field-responsive hydrogels depend on the composition, concentration, size, and uniformity of the magnetic particles in the hydrogel. When affected by the magnetic field effect, magnetic particles immediately gather, the hydrogel network contracts, and the solvent is "crowded out", causing the hydrogel shape to change rapidly [75]. Current studies have focused on the application of hydrogels containing magnetic nanoparticles for bone repair. Iqbal et al., synthesized magnetically modified Fe_2O_3 nanoparticles (m-nHAP) and added them to a polyvinyl alcohol (PVA) solution to prepare a m-nHAP/PVA hydrogel [76,77]. PVA, with excellent biocompatibility, high mechanical properties, and slow biodegradability, was essential for its application in bone repair. With the increase in m-nHAP content, the pore size inside the hydrogel gradually increased, which facilitated nutrient exchange and significantly increased the adhesion and proliferation of osteoblasts. Mahdavinia et al., recombined chitosan and magnetic Fe_3O_4 to produce magnetically responsive gel microspheres for the efficient adsorption of protein [78]. They fixed the Fe_3O_4 in situ in an inorganic thickening liquid, mixed the magnetic solution with polyvinyl alcohol and chitosan solution, and then repeatedly froze and melted the mixing solution to obtain the final gel sample. The results showed that, due to the introduction of the magnetic material into the gel microspheres, the maximum adsorption capacity was 240.5 mg/g. Isothermal adsorption data simulation showed that the sample adsorption process was more consistent with the Langmuir model than the Freundlich model. The research group also mixed Fe_3O_4 with carrageenan in situ to obtain magnetic nanoparticles, which were then crosslinked with chitosan as a crosslinker [79]. The introduction of magnetic particles into the carrageenan/chitosan complex had significant effects on the multiple properties of the hydrogel, and the results showed that with the increase in magnetic particles, the water absorption and encapsulation rate of the gel to the model drug increased significantly. The combination of xanthan gum and chitosan is also a new exploration. They can self-assemble in the presence of magnetic nanoparticles and form a magnetically responsive polyelectrolyte hydrogel under the action of glucuronic acid [80]. The results showed that the proliferation and adhesion capacity of fibroblasts on the gel were significantly enhanced under the influence of an external magnetic field. The addition of magnetic nanoparticles can also significantly improve the gel's mechanical strength and improve its rheological energy (increased energy storage modulus). Therefore, it is expected that this magnetic-responsive hydrogel will have potential applications in the field of bone defect repair.

It should be mentioned that magnetic materials have been proposed as potential agents to provide hydrogels with the anisotropy required for their use in tissue engineering. The intrinsic properties of magnetic nanoparticles enable their use as magnetomechanic remote actuators to control the behavior of the cells encapsulated within the hydrogels under the application of external magnetic fields. The incorporation of magnetic materials and the subsequent application of magnetic fields may present different advantages for bone tissue engineering strategies. First, magnetic nanoparticles can be used to provide the biomaterials with the visually anisotropic hierarchical architecture in native bone tissues, thus allowing the controlled design of anisotropic magnetically responsive scaffolding materials. In addition, the magnetic forces at the interface between cells and hybrid composites display the capacity for activating the sensitive receptors of the cell's surface, enhancing cell activity and promoting the bone formation process and the integration of scaffolds into the host bone [81–83]. In summary, magnetic hydrogels with anisotropic architectures provide not only an ordered 3D template in which the complex architectural properties of native tissues can be replicated but also add control over cell behavior. Therefore, there are specific strategies in the development of magnetic-responsive hydrogels with the required architectural properties to properly mimic different anisotropic tissues, such as tendons, bone, or cartilage (Figure 3) [84]. However, it has been mentioned that when the magnetic field-responsive hydrogels loaded with concentrated drugs are fixed to a body tissue that is in a pathological state, although the drugs can be accurately released from the hydrogels to achieve targeted controlled release for bone regeneration, the potential toxicity of magnetic nanoparticles derived from the degradation of the hydrogel-based scaffolds should be carefully considered, and the suitable usage concentration, purity, and geometry of the magnetic nanoparticles, as well as the rational fabrication of magnetically responsive hydrogels, remain challenging.

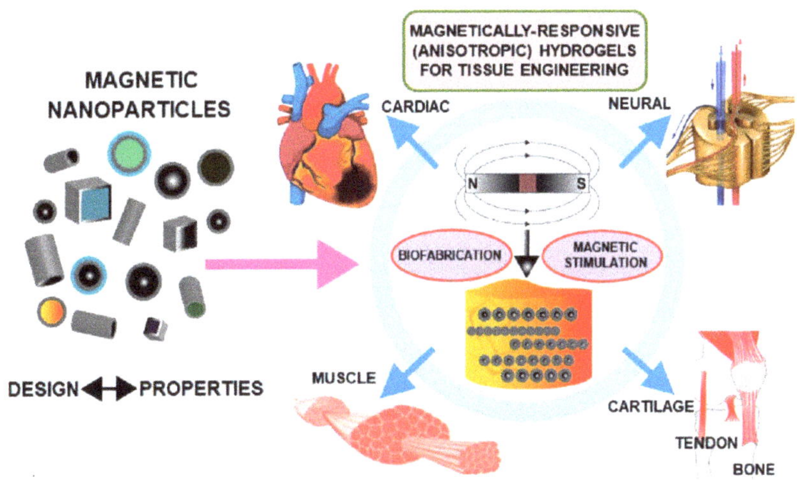

Figure 3. Schematic illustration: use of magnetic hydrogels to engineer different tissues of the human body. Reproduced from [84] with permission. Copyright 2021 American Chemical Society.

These stimulus-responsive hydrogels can also be used to fabricate constructs that replicate the main physicochemical of other tissues, such as tendons or tendon-to-bone interfaces. For example, Echave et al., developed a gelatin-based multiphasic hydrogel system with a distinct composition and microstructure [85]. In each phase, HAP particles or cellulose nanocrystals (CNC) were incorporated into an enzymatically crosslinked gelatin network to mimic bone or tendon tissue, respectively. Stiffer hydrogels were produced with the incorporation of mineralized particles, and magnetic alignment of CNC resulted in anisotropic structure formation. After the evaluation of biological commitment with human adipose-derived stem cells toward the tendon-to-bone interface, the results revealed aligned cell growth and higher synthesis and deposition of tenascin in the anisotropic phase, which indicated the potential versatility offered by the gelatin–transglutaminase tandem enzyme for the development of strategies to mimic the graded, composite, and complex intersections of the connective tissues.

2.6. Enzyme-Responsive Hydrogels

Enzyme-responsive hydrogels generally contain enzyme-responsive polypeptides, which can make structural changes in response to specific enzymes, thus promoting the formation or degradation of the hydrogel network. At the design level, enzyme-responsive hydrogels often use natural enzymes present in organisms or abnormally highly expressed in the lesion, such as matrix metalloproteinase (MMP) [86], phosphatase [87], and tyrosinase [88]. In the field of tissue engineering, enzyme-responsive hydrogels, acting as cell and protein carriers, can catalyze the degradation of the hydrogel cytoskeleton, thereby promoting the release of cell growth factors or providing an environment for cell proliferation and differentiation. Anjum et al., prepared a dual-responsive hydrogel based on the ECM sugar 695 aminoglycan (GAG) to regulate the delivery of cell growth factors and stem cell differentiation [89]. Using an MMP-sensitive glutamine transaminase factor XIII (FXIIIa)-specific lysine polypeptide sequence (TG-MMP-Lys) to functionalize the chondroitin sulfate (CS), CS was crosslinked with eight-arm PEG (PEG-Gln) modified by glutamine polypeptide to form a hydrogel. The cell adhesion polypeptide as a model ligand (TG-RGD-Lys) could improve the cell adhesion capacity of the hydrogel. After the encapsulation of bone-forming protein (BMP-2) and BM-MSCs into the hydrogels, BMSCs could maintain cell viability and realize proliferation and migration, and the released BMP-2 could induce the osteogenic differentiation of BMSCs (Figure 4). Therefore, a CS-PEG composite hydrogel can flexibly integrate the molecular tools that are needed to induce various tissues to simulate the characteristics of the extracellular environment, thus achieving the control of cell differentiation and tissue regeneration.

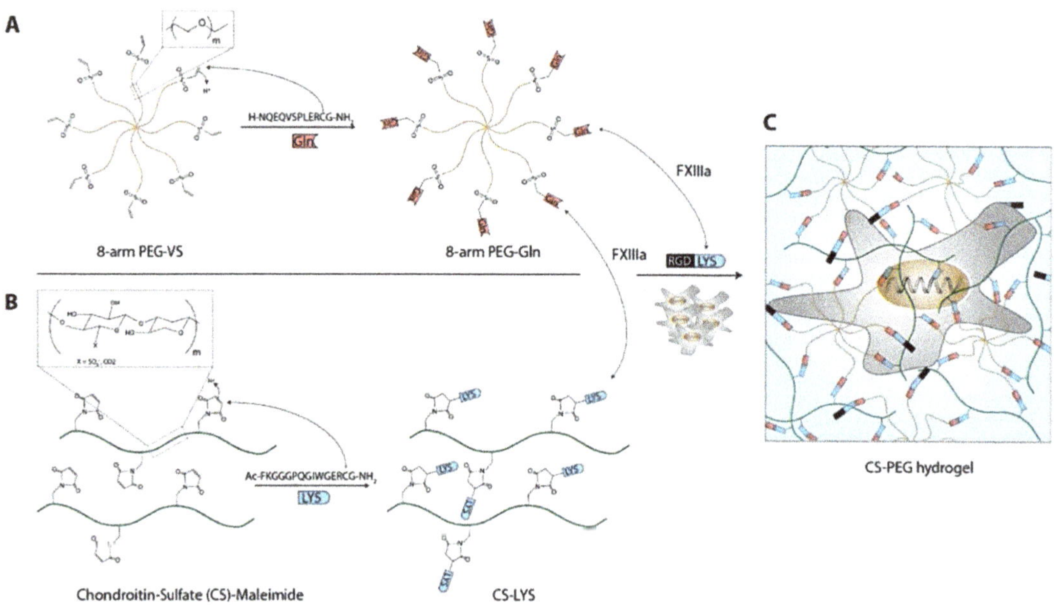

Figure 4. Factor XIII mediated CS-PEG hydrogel formation. The functionalization of (**A**) 8-PEG-VS with TG-Gln peptides resulted in 8-PEG-Gln and (**B**) CS-maleimide with CS-MMP-Lys in CSn-MMP-Lys. (**C**) Formulations of 8-PEG-Gln, CS-MMP-Lys, and the cell adhesion ligand TG-RGD-Lys were crosslinked by FXIIIa in presence of cells to form 3D cellular microenvironments. Reproduced from [89] with permission. Copyright 2016 Elsevier.

3. Conclusions and Perspectives

At present, stimulus responsive hydrogels are emerging as a new type of smart biomaterial that can sense external physical stimuli (light, temperature, pressure, electric field, magnetic field, etc.), chemical stimuli (pH, redox reaction, ions, etc.), biochemical stimuli (glucose, enzymes, etc.), and other different stimuli. They can respond to stimuli such as the characteristics of the 3D shape and solid–liquid phase state, and exhibit special properties, such as injection ability, self-repair, and shape memory, thus becoming an ideal material to provide tissue engineering scaffold adhesion, cell proliferation, and differentiation, and achieve precise bone defect repair. Nowadays, stimulus-responsive hydrogels not only have the advantages of being highly hydrophilic and good biocompatibility, but also possess regulated stimulus responsiveness, which makes them emerging scaffold materials, widely used for bone defect repair in tissue engineering. However, although many scholars have affirmed the efficacy of bone defect repair through stimulus-responsive hydrogels, their therapeutic effect on large bone defects has not been demonstrated due to the minimally invasive treatment of bone defects, the mechanical strength, and the degradation rate of bone regeneration. Thus, stimulus-responsive hydrogels with biocompatible, osteoconductive, osteoinductive, and osteogenic effects should have mechanical properties matching their degradation rate to meet the complex requirements of large bone defect repair.

During the last few decades, although significant progress ranging from cell biology up to advanced biomaterials has been made for tissue engineering in bone, we still have a long way to go to achieve functional bone tissue with the promise of revolutionizing healthcare by providing artificially engineered functional tissue and organ substitutes. In particular, the following three aspects still need further attention in the future.

First, although hydrogels with biocompatible composition and stimulus response are achievable for tissue-engineered scaffolds, a single response often cannot achieve the ideal treatment effect because of the complexity of the human physiological environment and the diversity of the lesion site environment; thus, a new kind of intelligent nanofiller-loaded hydrogel, such as magnetoelectric nanoparticles with multi-responsive performance, could be combined in the same carrier [90–92]. The hydrogel can choose the appropriate means of response according to the characteristics of the environment to achieve the ideal effect.

Second, the extensive development of precise fabrication and personalized medicine treatments for individual complexities will enable the engineering of biomaterials with precise structures and specific functions. Therefore, an in-depth study of the processing technologies and application of additive manufacturing in the biomedical field is needed, owing to its potential to provide personalized solutions for patients. For example, biofabrication is an emerging and rapidly growing research field in which additive manufacturing has been merged with tissue engineering to generate hierarchical tissue-like and personalized constructs. On the basis of 3D bioprinting technology, bioinks combining high-resolution printability with cytocompatibility have been developed as one ideal scaffold material for clinical translation, which will continue to be an active participant in the process of bone regeneration, not only as cells and molecular carriers, but also playing an important role in controlling delivery efficiency and delivery rate, thus making bioprinting hydrogel systems appealing alternatives for tissue engineering and drug delivery purposes, among others [93–96]. Therefore, it is of great significance to construct osteoblast and osteoclast co-culture models based on 3D printing techniques for promoting bone regeneration and studying the interaction between cells. In addition, the conception of green nanofabrication and the development of microfluidic particles with improved functionalities are essential for the creation of granular hydrogels, which will be also an innovative, green, sustained, and highly promising solution for different therapies in regenerative medicine areas [97].

Finally, further efforts to research hydrogel scaffolds with suitable degradation performance, mechanical properties, and vascular functionalization will be the main aspect and major challenge in the field of bone defect repair—for example, the design and development of extremely sensitive hydrogels that can be manipulated by using extremely weak external stimuli after implantation, thus avoiding the potential toxicity risks associated with some functional but harmful nanofillers (Au NPs, Ag NPs, MNPs) under strong magnetic radiations [88]. Another major challenge is the degree to which biomaterials bind to the local microenvironment in vivo. After biomaterial implantation, the altered microenvironment has a great impact on bone formation, so monitoring the material changes produced in the body in real time is crucial. It is believed that with the continuous progress of bone tissue engineering, stimulus-responsive hydrogels will develop rapidly to provide more solutions for the clinical treatment of bone defects and realize the transformation of clinical results.

Author Contributions: S.C., Z.L. and X.W. initiated the project. S.C., S.W., Z.L. and X.W. searched the database, and wrote and finalized the manuscript. All authors have read and agreed to the published version of the manuscript.

Funding: This work was supported by the National Natural Science Foundation of China (51973226) and Youth Innovation Promotion Association CAS (2019031).

Institutional Review Board Statement: Not applicable.

Informed Consent Statement: Not applicable.

Conflicts of Interest: The authors declare that they have no conflict of interest.

References

1. Tang, D.; Tare, R.S.; Yang, L.Y.; Williams, D.F.; Ou, K.L.; Oreffo, R.O. Biofabrication of bone tissue: Approaches, challenges and translation for bone regeneration. *Biomaterials* **2016**, *83*, 363–382. [CrossRef] [PubMed]
2. Fernandez-Yague, M.A.; Abbah, S.A.; McNamara, L.; Zeugolis, D.I.; Pandit, A.; Biggs, M.J. Biomimetic approaches in bone tissue engineering: Integrating biological and physicomechanical strategies. *Adv. Drug Deliv. Rev.* **2015**, *84*, 1–29. [CrossRef] [PubMed]
3. Zhang, D.W.; George, O.J.; Petersen, K.M.; Jimenez-Vergara, A.C.; Hahn, M.S.; Grunlan, M.A. A bioactive "self-fitting" shape memory polymer scaffold with potential to treat cranio-maxillo facial bone defects. *Acta Biomater.* **2014**, *10*, 4597–4605. [CrossRef] [PubMed]
4. Langer, R.; Vacanti, J. Tissue engineering. *Science* **1993**, *260*, 920–926. [CrossRef]
5. Xu, B.B.; Ye, J.; Fan, B.S.; Wang, X.J.; Zhang, J.Y.; Song, S.T.; Song, Y.F.; Jiang, W.B.; Wang, X.; Yu, J.K. Protein-spatiotemporal partition releasing gradient porous scaffolds and anti-inflammatory and antioxidant regulation remodel tissue engineered anisotropic Meniscus. *Bioact. Mater.* **2023**, *20*, 194–207. [CrossRef]
6. Fu, J.N.; Wang, X.; Yang, M.; Chen, Y.R.; Zhang, J.Y.; Deng, R.H.; Zhang, Z.N.; Yu, J.K.; Yuan, F.Z. Scaffold-based tissue engineering strategies for osteochondral repair. *Front. Bioeng. Biotechnol.* **2022**, *9*, 812383. [CrossRef]
7. Cheng, J.Y.; Liu, J.H.; Wu, B.; Liu, Z.Y.; Li, M.; Wang, X.; Tang, P.F.; Wang, Z. Graphene and its derivatives for bone tissue engineering: In vitro and in vivo evaluation of graphene-based scaffolds, membranes and coatings. *Front. Bioeng. Biotechnol.* **2021**, *9*, 734688. [CrossRef]
8. Chen, Y.; Li, W.L.; Zhang, C.; Wu, Z.Y.; Liu, J. Recent developments of biomaterials for additive manufacturing of bone scaffolds. *Adv. Healthc. Mater.* **2020**, *9*, 2000724. [CrossRef]
9. Rial, R.; González-Durruthy, M.; Liu, Z.; Ruso, J.M. Advanced materials based on nanosized hydroxyapatite. *Molecules* **2021**, *26*, 3190. [CrossRef]
10. Qu, H.W.; Fu, H.Y.; Han, Z.Y.; Sun, Y. Biomaterials for bone tissue engineering scaffolds: A review. *RSC Adv.* **2019**, *9*, 26252–26262. [CrossRef]
11. Jahan, K.; Tabrizian, M. Composite biopolymers for bone regeneration enhancement in bony defects. *Biomater. Sci.* **2016**, *4*, 25–39. [CrossRef]
12. Gong, T.; Xie, J.; Liao, J.F.; Zhang, T.; Lin, S.Y.; Lin, Y.F. Nanomaterials and bone regeneration. *Bone Res.* **2015**, 3, 15029. [CrossRef] [PubMed]
13. Alford, A.I.; Kozloff, K.M.; Hankenson, K.D. Extracellular matrix networks in bone remodeling. *Int. J. Biochem. Cell Biol.* **2015**, *65*, 20–31. [CrossRef] [PubMed]
14. Zhang, C.; Xia, D.D.; Li, J.J.; Zheng, Y.N.; Weng, B.W.; Mao, H.J.; Mei, J.; Wu, T.; Li, M.; Zhao, J.Y. BMSCs and osteoblast-engineered ECM synergetically promotes osteogenesis and angiogenesis in an ectopic bone formation model. *Front. Bioeng. Biotechnol.* **2022**, *10*, 818191. [CrossRef] [PubMed]
15. Lin, X.; Patil, S.; Gao, Y.G.; Qian, A.R. The bone extracellular matrix in bone formation and regeneration. *Front. Pharmacol.* **2020**, *11*, 757. [CrossRef]
16. Cao, H.F.; Wang, X.Y.; Chen, M.Y.; Liu, Y.H.; Cui, X.L.; Liang, J.; Wang, Q.G.; Fan, Y.J.; Zhang, X.D. Childhood cartilage ECM enhances the chondrogenesis of endogenous cells and subchondral bone repair of the unidirectional collagen–dECM scaffolds in combination with microfracture. *ACS Appl. Mater. Interfaces* **2021**, *13*, 57043–57057. [CrossRef]

17. Tang, G.K.; Tan, Z.H.; Zeng, W.S.; Wang, X.; Shi, C.G.; Liu, Y.; He, H.L.; Chen, R.; Ye, X.J. Recent advances of chitosan-based injectable hydrogels for bone and dental tissue regeneration. *Front. Bioeng. Biotechnol.* **2020**, *8*, 587658. [CrossRef]
18. Ribeiro, J.P.; Domingues, R.M.A.; Babo, P.S.; Nogueira, L.P.; Reseland, J.E.; Reis, R.L.; Gomez-Florit, M.; Gomes, M.E. Highly elastic and bioactive bone mimetic scaffolds based on platelet lysate and biomineralized cellulose nanocrystals. *Carbohy. Polym.* **2022**, *292*, 119638. [CrossRef]
19. Yu, T.T.; Wang, H.F.; Zhang, Y.F.; Wang, X.; Han, B. The delivery of RNA-interference therapies based on engineered hydrogels for bone tissue regeneration. *Front. Bioeng. Biotechnol.* **2020**, *8*, 445. [CrossRef]
20. Wang, X.; Gao, P.Y.; Yang, Y.Y.; Guo, H.X.; Wu, D.C. Dynamic and programmable morphology and size evolution via a living hierarchical self-assembly strategy. *Nat. Commun.* **2018**, *9*, 2772. [CrossRef]
21. Dou, X.Y.; Cao, Q.C.; Sun, F.F.; Wang, Y.Q.; Wang, H.F.; Shen, H.; Yang, F.; Wang, X.; Wu, D.C. Synergistic control of dual cross-linking strategy toward tailor-made hydrogels. *Sci. China Chem.* **2020**, *63*, 1793–1798. [CrossRef]
22. Wang, L.L.; Cao, Q.C.; Wang, X.; Wu, D.C. Visible light triggered controlled formation of rapidly self-healing hydrogels based on thiol-disulfide exchange. *Soft Matter* **2022**, *18*, 3004–3012. [CrossRef] [PubMed]
23. Hsu, R.S.; Chen, P.Y.; Fang, J.H.; Chen, Y.Y.; Chang, C.W.; Lu, Y.J.; Hu, S.H. Adaptable microporous hydrogels of propagating NGF-gradient by injectable building blocks for accelerated axonal outgrowth. *Adv. Sci.* **2019**, *6*, 1900520. [CrossRef] [PubMed]
24. Xin, S.; Chimene, D.; Garza, J.E.; Gaharwar, A.K.; Alge, D.L. Clickable PEG hydrogel microspheres as building blocks for 3D bioprinting. *Biomater. Sci.* **2019**, *7*, 1179–1187. [CrossRef]
25. Zhan, H.; Lwik, D.W.P.M. A hybrid peptide amphiphile fiber PEG hydrogel matrix for 3D cell culture. *Adv. Funct. Mater.* **2019**, *29*, 1808505. [CrossRef]
26. Leone, G.; Consumi, M.; Lamponi, S.; Bonechi, C.; Tamasi, G.; Donati, A.; Rossi, C.; Magnani, A. Thixotropic PVA hydrogel enclosing a hydrophilic PVP core as nucleus pulposus substitute. *Mater. Sci. Eng. C* **2019**, *98*, 696–704. [CrossRef]
27. Li, Y.; Zhu, C.; Fan, D.; Fu, R.Z.; Ma, P.; Duan, Z.G.; Li, X.; Lei, H.; Chi, L. A bi-layer PVA/CMC/PEG hydrogel with gradually changing pore sizes for wound dressing. *Macromol. Biosci.* **2019**, *19*, 1800424. [CrossRef]
28. Das, S.; Subuddhi, U. Controlled delivery of ibuprofen from poly(vinyl alcohol)-poly(ethylene glycol) interpenetrating polymeric network hydrogels. *J. Pharm. Anal.* **2019**, *9*, 108–116. [CrossRef]
29. Valentin, T.M.; DuBois, E.M.; Machnicki, C.E.; Bhaskar, D.; Cui, F.R.; Wong, L.Y. 3D printed self-adhesive PEGDA–PAA hydrogels as modular components for soft actuators and microfluidics. *Polym. Chem.* **2019**, *10*, 2015–2028. [CrossRef]
30. James, J.; Thomas, G.V.; Thomas, S. Fabrication of interpenetrating polymer networks of sbr and pmma with nano domains. *Mater. Today Proc.* **2019**, *9*, 77–84. [CrossRef]
31. Wang, X.; Yang, Y.Y.; Shi, Y.; Jia, F. Editorial: Smart hydrogels in tissue engineering and regenerative medicine. *Front. Chem.* **2020**, *8*, 245. [CrossRef] [PubMed]
32. Gomez-Florit, M.; Pardo, A.; Domingues, R.M.A.; Graça, A.L.; Babo, P.S.; Reis, R.L.; Gomes, M.E. Natural-based hydrogels for tissue engineering applications. *Molecules* **2020**, *25*, 5858. [CrossRef] [PubMed]
33. Tang, G.K.; Liu, Z.Q.; Liu, Y.; Yu, J.M.; Wang, X.; Tan, Z.H.; Ye, X.J. Recent trends in the development of bone regenerative biomaterials. *Front. Cell Dev. Biol.* **2021**, *9*, 665813. [CrossRef] [PubMed]
34. Gholizadeh, H.; Cheng, S.; Pozzoli, M.; Messerotti, E.; Traini, D.; Young, P.; Kourmatzis, A.; Ong, H.X. Smart thermosensitive chitosan hydrogel for nasal delivery of ibuprofen to treat neurological disorders. *Expert Opin. Drug Deliv.* **2019**, *16*, 453–466. [CrossRef]
35. Wu, B.Y.; Le, X.X.; Jian, Y.K.; Lu, W.; Yang, Z.Y.; Zheng, Z.K.; Théato, P.; Zhang, J.W.; Zhang, A.; Chen, T. pH and thermo dual-responsive fluorescent hydrogel actuator. *Macromol. Rapid Commun.* **2019**, *40*, e1800648. [CrossRef]
36. Lei, J.; Li, X.; Wang, S.; Yuan, L.; Ge, L.M.; Li, D.F.; Mu, C.D. Facile fabrication of biocompatible gelatin-based self-healing hydrogels. *ACS Appl. Polym. Mater.* **2019**, *1*, 1350–1358. [CrossRef]
37. Fan, Z.P.; Cheng, P.; Zhang, D.M.; Wang, W.L.; Han, J. Progress on stimulus responsive smart hydrogels based on natural polymers. *Mater. Rep.* **2020**, *34*, 21012–21025.
38. Wichterle, O.; Lim, D. Hydrophilic gels for biological use. *Nature* **1960**, *185*, 117–118. [CrossRef]
39. Wu, Y.; Wang, L.; Zhao, X.; Hou, S.; Guo, B.L.; Ma, P.X. Self-healing supramolecular bioelastomers with shape memory property as a multifunctional platform for biomedical applications via modular assembly. *Biomaterials* **2016**, *104*, 18–31. [CrossRef]
40. Koetting, M.C.; Peters, J.T.; Steichen, S.D.; Peppas, N.A. Stimulus-responsive hydrogels: Theory, modern advances, and applications. *Mater. Sci. Eng. R Rep.* **2015**, *93*, 1–49. [CrossRef]
41. Lei, L.J.; Bai, Y.J.; Qin, X.Y.; Liu, J.; Huang, W.; Lv, Q.Z. Current understanding of hydrogel for drug release and tissue engineering. *Gels* **2022**, *8*, 301. [CrossRef]
42. Samchenko, Y.; Ulberg, Z.; Korotych, O. Multipurpose smart hydrogel systems. *Adv. Colloid Interface Sci.* **2011**, *168*, 247–262. [CrossRef]
43. Wang, H.; Cai, L.J.; Zhang, D.G.; Shang, L.S.; Zhao, Y.J. Responsive janusstructural color hydrogel micromotors for label-free multiplex assays. *Research* **2021**, *1*, 1183–1191.
44. Lavrador, P.; Esteves, M.R.; Gaspar, V.M.; Mano, J.F. Stimuli-responsive nanocomposite hydrogels for biomedical applications. *Adv. Funct. Mater.* **2021**, *31*, 2005941. [CrossRef]
45. Burdick, J.A.; Murphy, W.L. Moving from static to dynamic complexity in hydrogel design. *Nat. Commun.* **2012**, *3*, 1269. [CrossRef] [PubMed]

46. Tomatsu, I.; Peng, K.; Kros, A. Photoresponsive hydrogels for biomedical applications. *Adv. Drug Deliv. Rev.* **2011**, *63*, 1257–1266. [CrossRef] [PubMed]
47. Zhao, H.; Sterner, E.S.; Coughlin, E.B.; Theato, P. O-nitrobenzyl alcohol derivatives: Opportunities in polymer and materials science. *Macromolecules* **2012**, *45*, 1723–1736. [CrossRef]
48. Khetan, S.; Guvendiren, M.; Legant, W.R.; Cohen, D.M.; Chen, C.S.; Burdick, J.A.A. Degradationmediated cellular traction directs stem cell fate in covalently crosslinked three-dimensional hydrogels. *Nat. Mater.* **2013**, *12*, 458–465. [CrossRef]
49. Fairbanks, B.D.; Schwartz, M.P.; Bowman, C.N.; Anseth, K.S. Photoinitiated polymerization of PEG-diacrylate with lithium phenyl-2,4,6-trimethylbenzoylphosphinate: Polymerization rate and cytocompatibility. *Biomaterials* **2009**, *30*, 6702–6707. [CrossRef]
50. DeForest, C.A.; Anseth, K.S. Photoreversible patterning of biomolecules within click-based hydrogels. *Angew. Chem. Int. Ed.* **2012**, *51*, 1816–1819. [CrossRef] [PubMed]
51. Katz, J.S.; Burdick, J.A. Light-responsive biomaterials: Development and applications. *Macromol. Biosci.* **2010**, *10*, 339–348. [CrossRef] [PubMed]
52. Feringa, B.L.; van Delden, R.A.; Koumura, N.; Geertsema, E.M. Chiroptical molecular switches. *Chem. Rev.* **2000**, *100*, 1789–1816. [CrossRef] [PubMed]
53. You, J.; Heo, J.S.; Kim, J.; Park, T.; Kim, B.; Kim, H.S.; Choi, Y.; Kim, H.O.; Kim, E. Noninvasive photo detachment of stem cells on tunable conductive polymer nano thin films: Selective harvesting and preserved differentiation capacity. *ACS Nano.* **2013**, *7*, 4119–4128. [CrossRef] [PubMed]
54. Lee, I.N.; Dobre, O.; Richards, D.; Ballestrem, C.; Curran, J.M.; Hunt, J.A.; Richardson, S.M.; Swift, J.; Wong, L.S. Photoresponsive hydrogels with photoswitchable mechanical properties allow time-resolved analysis of cellular responses to matrix stiffening. *ACS Appl. Mater. Interfaces* **2018**, *10*, 7765–7776. [CrossRef] [PubMed]
55. Helmchen, F.; Denk, W. Deep tissue two-photon microscopy. *Nat. Methods* **2005**, *2*, 932–940. [CrossRef]
56. Ullah, F.; Othman, M.B.; Javed, F.; Ahmad, Z.; Akil, H.M. Classification, processing and application of hydrogels: A review. *Mater. Sci. Eng. C* **2015**, *57*, 414–433. [CrossRef]
57. Yoshimatsu, K.; Lesel, B.K.; Yonamine, Y.; Beierle, J.M.; Hoshino, Y.; Shea, K.J. Temperature-responsive "catch and release" of proteins by using multifunctional polymer-based nanoparticles. *Angew. Chem. Int. Ed.* **2012**, *51*, 2405–2408. [CrossRef]
58. Nafee, N.; Zewail, M.; Boraie, N. Alendronate-loaded biodegradable smart hydrogel: A promising injectable depot formulation for osteoporosis. *J. Drug Target.* **2017**, *26*, 563–575. [CrossRef]
59. Ni, P.Y.; Ding, Q.X.; Fan, M.; Liao, J.F.; Qian, Z.Y.; Luo, J.C.; Li, X.Q.; Luo, F.; Yang, Z.M.; Wei, Y.Q. Injectable thermosensitive PEG-PCL-PEG hydrogel/acellular bone matrix composite for bone regeneration in cranial defects. *Biomaterials* **2014**, *35*, 236–248. [CrossRef]
60. Fu, S.Z.; Ni, P.Y.; Wang, B.Y.; Chu, B.Y.; Zheng, L.; Luo, F.; Luo, J.C.; Qian, Z.Y. Injectable and thermo-sensitive PEG-PCL-PEG copolymer/collagen/n-HA hydrogel composite for guided bone regeneration. *Biomaterials* **2012**, *33*, 4801–4809. [CrossRef]
61. Cai, L.; Dewi, R.E.; Heilshorn, S.C. Injectable hydrogels with in situ double network formation enhance retention of transplanted stem cells. *Adv. Funct. Mater.* **2015**, *25*, 1344–1351. [CrossRef] [PubMed]
62. Yang, X.F.; Liu, G.Q.; Peng, L.; Guo, J.H.; Tao, L.; Yuan, J.Y.; Chang, C.Y.; Wei, Y.; Zhang, L.N. Highly efficient self-healable and dual responsive cellulose-based hydrogels for controlled release and 3D cell culture. *Adv. Funct. Mater.* **2017**, *27*, 1703174. [CrossRef]
63. Kushwaha, S.K.; Saxena, P.; Rai, A. Stimuli sensitive hydrogels for ophthalmic drug delivery: A review. *Int. J. Pharm. Investig.* **2012**, *2*, 54–60. [CrossRef] [PubMed]
64. Hu, Y.H.; You, J.O.; Auguste, D.T.; Suo, Z.G.; Vlassak, J.J. Indentation: A simple, nondestructive method for characterizing the mechanical and transport properties of pH-sensitive hydrogels. *J. Mater. Res.* **2012**, *27*, 152–160. [CrossRef]
65. Auguste, D.T.; Armes, S.P.; Brzezinska, K.R.; Deming, T.J.; Kohn, J.; Prud'homme, R.K. pH triggered release of protective poly(ethylene glycol)-b-polycation copolymers from liposomes. *Biomaterials* **2006**, *27*, 2599–2608. [CrossRef] [PubMed]
66. Guix, M.; Meyer, A.K.; Koch, B.; Schmidt, O.G. Carbonate-based Janus micromotors moving in ultra-light acidic environment generated by HeLa cells in situ. *Sci. Rep.* **2016**, *6*, 21701. [CrossRef]
67. Yoshikawa, H.Y.; Rossetti, F.F.; Kaufmann, S.; Kaindl, T.; Madsen, J.; Engel, U.; Lewis, A.L.; Armes, S.P.; Tanaka, M. Quantitative evaluation of mechano-sensing of cells on dynamically tunable hydrogels. *J. Am. Chem. Soc.* **2011**, *133*, 1367–1374. [CrossRef]
68. Rogina, A.; Ressler, A.; Matić, I.; Ferrer, G.G.; Marijanović, I.; Ivanković, M.; Ivanković, H. Cellular hydrogels based on pH-responsive chitosan-hydroxyapatite system. *Carbohydr. Polym.* **2017**, *166*, 173–182. [CrossRef]
69. Lundberg, P.; Lynd, N.A.; Zhang, Y.N.; Zeng, X.H.; Krogstad, D.V.; Paffen, T.; Malkoch, M.; Nyström, A.M.; Hawker, C.J. pH-triggered self-assembly of biocompatible histamine-functionalized triblock copolymers. *Soft Matter* **2013**, *9*, 82–89. [CrossRef]
70. Greene, A.F.; Danielson, M.K.; Delawder, A.O.; Liles, K.P.; Li, X.S.; Natraj, A.; Wellen, A.; Barnes, J.C. Redox responsive artificial molecular muscles: Reversible radical-based self-assembly for actuating hydrogels. *Chem. Mater.* **2017**, *29*, 9498–9508. [CrossRef]
71. Papanikolaou, G.; Pantopoulos, K. Iron metabolism and toxicity. *Toxicol. Appl. Pharmacol.* **2005**, *202*, 199–211. [CrossRef]
72. Degtyar, E.; Harrington, M.J.; Politi, Y.; Fratzl, P. The mechanical role of metal ions in biogenic proteinbased materials. *Angew. Chem. Int. Ed.* **2014**, *53*, 12026. [CrossRef] [PubMed]
73. Niki, Y.; Matsumoto, H.; Suda, Y.; Otani, T.; Fujikawa, K.; Toyama, Y.; Hisamori, N.; Nozue, A. Metal ions induce bone-resorbing cytokine production through the redox pathway in synoviocytes and bone marrow macrophages. *Biomaterials* **2003**, *24*, 1447–1457. [CrossRef]

74. Zhang, J.Z.; Shang, Z.Z.; Jiang, Y.B.; Zhang, K.; Li, X.G.; Ma, M.L.; Li, Y.J.; Ma, B. Biodegradable metals for bone fracture repair in animal models: A systematic review. *Regen. Biomater.* **2021**, *8*, rbaa047. [CrossRef] [PubMed]
75. Ceylan, H.; Yasa, I.C.; Yasa, O.; Tabak, A.F.; Giltinan, J.; Sitti, M. 3D-printed biodegradable microswimmer for theranostic cargo delivery and release. *ACS Nano* **2019**, *13*, 3353–3362. [CrossRef]
76. Iqbal, H.; Khan, B.A.; Khan, Z.U.; Razzaq, A.; Khan, N.U.; Menaa, B.; Menaa, F. Fabrication, physical characterization sand in vitro antibacterial activity of cefadroxilloaded chitosan/poly(vinyl alcohol) nanofibers against Staphylococcus aureus clinical isolates. *Int. J. Biol. Macromol.* **2020**, *144*, 921–931. [CrossRef]
77. Hou, R.X.; Zhang, G.H.; Du, G.L.; Zhan, D.X.; Cong, Y.; Cheng, Y.J.; Fu, J. Magnetic nanohydroxyapatite/PVA composite hydrogels for promoted osteoblast adhesion and proliferation. *Colloids Surf. B Biointerfaces* **2013**, *103*, 318–325. [CrossRef]
78. Mahdavinia, G.R.; Soleymani, M.; Etemadi, H.; Sabzi, M.; Atlasi, Z. Model protein BSA adsorption onto novel magnetic chitosan/PVA/laponite RD hydrogel nanocomposite beads. *Int. J. Biol. Macromol.* **2018**, *107*, 719–729. [CrossRef]
79. Mahdavinia, G.R.; Mosallanezhad, A.; Soleymani, M.; Sabzi, M. Magnetic- and pH-responsive k-carrageenan/chitosan complexes for controlled release of methotrexate anticancer drug. *Int. J. Biol. Macromol.* **2017**, *97*, 209–217. [CrossRef]
80. Rao, K.M.; Kumar, A.; Han, S.S. Polysaccharide-based magnetically responsive polyelectrolyte hydrogels for tissue engineering applications. *J. Mater. Sci. Technol.* **2018**, *34*, 1371–1377. [CrossRef]
81. Yun, H.M.; Lee, E.S.; Kim, M.J.; Kim, J.J.; Lee, J.H.; Lee, H.H.; Park, K.R.; Yi, J.K.; Kim, H.W.; Kim, E.C. Magnetic nanocomposite scaffold-induced stimulation of migration and odontogenesis of human dental pulp cells through integrin signaling pathways. *PLoS ONE* **2015**, *10*, e0138614. [CrossRef] [PubMed]
82. Arjmand, M.; Ardeshirylajimi, A.; Maghsoudi, H.; Azadian, E. Osteogenic differentiation potential of mesenchymal stem cells cultured on nanofibrous scaffold improved in the presence of pulsed electromagnetic field. *J. Cell. Physiol.* **2018**, *233*, 1061–1070. [CrossRef] [PubMed]
83. Ganesh, N.; Ashokan, A.; Rajeshkannan, R.; Chennazhi, K.; Koyakutty, M.; Nair, S.V. Magnetic resonance functional nanohydroxyapatite incorporated poly(caprolactone) composite scaffolds for in situ monitoring of bone tissue regeneration by MRI. *Tissue Eng. Part A* **2014**, *20*, 2783–2794. [CrossRef] [PubMed]
84. Pardo, A.; Gomez-Florit, M.; Barbosa, S.; Taboada, P.; Domingues, R.M.A.; Gomes, M.E. Magnetic nanocomposite hydrogels for tissue engineering: Design concepts and remote actuation strategies to control cell fate. *ACS Nano* **2021**, *15*, 175–209. [CrossRef] [PubMed]
85. Echave, M.C.; Domingues, R.M.A.; Gómez-Florit, M.; Pedraz, J.L.; Reis, R.L.; Orive, G.; Gomes, M.E. Biphasic hydrogels integrating mineralized and anisotropic features for interfacial tissue engineering. *ACS Appl. Mater. Interfaces* **2019**, *11*, 47771–47784. [CrossRef]
86. Yu, S.; Duan, Y.Y.; Zuo, X.G.; Chen, X.Y.; Mao, Z.W.; Gao, C.Y. Mediating the invasion of smooth muscle cells into a cell-responsive hydrogel under the existence of immune cells. *Biomaterials* **2018**, *180*, 193–205. [CrossRef]
87. Toda, H.; Yamamoto, M.; Uyama, H.; Tabata, Y. Fabrication of hydrogels with elasticity changed by alkaline phosphatase for stem cell culture. *Acta Biomater.* **2016**, *29*, 215–227. [CrossRef]
88. Liu, H.Y.; Korc, M.; Lin, C.C. Biomimetic and enzyme-responsive dynamic hydrogels for studying cell-matrix interactions in pancreatic ductal adenocarcinoma. *Biomaterials* **2018**, *160*, 24–36. [CrossRef]
89. Anjum, F.; Lienemann, P.S.; Metzger, S.; Biernaskie, J.; Kallos, M.S.; Ehrbar, M. Enzyme responsive GAG-based natural-synthetic hybrid hydrogel for tunable growth factor delivery and stem cell differentiation. *Biomaterials* **2016**, *87*, 104–117. [CrossRef]
90. Wang, P.; Zhang, E.; Toledo, D.; Smith, I.T.; Navarrete, B.; Furman, N.; Hernandez, A.F.; Telusma, M.; McDaniel, D.; Liang, P.; et al. Colossal Magnetoelectric effect in core-shell magnetoelectric nanoparticles. *Nano. Lett.* **2020**, *20*, 5765–5772. [CrossRef]
91. Mehdipour, M.; Gloag, L.; Bennett, D.T.; Hoque, S.; Pardehkhorram, R.; Bakthavathsalam, P.; Gonçales, V.R.; Tilley, R.D.; Gooding, J.J. Synthesis of gold-coated magnetic conglomerate nanoparticles with a fast magnetic response for biosensing. *J. Mater. Chem. C* **2021**, *9*, 1034–1043. [CrossRef]
92. Wang, X.; Wang, J.; Yang, Y.Y.; Fei Yang, F.; Wu, D.C. Fabrication of multi-stimuli responsive supramolecular hydrogels based on host–guest inclusion complexation of a tadpole-shaped cyclodextrin derivative with the azobenzene dimer. *Polym. Chem.* **2017**, *8*, 3901–3909. [CrossRef]
93. Bakht, S.M.; Pardo, A.; Gómez-Florit, M.; Reis, R.L.; Domingues, R.M.A.; Gomes, M.E. Engineering Next-generation bioinks with nanoparticles: Moving from reinforcement fillers to multifunctional nanoelements. *J. Mater. Chem. B* **2021**, *9*, 5025. [CrossRef]
94. Ashammakhi, N.; Hasan, A.; Kaarela, O.; Byambaa, B.; Sheikhi, A.; Gaharwar, A.K.; Khademhosseini, A. Advancing frontiers in bone bioprinting. *Adv. Healthcare Mater.* **2019**, *8*, 1801048. [CrossRef] [PubMed]
95. Tavares, M.T.; Gaspar, V.M.; Monteiro, M.V.; Farinha, J.P.S.; Baleizão, C.; Mano, J.F. GelMA/bioactive silica nanocomposite bioinks for stem cell osteogenic differentiation. *Biofabrication* **2021**, *13*, 035012. [CrossRef]
96. Chimene, D.; Miller, L.; Cross, L.M.; Jaiswal, M.K.; Singh, I.; Gaharwar, A.K. Nanoengineered osteoinductive bioink for 3D bioprinting bone tissue. *ACS Appl. Mater. Interfaces* **2020**, *12*, 15976–15988. [CrossRef]
97. Rial, R.; Hassan, N.; Liu, Z.; Ruso, J.M. The Design and green nanofabrication of noble hydrogel systems with encapsulation of doped bioactive hydroxyapatite toward sustained drug delivery. *J. Mol. Liq.* **2021**, *343*, 117598. [CrossRef]

gels

MDPI

Review

Self-Assembling Peptide Hydrogels as Functional Tools to Tackle Intervertebral Disc Degeneration

Cosimo Ligorio [1,2,3,*], Judith A. Hoyland [3] and Alberto Saiani [1,2]

1 Department of Materials, School of Natural Sciences, Faculty of Science and Engineering, The University of Manchester, Manchester M1 3BB, UK; a.saiani@manchester.ac.uk
2 Manchester Institute of Biotechnology (MIB), The University of Manchester, Manchester M1 7DN, UK
3 Division of Cell Matrix Biology and Regenerative Medicine, School of Biological Sciences, Faculty of Biology, Medicine and Health, The University of Manchester, Manchester M13 9PG, UK; judith.a.hoyland@manchester.ac.uk
* Correspondence: cosimo.ligorio@manchester.ac.uk

Abstract: Low back pain (LBP), caused by intervertebral disc (IVD) degeneration, is a major contributor to global disability. In its healthy state, the IVD is a tough and well-hydrated tissue, able to act as a shock absorber along the spine. During degeneration, the IVD is hit by a cell-driven cascade of events, which progressively lead to extracellular matrix (ECM) degradation, chronic inflammation, and pain. Current treatments are divided into palliative care (early stage degeneration) and surgical interventions (late-stage degeneration), which are invasive and poorly efficient in the long term. To overcome these limitations, alternative tissue engineering and regenerative medicine strategies, in which soft biomaterials are used as injectable carriers of cells and/or biomolecules to be delivered to the injury site and restore tissue function, are currently being explored. Self-assembling peptide hydrogels (SAPHs) represent a promising class of *de novo* synthetic biomaterials able to merge the strengths of both natural and synthetic hydrogels for biomedical applications. Inherent features, such as shear-thinning behaviour, high biocompatibility, ECM biomimicry, and tuneable physiochemical properties make these hydrogels appropriate and functional tools to tackle IVD degeneration. This review will describe the pathogenesis of IVD degeneration, list biomaterials requirements to attempt IVD repair, and focus on current peptide hydrogel materials exploited for this purpose.

Keywords: self-assembling peptide hydrogels; intervertebral disc; tissue engineering

1. Introduction

1.1. Structure and Function of the IVD

The IVD, the building block of the spinal motion segment, was anatomically described for the first time at the University of Padua by the famous Italian anatomist Andreas Vesalius in 1555, who introduced the intervertebral body to early anatomists [1]. The human spine is composed of a total of at least 23 IVDs interposed between adjacent vertebrae, joining them from the axis to the sacrum [2]. The IVDs constitute one-third of the total spinal column's length, with 6 IVDs present in the cervical region, 12 in the thoracic region, 5 in the lumbar region, and 1 between the sacrum and the coccyx [2]. Each IVD is a pad of fibrocartilage, 7–10 mm thick and with an average diameter of 4 cm, which is composed of three integrated tissues: the nucleus pulposus (NP), the annulus fibrosus (AF), and the cartilage endplates (CEP) [3] (Figure 1). Schematically, the AF represents a thick outer ring of fibrous cartilage comprised of concentric layers of type I collagen that circumferentially surrounds a proteoglycan (PG)-rich gelatinous core (the NP tissue) to constitute a composite tissue sandwiched inferiorly and superiorly by two thin layers of cartilage, better known as CEP [3]. Fibres from the inner AF tissue bridge this tissue to CEP, while fibres coming from the outer layer of AF are anchored within the bony endplates of vertebrae to form a continuous vertebral body. Longitudinal ligaments (anterior and posterior) laying at the front and at the back of the vertebral body, along with the spinous processes, constitute a 'three-joint complex' called the vertebral unit (Figure 1). As the name suggests, the vertebral unit represents the unit of the column, dictating its biomechanics and biological function [2].

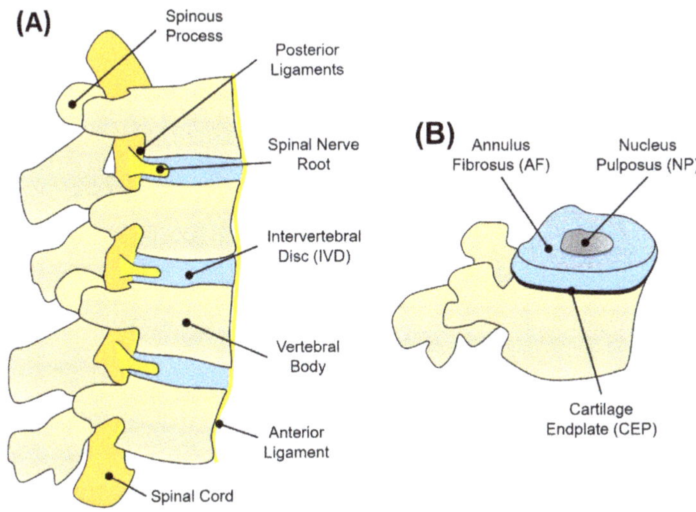

Figure 1. (**A**) Graphical representation of adjacent vertebral units in sagittal view. Each unit consists of vertebral bodies surrounding an IVD. The nerve supply of the IVD and vertebral bodies consists of the spinal cord disposed longitudinally along the vertebrae and passing through the intervertebral foramen. Vertebrae are kept in place by anterior and posterior ligaments. (**B**) Schematic representation of the IVD, showing the NP, AF, and CEP regions.

Biomechanically, the IVD acts as a shock absorber by transmitting loads along the spine. External loads exerted on the spine are mainly related to body weight and muscle traction to maintain equilibrium, but also different postures, body motion, and carrying weights are referred to as stress components acting on the column [4]. It has been estimated that IVDs are able to tolerate multiple cycles of loading such as compression, tension, torsion, and bending up to 2.5 MPa, from a minimum of 0.1 MPa for simple resting to a maximum of 2.3 MPa for carrying a weight of 20 kg in a flexed position [5]. The biomechanics of the IVD can be approximated as a mechanical interplay existing between the NP and AF. Under compression, hydrostatic pressure is created within the gelatinous NP, which in turn generates tensile stress (or 'hoop' stress) in the surrounding cartilaginous layers of the AF to counteract the original load [6]. The success behind this biomechanical response relies on the NP's high water content, which is needed to generate the hydrostatic pressure that increases in response to compressive loading, along with a high shear resistance and complex collagen network between AF's lamellae, which enable the AF to develop tensile stress opposite to the NP's pressure [6]. This biomechanical balance is essential to absorb mechanical forces and dissipate their energy. Moreover, under compression, the limited swelling of the CEP and its elasticity are fundamental to reducing the rate of water expelled from the pressurised NP, allowing the NP to swell and expand [6].

1.2. Nucleus Pulposus (NP)

The NP is a soft, highly hydrated and gelatinous tissue that occupies the central region of the IVD (Figure 1B). This tissue appears translucent and gel-like in the healthy state, with a heterogeneous structure composed of water, glycosaminoglycans (GAGs), collagen fibres, mineral salts, and cellular components [7]. The amount of water in the NP is around 80–90% in the early life, with a progressive decrease up to 70% at ages over 60, when the NP becomes white in colour, fibrotic, and amorphous [2,7]. Structurally, the NP can be considered as a three-dimensional (3D) viscoelastic network mainly composed of a PG-rich gel, in which multiple populations of collagen and non-collagenous proteins are dispersed. PGs represent up to 65% of the NP's dry weight in the early stages, and they are responsible for binding water within the tissue due to their negatively charged residues [8]. Although many comparisons are made between the NP and the articular cartilage (AC), in the NP, the proteoglycans (PGs) to fibrillar collagen ratio has been reported to be 27:1 compared to the 2:1 ratio for AC, making the NP a more hydrated and softer tissue compared to AC [9]. Aggrecan represents the most abundant PG in the NP, although traces of other GAGs, including hyaluronan, chrondroitin sulphates, decorin, lumican, biglycan, fibromodulin, and versican, have also been reported in the NP tissue [10–12]. In this PG-rich network, collagen fibres (principally collagen type II) are randomly dispersed and highly interconnected with PG fibres, constituting 15–20% of the dry weight (Figure 2). Collagen fibres, including type II, VI, IX, and XI, form a loose, secondary interpenetrating network, which has the scope with PGs to hold the NP together [13,14]. The remaining dry weight, accounting for 10–25%, is represented by non-collagenous fibres (mainly elastin up to 150 μm in length), which are arranged radially across the NP [12].

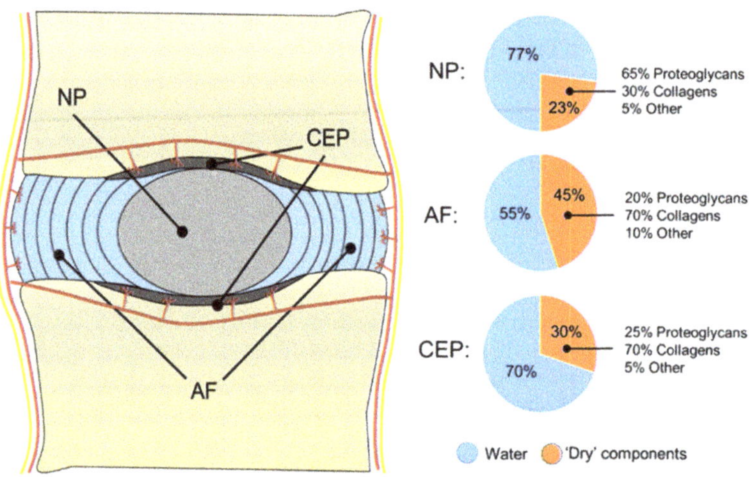

Figure 2. Schematic representation of the human IVD, showing its water and ECM content. The IVD is surrounded by blood and nerve vessels. Capillaries penetrate a few millimetres into the outer AF to provide nutrients and waste exchange. Cells from the avascular NP and inner AF receive nutrients and are able to exchange waste products through a bidirectional flow occurring via blood capillaries that penetrate the subchondral plate and reach the CEP.

In this ECM, the cell population of the NP is derived from the notochord during embryonic development, with large vacuolated embryonic notochordal cells (25–85 μm in size) that are gradually replaced by smaller and rounded 'chondrocyte-like' NP cells (10 μm in size) in juvenile discs, which have a lower metabolic activity than notochordal cells [15]. NP cells are sparse within the tissue, with an average cell density (4000 cells mm^{-3}) lower than the AF and CEP, as well as three-fold lower than the average number in AC [16,17]. Although NP cells have a similar morphology to AC chondrocytes, an extensive gene expression profiling conducted by Minogue et al. on human NP cells has identified a pool of specific markers upregulated in NP cells but not articular chondrocytes. These studies have allowed an improved characterisation of NP cells, making the comparison between the two cell types clearer [18,19]. The number of NP cells within the tissue tends to decrease with age, when an increased number of macrophages, T-cells, and lymphocytes start to populate the tissue due to blood vessel infiltration, contributing to the pro-inflammatory milieu that affects NP tissue during fibrosis degeneration and herniation [20,21]. Moreover, the normal NP is an avascular and aneural tissue, where NP cells are far at least 8 mm from the nearest blood supply [22], which makes nutrients and waste exchange based solely on diffusion and self-healing extremely difficult to achieve [23]. In this scenario, NP cells rely on anaerobic cellular respiration to produce adenosine triphosphate (ATP) for their vital functions, so lactate as the by-product of glycolysis is largely accumulated in the NP. This makes the NP a relatively hostile microenvironment, with a relatively low pH (pH < 6.5) and low oxygen content (O_2 levels < 5%) [24].

1.3. Annulus Fibrosus (AF)

The NP is circumferentially surrounded by the AF, which is composed of 15–25 concentric rings made of alternating lamellae of collagen type I inclined by 60° with respect to the main axis of the column [25] (Figure 1B). Although a clear demarcation region between NP and AF was drawn historically [2], it has been proved that the two tissues are extremely integrated, with a thin region between the two called the transition zone (TZ) having increased metabolic activity and sensitivity to normal forces and growth factor (GF) modulation [26]. This region is fundamental for the NP to expand at the expense of the AF, while for the AF, this region is necessary to sense the hydrostatic pressure coming from the NP and exert a mechanical resistance to counteract it [26]. The building blocks of the AF are the lamellae, which are 50–500 μm-thick collagen layers with thickness increasing from the inside to the outside [14,25]. Each lamella is composed of 100–200 nm collagen type I fibre bundles, uniformly oriented at 60° to the vertical axis, alternating to the left and right between adjacent lamellae [14,25]. The interconnection between lamellae is little, and elastin fibres have been found among collagen fibrils within the same lamellae [12]. This matrix organisation makes the lamellae prone to expand during compression, allowing different degrees of movement, still ensuring elastic recoil to its original position when the stress (e.g., bending, tension, and compression) is suppressed. The AF can be divided into the inner and outer annulus according to the distance to the NP. Moving from the inner part to the outer, the content of PGs, water, and collagen type II decreases, while the content of collagen type I increases, making the outer rings of AF stiffer and elastic, while the inner rings are softer and more deformable. As also occurs for tendons, collagen type I provides tension resistance in the AF, while collagen type II fibres intertwined with PGs help to bind and keep water molecules together in order to keep the AF stable against large compressive forces [13,27]. In the AF, all the collagen types represent 50–70% of the total dry weight (with collagen I being the dominant type), while 20–25% is provided by PGs and the remaining 5–10% by elastin fibrils [7] (Figure 2).

Regarding the cellular component, the AF is populated by elongated, fibroblast-like cells that tend to align to the collagen fibres within the lamellae, especially in the outer rings of the tissue [2]. The cells of the AF present large protrusions (up to 30 μm long) that have been not observed in AC, which have been suggested to act as mechanical strain sensors within the tissue to regulate shock absorbance [28]. Like the NP, the AF is largely avascular and aneural, although evidence of penetrating blood vessels and sensory nerve fibres up to a few millimetres in the outer rings have been found, which may contribute to nutrient transport and waste removal [23]. However, substantial nerve and blood vessel infiltration occur during IVD degeneration, which, coupled with a limited self-healing capability, makes AF injuries difficult to repair without surgical intervention [29,30].

1.4. Cartilage Endplate (CEP)

The third component of the IVD is represented by the CEP (Figure 1B). These endplates are thin layers of hyaline cartilage (~0.6 mm) that interface the avascular AF and NP to the highly vascularised vertebrae, specifically with vertebral bodies' endplates called bony endplates [31]. The thickness of the CEP is thinner at the centre and larger at the periphery so that NP tissue that stands at the core of the IVD can swell and expand easily by pushing on the CEP's inner surface [32]. The CEP is fundamental to maintaining the IVD and the vertebral compartments as distinct mechanically entities, but at the same time, it allows them to communicate biochemically by ensuring nutrition and waste exchange through capillaries and blood vessels that pass through the endplates into the IVD. Interestingly, Nachemson et al. showed that only the central part of the CEP is permeable to the diffusion of dyes or radioactive substances, leaving the rest of the nutrients mainly supplied by a network of vascular channels within the vertebral bodies [33].

Being constituted of hyaline cartilage, the CEP's dry weight is mainly composed of collagen fibres (75% of the dry weight), e.g., collagen type II, but also type III, VI, and X, as well as PGs such as decorin and biglycans (25% of the dry weight, Figure 2). The collagen content is higher at the periphery, while the PG content is higher at the centre of the CEP to allow mechanical compliance [34,35]. Collagen fibres within the CEP run horizontally to its surface in contact with the IVD, but some fibres also continue to the disc merging with fibres coming from the AF, as well as anchoring with the bone trabeculae of the bone endplates to ensure connections along the column [31]. The CEP is partly mineralised, but it is also populated by a family of chondrocytes that can become hypertrophic, especially during IVD degeneration, when CEP's calcification is increased, and the flow of nutrients and waste exchange is highly affected [36].

1.5. Anabolic Growth Factors Present in the IVD

As defined by Hynes, native ECMs are not 'just pretty fibrils', but they are a reservoir of growth factors (GFs), which are bound to the ECM proteins or dispersed in the ECM as soluble factors [37]. Indeed, every tissue for its differentiation and maturation require specific GFs, which are the master regulators of intracellular pathways. A broad range of GFs is known to be present in the IVD, including many members of the transforming growth factor β (TGF-β) superfamily. In particular, TGF-β molecules play an important role in IVD growth and homeostasis. TGF-β signalling orchestrates tissue formation at the embryonic stage via notochord condensation and supports tissue homeostasis at postnatal stages by promoting the synthesis of PGs and collagens, which dictate the correct structure and functionality of the native NP [38]. Gruber et al., for example, demonstrated that TGF-β1 could induce cell proliferation and reduce disc cell apoptosis, as well as promote ECM synthesis in vivo [39]. Similarly, Jin and co-workers observed that the suppression of TGF-β3 signalling in mice causes a notable reduction in the area and length of CEP tissue, resulting in NP cells being deprived of nutrients and prone to adopt a catabolic, pro-inflammatory phenotype [40]. Along with TGF-β molecules, growth differentiation factor (GDF) molecules are also highly involved in tissue homeostasis and matrix turnover in the IVD. Recombinant GDF-5, for example, has shown increased PGs and collagen type II gene and protein expression *in vitro*, both with mouse and bovine NP cells [41]. Similarly, GDF-6 has been shown to play a crucial role in the development of bones and joints by forming morphogenetic gradients during development, while knockout of GDF-6 in the disc reveals spine defects and reduced PG content in mice [41]. Due to their biological function, GFs, along with ECM components, should be considered active players and regulators of IVD during health and disease.

2. Pathophysiology of Intervertebral Disc Degeneration

2.1. Alterations in the ECM of Degenerated Discs

In the IVD, resident cells are responsible for maintaining a tight balance between anabolic and catabolic processes so that any imbalance towards degradative processes can lead to matrix catabolism and tissue breakdown in the IVD. Although the exact causes of this imbalance are not fully elucidated, new evidence has highlighted several factors that may contribute to, or may be responsible for, IVD degeneration. ECM breakdown, altered matrix synthesis, and changes in cell number, phenotype, and behaviour are characteristic features of the IVD [42]. During the early stages of degeneration, disc cells adopt a defensive mechanism by increasing their secretion of collagen type II molecules in an attempt to counteract ECM degradation and start a repair mechanism [43]. This phenomenon is only temporary since, with advancing degeneration, collagen type II synthesis shifts dramatically to collagen type I, specifically in the NP and in the inner AF [7,44]. In the NP, the collagen fibres shift more to the type I form and become less packed and looser due to a decrease in the pyridinoline cross-links between collagen fibres. This conformational change affects the biomechanics of the NP, which starts to become stiffer and granular [43]. A decrease in collagen type II synthesis is also accompanied by an increase in the production of collagen type X, which causes NP's hypertrophy and CEP calcification [45]. In particular, calcified CEP results in a limited diffusion of nutrients and waste into and out of the NP so that lactate accumulates, leading to tissue acidification. In healthy IVD, the pH has been shown to be ~7.1, while during degeneration, it can drop to 6.8 (mildly degenerated discs) or 6.5 (severely degenerated discs) [46]. Low pH values have been associated with reduced cell viability, as well as the reduced production of collagen and PGs in several 2D and 3D studies [47]. Gilbert and co-workers, for example, showed that human NP cells exposed to acidic pH (6.8 and 6.5) are able to sense low pH via acid-sensing ion channel-3 proteins (ASICs) proteins, resulting in a catabolic and degenerated phenotype for NP cells. Interestingly, when ASICs were inhibited, proinflammatory and pain-related makers in NP cells were downregulated, suggesting a correlation between low pH and catabolic phenotype [48].

In addition to changes in collagen expression, the disc also experiences a decrease in PGs [49]. Indeed, although degenerated NP cells are still capable of producing aggrecan and versican, these molecules become shorter and fragmented so that the overall content of PGs dramatically decreases [50] (Figure 3). As the aggrecan molecules of degenerated discs are disrupted and are in the form of fragments, they can only retain smaller quantities of water than their fully formed version, and they can easily leach out of the tissue due to their smaller dimensions [3]. During degeneration, as the content of aggrecan decreases, the synthesis of versican, biglycan, and decorin increases [44]. However, versican contains fewer chondroitin sulphate chains than aggrecan, so they cannot compensate for the water binding and high osmotic pressure lost with aggrecan degradation [3]. The result is a tissue that is less able to retain water and becomes dehydrated over time. Moreover, increased production of collagen type I leads the NP to become more fibrous and less compliant as the degeneration progresses. Among the other ECM molecules, fibronectin is another ECM component whose content as a fragmented molecule increases during disc degeneration [51]. It has been demonstrated, in fact, that fragments of fibronectin (~30 kDa) can induce catabolic cytokine production and stop aggrecan synthesis in vitro [52], while *in vivo*, they have been associated with the loss of PGs and osteophyte formation [53].

2.2. ECM Degradation in Degenerated Discs

A range of degradative enzymes and catabolic cytokines are released during disc degeneration. An increased expression of matrix metalloproteinases (MMPs) and 'a disintegrin and metalloproteinase with thrombospondin motifs' (ADAMTS), which are involved in the cleavage of the major components of the ECM, have been reported during degeneration. MMP-1, MMP-8, and MMP-13, for example, are able to degrade collagen type I and II, while MMP-2 and MMP-9 have been more generally associated with the degradation of triple helices along with gelatinases [54]. MMPs are also able to degrade aggrecan molecules at specific sites (G1–G2 and G2–G3 interglobular domains), although their activity is less pronounced than ADAMTS [55]. An impairment between ADAMTS and the tissue inhibitors of MMPs (TIMPs) can lead to matrix degradation, in particular through aggrecan fragmentation, which is a feature of degenerated discs [56].

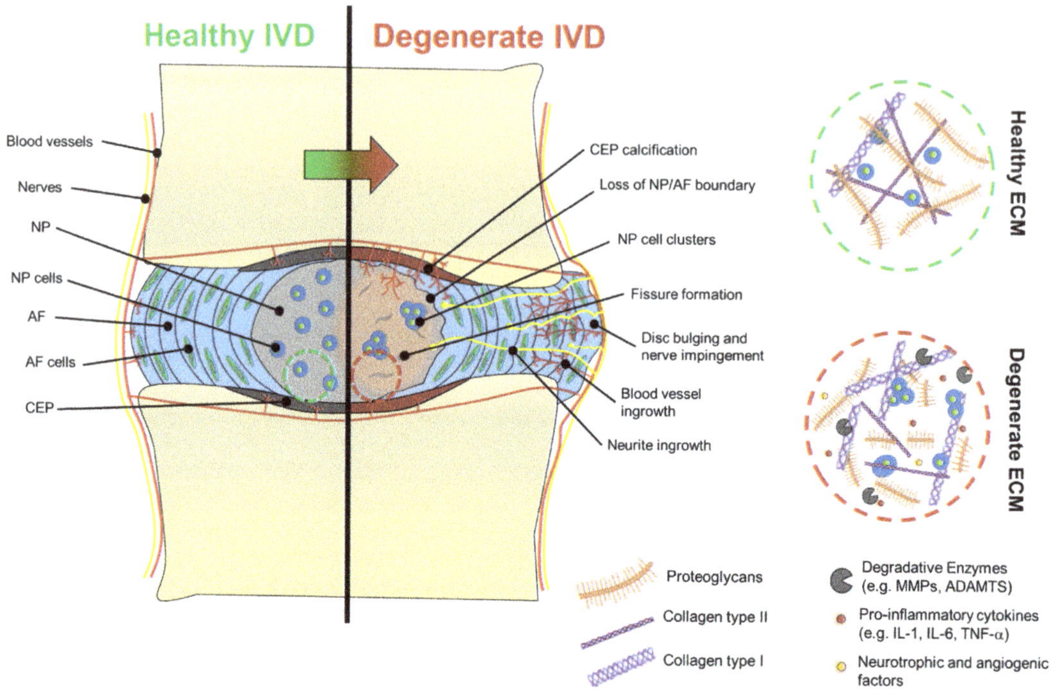

Figure 3. Macroscopic and microscopic changes of the IVD during degeneration. Characteristic changes of the NP's ECM after degeneration are illustrated in the dashed circles. ECM of degenerated discs (dashed red circle) shows shorter aggrecan macromolecules and more collagen type I fibres (thicker fibre bundles) than collagen type II (thinner collagen bundles), which are largely abundant in the ECM of healthy discs (dashed green circle).

It has been hypothesised that an increased expression of ADAMTS, especially ADAMTS-1, ADAMTS-4, and ADAMTS-9 in degenerated discs, could be potentially regulated by a series of proinflammatory cytokines and inflammatory mediators, such as interleukins (ILs), tumour necrosis factor-α (TNF-α), and nitric oxide, which are released by NP cells in the cellular microenvironment during degeneration [57]. These soluble regulators of cellular function are involved together, with some of them acting as master regulators of others. IL-6, for example, is thought to play indirectly by inducing NP cells to respond to IL-1 and TNF-α [58]. IL-1, in its two isoforms (IL-1α and IL-1β), along with its receptors (IL-1RI and IL-1RII) and their natural receptor antagonist (IL-1Ra), have all been identified in the IVD. In particular, IL-1 and IL-1Ra are in equilibrium during homeostasis, while an increase in the former compared to the latter has been observed in degenerated discs [59]. Strong evidence suggests that increased production of IL-1β by NP cells represents a master regulator of disc degeneration. Indeed, Le Maitre et al. showed that the administration of recombinant IL-1 to NP cells in vitro induced an increased production of MMP-3, MMP-13, and ADAMTS-4, along with a collagen II-to-collagen I shift and a decrease in aggrecan production, which are all signs of matrix breakdown and progressive degeneration [60]. On the other hand, IL-1Ra directly delivered into explants of degenerated IVDs has been shown to halt matrix degradation and start to reverse the typical signs of degeneration, highlighting the importance of the IL-1/IL-1Ra balance to keep the NP in its healthy state [61,62].

2.3. Nerve and Blood Vessel Infiltration in Degenerated Discs

One of the aspects that makes disc degeneration painful and debilitating is associated with vascular and nerve ingrowth during degeneration. At birth, only AF and CEP possess vascular networks, which recede with age so that only the endplates are vascularised, while NP and AF remain avascular and aneural [2]. During disc degradation, a decreased aggrecan content has been associated with the increased severity of degradation and neovascularisation, suggesting that blood vessels infiltration is avoided in a healthy state by aggrecan molecules themselves, which are thought to inhibit endothelial cell adhesion and migration [63]. Moreover, neovascularisation and innervation are also associated with the formation of fissures within the AF and NP, suggesting that tissue breakdown allows blood and nerve vessels to interdigitate within these tissues. In the last decade, biochemical links between neovascularisation and inflammation have been identified. An increase in IL-1β has been associated with the stimulation of angiogenic and neurogenic factors, such as vascular endothelial GF (VEGF), nerve GF (NGF), and brain-derived neurotrophic factor (BDNF) [42]. Both NP and AF, in their healthy state, have low levels of NGF and BDNF, but their levels increase notably during degeneration [30]. To support this, Purmessur et al. found that levels of NGF and BDNF can be increased by the addition of IL-1β and TNF-α to cultured human NP cells and that substance P, a molecule involved in pain sensation or nociception, can be highly expressed [30]. Driven by neurotrophic factors, neurites start to infiltrate NP and the AF tissues. In degenerated discs, nerve vessels are found alongside blood vessels, probably formed because of infiltrating endothelial cells that, during angiogenesis, secrete NGF [42]. Neurites found in the NP show a similar expression profile to neurites found in the AF, suggesting that neoinnervation may have a common source. The presence of NGF and BDNF in the cell microenvironment can also activate the NF-kB pathway, which is responsible for the secretion of further pro-inflammatory cytokines and degradative enzymes that accelerate disc degradation [64,65]. In addition to the pain caused by neurite infiltration (discogenic pain), degenerated discs often collapse and bulge out of the vertebral unit, pushing on the spinal nerves exiting the IVD foramen, causing nerve compression and additional pain [66]. Macroscopic and molecular features of IVD degeneration are shown in Figure 3.

2.4. The Effect of Ageing on Disc Degeneration

During development, NP tissue is populated by large vacuolated cells known as notochordal cells because of their origin, which are metabolically active and produce large amounts of PGs [67]. By the age of 10, notochordal cells are replaced by smaller chondrocyte-like cells known as NP cells, which are less metabolically active. The period within which notochordal cells start to disappear (3–10 years old) is accompanied by a high level of cell death [67]. In this period, the NP tissues lose their characteristic gel-like structure of the early years and become less hydrated, cartilaginous, and fibrous. These years also coincide with the earliest identifiable signs of disc degeneration ever reported through magnetic resonance imaging (MRI) [3]. It is clear that the passage from a cell population predominantly of notochordal cells to a population of NP cells and the changes in matrix composition and mechanics are the earliest signs of disc changes occurring with ageing. Cell death, occurring through apoptosis [68] and autophagy [69,70], comes in cycles followed by proliferation. Between 11 and 16 years, NP cells undergo a notable cell proliferation, which is thought to follow the massive cell death that occurs between ages 3 and 10 [67].

Cells during proliferation tend to create clusters, which, in turn, predispose the NP cells to replicate senescence traits [71]. In the last decade, several markers of cell senescence have been identified in degenerated discs [68,72,73]. Le Maitre and co-workers, for example, found that during degeneration, there is an increase in the protein and gene levels of $p16^{INK4A}$, which is a cell cycle inhibitor, accompanied by a decrease in the average telomere length (characteristic of ageing) and an increase in senescence-associated β-gal staining [72]. Similarly, Kim et al. a few years later, reported that the telomere-based p53-

p21-pRB pathway plays a crucial role in premature senescence in NP cells, confirming that with degeneration, senescent cells accumulate in the disc [73]. Senescence has been shown to also be associated with an increase in matrix catabolism since elevated gene expression of MMPs and ADAMTS were recorded in aged discs [74]. Although these studies and more suggest that there is a link between senescence, ageing, and disc degeneration, it is still not clear if the link is causative, i.e., if cell senescence causes IVD degeneration. However, evidence suggests that degeneration is accompanied by cell senescence, autophagy, and apoptosis, which are features of tissue ageing [75].

2.5. Mechanical Factors for Disc Degeneration

Although it has been reported that intense exercise has no adverse effect on the disc, experimental overloading and injury to the disc can be considered inductive factors for disc degeneration. In a study by Puustjärvi et al., for example, beagles were used as animal models to study the changes in PGs in two IVDs (cervical C5 and thoracic T6) after one year of running exercise on a treadmill (40 km per day) compared to non-running control dogs [76]. The animals subjected to extensive exercise showed an increase in PGs in the C5 disc and a reduction in T6 discs, in particular within the AF tissue. Although PG content is related to disc hydration and viscoelasticity, this study highlighted that PG content might be affected by a spine-location dependency after exercise, which can alter the biomechanics of the entire spinal column [76]. In another animal model, Iatridis et al. applied external compressive forces to the tails of rats to recreate spine overloading [77]. Chronically applied compressive forces resulted in a decreased disc height and loss of spine flexibility. The same effects were observed with rats' tail immobilisation, but earlier and with a larger magnitude. Similarly to Puustjärvi's work, the content of PGs in the overloading-induced group was increased, but the original biomechanical features of the disc, such as flexibility and angular laxity, were lost [77].

Along with macroscopic effects on spine mobility, mechanical loadings can also have a profound effect on the ECM organisation and phenotype of disc cells, which are ultimately the functional units of the IVD. Different in vitro studies have demonstrated that the magnitude, frequency, and duration of loading play a pivotal role in determining cell response and inducing consequent degeneration [78–80]. In two separate studies, Handa first [81] and Neidlinger-Wilke et al. later [82] showed that low-magnitude loadings could increase PG and collagen synthesis (aggrecan and collagen type II) in human disc cells and human nucleus explants, while the increased production of matrix degradative enzymes (mainly metalloproteinases-3, MMP-3) was reported with high-magnitude loadings. Once disc cells initiated a catabolic response and were degenerated, their mechanical response was largely affected, probably due to an alteration at the mechanotransduction level. Le Maitre and co-workers, for example, encapsulated AF and NP cells in alginate beads and applied dynamic hydrostatic pressure to the hydrogels (>2 MPa) to mimic the pressure experienced by cells *in vivo*. In non-degenerated cells, typical markers of disc matrix production such as SRY-box transcription factor-9 (SOX-9), collagen type II, and aggrecan were upregulated, with no effect on degradative enzymes [83]. On the contrary, degenerated cells showed no response to mechanical load, with no effect on the same gene target [83]. Similarly, Gilbert et al. showed that cyclic tensile strain applied at 1 Hz could reduce the anabolic response of AF cells derived from degenerated discs, while the same loading at 0.33 Hz could promote matrix catabolism without the involvement of any external cytokine [84,85]. The authors suggested that the response of human AF cells to cyclic tensile strain is frequency-dependent, affected by degeneration [84,85].

2.6. Genetic Influences on Disc Degeneration

IVD degeneration is a multifactorial, progressive disease, and recent evidence has also correlated disc degeneration with genetic mutations. Disc degeneration and herniation, for example, have been linked with a high genetic predisposition, which in some cases can have a hereditability of nearly 60% [86,87]. The most crucial genetic mutations are genes that code for ECM components, such as collagen and aggrecan [88–90]. Mutations of collagen type II (specifically COL2A1) led to highly disorganised ECM deposition in transgenic mice, along with the total absence of endochondral bone or the epiphyseal growth plate [88]. Similarly, a genetic mutation of collagen IX in a mouse model was associated with the shrinkage and disappearance of NP tissue and fissure formation in the AF, which are both early signs of disc degeneration [89]. Aggrecan, like collagen, is another important ECM component for IVD, and therefore, mutations of aggrecan-associated genes have been linked with degeneration. In a study by Watanabe et al., for example, it was shown that mouse cartilage deficiency known as 'cmd' was associated with a genetic defect of aggrecan synthesis, causing heterozygotes mice to have cervical spine misalignment and herniation, which is a cause of premature death [90]. On the other hand, single-nucleotide polymorphisms associated with ADAMTS-5, one of the major aggrecanases in the IVD, were associated with the formation of cracks and tears in the AF [91]. More recently, mutations and depletions of other genes, such as Paired box-1 (PAX1) and Forkhead box F1 (FOXF1), were also associated with spine deformations in the foetal disc [92,93]. Deletion of the FOXF1 gene in mice, for example, was associated with spinal deformations and fusion of vertebrae [92], while deletions of PAX1 were associated with aberrant vertebrae formation during axial skeleton development due to the key role played by PAX1 in sclerotome formation [94].

3. Impact and Current Treatments of Disc Degeneration

3.1. Socio-Economic Burden

Low back pain (LBP) is one of the most common degenerative diseases in Western countries, whose prevalence is increasing dramatically in recent decades and with the highest number of years lived with disabilities (YLDs) [95,96]. Although it is recognised that LBP is a multifactorial condition, 40% of cases affected have been associated with degeneration of the IVD [66]. Recently, de Schepper and co-workers, in a cross-sectional, population-based study, confirmed a strong association between LBP and IVD degeneration by finding a high prevalence of disc narrowing and osteophyte formation (two morphological features of disc degeneration) in people affected by LBP [97]. It has been estimated that nearly 84% of the world population will experience LBP during their lifetime, with the highest impact in Europe and the lowest impact in Africa, followed by South America and India [96,98]. The prevalence of LBP seems to increase steeply after the third decade of life, with a similar growth rate between men and women and with a peak over the age of 65 [99]. Among people affected by IVD degeneration, it has been assessed that at least 10% remain chronically disabled [99]. It has been estimated that in the UK alone, the socioeconomic burden of LBP costs around GBP 12 billion per year, while in the USA, the total cost associated with LBP has been estimated to be nearly USD 85 billion per year [95,100]. However, these estimates are conservative since LBP does not represent an economic burden only for the patient, but it has an enormous impact on carers, families, communities, and the healthcare system [101]. A deeper analysis of the socio-economic costs of LBP showed that indirect costs are often higher than direct medical costs. In Australia, for example, total costs for LBP are estimated at AUD 9 billion, where only 11% of the total amount accounts for direct costs. Similar amounts and proportions have also been reported for Netherlands and UK [96]. Moreover, due to the relation between ageing, spine overloading due to obesity, and LBP, it has been estimated that the total costs associated with LBP will increase steeply in future decades unless regenerative therapies to restore tissue function are implemented [99].

3.2. Grading Scales for the Assessment of Disc Degeneration

To date, magnetic resonance imaging (MRI) is the most used technique to clinically assess the onset of IVD degeneration associated with LBP. MRI is a powerful, non-invasive technique through which it is possible to obtain not only information about the disc height but also about the hydration level of the NP. Based on MRI images, in 2001, Pfirrmann and colleagues developed a disc-degeneration grading system able to estimate the water content signal in degenerated discs and correlate it with morphological parameters on a scale from I to V [102,103]. On MRI images, healthy discs appear bright for the NP and for the inner part of the AF due to their high level of hydration, while the outer AF appears dark. With the progression of IVD degeneration, the intensity of the NP is drastically reduced, and the distinction between the inner and outer AF is lost [102]. Taken together, the colour of the NP, demarcation between NP and AF, signal intensity, and disc height are used as morphological features to describe the grade. The Pfirrmann grading scale shows the changes in IVD degeneration from one stage to the other, with degeneration becoming more destructive as one moves from grade I to V (Table 1). In particular, grades I–III are usually associated with early degeneration, while grades IV–V are reported as late-stage or advanced disc degeneration [102] (Figure 4). Although the Pfirrmann grading scale has been widely accepted, it should be used carefully since it shows some inherent limitations. First, the grading system proved to be relatively non-discriminatory when elderly subjects were examined since, due to ageing, the NP becomes dehydrated, and the distinction between aged and degenerated discs becomes hard [104]. Second, the system is based on subjective and often ambiguous visual assessments of MRI images, so degenerated discs may be categorised into different levels of degeneration depending on the assessor. As such, very similar contiguous grades, such as grade III and grade IV, are difficult to distinguish, so highly degenerated discs may be underestimated [104]. Recently, Griffith et al. have slightly modified the five-level Pfirrmann grading scale into an eight-level scale to also discriminate IVDD severity in elderly subjects, where the Pfirrmann grading scale lacked accuracy [104]. Other parameters, such as the 'MRI Index', which is the product of the NP pixel area (or volume) with the intensity of the signal, are also being explored as more objective tools [105]. However, Pfirrmann's grading scale results are the most widely accepted [106,107].

Table 1. Morphological features and Pfirrmann grading system of IVD degeneration.

Grade	Structure Colour	NP/AF Distinction	Signal Intensity	IVD Height
I	Homogenous, bright white	Clear	Hyperintense	Normal
II	Inhomogeneous, opaque	Clear	Hyperintense	Normal
III	Inhomogeneous and grey	Unclear	Intermediate	Normal to slightly decreased
IV	Inhomogeneous, grey to black	Lost	Intermediate to Hypointense	Normal to moderately decreased
V	Inhomogeneous, black	Lost	Hypointense	Collapse of disc space

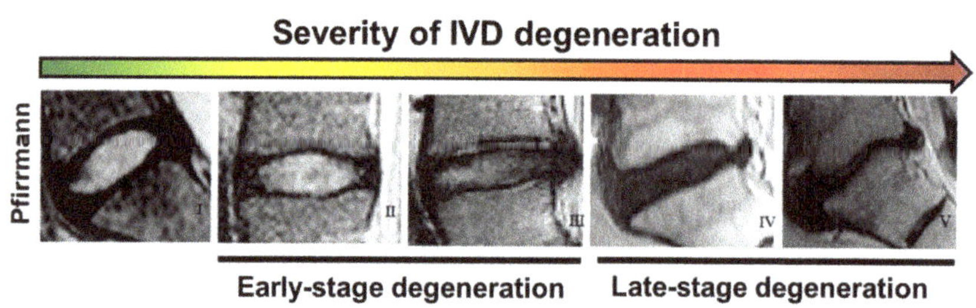

Figure 4. Comparison between different stages of intervertebral disc degeneration (**I** to **V**, left to right) according to the Pfirrmann grade scale based on MRI images. Reprinted and adapted with permission from Ref. [103]. Copyright 2022, Elsevier.

3.3. Conservative Therapies for Early Stage Disc Degeneration

Current treatments for IVD repair are divided into conservative and surgical therapies depending on the Pfirrmann grade of degeneration. For early stage degeneration (Pfirrmann grade < III), a series of conservative approaches is recommended before employing any invasive therapy. Conservative approaches are entitled 'palliative care' and can include, depending on the case, imposed rest, physical exercises to strengthen back muscles, physiotherapy, painkillers, and vitamin supplements [108]. Physical exercise has been shown to correlate positively with cell proliferation in several animal studies, while strengthening of the paraspinal muscles has been shown to have a positive effect on pain reduction [76,109]. Along with physical exercise, severe cases of LBP require the administration of non-steroidal anti-inflammatory drugs (NSAIDs, such as ibuprofen and paracetamol), corticosteroids, opioids, and muscle relaxants. These medications can be supplied via oral administration or via intradiscal injections, whose aim is to decrease the local inflammation and provide temporary anaesthesia to relieve the pain [108]. Among supplements, omega-3 fatty acids, glucosamine, and chondroitin are usually administered due to their minimal side effects, but there is a lack of evidence and conflicting findings on their therapeutic effect [110,111]. Although these treatments can offer temporary pain relief, these medications do not target the process of the disease (NP matrix degradation, NP cell senescence and death, or altered cell function), but only target the symptoms of LBP. Moreover, to ensure a therapeutic effect and overcome poor targeting, multiple doses are often required, without any guarantee of their sustained effect. In particular, NSAIDs, which are mainly based on COX inhibitors, can increase susceptibility to mucosal injury in the gastrointestinal tract, forcing the patients to take other medications, such as proton pump inhibitors, to cope with these complications [112–114].

3.4. Surgical Approaches for Late-Stage Disc Degeneration

When patients have not achieved pain relief after 6 months of non-surgical procedures and they are still physically constrained in daily activities, surgery is the only remedy [108]. Surgical operations are only performed to tackle late-stage disc degeneration (Pfirrmann grade > III) when disc function and mobility are irremediably compromised. Although different surgical procedures are currently available, their common goal is to separate nerve vessels from the disc by reducing the size of the disc bulging on the neural arch and restoring normal disc function [108,115]. Reducing disc bulging is often achieved through decompression or discectomy (mechanical, thermal, or chemical), in which a portion of IVD is removed to decompress the spinal canal and destroy the blood and nociceptive fibres that could have formed in the periphery of the disc during degeneration [108]. When the whole disc is compromised (Pfirrmann grade > IV), total disc replacement (TDR) and spinal fusions are the only options available to treat this disability. In TDR, the entire IVD (NP plus AF and CEPs) is substituted by an artificial implant made of polymers (for NP and AF bodies) and metals (for the CEP to anchor on the vertebral bodies) [116]. Commercially available products for TDR include InMotion/Charité®, ProDisc®, and Maverick™, while BioDisc™, GelStix™, and PerQdisc™ have recently been explored for NP replacement [117–120]. Metal and plastic materials are chosen because they are bioinert and, because of their high mechanical properties, they are suitable to support the biomechanics of the column. In spinal fusion, two or more vertebrae are merged into one functional unit [115]. During this procedure, vertebrae enclosing the damaged disc are fused together by the addition of tissue graft containing osteoinductive materials (e.g., bone allografts, autografts, and cements) to allow vertebrae fusion. Finally, the vertebral joint is stabilised through external fixators, such as surgical plates and screws [116].

The introduction of an inert biomaterial or the removal of a damaged disc is often a very invasive procedure accompanied by medical complications and poor efficacy in the long term. Inert biomaterials, in fact, are not living materials, so the damaged tissue is actually *substituted* rather than *repaired*. The incorporation of a solid implant can cause a mechanical mismatch between the implant and the adjacent vertebrae, which can result in hypermobility and the onset of IVDD in the adjacent discs [121]. Any abnormal process that occurs in the mobile segment adjacent to a spinal fusion is clinically defined as Adjacent Segment Disease (ASD). The list of abnormal processes occurring at the surrounding vertebrae and IVDs is long and includes herniated NPs, stenosis, hypertrophy of facet joints, and instability of the column [122]. From the several retrospective clinical reports that have been published on ASD, it emerged that the incidence of this medical complication is broad, ranging from 7% up to 52% [122]. In most cases, as reported by Leong [123], Guigui [124] and more recently by Kuslich's groups [125], ASD causes degenerative changes of the IVDs next to fused segments. In general, when post-fusion patients present pathological symptoms referable to ASD, conservative approaches are used to relieve the pain, but in most cases, another surgical intervention is required. Kim et al., for example, in a follow up of 1856 patients who underwent surgery for degenerated IVDs, reported a cumulative recurrence rate of up to 16% in 10 years [108]. Moreover, these implants require periodic revisions and are often accompanied by local infections arising near the fixators (surgical plates or bone screws), which may force the patient to undergo a second surgical operation [126].

3.5. Cell-Based Therapies for NP Regeneration: Cell Sources

As stated in the previous section, commercially available IVD or NP replacement materials are poorly efficient in the long term since they partially restore column segment mobility and do not biologically repair or regenerate the affected tissue. Moreover, they are often accompanied by a broad range of potential adverse effects and a high rate of recurrence. Hence, it is critical to develop novel therapeutic strategies that can promote IVD regeneration and can be translated into clinics to improve patients' quality of life. Since IVD degeneration initiates in the nucleus, the NP represents a promising tissue target for the design of novel therapies. In particular, tissue engineering (TE) and regenerative medicine can offer potential alternative therapies to current treatments by the use of instructive and functional biomaterials in conjunction with cells and biomolecules (e.g., GFs) as therapeutic agents. Indeed, biomaterials can help to restore disc height and IVD biomechanics as well as provide a vehicle for the retention and delivery of cells and relevant biomolecules to injured discs to re-establish healthy tissue.

The use of cells as therapeutic agents, with or without encapsulation in a biomaterial, is referred to as cell-based therapy. Cell-based approaches may involve the delivery of autologous, allogeneic, or xenogeneic sources of primary or stem cells, which can be injected intravenously, transplanted at a defect site within a scaffold, or recruited from a patient's own tissue to exploit the self-repair processes [127]. Different cell sources and cell types have been explored for cell-based therapies. Autologous cells are usually preferred since they are harvested from the patient's body to avoid transplant rejection. Mesenchymal stem cells (MSCs), for example, are receiving a significant amount of attention for the treatment of NP defects since they have been shown to undergo differentiation into NP-like cells, with high expression and deposition of proteins specific to the NP matrix [128]. Moreover, MSCs can be isolated from a large number of adult connective tissues, including bone marrow [129], umbilical cord [130], trabecular bone [131], adipose tissue [132], dermis [133], skeletal muscle [134], and AC [135]. In particular, the possibility to acquire MSCs from tissues such as fat or skin has limited the morbidities associated with stem cells harvesting, making cell isolation less invasive.

Recently, Clarke and co-workers tested the suitability of bone-marrow-derived MSCs (BM-MSCs) and adipose-derived MSCs (AD-MSCs) in conjunction with GDF-6 and TGF-β3 for the differentiation into NP-like cells for the treatment of IVD degeneration [136]. The authors found that GDF-6 stimulation of both types of stem cells induced a high expression of NP-marker genes and GAG production, with a higher effect using AD-MSCs rather than BM-MSCs. Moreover, GDF-6-stimulated AD-MSCs resulted in a softer ECM deposition, more akin to the native NP tissue [136]. MSCs are also explored for their ability to release bioactive factors and anti-inflammatory agents that can influence tissue regeneration and reduce the immune response and local inflammation. Moreover, cell-to-cell contact between MSCs and endogenous primary cells has shown to be beneficial for cell differentiation and ECM production in co-culture studies, suggesting that the injection of MSCs in degenerated discs may also be beneficial for resident NP cells [128]. For example, in a study by Strassburg et al. the co-culture of MSCs and NP-cells increased the expression of anabolic GFs by MSCs, differentiation towards the discogenic lineage, and improvement of the NP cell phenotype through MSCs-NP cell communication, which was possibly mediated by the bi-directional intercellular transfer of membrane components [137,138]. Taken together, these features confer MSCs' high clinical translatability for degenerated discs.

As shown recently by a systematic review by Panebianco et al., along with MSCs (59.5% of peer-reviewed articles screened), NP cells (40.5% of articles screened), patient or animal-derived, have also been employed in clinical studies and in preliminary in vitro studies for NP repair [139]. The transplantation of NP cells, for example, has been shown to retard the degenerative disease in a dog model, whereas a significant decrease in LBP score with the retention of disc height and hydration levels was observed in a randomised human clinical trial [140,141]. Despite these positive results, the senescent and catabolic phenotype of autologous cells harvested from degenerated discs introduces doubts to its use for cell therapies, in which normal or appropriate cell function is a strict requirement. Moreover, the acquisition of these cells may lead to further complications that can accelerate degeneration [142]. Nevertheless, the use of NP cells extracted from animal models is strongly encouraged when they are tested in conjunction with novel biomaterials, whose properties of biocompatibility and cell response are still unknown. In particular, many animal-sourced NP cells are derived from preclinical models of IVD degeneration widely accepted in the literature, which enables rapid progression from in vitro to in vivo testing [143]. Among the animal sources used for NP repair strategies, rabbit NP cells are the most commonly used (34.4%), followed by larger animals such as cows (25.5%) and pigs (21.1%) [139]. Bovine and porcine NP cells are largely preferred because they are cost-effective, and they are ideal for simulating in vitro force applications and transport studies, similar to those experienced by the human IVD [144].

3.6. Cell-Based Therapies for NP Regeneration: Biomaterial Requirements

For the biomaterial design aspect, biomaterials should fulfil specific requirements, which can be divided into 'mechanical', 'biological', and 'clinical'. Although the biomaterial should match the mechanical properties of the native NP on the one hand, on the other hand, it should be applicable and translatable into a clinical setting to be effectively used as an alternative therapy to current treatments.

Mechanically speaking, the biomaterial should restore the disc height and ensure biomechanical flexibility. To meet this requirement, the biomaterial should possess rheological properties similar to those of the NP, such as shear modulus, swelling behaviour, and permeability. As detailed by Iatridis et al., the NP tissue is a gelatinous material, with an average storage modulus at 1 Hz (G', accounting for the material's elasticity) of 10 kPa and a loss modulus (G'', accounting for the material's viscous properties) of 4 kPa [145]. Swelling properties are also important since the NP will swell over time under cyclic loading, and this is usually reported as a ~1.5% change of its weight for the human NP [146]. NP-mimicking biomaterials should also have values of permeability (k_a) similar to those of the human NP ($k_a = 6 \times 10^{-16}$ m^4 N s) to ensure a physiological exchange of nutrients and waste with the surrounding tissues, i.e., the AF and the CEP [146]. In addition, the biomaterial should be hydrophilic and highly hydrated to mimic the high water content of the native NP tissue (water content: 80–90%) [7].

Biologically speaking, ideally, the biomaterial should be able to inhibit the degenerative processes, prevent the formation of nerve and blood vessels, and stimulate tissue regeneration. To this extent, cell-based therapies in which the biomaterial is used as a vehicle for cells, proteins, and/or GFs has proved to be a successful strategy in numerous preclinical and clinical studies for NP repair [147,148]. Indeed, the incorporation of exogenous cells into an implanted biomaterial can promote ECM synthesis as well as secrete paracrine signals that may stimulate resident cells to deposit new ECM and release immunomodulatory signals to halt disc inflammation [149]. In line with Langer and Vacanti's principles of TE [150], to be biologically successful, biomaterials designed as cell carriers should retain implanted cells at the injury site and provide them with an instructive microenvironment that can support cell viability, proliferation, and differentiation [151]. At the same time, biomaterials should allow tissue remodelling, with a degradation rate that ideally should match the rate of tissue ingrowth during regeneration [151].

Clinically speaking, the designed biomaterial should be deliverable via minimally invasive tools (e.g., small-needle syringes, catheters) and be non-immunogenic in vivo in order to interdigitate and fully integrate with the NP ECM after implantation [152].

4. Injectable Hydrogels for NP Regeneration

4.1. Advantages of Using Injectable Hydrogels for NP TE

Hydrogels and, in particular, injectable ones, represent an ideal class of biomaterials to serve as cell carriers and/or as acellular scaffolds for NP TE for several reasons. Firstly, hydrogels are hydrated materials with a gelatinous and amorphous structure that closely mimics the viscosity and level of hydration of the NP. Secondly, hydrogels can be designed to be injected via minimally invasive techniques, such as syringe injections and catheters. With this approach, hydrogels can easily reach the NP, which is known to be avascular and therefore difficult to reach systemically. Hydrogels can be injected via small punctures (e.g., 26G–21G needles), which reduces the damage occurring via annular puncture to a minimum and allows faster recovery after the injection [151]. Due to their viscoelastic properties, hydrogels can flow easily and interdigitate within the NP to fill tissue defects and fissures. Moreover, injectable hydrogels can be designed to incorporate a range of biophysical and biochemical cues (e.g., GFs or anti-inflammatory cytokines) so that they can be used at different stages of disc degeneration. Indeed, hydrogels may be applicable to symptomatic early to moderate disc degeneration (Pfirrmann grade < 3) to attempt regeneration, while they may also be used as a prophylactic approach in segments adjacent to the degenerated discs of patients undergoing disc fusion to prevent the onset of adjacent vertebrae's degeneration [122].

Finally, hydrogels can serve as protection to retain their cargo (e.g., cells and/or biomolecules) at the injury site to avoid its leakage as well as to protect it against harsh environments. Degenerated discs are an example of such harsh microenvironments, being full of catabolic cytokines, with minimum levels of glucose, low oxygen, and acidic pH (pH < 6.2) [46,48]. Direct injection of cells without any cell carrier is usually accompanied by poor engraftment at the defect site (less than 3% of the total transplanted cells) along with a loss of control over cell fate after transplantation [153]. In a study by Vadalà et al., for example, injections of MSCs into IVD without any cell carrier were accompanied by cell leakage from the injection site, with leaked stem cells being involved in the formation of osteophytes on the vertebral bodies [154]. Similarly, it has been observed in vitro that MSC proliferation and matrix deposition is severely compromised when cells are exposed directly to low glucose, high osmolality, and acidic pH, which represent the environmental conditions of degenerated discs [155].

Due to their properties, injectable hydrogels exploited as cell carriers can represent an alternative therapy to current treatments by bridging the gap between mechanical (provided so far by disc prostheses) and biological needs (provided by cell transplants). Once injected, hydrogels may be used to provide high initial biomechanical efficiency to degenerated discs in the short term while still promoting tissue repair in the long term.

4.2. Naturally Derived Injectable Hydrogels

Based on the origin of their building blocks, hydrogels can be classified into natural (or naturally sourced) and synthetic. Most natural hydrogels used for IVD repair include hyaluronic acid, alginate, and chitosan [156]. These hydrogels are thought to have an advantage over synthetic counterparts since their building blocks are naturally presented *in vivo*, showing intrinsic ECM mimicry and inherent biocompatibility. Hyaluronic acid (HA) is an unbranched high-molecular-weight polysaccharide existing in the native ECM of most hydrated tissues, such as AC, synovium, and IVD, where it plays a key role in maintaining high levels of water content [157]. When dissolved in water, HA can create highly viscous solutions, which are also used in clinics as injectable viscosupplement to replenish the human synovium of HA during osteoarthritis [158]. For IVD research, Pandit's group showed that cross-linked HA hydrogels were able to induce a downregulation of inflammatory receptors and neurotrophins in NP cells exposed to IL-1β in an in vitro inflammation model [159]. The exact mechanism of HA seems to be related to the binding of HA chains to the cells' CD44 receptor, which prevents NP cells from undergoing further inflammation [159]. Similarly, injection of MSC-laden 15% HA hydrogels in rat IVDs showed cell proliferation and an increase in disc height, as well as a reduction in pain and inflammation [160]. However, HA is prone to undergoing rapid hydrolytic degradation *in vivo*, which results in generally poor mechanical properties [161]. Moreover, degradation of HA in small fragments has been shown to induce mRNA expression of key catabolic and pro-inflammatory cytokines by IVD cells, which is not clinically preferred for the treatment of degenerated discs [162].

Along with HA, alginate represents another popular material choice for NP TE. Alginate is a polysaccharide extracted from marine algae, with high biocompatibility and low cytotoxicity. Unfortunately, the applications of alginate hydrogels in the field of IVD repair are limited so far due to alginate's physiochemical properties, such as long gelation times and high degradability. Alginate monomers undergo gelation when exposed to divalent calcium ions that induce 'eggbox' chelation of its monomers, mannuronic and glucuronic acid. However, alginate's gelation is a quick process, during which the hydrogel boundaries bond tightly while the rest of the hydrogel remains physically weak. In a study by Kalaf et al., for example, 'slowly cross-linked' alginate gels were implanted in an ex vivo bovine IVD model and showed the ability to maintain disc height over cyclic axial loads better than alginate hydrogels that were 'instantaneously gelled' [163]. The elution of calcium ions over time with a decrease in mechanical properties is another major drawback of this system. Baer and colleagues, for example, showed that although AF and NP cells cultured in alginate gels were able to express collagen type I and II, respectively, the mechanical properties of the cell-laden hydrogel constructs decreased over 21 days, up to 50% of their initial strength [164].

Like alginate, chitosan has also been used extensively in preliminary studies of NP TE, but its long gelation time and weak mechanical properties make it poorly applicable as an injectable hydrogel for IVD repair. Chitosan hydrogels are formed by cationic polymeric chains of chitin, which is largely abundant in the shells of crustaceans and in the wings of arthropods. Chitosan has excellent biodegradability, water dispersibility, promotes cell adhesion, and its cationic chains can be exploited to retain negatively charged PGs in the hydrogel network [156]. Chitosan can be mixed with beta-glycerophosphate (BCP) to make its gelation thermosensitive and faster so that chitosan hydrogels can be injected into degenerated discs and undergo gelation in situ [165]. Unfortunately, although high levels of BCP are beneficial for chitosan's gelation, BCP in high concentrations has been revealed to be highly cytotoxic [166]. Alini's group, for example, successfully cultured bovine NP cells in chitosan gel and showed a high retention of NP-synthesised PGs within the hydrogel, but the gel became cytotoxic when exposed to AF cells [167]. In addition, chitosan gels present relatively low mechanical properties and poor cell adhesion, which are not ideal for load-bearing tissues, such as the NP [156].

Other natural hydrogels, such as fibrin, gelatin, cellulose, and gellan gum, have also been explored as IVD injectable systems. However, as occurs for HA, alginate, and chitosan, typical natural hydrogels show poor mechanical properties, a high degradation rate, and low adhesion properties, so extensive cross-linking or co-assembly with another hydrogel network is usually required to overcome these issues [156]. In some cases, hydrogel cross-linking approaches involve the use of chemicals (e.g., glutaraldehyde and glyceraldehyde) that are highly cytotoxic and whose by-products should be removed completely before cell encapsulation, while some cross-linking mechanisms rely on UV-photocurable materials, which is not practical for NP injection [168]. Finally, due to their animal or vegetable source, many natural hydrogels suffer from inherent batch-to-batch variability, which makes them poorly reproducible for clinical studies [169].

4.3. Synthetic Injectable Hydrogels

Some of the limits encountered with natural hydrogels, such as batch-to-batch variability, low mechanical properties, and high degradation rates, can be overcome with the use of purely synthetic materials. Indeed, synthetic hydrogels allow easier large-scale production, high consistency across batches, and highly tuneable properties. Among synthetic injectable hydrogels used as cell-laden scaffolds for NP repair, poly(N-isopropylacrylamide) (pNIPAM), poly(ethylene glycol) (PEG), and poly(vinyl alcohol) (PVA) hydrogels are the most used in the field [156]. PNIPAM hydrogels have a thermosensitive network with hydrophilic and hydrophobic building blocks. When the temperature of the pNIPAM solution is below its lower critical solution temperature (LCST), the hydrophilic building blocks of this polymer interact with water, and they are able to form a cross-linked hydrated network. Once the temperature is raised above the LCST, the hydrophilic building blocks dehydrate and collapse into a globular system. The transition temperature between the two states can be set to be the human body temperature, allowing pNIPAM to undergo gelation in situ once injected. To this extent, Le Maitre and Sammon's groups developed a pNIPAM-based hydrogel that is liquid ex vivo at 39 °C while undergoing gelation in vivo at 37 °C, triggered by body temperature, when it is injected into the IVD [170]. The authors showed that a pNIPAM-clay hydrogel composite promoted the differentiation of MSCs into NP-like cells without additional GFs in hypoxic conditions mimicking degenerated discs [171]. However, the biggest disadvantage of this system is that pNIPAM hydrogels are not degradable; therefore, after gelation, cells are entrapped in the hydrogel and tissue remodelling by encapsulated cells may be hindered [172].

Due to its biocompatibility, non-immunogenicity, and ease of functionalisation, PEG hydrogels are widely studied in TE and for NP repair. However, due to the lack of their recognition site by encapsulated cells and non-degradability, PEG is largely used either as an interpenetrating network in hydrogel composites or as a non-toxic hydrogel cross-linker. Similarly, PVA hydrogels are usually based on composite mixtures of PVA and other polymeric matrices, either natural (e.g., silk fibroin, cellulose) or synthetic (e.g., poly(vinyl pyrrolidone) or PVP) [156]. PVA-based hydrogels, like other polyesters, start to degrade when hydrolytically labile linkages are exposed to water; therefore, the degradation rate and time of the final hydrogel can be tuned for specific applications and environments. However, the use of polyesters is not recommended for the repair of degenerated discs since hydrogel degradation leads to the formation of acidic by-products that cause further acidification of the hydrogel at the injection site. Moreover, acidic by-products have also been shown to have an 'auto-catalytic' effect on hydrogel degradation [173]. In the case of degenerated NP, where the pH can go down to 6.5, further acidification at the injection site would be deleterious for the resident cells, promoting a catabolic phenotype.

Despite the several advantages of synthetic hydrogels, their lack of biocompatibility and the need for chemical cross-linking with toxic reagents represent issues that need to be overcome before any biological translation. In particular, although many synthetic hydrogels show excellent mechanical properties and ease of injectability, these systems lack biological cues that could promote cell adhesion and proliferation. In many cases, in fact, synthetic hydrogels are coated with ECM proteins or decorated with corresponding biological motifs for TE applications [174,175]. Table 2 summarises the major advantages and disadvantages of natural and synthetic hydrogels.

Table 2. Summary of advantages and disadvantages of natural vs. synthetic hydrogels.

Hydrogels	Relevant Examples	Main Advantages	Main Disadvantages
Natural	Hyaluronic Acid, alginate, chitosan, fibrin, gelatin, cellulose and gellan gum	Biodegradable Cell-binding sites Low immunogenicity Relatively cheap	Poor mechanical properties High degradation rates Batch-to-batch variability
Synthetic	poly(N-isopropylacrylamide), poly(ethylene glycol), poly(vinyl alcohol), poly(vinyl pyrrolidone)	Tuneable mechanical properties and shapes Tuneable degradation Ease of scalability	Lack of cell-recognition sites Toxic degradation by-products Toxic cross-linkers

5. Hydrogels Based on Self-Assembling Peptides

A versatile class of hydrogels that combine the advantages of both natural and synthetic hydrogels is represented by self-assembling peptide hydrogels (SAPHs). Peptide hydrogels offer great potential for their use as a 3D scaffold and 3D cell carriers for NP TE since they satisfy most of the criteria cited above. These systems undergo a sol–gel transition in hydrophilic environments through the mechanism of molecular self-assembly, without the need for any toxic cross-linkers that are usually needed for natural and synthetic hydrogels [176]. Molecular self-assembly, in fact, is a kinetically and thermodynamically driven, bottom-up approach, which is ubiquitous in nature, allowing the formation of different structures from nanoparticles to nanofibres and supramolecular constructs present in biology, such as virus capsids and ribosomal units [177].

SAPHs are water-rich (>95%) and have a nanofibrous microenvironment that mimics the native ECM and creates a suitable platform for cell encapsulation and 3D culture [176,178]. Moreover, they can be designed to show specific peptide length and bioactive motifs and affect the network elasticity, which can be exploited as a physiochemical regulator of cellular fate [179]. Along with their biocompatibility, peptide hydrogels are also shear-thinning and are easily injected via syringes or catheters, with the ability to recover their bulk properties after injection/transplant, which makes them an ideal candidate for minimally invasive therapies [180]. The majority of SAPHs are based on L-amino acids, which are naturally found and processed by the human body, making them low in terms of immunogenicity and inflammatory properties [181,182]. Peptides are typically synthesised using fluorenylmethoxycarbonyl (Fmoc) solid-phase synthesis, which results in the production of short peptide sequences with high purity (>95%) [183]. This high purity makes peptides more reproducible and reduces the batch-to-batch variability encountered with naturally derived hydrogels. Indeed, many peptide systems are now readily available in the market, including PuraMatrix (from Corning, US), HydroMatrix (from Sigma-Aldrich, UK), and PeptiGels (from Manchester BIOGEL, UK), which highlights the ability of these systems to be scaled up for large production. Moreover, some SAPHs have also started to find their way into approved clinical products over the last decade [184].

5.1. Amino Acids as Molecular Building Blocks

At the base of molecular self-assembly, there is the precise choice of building blocks, which will adopt the most energetically favourable conformation to produce highly organised structures with defined properties. In nature, twenty amino acids are available as building blocks for the synthesis of peptides and proteins. All amino acids, except for glycine (G), are chiral macromolecules that exist in biology in their L-form, with a common structure consisting of a central carbon atom (alpha carbon) covalently bound to a carbonyl (-COOH) and an amine group (-NH2), along with a specific R-group. The configuration of a peptide sequence strongly depends on the nature of the R-groups that are present between two adjacent amino acids. Indeed, according to the properties of the R-group, amino acids can be divided into hydrophobic, hydrophilic, charged, and 'others' [185] (Figure 5). Amino acids interact with each other in the same peptide sequence or with adjacent peptides through a vast range of physiochemical interactions, including H-bonding, ionic bonds, pi–pi stacking, and electrostatic and hydrophobic interactions.

Typically, aromatic and hydrophobic amino acids are involved in pi–pi stacking interactions, while hydrophilic residues are involved in charge–charge interactions and H-bonding [185]. Inversely, cysteine is typically exploited for chemical functionalisation, glycine to add peptide flexibility, while proline is used for chemical hindrance [185].

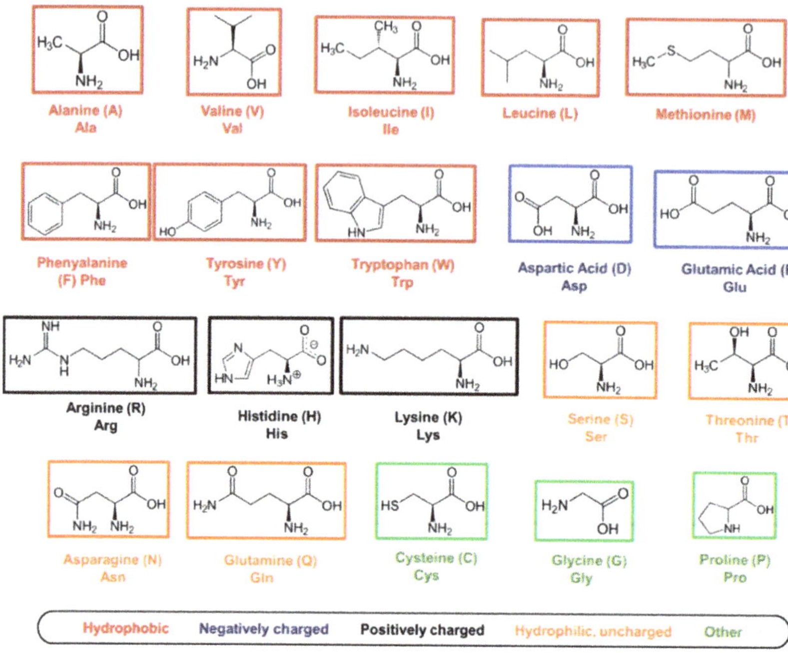

Figure 5. List of 20 natural amino acids. For each amino acid, its 3-letter abbreviation, 1-letter code, and property ('hydrophobic', 'negatively charged', 'positively charged', 'hydrophilic, uncharged', and 'other') are provided. Cys, Gly, and Pro are listed as 'other' since they have specific roles in peptide self-assembly that cannot be associated with the other properties listed.

As will be discussed in the following sections, during peptide synthesis, amino acids are linked together one after the other in a linear chain, and the position of amino acids plays a key role in both peptide self-assembly and in the interaction of peptides with the surrounding environment [178]. Using a bottom-up approach, amino acids can be exploited to recreate secondary and tertiary structures observed in cellular proteins. A linear chain of amino acids (or peptides) represents the primary structure of amino acids, while secondary and tertiary structures describe how peptide chains are folded into a 3D conformation. Typically, the length of a peptides chain is usually reported as between 2 (dipeptides) and <50 amino acids (polypeptides). Depending on the constituent amino acids, peptides acquire secondary structures in solution, such as β-sheets, β-hairpins, α-helices, and random coils, which can further self-assemble into supramolecular structures in response to a change in ionic strength, pH shift, enzymes' activation, peptide concentration, light, and temperature [185]. SAPHs are, therefore, the result of secondary structures assembling/entangling over a critical gelation concentration into self-supporting water-swollen networks, i.e., 3D hydrogels [186]. According to the secondary structures acquired, SAPHs are divided into nature-mimicking structures (e.g., α-helices and β-sheets) and newly designed derivatives, such as amphiphilic and short aromatic peptides.

5.2. Self-Assembling Peptide Structures: β-Sheets and β-Hairpins

The first example of SAPHs was introduced in 1989 by Zhang et al., who discovered that a peptide sequence of zuotin (AEAEAKAKAEAEAKAK or EAK16), a left-handed Z-DNA binding protein found in yeast, was able to self-assemble into nanofibres and form hydrogel networks in the presence of electrolytes. Inspired by this discovery, the first two sequences of complementary ionic polypeptides (16 amino acids), EAK16, and RADARADARADARADA (RADA16), were designed [184]. These β-sheet-rich fibrillar matrices proved to be highly biocompatible for 2D and 3D cell cultures of different cell lines [184]. Zhang's discovery opened the route for the design of β-sheet forming peptides, in which charged hydrophilic amino acids are alternated with hydrophobic residues to form β-sheets units, which display one hydrophilic side and one hydrophobic site. When β-sheets are exposed to a hydrophilic environment, upon external stimuli, such as pH, temperature, enzymes, or light, the hydrophobic sides stack together, leading to the formation of different supramolecular structures, such as tapes, ribbons, fibrils, and fibres (Figure 6A,B) depending on the peptide concentration [185,186]. RADA16 and EAK16 have been extensively used in TE, both as pure peptides as well as combined with bioactive motifs, enzymes, and GFs. Bioactive motifs introduced in RADA16 systems include, but are not limited to, integrin-binding motifs (e.g., RGDS) and laminin-mimetic motifs (e.g., IKVAV) for enhanced cell adhesion and migration, VEGF-mimetic for angiogenesis, and BDNF-mimetic for neurogenesis [187]. Along with RADA16 and EAK16, dodecapeptides, such as KLD12 (KLDLKLDLKLDL), have also proved to self-assemble into fibrillar hydrogels with high biocompatibility and anti-microbial properties, which have been used extensively for cartilage, bone, and IVD applications [188–191].

In the last two decades, based on Zhang's design, Saiani's group developed a class of de novo β-sheet forming peptides based on short sequences (8 to 10 amino acids), which were able to entangle/assemble into nanofibrillar 3D network (fibres being 3–5 nm) and self-supporting hydrogels [192]. One of the earliest studies on these materials explored four octapeptides based on the alternation of alanine (A) and phenylalanine (F) with glutamic acid (E) and lysine (K), namely, AEAEAKAK, AEAKAEAK, FEFEFKFK, and FEFKFEFK. In that study, Saiani et al. showed that F-based octapeptides were able to form defined β-sheet fibres (~3 nm thick) and 3D fibrillar hydrogels regardless of the position of E and K in the sequence, while within the A-based systems, only AEAEAKAK self-assembled into rigid aggregated fibres [192]. From that study, different phenylalanine-based sequences of 8 to 10 amino acids were formulated and proved to be highly biocompatible, injectable, and low in terms of immunogenicity for a wide range of biomedical applications. Indeed, many cell lines were explored within this system for different biomedical applications, both in 2D and 3D settings, including osteoblasts [193], chondrocytes [194], NP cells [195], cardiomyocytes [196], and synoviocytes [197]. Moreover, this system is highly shear-thinning and able to recover after different cycles of shear strain, providing a versatile material platform for printable, injectable, and sprayable strategies [197–199]. During the self-assembly of β-sheets, two or more monomers, called β-strands, are joined together by at least two or three backbone H-bonds to form a twisted, pleated sheet. β-strands can join as parallel strands oriented in the same direction, such as N- to C-terminus, to form parallel β-sheets, or they can join in opposite directions, forming anti-parallel β-sheets. Based on this structure, Schneider and Pochan created a new class of peptide hydrogels by linking together two anti-parallel β-sheets via a β-turn to form a β-hairpin. In 2002, the first β-hairpin-based hydrogels for biomedical applications were formulated, namely, MAX1 and MAX8, in which a proline-based tetra peptide (-V^{D}PPT-) was designed as a β-turn link to induce a type II′ turn structure [200] (Figure 6C,D). These hydrogels have gained interest due to their biocompatibility, their shear-thinning behaviour, and good mechanical properties [201]. Within these systems, in fact, several cell types (e.g., osteosarcoma cells, human trabecular cells, and fibroblasts) and biomolecules/drugs (e.g., curcumin, NGF, BDNF, and vincristine) were encapsulated and delivered successfully [202,203].

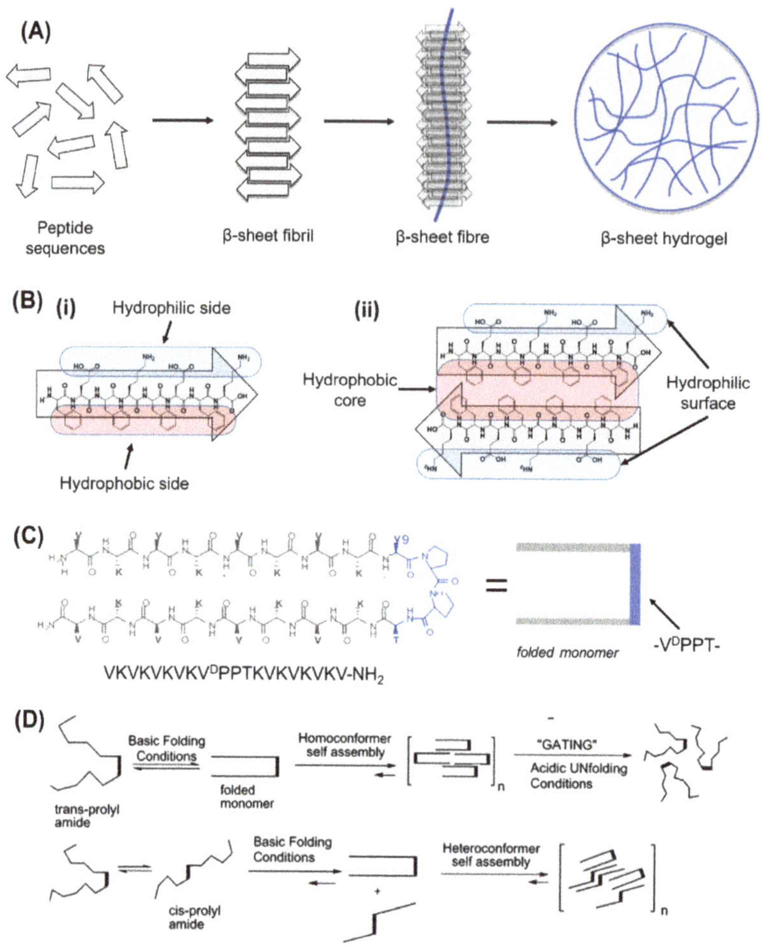

Figure 6. Beta-sheet and beta-hairpin peptide system designs. (**A**) Schematic representation of β-sheet-forming peptide hydrogel formation. Peptide sequences under external stimuli (e.g., pH, enzyme, temperature, light, time) self-assemble into β-sheet fibrils and fibres, which above a critical gelation concentration entrap water in water-swollen networks, i.e., hydrogels. (**B**) Detail of a β-sheet-forming polypeptide (**i**) reacting in water with another polypeptide via stacking of hydrophobic regions (**ii**). (**C**) Design of a β-hairpin sequence (i.e., MAX1), in which two valine-based peptides are linked together by a -V^{D}PPT- turn. This tetrapeptide based on D-isomer valine induces a trans-prolyl amide bond re-arrangement favouring the β-hairpin formation. (**D**) Folding/self-assembly pathways of β-hairpins. Adapted with permission from Ref. [200]. Copyright 2022, American Chemical Society.

5.3. Self-Assembling Peptide Structures: α-Helices and Coiled Coils

An alternative to the use of β-sheets as a secondary structure to form fibrous hydrogels is provided by the α-helices, which are used as building blocks to form coiled-coil structures, as was proposed initially by Pauling and Corey [204]. The design of these structures relies on seven coil residues (*abcdefg* heptad), which possess a hydrophobic and a hydrophilic face according to the residues presented within the helix structure. In particular, the carbonyl group of the first amino acid of the heptad interacts with the fourth amino acid of the same heptad so that there are two turns within the same heptad [185]. A coiled-coil structure is the result of the interactions of two α-helices burying their hydrophobic sides within a hydrophilic environment (Figure 7A). The design rules of coiled-coil structures have been extensively studied by Woolfson's group and are well understood [205]. Typically, *a* and *d* residues are hydrophobic, providing the hydrophobic core of the coil; *e* and *g* are generally charged amino acids (with opposite charge) used to stabilise the coil formation, while the triad *bfc* can be used to introduce bioactive motifs [185,206]. Examples of helical coiled-coil fibres are also present in numerous natural proteins, such as myosin and tropomyosin, fibrin, and keratin [207]. Variations of these systems are based on coiled coils with 'sticky ends', which stack together in a staggered manner along the axis of the coils. Larger coiled-coil systems can also form α-helical barrels with accessible channels, appealing as functional materials for the catalysis, binding, and transport of molecules (Figure 7B) [208]. These systems, as demonstrated by Woolfson's group, offer a valid platform for regenerative medicine applications [209]. For example, Mehrban et al. showed that RGDS-decorated coiled-coil peptide hydrogels were able to sustain cell growth and differentiation of murine embryonic neural stem cells in an in vitro study [210]. The same system injected into a rat with a partial-thickness abdominal wall defect showed no foreign body reaction and good biocompatibility [211].

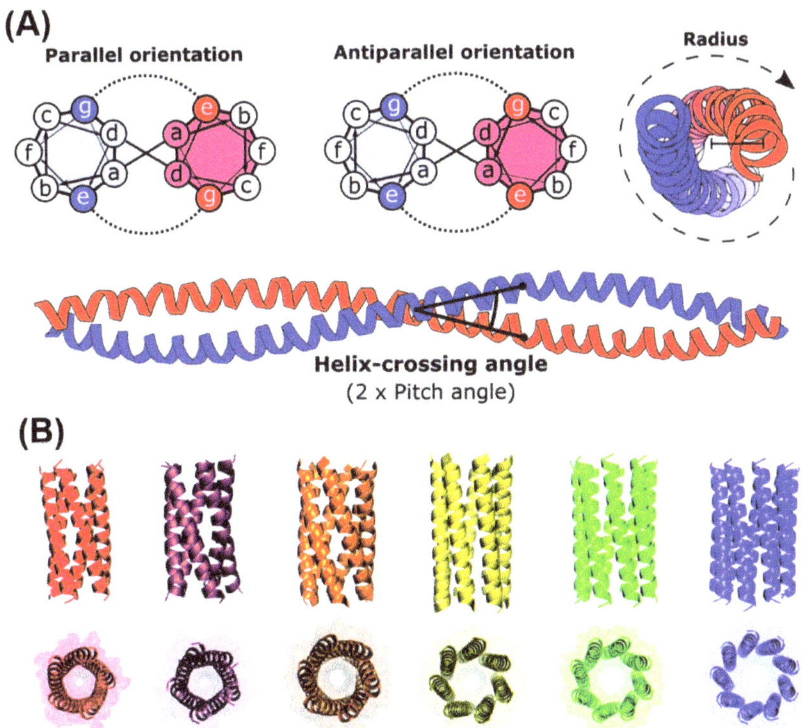

Figure 7. Coiled-coil and α-helix barrel peptide system designs. (**A**) Heptad wheel representation of parallel and anti-parallel coiled coils. Supramolecular structures form due to the interactions occurring between 'a' and 'd' (usually hydrophobic), while 'g' and 'e' (usually charged) stabilise the assembly. (**B**) Larger coiled-coil structures form α-helical barrels (αHBs) with accessible central channels. From right (red) to left (blue), X-ray crystal structures of multiple pentameric to nonameric αHBs. Images were adapted with permission from Ref. [206]. Copyright 2022, Royal Society of Chemistry.

5.4. Short Aromatic Peptides and Peptide Amphiphiles

Short aromatic peptide derivatives have been designed to create hydrogels based on π–π stacking interactions. In these systems, short peptides (typically up to five amino acids) exploit the presence of aromatic residues such as phenylalanine (F), tyrosine (Y), and/or tryptophan (W) as sticky points to promote self-assembly. The first example of short aromatic peptides was reported by Reches and Gazit, who found that diluting a hexafluoroisopropanol solution of Fmoc-FF in water caused self-assembly into amyloid-like fibrillar structures [212]. Since then, most common designs have been based on di-phenylalanine residues (FF) or di-phenylalanine coupled with Fmoc groups, which can also be extended to contain bioactive motifs such as RGD and serine residues (S) [185]. Fmoc-FF-based systems allow the creation of stable β-sheet-forming hydrogels in which Fmoc acts as an interlocking zipper to bring together different sheets along the nanofibre axis. In particular, the supramolecular unit of Fmoc-FF is thought to be composed of anti-parallel β-sheets organised into cylindrical fibres, held together by π–π stacking of adjacent sheets [185,213]. Two remarkable examples in this field are those by Jayawarna and Zhou, who designed FF-modified peptides able to create ECM-mimicking nanofibrous structures for the encapsulation of chondrocytes and fibroblasts in 3D cell cultures [214,215]. In particular, Zhou et al. showed that a mixture of Fmoc-RGD and Fmoc-FF self-assembled into RGD-displaying peptides that were able to promote fibroblast viability and provide a highly hydrated, stiff nanofibrous network [215] (Figure 8A). Jayawarna et al. extended the library of Fmoc-FF hydrogels by introducing serine (S), lysine (K), and aspartic acid (D) to form bioactive hydrogels for different cell lines [216].

Peptide amphiphiles (PAs) are represented by polypeptides with both hydrophobic and hydrophilic ends. Each PA sequence is composed of three parts: (1) a hydrophobic alkyl tail, (2) a β-sheet forming segment, and (3) a hydrophilic peptide head [185] (see Figure 8B). When exposed to water, PA sequences assemble into high-aspect-ratio rods having a hydrophobic core composed of the alkyl tails and a hydrophilic surface provided by the PA's head. The glycine linker region is used to promote H bonding between adjacent PA monomers assembling within the same rod. By varying the alkyl tail length (12 to 16 carbon atoms), it is possible to change the fibre diameter, their flexibility, and, in turn, the PA's network mechanical properties [185]. The peptide sequences of the hydrophilic head can be designed to show bioactive motifs to mimic the native ECM. This design strategy allows functional groups, such as cell attachment or GF-binding sites, to be 'micro-patterned' directly on the surface of peptide nanofibres, which could be used to regulate cell behaviour [217,218]. The possibility to introduce ECM-mimetic sequences directly exposed on the surface of the PA fibres without disrupting the process of self-assembly (which happens through the hydrophobic tail) represents one of the major advantages of this peptide-based system. Moreover, different 'families' of PA can be mixed to 'dilute' or 'concentrate' the density of the bioactive site on the surface of the fibres [218]. Firstly, introduced by Tirrell's group in 1995 [219], PAs were extensively designed and used for TE by Stupp and co-workers, who applied PAs decorated with RGD, IKVAV, and GF-binding motifs for different applications in TE, including angiogenesis, neurogenesis, chondrogenesis, and enamel and bone regeneration [220–223].

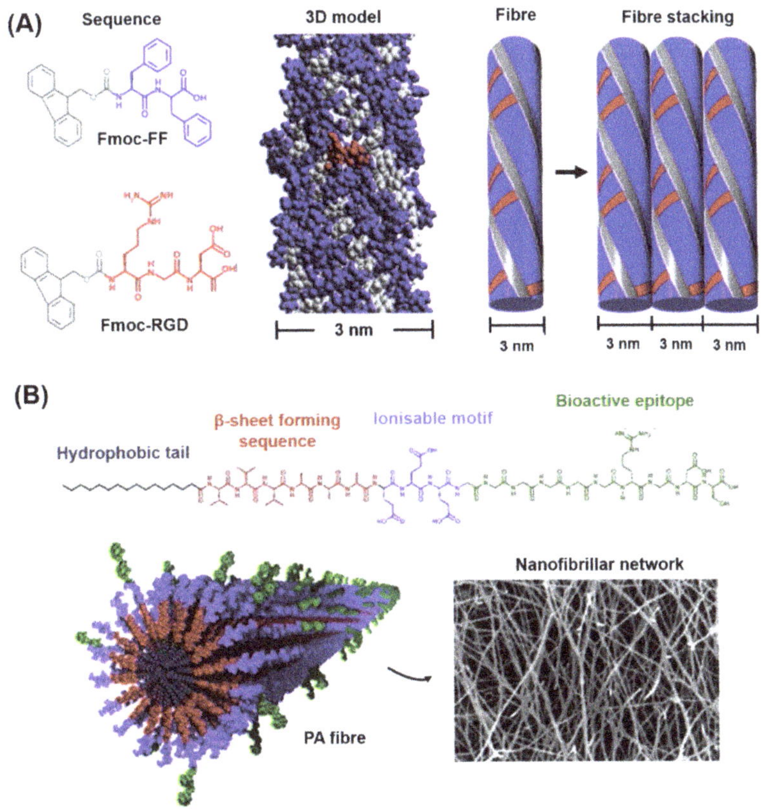

Figure 8. Examples of short aromatic and peptide amphiphile designs. (**A**) Chemical structures of Fmoc-FF and Fmoc-RGD assembling into nanofibrils with RGD sequences on the fibre surface. (**B**) Example of a PA sequence consisting of a hydrophobic tail, a β-forming segment, an ionisable motif, and a hydrophilic bioactive epitope. Self-assembled PAs show bioactive epitope on the surface (green appendages) and hydrophobic tails in the core (dark grey). Reprinted and adapted with permission from Ref. [223]. Copyright 2022, Elsevier.

6. Peptide-Based Hydrogels for IVD Repair

The good biocompatibility, tuneable physiochemical properties, and ease of injectability made SAPHs an attractive soft biomaterial for NP tissue engineering. Indeed, with the recent attention on NP regeneration as a therapeutic approach to treat early stage IVD degeneration, an increasing number of peptide materials have been explored as injectable NP hydrogel scaffolds. RADA16 systems, first introduced by Zhang et al. [184], represent the most used class of peptide hydrogels for NP regeneration. Li et al., for example, designed a new RADA16-based peptide (RKP) by attaching KPSS (KPSSAPTQLN), which is a short bioactive motif of bone morphogenetic protein-7 (BMP-7), to the C-terminal of a RADA16 for IVD TE applications [224]. KPSS was chosen since BMP-7 has been shown to stimulate the expression and deposition of PGs and collagen type II in vitro as well as induce an increase in disc height in degenerated discs in rabbits [225]. Dike's group, for example, showed that transduced NP cells with human BMP-7, via adeno-associated virus-2 and injected into a canine model of disc degeneration, were able to effectively preserve the structural integrity, ECM, and biomechanical properties of the canine's degenerated discs [226]. In a work by Li and colleagues, human NP-derived stem cells (NDSCs) were cultured in RKP and bare RADA16 peptide hydrogels in an apoptosis-inducing environment generated by the addition of TNF-α [224]. After 48 h, RKP effectively protected NDSCs from the pro-inflammatory TNF-α cytokine, promoting high viability and cell proliferation. Moreover, the hydrogel stimulated PG and collagen type II synthesis *in vitro*; it promoted upregulation of NP-marker genes, such as collagen type II, SOX9, and aggrecan and, more importantly, it restored disc height in rabbits affected by IVD degeneration [224]. Using the KPSS motif, Ruan's groups showed that a mixture of RADA16 with KPSS-containing RADA16 in a 1:1 ratio promoted cell proliferation and chemotaxis of bone-marrow-derived MSCs as well as kept cell viability as a cell carrier up to 14 days once injected into an ex vivo cultured disc model [227]. In another two studies, Tao et al. took a further step in this direction with the use of BMP-mimetic sequences co-assembling RADA16 with RKP as well as RADA16 mixed with another two BMP-2 motifs, SNVILKKYRN and KAISVLYFDDS, and improved the therapeutic effect of RADA16, leading to higher gene and protein expression of NP cells [228].

Besides BMPs, Link N (peptide sequence: DHLSKNYTLDHDRAIH) represents another popular bioactive motif that has been used for NP TE applications. Link N has been shown to stabilise PGs in the ECM and to induce a GF-like effect on chondrocytes in vitro by stimulating the synthesis of PG and collagen type II [229]. Due to the presence of PG and collagen type II as major components of the NP's ECM, it is not surprising that Link N has also been explored for NP studies. Wang et al., for example, conjugated Link N to RADA16 peptide hydrogel to form RLN and further co-assembled this with RADA16 to create a multicomponent peptide scaffold called LN-NS [230]. The authors showed that rabbit NP cells seeded on LN-NS migrated into the hydrogel and remained viable over time, while they did not with RADA16. Moreover, LN-NS hydrogels promoted a nearly two-fold increase in collagen type II and aggrecan gene expression compared with RADA16 only used as control [230]. Similarly, Ma et al. mixed RADA16 with Link N in a 1:1 ratio and showed that the new mixture promoted rabbit notochordal cell adhesion and the deposition of aggrecan and collagen type II after 14 days [231].

Besides RADA-based systems, other β-sheet forming peptides have been explored for NP repair. Bian and colleagues, for example, formulated a physical mixture of KLD12 (dodecapeptide) and TGFβ1 for the culture of rabbit MSCs for disc regeneration, which showed good MSC viability and proliferation after 14 days of culture within the KLD12/TGFβ1 complex [190]. In this case, peptide hydrogels were able to induce MSCs differentiation into NP-like cells, and this was accompanied by an upregulation of aggrecan and collagen type II gene expression [190]. The same system alone was studied as an injectable scaffold for NP cells by Sun and co-workers. KLD12 self-assembled into long nanometric fibres (10–30 nm in diameter), which sustained rabbit NP cell viability for up to 14 days and increased the production of GAGs and collagen type II over time [232]. Shorter than 12 amino acids, we explored the use of the octapeptide FEFEFKFK for the 3D culture of bovine NP cells (BNPCs), as shown in Figure 9. A rheological study of FEFEFKFK revealed that the hydrogel stiffness could be tuned by increasing the peptide concentration (up to 35 mg mL^{-1}) to mimic the native NP rheological properties [195]. BNPCs encapsulated in FEFEFKFK showed an upregulation of aggrecan, collagen type II, and novel NP-marker genes, such as FOXF1, KRT8, and KRT18, over 14 days of 3D cell culture [195]. Based on a similar sequence, recently, our group developed an injectable graphene oxide (GO) self-assembling peptide nanocomposite based on the β-sheet-forming peptide FEFKFEFK (F8) [233]. The octapeptide is a short octa-peptide that readily self-assembles into ~3 nm-diameter β-sheet-rich fibres, which above a critical gelation concentration entangle/associate to form self-supporting, transparent, and injectable hydrogels. The incorporation of GO had a two-pronged objective. On the one side, GO flakes effectively increased the mechanical properties of peptide hydrogels to mimic the value of native NP tissue so that higher mechanical properties were achieved even with lower peptide concentrations (10–20 mg mL^{-1}). On the other side, GO flakes could promote cell attachment and be decorated with relevant biomolecules to support cell viability and function. In a preliminary 3D cell culture study of 7 days, GO-F8 hydrogels were conducive to NP cell proliferation and preserved anabolic cellular activity, similar to those experienced by NP cells in vivo [233]. Due to the molecular design of F8, at a low pH, strong electrostatic interactions between positively charged F8 and negatively charged GO flakes drove the assembly of peptide nanofibres onto GO flakes, while at a physiological pH, hydrophobic interactions prevailed to create further hydrogel stiffening. These hydrogels were designed to shear-thin and recover their mechanical property instantaneously upon the removal of the injection's shear forces and undergo further stiffening in contact with cell culture media [233]. In a subsequent study, GO flakes were decorated with TGF-β3, an important anabolic GF involved in disc formation and homeostasis [38]. Decorated GO flakes were incorporated in F8 hydrogels and compared with systems with TGF-β3 mixed within the hydrogel and TGF-β3 supplied exogenously. When TGF-β3 was anchored on GO's surfaces, NP cells showed optimal matrix deposition after 21 days of culture, with a significant increase in gene expression (i.e., ACAN, COL2A1, FOXF1 and PAX1) and activation of SMAD signalling compared to peptide hydrogels with TGF-β3 exogenously added in the culture media or 'free' in the peptide hydrogels [234]. These two studies represent the first example of the potential advantages of merging peptide hydrogels with carbon-based nanofillers to achieve IVD repair and NP regeneration. Shorter than octapeptides, ultrasmall peptides (three to seven aminoacids) were also exploited by Hauser's group and tested with porcine NP cells (pNPCs), rabbit erythrocytes, and human MSCs for IVD TE applications. The peptides in aqueous solution self-assembled via transitional α-helices that transformed into β-turn supramolecular structures. The formulated hydrogels resulted in NP tissue of two orders of magnitude stiffer than native porcine, but comparable with human NP. When incubated with the peptides, minimal hemolysis was observed, and when the peptides were neutralised to pH 7.4, pNPCs and hMSCs showed good viability and characteristic morphology [235].

Less explored for IVD regeneration, PA-based hydrogels were recently tested by Uysal et al. for NP-repair applications. In their study, Uysal et al. cultured rat MSCs in a PA, presenting the collagen-mimetic sequence Pro-Hyp-Gly (Figure 9) [236]. The hydrogels were reported in vitro to enhance MSC differentiation into NP-like cells, as observed by an increase in SOX9 over time. When injected into degenerated rabbit IVDs, the PA hydrogel induced more GAG and collagen deposition compared to saline solution and to the PA hydrogel lacking the POG motif [236]. The injected hydrogel provided a functional recovery of the rabbit IVDs, as shown by the degeneration index score obtained [236]. Among commercially available peptide hydrogels, Barreto-Henriksson et al. showed that PuraMatrix® was able to support encapsulated MSCs' GAG deposition over time as well as provide mechanical properties mimicking the native NP tissue [237]. As a potential tissue-engineering scaffold to reconstitute the NP in early degeneration, Moss et al. formulated a hydrogel composite based on an elastin-like polypeptide and thiol-modified hyaluronan [238]. The composite resulted in an injectable stiff hydrogel with time-dependant features desirable for injectable applications. Biologically, the hybrid hydrogel maintained high cell viability and maintenance of anabolic phenotype based on morphologic and immunohistochemical data [238]. Moreover, the hydrogels showed good biocompatibility in a preclinical rabbit annular puncture model [238]. More recently, Wilcox's group formulated a peptide:GAG nanocomposite, where GAGs were vital components of the IVD's ECM, regulating hydration and swelling pressure in vivo [239]. Peptide:GAG hybrid hydrogels were obtained by mixing four β-sheet tape-forming peptides (P_{11}-4, P_{11}-8, P_{11}-9, and P_{11}-12) with varying net charge and hydrogen bonding capacities with chondroitin sulphate, a proof-of-concept GAG present in the native NP. Specific peptides and specific peptide:GAG ratios were found to be key parameters to control gel rheological properties. In particular, peptide:GAG ratios of 1:2 and 1:10 matched human NP tissue properties and allowed injection in an ex vivo caudal disc model with minimal leakage from the injection site and fast in situ gelation [239]. Representative studies using self-assembling peptide hydrogels for NP tissue repair are collated in Table 3.

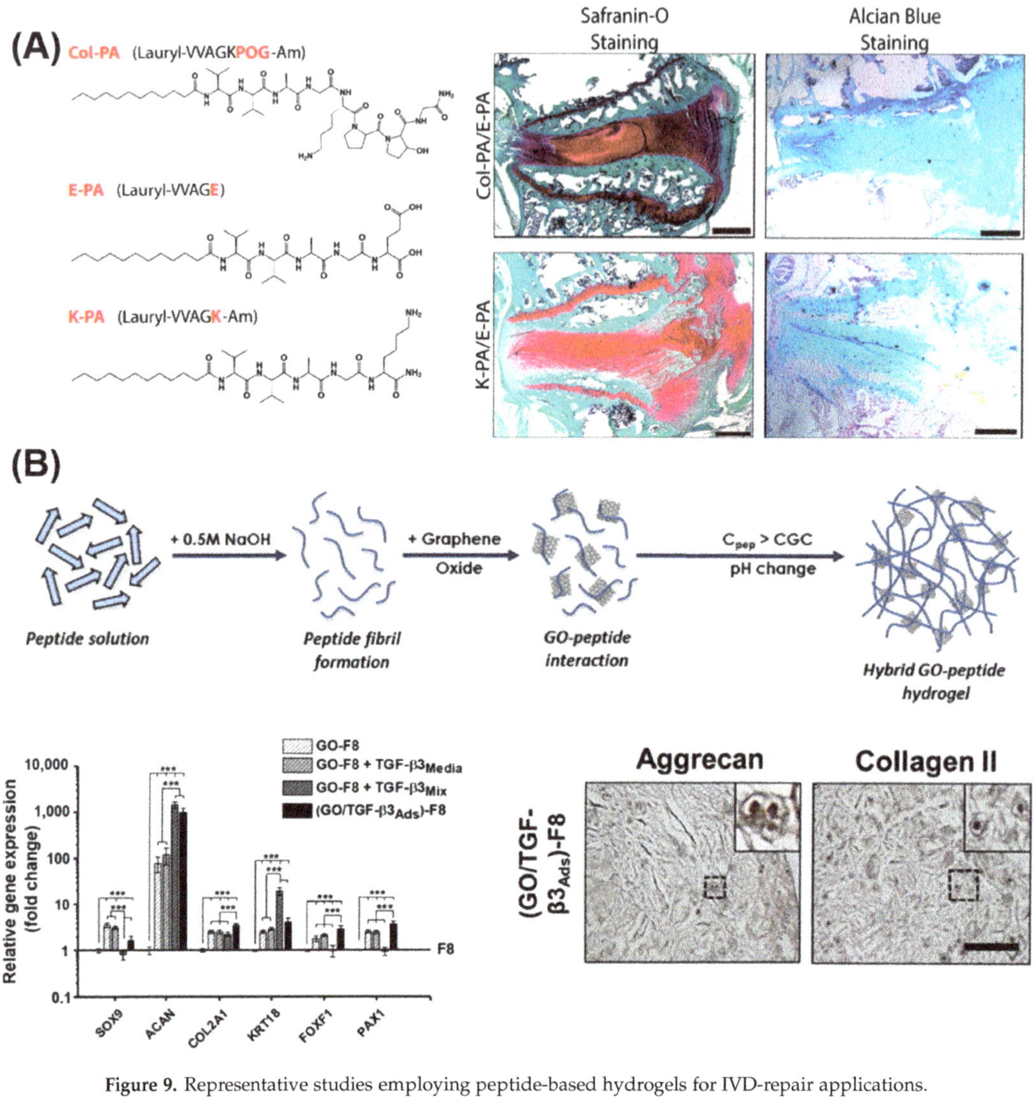

Figure 9. Representative studies employing peptide-based hydrogels for IVD-repair applications. (**A**) PA structures mimicking collagen fibres for IVD applications. MSCs cultured on PA nanofibres showed successful differentiation after 14 days and abundant GAG deposition (Safranin O and Alcian Blue stainings shown). (**B**) Self-assembly and formulation of GO-containing F8 hydrogels. TGF-β3-coated GO-F8 hydrogels showed increased gene and protein expression of NP cells after 3 weeks of 3D culture; *** p-value < 0.001. Reprinted and adapted with permission from Refs. [233,234]. Copyright 2022, Elsevier. Reprinted and adapted with permission from [236]. Copyright 2022, American Chemical Society.

Table 3. Studies reporting the use of peptide hydrogels as 3D scaffolds for IVD repair.

Hydrogel	Injectable?	Cell Type	Duration	Outcomes	Refs.
RKP (RADA16 functionalised with KPSS)	N/A	Human NP-derived stem cells	2 days	High cell proliferation and upregulation of the gene expression of collagen II, aggrecan, and Sox-9 compared to controls	[224]
RADA16:RADA-KPSS in 1:1 ratio	Yes	Human MSCs	14 days	Hydrogels promoted cell proliferation and chemotaxis as well as kept cell viability up to 14 days once injected into an ex vivo cultured disc mode	[227]
RAD-RKP (co-assembly of RADA16 with RKP)	Yes	Degenerated human NP cells	28 days	High cell proliferation and migration. Collagen type II, SOX9, and aggrecan were upregulated, while collagen I was downregulated compared to controls	[228]
LN-NS (co-assembly of link N-conjugated RADA16 with RADA16)	Yes	Rabbit NP cells	14 days	High viability and cell adhesion. Increased gene expression of PGs and collagen type II compared to controls	[230]
LN-NS (1:1 mixture of link N-conjugated RADA16 with RADA16)	Yes	Rabbit Notochordal Cells	14 days	High cell survival rate. Increased gene expression of aggrecan and collagen type II compared to controls	[231]
KLD12	Yes	Rabbit NP cells	14 days	Increased GAG and collagen type II production over time. NP cells preserved high viability and characteristic shape	[232]
KLD12/TGF-β1	N/A	Rabbit MSCs	14 days	High cell viability over time. Increased gene expression and protein deposition of aggrecan and collagen type II compared to controls	[190]
FEFEFKFK	Yes	Bovine NP Cells	14 days	Characteristic morphology and high viability over time. Upregulation of collagen type II, aggrecan, cytokeratin-8, cytokeratin-18, SOX9, and FOXF1, CA12	[195]
FEFKFEFK-GO	Yes	Bovine NP cells	7 days	Increased viability of NP cells and stable metabolic activity over time	[233]
FEFKFEFK-(GO + TGF-β3)	Yes	Bovine NP cells	21 days	TGF-β3-decorated GO flakes induced increased gene expression and matrix deposition over time compared to TGF-β3 'free' in peptide hydrogels or added exogenously. SMAD signalling was preserved when TGF-β3 was anchored on GO flakes, inducing ECM production	[234]
Ac-ID_3, Ac-LD_6, Ac-AD_6	N/A	Porcine NP cells	2 days	Porcine NP cells showed good viability when incubated with peptides	[235]
Col-PA/E-PA (PA hydrogel decorated with collagen-mimic POG motif)	Yes	Rat MSCs	14 days	*In vitro* differentiation into NP-like cells. Increased GAGs and collagen deposition compared to controls. Functional recovery of rabbit IVDs after injection *in vivo*	[236]
PuraMatrix®	Yes	Human MSCs	21 days	High deposition of GAGs accompanied by fast stress relaxation and mechanical properties mimicking native NP tissue	[237]
Elastin-like polypeptide/thiol-modified hyaluronan	Yes	Human IVD cells	21 days	High cell viability and maintenance of anabolic cell phenotype. Good biocompatibility in a preclinical rabbit annular puncture model	[238]
P_{11}-4, P_{11}-8, P_{11}-9, P_{11}-12/chondroitin sulphate	Yes	Ex vivo bovine caudal spine model	1 day	Hybrid hydrogels showed mechanical properties similar to the human NP. Hydrogels injected in a caudal spine model undergo gelation in situ with minimal leakage over time	[239]

7. Summary and Conclusions

IVD degeneration is a degenerative disease that initiates within the NP, the gelatinous core of the disc, and ends up involving the entire vertebral segment. The disc loses its high level of hydration and, unable to withstand external stresses, it collapses while nerves and blood vessels start to infiltrate. Current treatments, based on either palliative care (e.g., steroids, NSAIDs) or artificial prostheses (e.g., discectomy), mainly target the symptoms but not the cause of the disease. This leads to poorly efficient treatments and co-morbidities. In the search for alternative therapy, peptide hydrogels could play a pivotal role in the treatment of NP defects, especially in the early stage of degeneration. Due to their inherent properties, peptide hydrogels can be mixed with cell suspensions (e.g., MSCs and/or NP cells), and quick and efficient gelation can be induced by mild external stimuli, such as contact with cell culture media. This approach allows the formulation of bioactive hydrogels with homogenous cell dispersion for downstream applications. Peptide hydrogels not only behave as protective carriers for encapsulated cells but, more importantly, they can be designed to serve as an instructive niche by depicting both mechanical (e.g., stiffness gradients) and biochemical cues (e.g., bioactive epitopes). The tunability of SAPHs allows them also to host nanofillers that can enhance or add functionality. In this context, the incorporation of graphene-based materials, for example, offers a valid example in which 2D nanomaterials could dramatically improve bulk mechanical properties, increase cell attachment, and act as a solid-phase presentation platform for relevant biomolecules, such as proteins and anabolic GFs [234,240,241]. Finally, the ability of SAPHs to be injected with excellent shear-thinning behaviour make them a suitable candidate for minimally invasive procedures and the spatial dispensing of hydrogels, for example, via minimally invasive punctures and bioprinting [199,242]. Taking these advantages together, we envisage that peptide hydrogels represent a promising tool to tackle IVD degeneration, especially when used as functional injectable fillers in the early stage of degeneration.

Author Contributions: C.L. wrote the original draft and drew the figures; C.L., J.A.H., and A.S. revised and approved the manuscript. All authors have read and agreed to the published version of the manuscript.

Funding: C.L. acknowledges financial support from the EPSRC through the Doctoral Prize Fellowship (grant n°: EP/T517823/1). A.S. acknowledges the financial support from the EPSRC Early Career Research Fellowship (grant n°: EP/K016210/1) and the MRC Acellular/Smart Materials—3D Architecture: UK RMP Hub (grant n°: MR/R015651/1).

Institutional Review Board Statement: Not applicable.

Informed Consent Statement: Not applicable.

Data Availability Statement: Not applicable.

Conflicts of Interest: The authors declare no conflict of interest.

References

1. Robinson, J.S. Sciatica and the lumbar disk syndrome: A historic perspective. *South. Med. J.* **1983**, *76*, 232–238. [CrossRef] [PubMed]
2. Humzah, M.D.; Soames, R.W. Human Intervertebral Disc: Structure and Function. *Anat. Rec.* **1988**, *220*, 337–356. [CrossRef] [PubMed]
3. Raj, P.P. Intervertebral Disc: Pathophysiology-Treatment. *Pain Pract.* **2008**, *8*, 18–44. [CrossRef]
4. Neidlinger-Wilke, C.; Galbusera, F.; Pratsinis, H.; Mavrogonatou, E.; Mietsch, A.; Kletsas, D.; Wilke, H.-J. Mechanical loading of the intervertebral disc: From the macroscopic to the cellular level. *Eur. Spine J. Off. Publ. Eur. Spine Soc. Eur. Spinal Deform. Soc. Eur. Sect. Cerv. Spine Res. Soc.* **2014**, *23* (Suppl. S3), S333–S343. [CrossRef] [PubMed]
5. Wilke, H.J.; Neef, P.; Caimi, M.; Hoogland, T.; Claes, L.E. New in vivo measurements of pressures in the intervertebral disc in daily life. *Spine (Phila. Pa. 1976)* **1999**, *24*, 755–762. [CrossRef] [PubMed]
6. Newell, N.; Little, J.P.; Christou, A.; Adams, M.A.; Adam, C.J.; Masouros, S.D. Biomechanics of the human intervertebral disc: A review of testing techniques and results. *J. Mech. Behav. Biomed. Mater.* **2017**, *69*, 420–434. [CrossRef]
7. Antoniou, J.; Steffen, T.; Nelson, F.; Winterbottom, N.; Hollander, A.P.; Poole, R.A.; Aebi, M.; Alini, M. The human lumbar intervertebral disc: Evidence for changes in the biosynthesis and denaturation of the extracellular matrix with growth, maturation, ageing, and degeneration. *J. Clin. Investig.* **1996**, *98*, 996–1003. [CrossRef]
8. Iatridis, J.C.; Weidenbaum, M.; Setton, L.A.; Mow, V.C. Is the nucleus pulposus a solid or a fluid? Mechanical behaviors of the nucleus pulposus of the human intervertebral disc. *Spine (Phila. Pa. 1976)* **1996**, *21*, 1174–1184. [CrossRef]

9. Mwale, F.; Roughley, P.; Antoniou, J. Distinction between the extracellular matrix of the nucleus pulposus and hyaline cartilage: A requisite for tissue engineering of intervertebral disc. *Eur. Cell. Mater.* **2004**, *8*, 54–58. [CrossRef]
10. Sztrolovics, R.; Alini, M.; Mort, J.S.; Roughley, P.J. Age-related changes in fibromodulin and lumican in human intervertebral discs. *Spine (Phila. Pa. 1976)* **1999**, *24*, 1765–1771. [CrossRef]
11. Melrose, J.; Ghosh, P.; Taylor, T.K. A comparative analysis of the differential spatial and temporal distributions of the large (aggrecan, versican) and small (decorin, biglycan, fibromodulin) proteoglycans of the intervertebral disc. *J. Anat.* **2001**, *198*, 3–15. [CrossRef] [PubMed]
12. Yu, J.; Peter, C.; Roberts, S.; Urban, J.P.G. Elastic fibre organization in the intervertebral discs of the bovine tail. *J. Anat.* **2002**, *201*, 465–475. [CrossRef] [PubMed]
13. Eyre, D.R.; Muir, H. Types I and II collagens in intervertebral disc. Interchanging radial distributions in annulus fibrosus. *Biochem. J.* **1976**, *157*, 267–270. [CrossRef]
14. Inoue, H.; Takeda, T. Three-dimensional observation of collagen framework of lumbar intervertebral discs. *Acta Orthop.* **1975**, *46*, 949–956. [CrossRef]
15. Risbud, M.V.; Schoepflin, Z.R.; Mwale, F.; Kandel, R.A.; Grad, S.; Iatridis, J.C.; Sakai, D.; Hoyland, J.A. Defining the phenotype of young healthy nucleus pulposus cells: Recommendations of the Spine Research Interest Group at the 2014 annual ORS meeting. *J. Orthop. Res.* **2015**, *33*, 283–293. [CrossRef] [PubMed]
16. Stockwell, R.A. The interrelationship of cell density and cartilage thickness in mammalian articular cartilage. *J. Anat.* **1971**, *109*, 411–421.
17. Maroudas, A.; Stockwell, R.A.; Nachemson, A.; Urban, J. Factors involved in the nutrition of the human lumbar intervertebral disc: Cellularity and diffusion of glucose in vitro. *J. Anat.* **1975**, *120*, 113–130. Available online: https://www.ncbi.nlm.nih.gov/pmc/articles/PMC1231728/ (accessed on 30 March 2022).
18. Minogue, B.M.; Richardson, S.M.; Zeef, L.A.H.; Freemont, A.J.; Hoyland, J.A. Characterization of the human nucleus pulposus cell phenotype and evaluation of novel marker gene expression to define adult stem cell differentiation. *Arthritis Rheum.* **2010**, *62*, 3695–3705. [CrossRef]
19. Minogue, B.M.; Richardson, S.M.; Zeef, L.A.H.; Freemont, A.J.; Hoyland, J.A. Transcriptional profiling of bovine intervertebral disc cells: Implications for identification of normal and degenerate human intervertebral disc cell phenotypes. *Arthritis Res. Ther.* **2010**, *12*, R22. [CrossRef] [PubMed]
20. Nerlich, A.G.; Weiler, C.; Zipperer, J.; Narozny, M.; Boos, N. Immunolocalization of phagocytic cells in normal and degenerated intervertebral discs. *Spine (Phila. Pa. 1976)* **2002**, *27*, 2484–2490. [CrossRef]
21. Shamji, M.F.; Setton, L.A.; Jarvis, W.; So, S.; Chen, J.; Jing, L.; Bullock, R.; Isaacs, R.E.; Brown, C.; Richardson, W.J. Proinflammatory cytokine expression profile in degenerated and herniated human intervertebral disc tissues. *Arthritis Rheum.* **2010**, *62*, 1974–1982. [CrossRef] [PubMed]
22. Urban, J.P.G. The role of the physicochemical environment in determining disc cell behaviour. *Biochem. Soc. Trans.* **2002**, *30*, 858–864. [CrossRef] [PubMed]
23. Holm, S.; Maroudas, A.; Urban, J.P.; Selstam, G.; Nachemson, A. Nutrition of the intervertebral disc: Solute transport and metabolism. *Connect. Tissue Res.* **1981**, *8*, 101–119. [CrossRef]
24. Mwale, F.; Ciobanu, I.; Giannitsios, D.; Roughley, P.; Steffen, T.; Antoniou, J. Effect of oxygen levels on proteoglycan synthesis by intervertebral disc cells. *Spine (Phila. Pa. 1976)* **2011**, *36*, E131–E138. [CrossRef] [PubMed]
25. Marchand, F.; Ahmed, A.M. Investigation of the laminate structure of lumbar disc anulus fibrosus. *Spine (Phila. Pa. 1976)* **1990**, *15*, 402–410. [CrossRef]
26. Taylor, T.K.; Ghosh, P.; Bushell, G.R. The contribution of the intervertebral disk to the scoliotic deformity. *Clin. Orthop. Relat. Res.* **1981**, *156*, 79–90. [CrossRef]
27. Schollmeier, G.; Lahr-Eigen, R.; Lewandrowski, K.U. Observations on fiber-forming collagens in the anulus fibrosus. *Spine (Phila. Pa. 1976)* **2000**, *25*, 2736–2741. [CrossRef]
28. Errington, R.J.; Puustjarvi, K.; White, I.R.; Roberts, S.; Urban, J.P. Characterisation of cytoplasm-filled processes in cells of the intervertebral disc. *J. Anat.* **1998**, *192 Pt 3*, 369–378. [CrossRef]
29. Freemont, A.J.; Peacock, T.E.; Goupille, P.; Hoyland, J.A.; O'Brien, J.; Jayson, M.I. Nerve ingrowth into diseased intervertebral disc in chronic back pain. *Lancet* **1997**, *350*, 178–181. [CrossRef]
30. Purmessur, D.; Freemont, A.J.; Hoyland, J.A. Expression and regulation of neurotrophins in the nondegenerate and degenerate human intervertebral disc. *Arthritis Res. Ther.* **2008**, *10*, R99. [CrossRef]
31. Roberts, S.; Menage, J.; Urban, J.P. Biochemical and structural properties of the cartilage end-plate and its relation to the intervertebral disc. *Spine (Phila. Pa. 1976)* **1989**, *14*, 166–174. [CrossRef] [PubMed]
32. Roberts, S.; McCall, I.W.; Menage, J.; Haddaway, M.J.; Eisenstein, S.M. Does the thickness of the vertebral subchondral bone reflect the composition of the intervertebral disc? *Eur. Spine J.* **1997**, *6*, 385–389. [CrossRef] [PubMed]
33. Nachemson, A.; Lewin, T.; Maroudas, A.; Freeman, M.A. In vitro diffusion of dye through the end-plates and the annulus fibrosus of human lumbar inter-vertebral discs. *Acta Orthop. Scand.* **1970**, *41*, 589–607. [CrossRef] [PubMed]
34. Roberts, S.; Menage, J.; Duance, V.; Wotton, S.F. Type III collagen in the intervertebral disc. *Histochem. J.* **1991**, *23*, 503–508. [CrossRef] [PubMed]

35. Roberts, S.; Menage, J.; Duance, V.; Wotton, S.; Ayad, S. 1991 Volvo Award in basic sciences. Collagen types around the cells of the intervertebral disc and cartilage end plate: An immunolocalization study. *Spine (Phila. Pa. 1976)* **1991**, *16*, 1030–1038. [CrossRef]
36. Moore, R.J. The vertebral endplate: Disc degeneration, disc regeneration. *Eur. Spine J.* **2006**, *15*, 333–337. [CrossRef]
37. Hynes, R.O. Extracellular matrix: Not just pretty fibrils. *Science* **2009**, *326*, 1216–1219. [CrossRef]
38. Chen, S.; Liu, S.; Ma, K.; Zhao, L.; Lin, H.; Shao, Z. TGF-β signaling in intervertebral disc health and disease. *Osteoarthr. Cartil.* **2019**, *27*, 1109–1117. [CrossRef]
39. Gruber, H.E.; Fisher, E.C.; Desai, B.; Stasky, A.A.; Hoelscher, G.; Hanley, E.N. Human intervertebral disc cells from the annulus: Three-dimensional culture in agarose or alginate and responsiveness to TGF-beta1. *Exp Cell Res.* **1997**, *235*, 13–21. [CrossRef]
40. Jin, H.; Shen, J.; Wang, B.; Wang, M.; Shu, B.; Chen, D. TGF-β signaling plays an essential role in the growth and maintenance of intervertebral disc tissue. *FEBS Lett.* **2011**, *585*, 1209–1215. [CrossRef]
41. Hodgkinson, T.; Shen, B.; Diwan, A.; Hoyland, J.A.; Richardson, S.M. Therapeutic potential of growth differentiation factors in the treatment of degenerative disc diseases. *JOR Spine* **2019**, *2*, e1045. [CrossRef] [PubMed]
42. Richardson, S.M.; Freemont, A.J.; Hoyland, J.A. Pathogenesis of Intervertebral Disc Degeneration. In *The Intervertebral Disc*; Shapiro, I.M., Risbud, M.V., Eds.; Springer: Vienna, Vienna, 2014; pp. 177–200. [CrossRef]
43. Duance, V.C.; Crean, J.K.; Sims, T.J.; Avery, N.; Smith, S.; Menage, J.; Eisenstein, S.M.; Roberts, S. Changes in collagen cross-linking in degenerative disc disease and scoliosis. *Spine (Phila. Pa. 1976)* **1998**, *23*, 2545–2551. [CrossRef] [PubMed]
44. le Maitre, C.L.; Pockert, A.; Buttle, D.J.; Freemont, A.J.; Hoyland, J.A. Matrix synthesis and degradation in human intervertebral disc degeneration. *Biochem. Soc. Trans.* **2007**, *35*, 652–655. [CrossRef]
45. Boos, N.; Nerlich, A.G.; Wiest, I.; von der Mark, K.; Aebi, M. Immunolocalization of type X collagen in human lumbar intervertebral discs during ageing and degeneration. *Histochem Cell Biol.* **1997**, *108*, 471–480. [CrossRef] [PubMed]
46. Nachemson, A. Intradiscal Measurements of pH in Patients with Lumbar Rhizopathies. *Acta Orthop. Scand.* **1969**, *40*, 23–42. [CrossRef]
47. Ohshima, H.; Urban, J.P. The effect of lactate and pH on proteoglycan and protein synthesis rates in the intervertebral disc. *Spine (Phila. Pa. 1976)* **1992**, *17*, 1079–1082. [CrossRef]
48. Gilbert, H.T.J.; Hodson, N.; Baird, P.; Richardson, S.M.; Hoyland, J.A. Acidic pH promotes intervertebral disc degeneration: Acid-sensing ion channel -3 as a potential therapeutic target. *Sci. Rep.* **2016**, *6*, 37360. [CrossRef]
49. Cs-Szabo, G.; Juan, D.R.; Turumella, V.; Masuda, K.; Thonar, E.J.; An, H.S. Changes in mRNA and protein levels of proteoglycans of the anulus fibrosus and nucleus pulposus during intervertebral disc degeneration. *Spine (Phila. Pa. 1976)* **2002**, *27*, 2212–2219. [CrossRef]
50. Sztrolovics, R.; Grover, J.; Cs-Szabo, G.; Shi, S.-L.; Zhang, Y.; Mort, J.S.; Roughley, P.J. The characterization of versican and its message in human articular cartilage and intervertebral disc. *J. Orthop. Res.* **2002**, *20*, 257–266. [CrossRef]
51. Oegema, T.R.J.; Johnson, S.L.; Aguiar, D.J.; Ogilvie, J.W. Fibronectin and its fragments increase with degeneration in the human intervertebral disc. *Spine (Phila. Pa. 1976)* **2000**, *25*, 2742–2747. [CrossRef]
52. Aota, Y.; An, H.S.; Homandberg, G.; Thonar, E.J.M.-A.; Andersson, G.B.J.; Pichika, R.; Masuda, K. Differential effects of fibronectin fragment on proteoglycan metabolism by intervertebral disc cells: A comparison with articular chondrocytes. *Spine (Phila. Pa. 1976)* **2005**, *30*, 722–728. [CrossRef] [PubMed]
53. Anderson, D.G.; Li, X.; Tannoury, T.; Beck, G.; Balian, G. A fibronectin fragment stimulates intervertebral disc degeneration in vivo. *Spine (Phila. Pa. 1976)* **2003**, *28*, 2338–2345. [CrossRef] [PubMed]
54. Nagase, H.; Kashiwagi, M. Aggrecanases and cartilage matrix degradation. *Arthritis Res. Ther.* **2003**, *5*, 94–103. [CrossRef] [PubMed]
55. Nagase, H.; Woessner, J.F.J. Matrix metalloproteinases. *J. Biol. Chem.* **1999**, *274*, 21491–21494. [CrossRef]
56. Risbud, M.V. The Intervertebral Disc. *Lancet* **1951**, *257*, 275–276. [CrossRef]
57. Pockert, A.J.; Richardson, S.M.; le Maitre, C.L.; Lyon, M.; Deakin, J.A.; Buttle, D.J.; Freemont, A.J.; Hoyland, J.A. Modified expression of the ADAMTS enzymes and tissue inhibitor of metalloproteinases 3 during human intervertebral disc degeneration. *Arthritis Rheum.* **2009**, *60*, 482–491. [CrossRef]
58. Studer, R.K.; Vo, N.; Sowa, G.; Ondeck, C.; Kang, J. Human nucleus pulposus cells react to IL-6: Independent actions and amplification of response to IL-1 and TNF-alpha. *Spine (Phila. Pa. 1976)* **2011**, *36*, 593–599. [CrossRef]
59. le Maitre, C.L.; Freemont, A.J.; Hoyland, J.A. Localization of degradative enzymes and their inhibitors in the degenerate human intervertebral disc. *J Pathol.* **2004**, *204*, 47–54. [CrossRef]
60. le Maitre, C.L.; Freemont, A.J.; Hoyland, J.A. The role of interleukin-1 in the pathogenesis of human Intervertebral disc degeneration. *Arthritis Res. Ther.* **2005**, *7*, R732. [CrossRef]
61. le Maitre, C.L.; Freemont, A.J.; Hoyland, J.A. A preliminary in vitro study into the use of IL-1Ra gene therapy for the inhibition of intervertebral disc degeneration. *Int. J. Exp. Pathol.* **2006**, *87*, 17–28. [CrossRef]
62. le Maitre, C.L.; Hoyland, J.A.; Freemont, A.J. Interleukin-1 receptor antagonist delivered directly and by gene therapy inhibits matrix degradation in the intact degenerate human intervertebral disc: An in situ zymographic and gene therapy study. *Arthritis Res. Ther.* **2007**, *9*, R83. [CrossRef] [PubMed]
63. Johnson, W.E.B.; Caterson, B.; Eisenstein, S.M.; Hynds, D.L.; Snow, D.M.; Roberts, S. Human intervertebral disc aggrecan inhibits nerve growth in vitro. *Arthritis Rheum.* **2002**, *46*, 2658–2664. [CrossRef]

64. Richardson, S.M.; Doyle, P.; Minogue, B.M.; Gnanalingham, K.; Hoyland, J.A. Increased expression of matrix metalloproteinase-10, nerve growth factor and substance P in the painful degenerate intervertebral disc. *Arthritis Res. Ther.* **2009**, *11*, R126. [CrossRef] [PubMed]
65. Wallach, D.; Arumugam, T.U.; Boldin, M.P.; Cantarella, G.; Ganesh, K.A.; Goltsev, Y.; Goncharov, T.M.; Kovalenko, A.V.; Rajput, A.; Varfolomeev, E.E.; et al. How are the regulators regulated? The search for mechanisms that impose specificity on induction of cell death and NF-kappaB activation by members of the TNF/NGF receptor family. *Arthritis Res.* **2002**, *4* (Suppl. S3), S189–S196. [CrossRef] [PubMed]
66. Cheung, K.M.C.; Karppinen, J.; Chan, D.; Ho, D.W.H.; Song, Y.-Q.; Sham, P.; Cheah, K.S.E.; Leong, J.C.Y.; Luk, K.D.K. Prevalence and pattern of lumbar magnetic resonance imaging changes in a population study of one thousand forty-three individuals. *Spine (Phila. Pa. 1976)* **2009**, *34*, 934–940. [CrossRef] [PubMed]
67. Boos, N.; Weissbach, S.; Rohrbach, H.; Weiler, C.; Spratt, K.F.; Nerlich, A.G. Classification of age-related changes in lumbar intervertebral discs: 2002 Volvo Award in basic science. *Spine (Phila. Pa. 1976)* **2002**, *27*, 2631–2644. [CrossRef] [PubMed]
68. Risbud, M.V.; Fertala, J.; Vresilovic, E.J.; Albert, T.J.; Shapiro, I.M. Nucleus pulposus cells upregulate PI3K/Akt and MEK/ERK signaling pathways under hypoxic conditions and resist apoptosis induced by serum withdrawal. *Spine (Phila. Pa. 1976)* **2005**, *30*, 882–889. [CrossRef]
69. Shen, C.; Yan, J.; Jiang, L.-S.; Dai, L.-Y. Autophagy in rat annulus fibrosus cells: Evidence and possible implications. *Arthritis Res. Ther.* **2011**, *13*, R132. [CrossRef]
70. Ye, W.; Xu, K.; Huang, D.; Liang, A.; Peng, Y.; Zhu, W.; Li, C. Age-related increases of macroautophagy and chaperone-mediated autophagy in rat nucleus pulposus. *Connect. Tissue Res.* **2011**, *52*, 472–478. [CrossRef] [PubMed]
71. Roberts, S.; Evans, E.H.; Kletsas, D.; Jaffray, D.C.; Eisenstein, S.M. Senescence in human intervertebral discs. *Eur. Spine J. Off. Publ. Eur. Spine Soc. Eur. Spinal Deform. Soc. Eur. Sect. Cerv. Spine Res. Soc.* **2006**, *15* (Suppl. S3), S312–S316. [CrossRef] [PubMed]
72. le Maitre, C.L.; Freemont, A.J.; Hoyland, J.A. Accelerated cellular senescence in degenerate intervertebral discs: A possible role in the pathogenesis of intervertebral disc degeneration. *Arthritis Res. Ther.* **2007**, *9*, R45. [CrossRef] [PubMed]
73. Kim, K.-W.; Chung, H.-N.; Ha, K.-Y.; Lee, J.-S.; Kim, Y.-Y. Senescence mechanisms of nucleus pulposus chondrocytes in human intervertebral discs. *Spine J.* **2009**, *9*, 658–666. [CrossRef]
74. Zhang, C.; Wang, B.; Zhao, X.; Li, X.; Lou, Z.; Chen, X.; Zhang, F. Iron deficiency accelerates intervertebral disc degeneration through affecting the stability of DNA polymerase epsilon complex. *Am. J. Transl. Res.* **2018**, *10*, 3430–3442.
75. Barbosa, M.C.; Grosso, R.A.; Fader, C.M. Hallmarks of Aging: An Autophagic Perspective. *Front. Endocrinol.* **2019**, *9*, 790. [CrossRef]
76. Puustjarvi, K.; Lammi, M.; Helminen, H.; Inkinen, R.; Tammi, M. Proteoglycans in the intervertebral disc of young dogs following strenuous running exercise. *Connect. Tissue Res.* **1994**, *30*, 225–240. [CrossRef]
77. Iatridis, J.C.; Mente, P.L.; Stokes, I.A.F.; Aronsson, D.D.; Alini, M. Compression-Induced Changes in Intervertebral Disc Properties in a Rat Tail Model. *Spine (Phila. Pa. 1976)* **1999**, *24*. [CrossRef]
78. MacLean, J.J.; Lee, C.R.; Alini, M.; Iatridis, J.C. The effects of short-term load duration on anabolic and catabolic gene expression in the rat tail intervertebral disc. *J. Orthop. Res.* **2005**, *23*, 1120–1127. [CrossRef] [PubMed]
79. Wuertz, K.; Godburn, K.; MacLean, J.J.; Barbir, A.; Donnelly, J.S.; Roughley, P.J.; Alini, M.; Iatridis, J.C. In vivo remodeling of intervertebral discs in response to short- and long-term dynamic compression. *J. Orthop. Res.* **2009**, *27*, 1235–1242. [CrossRef] [PubMed]
80. Maclean, J.J.; Lee, C.R.; Alini, M.; Iatridis, J.C. Anabolic and catabolic mRNA levels of the intervertebral disc vary with the magnitude and frequency of in vivo dynamic compression. *J. Orthop. Res.* **2004**, *22*, 1193–1200. [CrossRef] [PubMed]
81. Handa, T.; Ishihara, H.; Ohshima, H.; Osada, R.; Tsuji, H.; Obata, K. Effects of hydrostatic pressure on matrix synthesis and matrix metalloproteinase production in the human lumbar intervertebral disc. *Spine (Phila. Pa. 1976)* **1997**, *22*, 1085–1091. [CrossRef]
82. Neidlinger-Wilke, C.; Wurtz, K.; Urban, J.P.G.; Borm, W.; Arand, M.; Ignatius, A.; Wilke, H.-J.; Claes, L.E. Regulation of gene expression in intervertebral disc cells by low and high hydrostatic pressure. *Eur. Spine J. Off. Publ. Eur. Spine Soc. Eur. Spinal Deform. Soc. Eur. Sect. Cerv. Spine Res. Soc.* **2006**, *15* (Suppl. S3), S372–S378. [CrossRef] [PubMed]
83. le Maitre, C.L.; Frain, J.; Fotheringham, A.P.; Freemont, A.J.; Hoyland, J.A. Human cells derived from degenerate intervertebral discs respond differently to those derived from non-degenerate intervertebral discs following application of dynamic hydrostatic pressure. *Biorheology* **2008**, *45*, 563–575. [CrossRef]
84. Gilbert, H.T.J.; Hoyland, J.A.; Millward-Sadler, S.J. The response of human anulus fibrosus cells to cyclic tensile strain is frequency-dependent and altered with disc degeneration. *Arthritis Rheum.* **2010**, *62*, 3385–3394. [CrossRef] [PubMed]
85. Gilbert, H.T.J.; Hoyland, J.A.; Freemont, A.J.; Millward-Sadler, S.J. The involvement of interleukin-1 and interleukin-4 in the response of human annulus fibrosus cells to cyclic tensile strain: An altered mechanotransduction pathway with degeneration. *Arthritis Res. Ther.* **2011**, *13*, R8. [CrossRef] [PubMed]
86. Varlotta, G.P.; Brown, M.D.; Kelsey, J.L.; Golden, A.L. Familial predisposition for herniation of a lumbar disc in patients who are less than twenty-one years old. *J. Bone Jt. Surg. Am.* **1991**, *73*, 124–128. Available online: https://journals.lww.com/jbjsjournal/Abstract/1991/73010/Familial_predisposition_for_herniation_of_a_lumbar.16.aspx (accessed on 4 March 2022). [CrossRef]
87. MacGregor, A.J.; Andrew, T.; Sambrook, P.N.; Spector, T.D. Structural, psychological, and genetic influences on low back and neck pain: A study of adult female twins. *Arthritis Rheum.* **2004**, *51*, 160–167. [CrossRef] [PubMed]

88. Li, S.W.; Prockop, D.J.; Helminen, H.; Fassler, R.; Lapvetelainen, T.; Kiraly, K.; Peltarri, A.; Arokoski, J.; Lui, H.; Arita, M. Transgenic mice with targeted inactivation of the Col2 alpha 1 gene for collagen II develop a skeleton with membranous and periosteal bone but no endochondral bone. *Genes Dev.* **1995**, *9*, 2821–2830. [CrossRef]
89. Kimura, T.; Nakata, K.; Tsumaki, N.; Miyamoto, S.; Matsui, Y.; Ebara, S.; Ochi, T. Progressive degeneration of articular cartilage and intervertebral discs. An experimental study in transgenic mice bearing a type IX collagen mutation. *Int. Orthop.* **1996**, *20*, 177–181. [CrossRef]
90. Watanabe, H.; Nakata, K.; Kimata, K.; Nakanishi, I.; Yamada, Y. Dwarfism and age-associated spinal degeneration of heterozygote cmd mice defective in aggrecan. *Proc. Natl. Acad. Sci. USA* **1997**, *94*, 6943–6947. [CrossRef]
91. Rajasekaran, S.; Kanna, R.M.; Senthil, N.; Raveendran, M.; Cheung, K.M.C.; Chan, D.; Subramaniam, S.; Shetty, A.P. Phenotype variations affect genetic association studies of degenerative disc disease: Conclusions of analysis of genetic association of 58 single nucleotide polymorphisms with highly specific phenotypes for disc degeneration in 332 subjects. *Spine J.* **2013**, *13*, 1309–1320. [CrossRef]
92. Stankiewicz, P.; Sen, P.; Bhatt, S.S.; Storer, M.; Xia, Z.; Bejjani, B.A.; Ou, Z.; Wiszniewska, J.; Driscoll, D.J.; Maisenbacher, M.K.; et al. Genomic and genic deletions of the FOX gene cluster on 16q24.1 and inactivating mutations of FOXF1 cause alveolar capillary dysplasia and other malformations. *Am. J. Hum. Genet.* **2009**, *84*, 780–791. [CrossRef]
93. Smith, C.A.; Tuan, R.S. Functional involvement of Pax-1 in somite development: Somite dysmorphogenesis in chick embryos treated with Pax-1 paired-box antisense oligodeoxynucleotide. *Teratology* **1995**, *52*, 333–345. [CrossRef]
94. Wallin, J.; Wilting, J.; Koseki, H.; Fritsch, R.; Christ, B.; Balling, R. The role of Pax-1 in axial skeleton development. *Development* **1994**, *120*, 1109–1121. [CrossRef] [PubMed]
95. Stewart, W.F.; Ricci, J.A.; Chee, E.; Morganstein, D.; Lipton, R. Lost productive time and cost due to common pain conditions in the US workforce. *JAMA* **2003**, *290*, 2443–2454. [CrossRef] [PubMed]
96. Vlaeyen, J.W.S.; Maher, C.G.; Wiech, K.; van Zundert, J.; Meloto, C.B.; Diatchenko, L.; Battié, M.C.; Goossens, M.; Koes, B.; Linton, S.J. Low back pain. *Nat. Rev. Dis. Prim.* **2018**, *4*, 52. [CrossRef]
97. de Schepper, E.I.T.; Damen, J.; van Meurs, J.B.J.; Ginai, A.Z.; Popham, M.; Hofman, A.; Koes, B.W.; Bierma-Zeinstra, S.M. The association between lumbar disc degeneration and low back pain: The influence of age, gender, and individual radiographic features. *Spine (Phila. Pa. 1976)* **2010**, *35*, 531–536. [CrossRef] [PubMed]
98. Ravindra, V.M.; Senglaub, S.S.; Rattani, A.; Dewan, M.C.; Härtl, R.; Bisson, E.; Park, K.B.; Shrime, M.G. Degenerative Lumbar Spine Disease: Estimating Global Incidence and Worldwide Volume. *Glob. Spine J.* **2018**, *8*, 784–794. [CrossRef]
99. Hoy, D.; Brooks, P.; Blyth, F.; Buchbinder, R. The Epidemiology of low back pain. *Best Pract. Res. Clin. Rheumatol.* **2010**, *24*, 769–781. [CrossRef] [PubMed]
100. Maniadakis, N.; Gray, A. The economic burden of back pain in the UK. *Pain* **2000**, *84*, 95–103. [CrossRef]
101. Hoy, D.; March, L.; Brooks, P.; Blyth, F.; Woolf, A.; Bain, C.; Williams, G.; Smith, E.; Vos, T.; Barendregt, J.; et al. The global burden of low back pain: Estimates from the Global Burden of Disease 2010 study. *Ann. Rheum. Dis.* **2014**, *73*, 968–974. [CrossRef] [PubMed]
102. Pfirrmann, C.W.A.; Metzdorf, A.; Zanetti, M.; Hodler, J.; Boos, N. Magnetic Resonance Classification of Lumbar Intervertebral Disc Degeneration. *Spine (Phila. Pa. 1976)* **2001**, *26*, 1873–1878. [CrossRef]
103. Radek, M.; Pacholczyk-Sienicka, B.; Jankowski, S.; Albrecht, Ł.; Grodzka, M.; Depta, A.; Radek, A. Assessing the correlation between the degree of disc degeneration on the Pfirrmann scale and the metabolites identified in HR-MAS NMR spectroscopy. *Magn. Reson. Imaging* **2016**, *34*, 376–380. [CrossRef] [PubMed]
104. Griffith, J.F.; Wang, Y.-X.J.; Antonio, G.E.; Choi, K.C.; Yu, A.; Ahuja, A.T.; Leung, P.C. Modified Pfirrmann Grading System for Lumbar Intervertebral Disc Degeneration. *Spine (Phila. Pa. 1976)* **2007**, *32*, E708–E712. [CrossRef] [PubMed]
105. Sobajima, S.; Kompel, J.F.; Kim, J.S.; Wallach, C.J.; Robertson, D.D.; Vogt, M.T.; Kang, J.D.; Gilbertson, L.G. A slowly progressive and reproducible animal model of intervertebral disc degeneration characterized by MRI, X-ray, and histology. *Spine (Phila. Pa. 1976)* **2005**, *30*, 15–24. [CrossRef] [PubMed]
106. Schenk, P.; Läubli, T.; Hodler, J.; Klipstein, A. Magnetic resonance imaging of the lumbar spine: Findings in female subjects from administrative and nursing professions. *Spine (Phila. Pa. 1976)* **2006**, *31*, 2701–2706. [CrossRef] [PubMed]
107. Waldenberg, C.; Hebelka, H.; Brisby, H.; Lagerstrand, K.M. MRI histogram analysis enables objective and continuous classification of intervertebral disc degeneration. *Eur. Spine J.* **2018**, *27*, 1042–1048. [CrossRef]
108. Wu, P.H.; Kim, H.S.; Jang, I.-T. Intervertebral Disc Diseases PART 2: A Review of the Current Diagnostic and Treatment Strategies for Intervertebral Disc Disease. *Int. J. Mol. Sci.* **2020**, *21*, 2135. [CrossRef] [PubMed]
109. Luan, S.; Wan, Q.; Luo, H.; Li, X.; Ke, S.; Lin, C.; Wu, Y.; Wu, S.; Ma, C. Running exercise alleviates pain and promotes cell proliferation in a rat model of intervertebral disc degeneration. *Int. J. Mol. Sci.* **2015**, *16*, 2130–2144. [CrossRef]
110. Maroon, J.C.; Bost, J.W. Omega-3 fatty acids (fish oil) as an anti-inflammatory: An alternative to nonsteroidal anti-inflammatory drugs for discogenic pain. *Surg. Neurol.* **2006**, *65*, 326–331. [CrossRef] [PubMed]
111. Stuber, K.; Sajko, S.; Kristmanson, K. Efficacy of glucosamine, chondroitin, and methylsulfonylmethane for spinal degenerative joint disease and degenerative disc disease: A systematic review. *J. Can. Chiropr. Assoc.* **2011**, *55*, 47–55.
112. Lanas, A. Nonsteroidal antiinflammatory drugs and cyclooxygenase inhibition in the gastrointestinal tract: A trip from peptic ulcer to colon cancer. *Am. J. Med. Sci.* **2009**, *338*, 96–106. [CrossRef] [PubMed]

113. Brune, K.; Patrignani, P. New insights into the use of currently available non-steroidal anti-inflammatory drugs. *J. Pain Res.* **2015**, *8*, 105–118. [CrossRef]
114. Atchison, J.W.; Herndon, C.M.; Rusie, E. NSAIDs for musculoskeletal pain management:current perspectives and novel strategies to improve safety. *J. Manag. Care Pharm.* **2013**, *19*, S3–S19. [PubMed]
115. Errico, T.J. Lumbar disc arthroplasty. *Clin. Orthop. Relat. Res.* **2005**, *435*, 106–117. [CrossRef] [PubMed]
116. Lee, Y.C.; Zotti, M.G.T.; Osti, O.L. Operative Management of Lumbar Degenerative Disc Disease. *Asian Spine J.* **2016**, *10*, 801–819. [CrossRef] [PubMed]
117. Serhan, H.; Mhatre, D.; Defossez, H.; Bono, C.M. Motion-preserving technologies for degenerative lumbar spine: The past, present, and future horizons. *SAS J.* **2011**, *5*, 75–89. [CrossRef] [PubMed]
118. Zigler, J.; Delamarter, R.; Spivak, J.M.; Linovitz, R.J.; Danielson, G.O.I.I.I.; Haider, T.T.; Cammisa, F.; Zuchermann, J.; Balderston, R.; Kitchel, S.; et al. Results of the Prospective, Randomized, Multicenter Food and Drug Administration Investigational Device Exemption Study of the ProDisc®-L Total Disc Replacement Versus Circumferential Fusion for the Treatment of 1-Level Degenerative Disc Disease. *Spine (Phila. Pa. 1976)* **2007**, *32*, 1155–1162. [CrossRef] [PubMed]
119. Link, H.D. History, design and biomechanics of the LINK SB Charité artificial disc. In *Arthroplasty of the Spine*; Gunzburg, R., Mayer, H.M., Szpalski, M., Aebi, M., Eds.; Springer: Berlin/Heidelberg, Germany, 2004; pp. 36–43. [CrossRef]
120. Guerrero, J.; Häckel, S.; Croft, A.S.; Hoppe, S.; Albers, C.E.; Gantenbein, B. The nucleus pulposus microenvironment in the intervertebral disc: The fountain of youth? *Eur. Cell. Mater.* **2021**, *41*, 707–738. [CrossRef] [PubMed]
121. Chou, W.-Y.; Hsu, C.-J.; Chang, W.-N.; Wong, C.-Y. Adjacent segment degeneration after lumbar spinal posterolateral fusion with instrumentation in elderly patients. *Arch. Orthop. Trauma Surg.* **2002**, *122*, 39–43. [CrossRef] [PubMed]
122. Park, P.; Garton, H.J.; Gala, V.C.; Hoff, J.T.; McGillicuddy, J.E. Adjacent Segment Disease after Lumbar or Lumbosacral Fusion: Review of the Literature. *Spine (Phila. Pa. 1976)* **2004**, *29*, 1938–1944. [CrossRef] [PubMed]
123. Leong, J.C.; Chun, S.Y.; Grange, W.J.; Fang, D. Long-term results of lumbar intervertebral disc prolapse. *Spine (Phila. Pa. 1976)* **1983**, *8*, 793–799. [CrossRef] [PubMed]
124. Guigui, P.; Wodecki, P.; Bizot, P.; Lambert, P.; Chaumeil, G.; Deburge, A. Long-term influence of associated arthrodesis on adjacent segments in the treatment of lumbar stenosis: A series of 127 cases with 9-year follow-up. *Rev. Chir. Orthop. Reparatrice Appar. Mot.* **2000**, *86*, 546–557. [PubMed]
125. Kuslich, S.D.; Danielson, G.; Dowdle, J.D.; Sherman, J.; Fredrickson, B.; Yuan, H.; Griffith, S.L. Four-year follow-up results of lumbar spine arthrodesis using the Bagby and Kuslich lumbar fusion cage. *Spine (Phila. Pa. 1976)* **2000**, *25*, 2656–2662. [CrossRef] [PubMed]
126. Deyo, R.A.; Mirza, S.K.; Martin, B.I.; Kreuter, W.; Goodman, D.C.; Jarvik, J.G. Trends, major medical complications, and charges associated with surgery for lumbar spinal stenosis in older adults. *JAMA* **2010**, *303*, 1259–1265. [CrossRef] [PubMed]
127. Buzhor, E.; Leshansky, L.; Blumenthal, J.; Barash, H.; Warshawsky, D.; Mazor, Y.; Shtrichman, R. Cell-based therapy approaches: The hope for incurable diseases. *Regen. Med.* **2014**, *9*, 649–672. [CrossRef] [PubMed]
128. Richardson, S.M.; Hoyland, J.A.; Mobasheri, R.; Csaki, C.; Shakibaei, M.; Mobasheri, A. Mesenchymal stem cells in regenerative medicine: Opportunities and challenges for articular cartilage and intervertebral disc tissue engineering. *J. Cell. Physiol.* **2010**, *222*, 23–32. [CrossRef] [PubMed]
129. Pittenger, M.F.; Mackay, A.M.; Beck, S.C.; Jaiswal, R.K.; Douglas, R.; Mosca, J.D.; Moorman, M.A.; Simonetti, D.W.; Craig, S.; Marshak, D.R. Multilineage potential of adult human mesenchymal stem cells. *Science* **1999**, *284*, 143–147. [CrossRef]
130. Erices, A.; Conget, P.; Minguell, J.J. Mesenchymal progenitor cells in human umbilical cord blood. *Br. J. Haematol.* **2000**, *109*, 235–242. [CrossRef] [PubMed]
131. Tuli, R.; Tuli, S.; Nandi, S.; Wang, M.L.; Alexander, P.G.; Haleem-Smith, H.; Hozack, W.J.; Manner, P.A.; Danielson, K.G.; Tuan, R.S. Characterization of multipotential mesenchymal progenitor cells derived from human trabecular bone. *Stem Cells.* **2003**, *21*, 681–693. [CrossRef] [PubMed]
132. Zuk, P.A.; Zhu, M.; Mizuno, H.; Huang, J.; Futrell, J.W.; Katz, A.J.; Benhaim, P.; Lorenz, H.P.; Hedrick, M.H. Multilineage cells from human adipose tissue: Implications for cell-based therapies. *Tissue Eng.* **2001**, *7*, 211–228. [CrossRef]
133. Young, H.E.; Steele, T.A.; Bray, R.A.; Hudson, J.; Floyd, J.A.; Hawkins, K.; Thomas, K.; Austin, T.; Edwards, C.; Cuzzourt, J.; et al. Human reserve pluripotent mesenchymal stem cells are present in the connective tissues of skeletal muscle and dermis derived from fetal, adult, and geriatric donors. *Anat. Rec.* **2001**, *264*, 51–62. [CrossRef]
134. Jankowski, R.J.; Deasy, B.M.; Huard, J. Muscle-derived stem cells. *Gene Ther.* **2002**, *9*, 642–647. [CrossRef] [PubMed]
135. Alsalameh, S.; Amin, R.; Gemba, T.; Lotz, M. Identification of mesenchymal progenitor cells in normal and osteoarthritic human articular cartilage. *Arthritis Rheum.* **2004**, *50*, 1522–1532. [CrossRef]
136. Clarke, L.E.; McConnell, J.C.; Sherratt, M.J.; Derby, B.; Richardson, S.M.; Hoyland, J.A. Growth differentiation factor 6 and transforming growth factor-beta differentially mediate mesenchymal stem cell differentiation, composition, and micromechanical properties of nucleus pulposus constructs. *Arthritis Res. Ther.* **2014**, *16*, R67. [CrossRef]
137. Strassburg, S.; Richardson, S.M.; Freemont, A.J.; Hoyland, J.A. Co-culture induces mesenchymal stem cell differentiation and modulation of the degenerate human nucleus pulposus cell phenotype. *Regen. Med.* **2010**, *5*, 701–711. [CrossRef] [PubMed]
138. Strassburg, S.; Hodson, N.W.; Hill, P.I.; Richardson, S.M.; Hoyland, J.A. Bi-Directional Exchange of Membrane Components Occurs during Co-Culture of Mesenchymal Stem Cells and Nucleus Pulposus Cells. *PLoS ONE* **2012**, *7*, e33739. [CrossRef] [PubMed]

139. Panebianco, C.J.; Meyers, J.H.; Gansau, J.; Hom, W.W.; Iatridis, J.C. Balancing biological and biomechanical performance in intervertebral disc repair: A systematic review of injectable cell delivery biomaterials. *Eur. Cell. Mater.* **2020**, *40*, 239–258. [CrossRef] [PubMed]
140. Hohaus, C.; Ganey, T.M.; Minkus, Y.; Meisel, H.J. Cell transplantation in lumbar spine disc degeneration disease. *Eur. Spine J.* **2008**, *17*, 492–503. [CrossRef] [PubMed]
141. Nomura, T.; Mochida, J.; Okuma, M.; Nishimura, K.; Sakabe, K. Nucleus pulposus allograft retards intervertebral disc degeneration. *Clin. Orthop. Relat. Res.* **2001**, *389*, 94–101. [CrossRef] [PubMed]
142. Nassr, A.; Lee, J.Y.; Bashir, R.S.; Rihn, J.A.; Eck, J.C.; Kang, J.D.; Lim, M.R. Does incorrect level needle localization during anterior cervical discectomy and fusion lead to accelerated disc degeneration? *Spine (Phila. Pa. 1976)* **2009**, *34*, 189–192. [CrossRef] [PubMed]
143. Daly, C.; Ghosh, P.; Jenkin, G.; Oehme, D.; Goldschlager, T. A Review of Animal Models of Intervertebral Disc Degeneration: Pathophysiology, Regeneration, and Translation to the Clinic. *Biomed Res. Int.* **2016**, *2016*, 5952165. [CrossRef] [PubMed]
144. Alini, M.; Eisenstein, S.M.; Ito, K.; Little, C.; Kettler, A.A.; Masuda, K.; Melrose, J.; Ralphs, J.; Stokes, I.; Wilke, H.J. Are animal models useful for studying human disc disorders/degeneration? *Eur. Spine J.* **2008**, *17*, 2–19. [CrossRef] [PubMed]
145. Iatridis, J.C.; Setton, L.A.; Weidenbaum, M.; Mow, V.C. The viscoelastic behavior of the non-degenerate human lumbar nucleus pulposus in shear. *J. Biomech.* **1997**, *30*, 1005–1013. [CrossRef]
146. Cortes, D.H.; Jacobs, N.T.; DeLucca, J.F.; Elliott, D.M. Elastic, permeability and swelling properties of human intervertebral disc tissues: A benchmark for tissue engineering. *J. Biomech.* **2014**, *47*, 2088–2094. [CrossRef]
147. Sakai, D. Stem cell regeneration of the intervertebral disk, Orthop. *Clin. North Am.* **2011**, *42*, 555–562. [CrossRef]
148. Schol, J.; Sakai, D. Cell therapy for intervertebral disc herniation and degenerative disc disease: Clinical trials. *Int. Orthop.* **2019**, *43*, 1011–1025. [CrossRef]
149. Clouet, J.; Fusellier, M.; Camus, A.; le Visage, C.; Guicheux, J. Intervertebral disc regeneration: From cell therapy to the development of novel bioinspired endogenous repair strategies. *Adv. Drug Deliv. Rev.* **2019**, *146*, 306–324. [CrossRef] [PubMed]
150. Langer, R.; Vacanti, J.P. Tissue engineering. *Science* **1993**, *260*, 920–926. [CrossRef]
151. Burdick, J.A.; Mauck, R.L.; Gerecht, S. To Serve and Protect: Hydrogels to Improve Stem Cell-Based Therapies. *Cell Stem Cell* **2016**, *18*, 13–15. [CrossRef]
152. Schmitz, T.C.; Salzer, E.; Crispim, J.F.; Fabra, G.T.; LeVisage, C.; Pandit, A.; Tryfonidou, M.; le Maitre, C.; Ito, K. Characterization of biomaterials intended for use in the nucleus pulposus of degenerated intervertebral discs. *Acta Biomater.* **2020**, *114*, 1–15. [CrossRef]
153. Mooney, D.J.; Vandenburgh, H. Cell Delivery Mechanisms for Tissue Repair. *Cell Stem Cell* **2008**, *2*, 205–213. [CrossRef] [PubMed]
154. Vadalà, G.; Sowa, G.; Hubert, M.; Gilbertson, L.G.; Denaro, V.; Kang, J.D. Mesenchymal stem cells injection in degenerated intervertebral disc: Cell leakage may induce osteophyte formation. *J. Tissue Eng. Regen. Med.* **2012**, *6*, 348–355. [CrossRef]
155. Wuertz, K.; Godburn, K.; Neidlinger-Wilke, C.; Urban, J.; Iatridis, J.C. Behavior of Mesenchymal Stem Cells in the Chemical Microenvironment of the Intervertebral Disc. *Spine (Phila. Pa. 1976)* **2008**, *33*, 1843–1849. [CrossRef] [PubMed]
156. Baumgartner, L.; Wuertz-Kozak, K.; le Maitre, C.L.; Wignall, F.; Richardson, S.M.; Hoyland, J.; Wills, C.R.; Ballester, M.A.G.; Neidlin, M.; Alexopoulos, L.G.; et al. Multiscale Regulation of the Intervertebral Disc: Achievements in Experimental, In Silico, and Regenerative Research. *Int. J. Mol. Sci.* **2021**, *22*, 703. [CrossRef]
157. Gupta, R.C.; Lall, R.; Srivastava, A.; Sinha, A. Hyaluronic Acid: Molecular Mechanisms and Therapeutic Trajectory. *Front. Vet. Sci.* **2019**, *6*, 192. [CrossRef] [PubMed]
158. Dagenais, S. Intra-articular hyaluronic acid (viscosupplementation) for knee osteoarthritis. *Issues Emerg. Health Technol.* **2006**, *98*, 1–4.
159. Isa, I.L.M.; Srivastava, A.; Tiernan, D.; Owens, P.; Rooney, P.; Dockery, P.; Pandit, A. Hyaluronic Acid Based Hydrogels Attenuate Inflammatory Receptors and Neurotrophins in Interleukin-1β Induced Inflammation Model of Nucleus Pulposus Cells. *Biomacromolecules* **2015**, *16*, 1714–1725. [CrossRef] [PubMed]
160. Isa, I.L.M.; Abbah, S.A.; Kilcoyne, M.; Sakai, D.; Dockery, P.; Finn, D.P.; Pandit, A. Implantation of hyaluronic acid hydrogel prevents the pain phenotype in a rat model of intervertebral disc injury. *Sci. Adv.* **2018**, *4*, eaaq0597. [CrossRef]
161. Tokita, Y.; Okamoto, A. Hydrolytic degradation of hyaluronic acid. *Polym. Degrad. Stab.* **1995**, *48*, 269–273. [CrossRef]
162. Quero, L.; Klawitter, M.; Schmaus, A.; Rothley, M.; Sleeman, J.; Tiaden, A.N.; Klasen, J.; Boos, N.; Hottiger, M.O.; Wuertz, K.; et al. Hyaluronic acid fragments enhance the inflammatory and catabolic response in human intervertebral disc cells through modulation of toll-like receptor 2 signalling pathways. *Arthritis Res. Ther.* **2013**, *15*, R94. [CrossRef] [PubMed]
163. Kalaf, E.A.G.; Pendyala, M.; Bledsoe, J.G.; Sell, S.A. Characterization and restoration of degenerated IVD function with an injectable, in situ gelling alginate hydrogel: An in vitro and ex vivo study. *J. Mech. Behav. Biomed. Mater.* **2017**, *72*, 229–240. [CrossRef] [PubMed]
164. Baer, A.E.; Wang, J.Y.; Kraus, V.B.; Setton, L.A. Collagen gene expression and mechanical properties of intervertebral disc cell-alginate cultures. *J. Orthop. Res.* **2001**, *19*, 2–10. [CrossRef]
165. Chenite, A.; Chaput, C.; Wang, D.; Combes, C.; Buschmann, M.D.; Hoemann, C.D.; Leroux, J.C.; Atkinson, B.L.; Binette, F.; Selmani, A. Novel injectable neutral solutions of chitosan form biodegradable gels in situ. *Biomaterials* **2000**, *21*, 2155–2161. [CrossRef]

166. Ahmadi, R.; de Bruijn, J.D. Biocompatibility and gelation of chitosan-glycerol phosphate hydrogels. *J. Biomed. Mater. Res. A.* **2008**, *86*, 824–832. [CrossRef] [PubMed]
167. Roughley, P.; Hoemann, C.; DesRosiers, E.; Mwale, F.; Antoniou, J.; Alini, M. The potential of chitosan-based gels containing intervertebral disc cells for nucleus pulposus supplementation. *Biomaterials* **2006**, *27*, 388–396. [CrossRef]
168. Sisson, K.; Zhang, C.; Farach-Carson, M.C.; Chase, D.B.; Rabolt, J.F. Evaluation of cross-linking methods for electrospun gelatin on cell growth and viability. *Biomacromolecules* **2009**, *10*, 1675–1680. [CrossRef] [PubMed]
169. Slaughter, B.V.; Khurshid, S.S.; Fisher, O.Z.; Khademhosseini, A.; Peppas, N.A. Hydrogels in Regenerative Medicine. *Adv. Mater.* **2009**, *21*, 3307–3329. [CrossRef] [PubMed]
170. Thorpe, A.A.; Dougill, G.; Vickers, L.; Reeves, N.D.; Sammon, C.; Cooper, G.; le Maitre, C.L. Thermally triggered hydrogel injection into bovine intervertebral disc tissue explants induces differentiation of mesenchymal stem cells and restores mechanical function. *Acta Biomater.* **2017**, *54*, 212–226. [CrossRef] [PubMed]
171. Thorpe, A.A.; Boyes, V.L.; Sammon, C.; le Maitre, C.L. Thermally triggered injectable hydrogel, which induces mesenchymal stem cell differentiation to nucleus pulposus cells: Potential for regeneration of the intervertebral disc. *Acta Biomater.* **2016**, *36*, 99–111. [CrossRef] [PubMed]
172. Haq, M.A.; Su, Y.; Wang, D. Mechanical properties of PNIPAM based hydrogels: A review. *Mater. Sci. Eng. C.* **2017**, *70*, 842–855. [CrossRef]
173. Woodard, L.N.; Grunlan, M.A. Hydrolytic Degradation and Erosion of Polyester Biomaterials. *ACS Macro Lett.* **2018**, *7*, 976–982. [CrossRef]
174. Tibbitt, M.W.; Anseth, K.S. Hydrogels as extracellular matrix mimics for 3D cell culture. *Biotechnol. Bioeng.* **2009**, *103*, 655–663. [CrossRef] [PubMed]
175. Peppas, N.A.; Hilt, J.Z.; Khademhosseini, A.; Langer, R. Hydrogels in Biology and Medicine: From Molecular Principles to Bionanotechnology. *Adv. Mater.* **2006**, *18*, 1345–1360. [CrossRef]
176. Gough, J.E.; Saiani, A.; Miller, A.F. Peptide hydrogels: Mimicking the extracellular matrix. *Bioinspired Biomim. Nanobiomaterials* **2012**, *1*, 4–12. [CrossRef]
177. Zhang, S. Fabrication of novel biomaterials through molecular self-assembly. *Nat. Biotechnol.* **2003**, *21*, 1171–1178. [CrossRef]
178. Zhang, S.; Holmes, T.C.; DiPersio, C.M.; Hynes, R.O.; Su, X.; Rich, A. Self-complementary oligopeptide matrices support mammalian cell attachment. *Biomaterials* **1995**, *16*, 1385–1393. [CrossRef]
179. Engler, A.J.; Sen, S.; Sweeney, H.L.; Discher, D.E. Matrix Elasticity Directs Stem Cell Lineage Specification. *Cell* **2006**, *126*, 677–689. [CrossRef]
180. Guvendiren, M.; Lu, H.D.; Burdick, J.A. Shear-thinning hydrogels for biomedical applications. *Soft Matter* **2012**, *8*, 260–272. [CrossRef]
181. Morris, O.; Elsawy, M.A.; Fairclough, M.; Williams, K.J.; Mcmahon, A.; Grigg, J.; Forster, D.; Miller, A.F.; Saiani, A.; Prenant, C. In vivo characterisation of a therapeutically relevant self-assembling (18) F-labelled β-sheet forming peptide and its hydrogel using positron emission tomography. *J. Labelled Comp. Radiopharm.* **2017**, *60*, 481–488. [CrossRef]
182. Markey, A.; Workman, V.L.; Bruce, I.A.; Woolford, T.J.; Derby, B.; Miller, A.F.; Cartmell, S.H.; Saiani, A. Peptide hydrogel in vitro non-inflammatory potential. *J. Pept. Sci.* **2016**, *23*, 148–154. [CrossRef] [PubMed]
183. Nagarkar, R.P.; Schneider, J.P. Synthesis and primary characterization of self-assembled peptide-based hydrogels. *Methods Mol. Biol.* **2008**, *474*, 61–77. [CrossRef] [PubMed]
184. Gelain, F.; Luo, Z.; Rioult, M.; Zhang, S. Self-assembling peptide scaffolds in the clinic. *npj Regen. Med.* **2021**, *6*, 9. [CrossRef] [PubMed]
185. Ulijn, R.V.; Smith, A.M. Designing peptide based nanomaterials. *Chem. Soc. Rev.* **2008**, *37*, 664. [CrossRef]
186. Aggeli, A.; Nyrkova, I.A.; Bell, M.; Harding, R.; Carrick, L.; McLeish, T.C.B.; Semenov, A.N.; Boden, N. Hierarchical self-assembly of chiral rod-like molecules as a model for peptide β-sheet tapes, ribbons, fibrils, and fibers. *Proc. Natl. Acad. Sci. USA* **2001**, *98*, 11857–11862. [CrossRef]
187. Ding, X.; Zhao, H.; Li, Y.; Lee, A.L.; Li, Z.; Fu, M.; Li, C.; Yang, Y.Y.; Yuan, P. Synthetic peptide hydrogels as 3D scaffolds for tissue engineering. *Adv. Drug Deliv. Rev.* **2020**, *160*, 78–104. [CrossRef] [PubMed]
188. Kisiday, J.; Jin, M.; Kurz, B.; Hung, H.; Semino, C.; Zhang, S.; Grodzinsky, A.J. Self-assembling peptide hydrogel fosters chondrocyte extracellular matrix production and cell division: Implications for cartilage tissue repair. *Proc. Natl. Acad. Sci. USA* **2002**, *99*, 9996–10001. [CrossRef]
189. Kopesky, P.W.; Vanderploeg, E.J.; Sandy, J.S.; Kurz, B.; Grodzinsky, A.J. Self-Assembling Peptide Hydrogels Modulate In Vitro Chondrogenesis of Bovine Bone Marrow Stromal Cells. *Tissue Eng. Part A* **2010**, *16*, 465–477. [CrossRef]
190. Bian, Z.; Sun, J. Development of a KLD-12 polypeptide/TGF-β1-tissue scaffold promoting the differentiation of mesenchymal stem cell into nucleus pulposus-like cells for treatment of intervertebral disc degeneration. *Int. J. Clin. Exp. Pathol.* **2015**, *8*, 1093–1103.
191. Tripathi, J.K.; Pal, S.; Awasthi, B.; Kumar, A.; Tandon, A.; Mitra, K.; Chattopadhyay, N.; Ghosh, J.K. Variants of self-assembling peptide, KLD-12 that show both rapid fracture healing and antimicrobial properties. *Biomaterials* **2015**, *56*, 92–103. [CrossRef]
192. Saiani, A.; Mohammed, A.; Frielinghaus, H.; Collins, R.; Hodson, N.; Kielty, C.M.; Sherratt, M.J.; Miller, A.F. Self-assembly and gelation properties of [small alpha]-helix versus [small beta]-sheet forming peptides. *Soft Matter* **2009**, *5*, 193–202. [CrossRef]

193. Diaz, L.A.C.; Saiani, A.; Gough, J.E.; Miller, A.F.; Diaz, L.A.C.; Saiani, A.; Gough, J.E.; Miller, A.F.; Diaz, L.A.C.; Saiani, A.; et al. Human osteoblasts within soft peptide hydrogels promote mineralisation in vitro. *J. Tissue Eng.* **2014**, *5*, 2041731414539344. [CrossRef]
194. Mujeeb, A.; Miller, A.F.; Saiani, A.; Gough, J.E. Self-assembled octapeptide scaffolds for in vitro chondrocyte culture. *Acta Biomater.* **2013**, *9*, 4609–4617. [CrossRef] [PubMed]
195. Wan, S.; Borland, S.; Richardson, S.M.; Merry, C.L.R.; Saiani, A.; Gough, J.E.; Wan, S.; Wan, S.; Borland, S.; Richardson, S.M.; et al. Self-Assembling Peptide Hydrogel for Intervertebral Disc Tissue Engineering. *Acta Biomater.* **2016**, *46*, 29–40. [CrossRef]
196. Burgess, K.A.; Frati, C.; Meade, K.; Gao, J.; Diaz, L.C.; Madeddu, D.; Graiani, G.; Cavalli, S.; Miller, A.F.; Oceandy, D.; et al. Functionalised peptide hydrogel for the delivery of cardiac progenitor cells. *Mater. Sci. Eng. C* **2021**, *119*, 111539. [CrossRef]
197. Imere, A.; Ligorio, C.; O'Brien, M.; Wong, J.; Domingos, M.; Cartmell, S. Engineering a cell-hydrogel-fibre composite to mimic the structure and function of the tendon synovial sheath. *Acta Biomater.* **2021**, *119*, 140–154. [CrossRef]
198. Wychowaniec, J.K.; Smith, A.M.; Ligorio, C.; Mykhaylyk, O.O.; Miller, A.F.; Saiani, A. Role of Sheet-Edge Interactions in β-sheet Self-Assembling Peptide Hydrogels. *Biomacromolecules* **2020**, *21*, 2285–2297. [CrossRef]
199. Chiesa, I.; Ligorio, C.; Bonatti, A.F.; de Acutis, A.; Smith, A.M.; Saiani, A.; Vozzi, G.; de Maria, C. Modeling the Three-Dimensional Bioprinting Process of β-Sheet Self-Assembling Peptide Hydrogel Scaffolds. *Front. Med. Technol.* **2020**, *2*, 4. [CrossRef]
200. Schneider, J.P.; Pochan, D.J.; Ozbas, B.; Rajagopal, K.; Pakstis, L.; Kretsinger, J. Responsive hydrogels from the intramolecular folding and self-assembly of a designed peptide. *J. Am. Chem. Soc.* **2002**, *124*, 15030–15037. [CrossRef]
201. Yan, C.; Altunbas, A.; Yucel, T.; Nagarkar, R.P.; Schneider, J.P.; Pochan, D.J. Injectable solid hydrogel: Mechanism of shear-thinning and immediate recovery of injectable β-hairpin peptide hydrogels. *Soft Matter* **2010**, *6*, 5143–5156. [CrossRef] [PubMed]
202. Lindsey, S.; Piatt, J.H.; Worthington, P.; Sönmez, C.; Satheye, S.; Schneider, J.P.; Pochan, D.J.; Langhans, S.A. Beta Hairpin Peptide Hydrogels as an Injectable Solid Vehicle for Neurotrophic Growth Factor Delivery. *Biomacromolecules* **2015**, *16*, 2672–2683. [CrossRef]
203. Altunbas, A.; Lee, S.J.; Rajasekaran, S.A.; Schneider, J.P.; Pochan, D.J. Encapsulation of curcumin in self-assembling peptide hydrogels as injectable drug delivery vehicles. *Biomaterials* **2011**, *32*, 5906–5914. [CrossRef] [PubMed]
204. Pauling, L.; Corey, R.B. Compound Helical Configurations of Polypeptide Chains: Structure of Proteins of the α-Keratin Type. *Nature* **1953**, *171*, 59–61. [CrossRef]
205. Banwell, E.F.; Abelardo, E.S.; Adams, D.J.; Birchall, M.A.; Corrigan, A.; Donald, A.M.; Kirkland, M.; Serpell, L.C.; Butler, M.F.; Woolfson, D.N. Rational design and application of responsive α-helical peptide hydrogels. *Nat. Mater.* **2009**, *8*, 596–600. [CrossRef] [PubMed]
206. Lapenta, F.; Aupič, J.; Strmšek, Ž.; Jerala, R. Coiled coil protein origami: From modular design principles towards biotechnological applications. *Chem. Soc. Rev.* **2018**, *47*, 3530–3542. [CrossRef] [PubMed]
207. Woolfson, D.N.; Ryadnov, M.G. Peptide-based fibrous biomaterials: Some things old, new and borrowed. *Curr. Opin. Chem. Biol.* **2006**, *10*, 559–567. [CrossRef]
208. Dawson, W.M.; Martin, F.J.O.; Rhys, G.G.; Shelley, K.L.; Brady, R.L.; Woolfson, D.N. Coiled coils 9-to-5: Rational de novo design of α-helical barrels with tunable oligomeric states. *Chem. Sci.* **2021**, *12*, 6923–6928. [CrossRef] [PubMed]
209. MacPhee, C.E.; Woolfson, D.N. Engineered and designed peptide-based fibrous biomaterials. *Curr. Opin. Solid State Mater. Sci.* **2004**, *8*, 141–149. [CrossRef]
210. Mehrban, N.; Zhu, B.; Tamagnini, F.; Young, F.I.; Wasmuth, A.; Hudson, K.L.; Thomson, A.R.; Birchall, M.A.; Randall, A.D.; Song, B.; et al. Functionalized α-Helical Peptide Hydrogels for Neural Tissue Engineering. *ACS Biomater. Sci. Eng.* **2015**, *1*, 431–439. [CrossRef] [PubMed]
211. Mehrban, N.; Molina, C.P.; Quijano, L.M.; Bowen, J.; Johnson, S.A.; Bartolacci, J.; Chang, J.T.; Scott, D.A.; Woolfson, D.N.; Birchall, M.A.; et al. Host macrophage response to injectable hydrogels derived from ECM and α-helical peptides. *Acta Biomater.* **2020**, *111*, 141–152. [CrossRef] [PubMed]
212. Reches, M.; Gazit, E. Self-assembly of peptide nanotubes and amyloid-like structures by charged-termini-capped diphenylalanine peptide analogues. *Isr. J. Chem.* **2005**, *45*, 363–371. [CrossRef]
213. Smith, A.M.; Williams, R.J.; Tang, C.; Coppo, P.; Collins, R.F.; Turner, M.L.; Saiani, A.; Ulijn, R.V. Fmoc-Diphenylalanine Self Assembles to a Hydrogel via a Novel Architecture Based on π–π Interlocked β-Sheets. *Adv. Mater.* **2008**, *20*, 37–41. [CrossRef]
214. Jayawarna, V.; Smith, A.; Gough, J.E.; Ulijn, R.V. Three-dimensional cell culture of chondrocytes on modified di-phenylalanine scaffolds. *Biochem. Soc. Trans.* **2007**, *35*, 535–537. [CrossRef]
215. Zhou, M.; Smith, A.M.; Das, A.K.; Hodson, N.W.; Collins, R.F.; Ulijn, R.V.; Gough, J.E. Self-assembled peptide-based hydrogels as scaffolds for anchorage-dependent cells. *Biomaterials* **2009**, *30*, 2523–2530. [CrossRef]
216. Jayawarna, V.; Richardson, S.M.; Hirst, A.R.; Hodson, N.W.; Saiani, A.; Gough, J.E.; Ulijn, R.V. Introducing chemical functionality in Fmoc-peptide gels for cell culture. *Acta Biomater.* **2009**, *5*, 934–943. [CrossRef]
217. Mata, A.; Hsu, L.; Capito, R.; Aparicio, C.; Henrikson, K.; Stupp, S.I. Micropatterning of bioactive self-assembling gels. *Soft Matter* **2009**, *5*, 1228–1236. [CrossRef] [PubMed]
218. Webber, M.J.; Tongers, J.; Renault, M.-A.; Roncalli, J.G.; Losordo, D.W.; Stupp, S.I. Development of bioactive peptide amphiphiles for therapeutic cell delivery. *Acta Biomater.* **2010**, *6*, 3–11. [CrossRef]
219. Berndt, P.; Fields, G.B.; Tirrell, M. Synthetic lipidation of peptides and amino acids: Monolayer structure and properties. *J. Am. Chem. Soc.* **1995**, *117*, 9515–9522. [CrossRef]

220. Chow, L.W.; Bitton, R.; Webber, M.J.; Carvajal, D.; Shull, K.R.; Sharma, A.K.; Stupp, S.I. A bioactive self-assembled membrane to promote angiogenesis. *Biomaterials* **2011**, *32*, 1574–1582. [CrossRef]
221. Shah, R.N.; Shah, N.A.; Lim, M.M.d.; Hsieh, C.; Nuber, G.; Stupp, S.I. Supramolecular design of self-assembling nanofibers for cartilage regeneration. *Proc. Natl. Acad. Sci. USA* **2010**, *107*, 3293–3298. [CrossRef] [PubMed]
222. Webber, M.J.; Stupp, S.I. Emerging Peptide Nanomedicine to Regenerate Tissues and Organs. *J. Intern. Med.* **2010**, *267*, 71–88. [CrossRef]
223. Zhou, S.; Hokugo, A.; McClendon, M.; Zhang, Z.; Bakshi, R.; Wang, L.; Segovia, L.A.; Rezzadeh, K.; Stupp, S.I.; Jarrahy, R. Bioactive peptide amphiphile nanofiber gels enhance burn wound healing. *Burns* **2019**, *45*, 1112–1121. [CrossRef] [PubMed]
224. Li, X.-C.; Wu, Y.-H.; Bai, X.-D.; Ji, W.; Guo, Z.-M.; Wang, C.-F.; He, Q.; Ruan, D. BMP7-Based Functionalized Self-Assembling Peptides Protect Nucleus Pulposus-Derived Stem Cells From Apoptosis *In Vitro*. *Tissue Eng. Part A* **2016**, *22*, 1218–1228. [CrossRef]
225. An, H.S.; Takegami, K.; Kamada, H.; Nguyen, C.M.; Thonar, E.J.-M.A.; Singh, K.; Andersson, G.B.; Masuda, K. Intradiscal administration of osteogenic protein-1 increases intervertebral disc height and proteoglycan content in the nucleus pulposus in normal adolescent rabbits. *Spine (Phila. Pa. 1976)* **2005**, *30*, 22–25. [CrossRef] [PubMed]
226. Gu, T.; Shi, Z.; Wang, C.; Chen, C.; Wu, J.; Wang, D.; Xu, C.; Qing, H.; Dike, R. Human bone morphogenetic protein 7 transfected nucleus pulposus cells delay the degeneration of intervertebral disc in dogs. *J. Orthop. Res. Off. Publ. Orthop. Res. Soc.* **2017**, *35*, 1311–1322. [CrossRef]
227. Wu, Y.; Jia, Z.; Liu, L.; Zhao, Y.; Li, H.; Wang, C.; Tao, H.; Tang, Y.; He, Q.; Ruan, D. Functional Self-Assembled Peptide Nanofibers for Bone Marrow Mesenchymal Stem Cell Encapsulation and Regeneration in Nucleus Pulposus. *Artif. Organs* **2016**, *40*, E112–E119. [CrossRef]
228. Tao, H.; Zhang, Y.; Wang, C.; Zhang, C.; Wang, X.; Wang, D.; Bai, X.; Wen, T.; Xin, H.; Wu, J.; et al. Biological Evaluation of Human Degenerated Nucleus Pulposus Cells in Functionalized Self-Assembling Peptide Nanofiber Hydrogel Scaffold. *Tissue Eng. Part A* **2014**, *20*, 1621–1631. [CrossRef] [PubMed]
229. McKenna, L.A.; Liu, H.; Sansom, P.A.; Dean, M.F. An N-terminal peptide from link protein stimulates proteoglycan biosynthesis in human articular cartilage in vitro. *Arthritis Rheum.* **1998**, *41*, 157–162. [CrossRef]
230. Wang, B.; Wu, Y.; Shao, Z.; Yang, S.; Che, B.; Sun, C.; Ma, Z.; Zhang, Y. Functionalized self-assembling peptide nanofiber hydrogel as a scaffold for rabbit nucleus pulposus cells. *J. Biomed. Mater. Res.—Part A* **2012**, *100*, 646–653. [CrossRef]
231. Ma, K.; Wu, Y.; Wang, B.; Yang, S.; Wei, Y.; Shao, Z. Effect of a synthetic link N peptide nanofiber scaffold on the matrix deposition of aggrecan and type II collagen in rabbit notochordal cells. *J. Mater. Sci. Mater. Med.* **2013**, *24*, 405–415. [CrossRef] [PubMed]
232. Sun, J.; Zheng, Q.; Wu, Y.; Liu, Y.; Guo, X.; Wu, W. Culture of nucleus pulposus cells from intervertebral disc on self-assembling KLD-12 peptide hydrogel scaffold. *Mater. Sci. Eng. C* **2010**, *30*, 975–980. [CrossRef]
233. Ligorio, C.; Zhou, M.; Wychowaniec, J.K.; Zhu, X.; Bartlam, C.; Miller, A.F.; Vijayaraghavan, A.; Hoyland, J.A.; Saiani, A. Graphene oxide containing self-assembling peptide hybrid hydrogels as a potential 3D injectable cell delivery platform for intervertebral disc repair applications. *Acta Biomater.* **2019**, *92*, 92–103. [CrossRef] [PubMed]
234. Ligorio, C.; O'Brien, M.; Hodson, N.W.; Mironov, A.; Iliut, M.; Miller, A.F.; Vijayaraghavan, A.; Hoyland, J.A.; Saiani, A. TGF-β3-loaded graphene oxide—Self-assembling peptide hybrid hydrogels as functional 3D scaffolds for the regeneration of the nucleus pulposus. *Acta Biomater.* **2021**, *127*, 116–130. [CrossRef]
235. Mishra, A.; Loo, Y.; Deng, R.; Chuah, Y.J.; Hee, H.T.; Ying, J.Y.; Hauser, C.A.E. Ultrasmall natural peptides self-assemble to strong temperature-resistant helical fibers in scaffolds suitable for tissue engineering. *Nano Today* **2011**, *6*, 232–239. [CrossRef]
236. Uysal, O.; Arslan, E.; Gulseren, G.; Kilinc, M.C.; Dogan, I.; Ozalp, H.; Caglar, Y.S.; Guler, M.O.; Tekinay, A.B. Collagen Peptide Presenting Nanofibrous Scaffold for Intervertebral Disc Regeneration. *ACS Appl. Bio Mater.* **2019**, *2*, 1686–1695. [CrossRef] [PubMed]
237. Barreto-Henriksson, H.; Llorente, M.; Larsson, A.; Brisby, H.; Gold, J.; Schuster, E.; Ström, A. Determination of mechanical and rheological properties of a cell-loaded peptide gel during ECM production. *Int. J. Pharm.* **2019**, *563*, 437–444. [CrossRef] [PubMed]
238. Moss, I.L.; Gordon, L.; Woodhouse, K.A.; Whyne, C.M.; Yee, A.J.M. A novel thiol-modified hyaluronan and elastin-like polypetide composite material for tissue engineering of the nucleus pulposus of the intervertebral disc. *Spine (Phila. Pa. 1976)* **2011**, *36*, 1022–1029. [CrossRef] [PubMed]
239. Miles, D.E.; Mitchell, E.A.; Kapur, N.; Beales, P.A.; Wilcox, R.K. Peptide:glycosaminoglycan hybrid hydrogels as an injectable intervention for spinal disc degeneration. *J. Mater. Chem. B.* **2016**, *4*, 3225–3231. [CrossRef]
240. Yoon, H.H.; Bhang, S.H.; Kim, T.; Yu, T.; Hyeon, T.; Kim, B.-S. Dual Roles of Graphene Oxide in Chondrogenic Differentiation of Adult Stem Cells: Cell-Adhesion Substrate and Growth Factor-Delivery Carrier. *Adv. Funct. Mater.* **2014**, *24*, 6455–6464. [CrossRef]
241. Zhou, M.; Lozano, N.; Wychowaniec, J.K.; Hodgkinson, T.; Richardson, S.M.; Kostarelos, K.; Hoyland, J.A. Graphene oxide: A growth factor delivery carrier to enhance chondrogenic differentiation of human mesenchymal stem cells in 3D hydrogels. *Acta Biomater.* **2019**, *96*, 271–280. [CrossRef]
242. Raphael, B.; Khalil, T.; Workman, V.L.; Smith, A.; Brown, C.P.; Streuli, C.; Saiani, A.; Domingos, M. 3D cell bioprinting of self-assembling peptide-based hydrogels. *Mater. Lett.* **2017**, *190*, 103–106. [CrossRef]

MDPI AG
Grosspeteranlage 5
4052 Basel
Switzerland
Tel.: +41 61 683 77 34

Gels Editorial Office
E-mail: gels@mdpi.com
www.mdpi.com/journal/gels

www.ingramcontent.com/pod-product-compliance
Lightning Source LLC
Chambersburg PA
CBHW061818210926
51850CB00006B/738